Variations in the Global Water Budget

Variations in the Global Water Budget

edited by

ALAYNE STREET-PERROTT
School of Geography, University of Oxford, U.K.

MAX BERAN
Institute of Hydrology, Wallingford, U.K.

and

ROBERT RATCLIFFE
Royal Meteorological Society, U.K.

D. REIDEL PUBLISHING COMPANY

A MEMBER OF THE KLUWER  ACADEMIC PUBLISHERS GROUP

DORDRECHT / BOSTON / LANCASTER

Library of Congress Cataloging in Publication Data
Main entry under title:

Variations in the global water budget.

    Papers presented at the Symposium on Variations in the Global Water
Budget held in Oxford, Aug. 1981 and sponsored by the Palaeoclimate Commis-
sion of the International Union for Quaternary Research . . . [et al.].
    Includes index.
    1. Water balance (Hydrology)–Congresses. I. Street-Perrott, Alayne,
1950–   . II. Beran, Max. III. Ratcliffe, Robert. IV. Symposium on
Variations in the Global Water Budget (1981 : Oxford, Oxfordshire)
V. International Union for Quaternary Research. Palaeoclimate Commission.
GB665.V27   1983   551.48   83–8687
ISBN-13: 978-94-009-6956-8   e-ISBN-13: 978-94-009-6954-4
DOI: 10.1007/ 978-94-009-6954-4

Published by D. Reidel Publishing Company,
P.O. Box 17, 3300 AA Dordrecht, Holland.

Sold and distributed in the U.S.A. and Canada
by Kluwer Boston Inc.,
190 Old Derby Street, Hingham, MA 02043, U.S.A.

In all other countries, sold and distributed
by Kluwer Academic Publishers Group,
P.O. Box 322, 3300 AH Dordrecht, Holland.

TABLE OF CONTENTS

SECULAR VARIABILITY: INTERACTIONS AND TELECONNECTIONS

LONG-TERM CHANGES

MODELLING AND PREDICTION

PREFACE

R.E. Newell

President, International Commission on Climate
International Association of Meteorology and
Atmospheric Physics

    Water is the active ingredient in the global climatic system,
its physical properties ensuring that it plays a major role. Its
high thermal capacity provides a mechanism for moderating mid-
latitude winter temperatures; solar energy is absorbed by the
surface layers of the middle latitude oceans in summer and is
released to the atmosphere in winter as the ocean cools. The
variation of saturation vapour pressure with temperature is the
factor which causes oceanic surface temperatures at low latitudes
to be limited by evaporation to values near 29°C, thereby
limiting tropical marine air temperatures to about the same value.
The substantial amount of energy involved in phase changes – the
latent heat – governs the passage of solar energy to the atmosphere; visible solar radiation is absorbed at the Earth's surface,
energy is supplied to evaporate water and the latent heat is
released to the atmosphere when and where condensation occurs,
which is often a considerable distance from the source of the
moisture. The infrared radiative characteristics of water vapour,
namely the broad vibration-rotation bands typical of a triatomic
molecule, permit it to act as the principal agent of energy loss
from the atmosphere, through infrared radiation to space.
    Whether climate is viewed on a time scale of a few months or
years, or many millennia, the energy balance of the atmosphere
remains broadly the same: about 85% of its energy input comes from
the liberation of latent heat when rainfall occurs; an even larger
fraction of the total energy lost by the atmosphere to space occurs
through infrared radiation from water vapour. It is this passage
of energy to and from the atmosphere, coupled with all the accom-
panying dynamical processes, that governs the free air temperature.
    Investigators in several fields of scientific enquiry (both
fundamental and applied) have a direct interest in the global

A. Street-Perrott et al. (eds.), Variations in the Global Water Budget, ix–xiv.
Copyright © 1983 by D. Reidel Publishing Company.

water cycle: hydrologists, agriculturalists, civil engineers, glaciologists, and climatologists, for example. In general, they are directly involved with only one aspect of the hydrological cycle, such as rainfall during the growing season or evaporation from reservoirs; nevertheless, all the processes investigated are influenced by its variability. This variability is linked in turn to fluctuations in the large-scale atmospheric circulation in ways which are only just beginning to be understood. During the seventeenth General Assembly of the International Union of Geodesy and Geophysics (IUGG) held in Canberra, Australia in December 1979, both the International Association of Meteorology and Atmospheric Physics (IAMAP) and the International Association of Hydrological Sciences (IAHS) held meetings which dealt directly or indirectly with various facets of the global water cycle. These groups clearly had some common problems, made use of some common material, and had some complementary ideas; the hydrologists were mostly concerned with what happened at and below the surface, while the meteorologists dealt more with what happened in the atmosphere. A major common problem was the variability of the cycle. It seemed that both groups could benefit by focusing attention jointly on this topic. As much of the basic material is not readily available for use by students and researchers, it was decided to convene a symposium which would be followed by publication of a book that could act as a state-of-the-art summary of material on the hydrological cycle, with emphasis on its variability. This was the underlying concept of the Symposium on Variations in the Global Water Budget (held in Oxford in August 1981).

From the outset, it was realized that the Palaeoclimate Commission of the International Union for Quaternary Research (INQUA) also had an active interest in this topic, and this body readily agreed to act as a co-sponsor of the symposium, together with the Royal Meteorological Society, the American Meteorological Society and UNESCO.

Several members of the sponsoring organizations work in the field of palaeoclimatology. From an examination of sediments on the ocean floor, in lakes, and on the land surface, and a knowledge of the present day interrelationships between surface characteristics, the biosphere and climate, they try to reconstruct the climate of the past. These scientists need to know how these linkages operated during the relatively small changes in climate that have occurred recently, say over the past century, before they can properly interpret data from the more distant past. Hence, they gave strong support to the idea of a symposium on the variability of the global water cycle on time scales long enough to encompass glacial-interglacial changes.

It should be stressed that, as yet, there is no satisfactory theory of climate. Although such a theory is being pursued through a combination of studies, some purely theoretical, some involving atmospheric and oceanic observations, and others

numerical modelling of the observed features, this goal is still
far distant. It is towards this end that a large part of the
research promoted by the World Meteorological Organization and
the International Council of Scientific Unions, through their
joint World Climate Research Programme, is directed. A major
prerequisite is clearly a detailed understanding of the hydrological cycle, including the distribution of water in its various
forms throughout the climatic system, the rate of transfer of
water from one phase to another and from one place to another, and
the factors which control the atmospheric source (evaporation) and
the atmospheric sink (precipitation).

Much of the variability in the hydrological cycle is introduced through the atmospheric branch. Certain phenomena, such as
the intensity and latitude of the rising and sinking motion
associated with the tropical Hadley cell circulation, vary from
year to year so that rainfall patterns and the frequency of clear
skies fluctuate concomitantly. Discussions of the atmospheric
branch of the cycle are scattered through various meteorological
and hydrological journals and texts. It was considered that an
integrated view of the present state of knowledge would make a
good starting point for the symposium and, hence, for this book.
Peixóto and Oort laboured hard to produce this view from the most
comprehensive set of atmospheric data that has yet been analysed,
covering a period of ten years. The relative roles of large-scale
quasi-horizontal eddies and the mean circulations are enumerated
for different latitudes and seasons, the actual paths of the
water vapour from source to sink are traced, substantial connections with the geography and surface properties are noted, and
the entire cycle is treated quantitatively.

This paper can be used by the lecturer to introduce students
to the subject of the hydrological cycle, or can be read by the
researcher new to the field. The tables will be quoted for many
years to come, and the fact that ten years of observations are
available gives scope for future studies of interannual variations.

A complementary approach is to focus on the regional water-
vapour budget as Alestalo does for Europe. He finds quite good
agreement between the water-vapour flux divergence deduced
entirely from radiosonde data and the surface water balance, the
latter being obtained either from observations of rainfall and
estimates of evaporation, or from river runoff. It would be
valuable to perform such studies for regions throughout the globe,
but the radiosonde data are not yet good enough over many land
regions and are, of course, scarce over the oceans. There is
some hope that global distributions of atmospheric water may be
monitored in the future by satellite. The papers by Gorodetsky
and Syachinov, Pathak, Taylor and others describe some of the
progress made in this field.

Palaeoclimatologists have made extensive use of stable
isotope measurements on ice cores. Fractionation of both hydrogen
and oxygen isotopes occurs during evaporation. Isotope ratios in

precipitation are governed by a series of processes, from first
evaporation to final fallout from the atmosphere. Often, water
vapour is involved in a series of condensation-evaporation cycles
as clouds form and decay. Due to changing circulation patterns,
precipitation at a particular place may originate from evaporation at a number of different sites. Sonntag *et al.* examine the
factors controlling stable isotope ratios in precipitation, based
on present-day knowledge of atmospheric moisture transport as set
out, for example, by Peixóto and Oort, and on current information
about the way in which water passes through a precipitating system.

Water vapour and latent heat are transferred from the surface
to the atmosphere by evaporation, a process which has always been
difficult to measure. Shuttleworth's article provides a vigorous
summary of the various models which have been used to estimate
evaporation from other variables, and shows what assumptions are
made and how they are related to the several processes involved.
Mastery of this article would be a valuable prerequisite to work
on the incorporation of evaporation into models of the atmospheric
general circulation.

Latent energy is only supplied to the atmosphere when and
where precipitation occurs; hence, the spatial and temporal
patterns of precipitation are of major importance in climatology.
Jaeger has made these patterns his special interest, and presents
the results of his latest synthesis, which are quite compatible
with the atmospheric observations of Peixóto and Oort. Little is
known about rainfall rates over the sea, as ships do not ordinarily report rainfall. Recently, however, it has been found
possible to deduce rainfall rates from satellite microwave
observations, and these are compared by Ruprecht *et al.* with ship-
based radar measurements made during the 1975 GATE expedition in
the tropical Atlantic. The agreement is encouraging, although
further work is needed. It would clearly be of great value to
have continuously available satellite data in order to monitor
the energy injected into the atmosphere over the sea.

At high latitudes, because the air is cold, it can only hold
small amounts of moisture. Water stored in vegetation thus takes
on a relatively important role, which is addressed by Miller.
Hitherto, this aspect of the hydrological cycle has received
little attention.

A substantial part of the book deals with the variability of
different components of the hydrological cycle over a time period
of a century or less. Rainfall variability is treated by
Nicholson and Chervin for Africa, by Mooley and Parthasarathy for
the Indian summer monsoon, by Kininmonth for Northern Australia,
by Zhang for summer conditions over China, and by Reiter for the
tropical Pacific. Variations in moisture and evaporation over the
tropical Pacific are also dealt with by Weare.

River levels in Germany are examined by Liebscher. Secular
changes in glacier mass balance are summarized by Reynaud for the
Alps and by Bhandari *et al.* for the Himalayas. Fluctuations of

Antarctic sea ice are described by Chiu. A common theme in the
papers by Weare, Reiter and Chiu is that a significant fraction of
the observed variability is associated with the Southern Oscillation. The papers by Zhang, Weare and Reiter all conclude that
rainfall is related to sea-surface temperature patterns.

Long-term fluctuations in the atmospheric circulation and
water budget are covered in the third section of the book.
Satellite photographs of the Earth's surface reveal aeolian
features, such as sand dunes, that can be used to assess past
wind directions. Wells' paper marks the first time that the
satellite approach has been used on a continental scale. Wells
shows that, during the last glaciation, the prevailing flow in
mid-latitudes followed a three-wave pattern; knowledge of the flow
can then be used to make deductions about possible moisture fluxes.
Papers by Street-Perrott and Roberts, and by Tetzlaff and Adams,
use fluctuations in lake levels as indicators of palaeoclimate.
Gac *et al*. employ biological indicators in estuaries to determine
how the discharge of tropical rivers has varied through time
during the Holocene. There are implications in these lake and
river studies concerning the possible latitudinal displacement
and intensity modulation of the tropical Hadley cell circulation.

Work on monsoon rainfall patterns during the late Pleistocene and early Holocene is summarized by Kutzbach who finds that
the enhanced monsoon rains of the early Holocene may be partly,
though not completely, explained in terms of the Earth's orbital
variation - the Milankovitch theory.

There is evidence from recent observations that the atmospheric CO_2 concentration increases when the equatorial ocean is
warm and evaporation is high and *vice versa*. Flohn applies this
recent work to glacial-interglacial data to show how the water
and carbon dioxide cycles are closely interrelated. This third
section of the book closes with a warning from Mörner. He
questions the interpretation of some of the observed sea-level
fluctuations in terms of ice-volume changes, pointing out that
geoidal changes provide an alternative explanation. He also
expresses scepticism about the use of isotope data from deep-sea
cores as a measure of palaeoglaciation. The objections need to
be answered and a debate instigated, perhaps with this paper as
the starting point.

The final section of the book deals with the modelling and
prediction of various parts of the water cycle. Lockwood and
Sellers have developed a model to assess the effects of changing
vegetation on surface hydroclimatology. Eventually, it will be
desirable to match this type of model to large-scale models in
order to study the influence of natural and anthropogenic changes
in vegetation on the general atmospheric circulation. Not only
would this be helpful in the interpretation of palaeovegetation
data in terms of palaeoclimate, but it would be of practical
value in the selection of measures to ameliorate drought and
desertification, and to study the possible consequences of large-

scale forest removal. Vegetation is also an important factor in
controlling runoff, a topic treated by Němec. Here again, it may
eventually be possible to apply Hutton's principle of uniformi-
tarianism to the present observed relationship between surface
cover and runoff in order to deduce some information about past
runoff patterns from the vegetation of the past.

The drought in northwest Europe in 1975-76 engendered much
debate, one question being whether the persistence of the drought
was partly due to the depletion of soil moisture. The modelling
work by Rowntree and Bolton suggests that soil moisture plays an
important role. Under clear skies, sea surface temperatures in
the surrounding region increased, which contributed both to the
maintenance of the drought during the summer of 1976 and to its
subsequent breakdown in the autumn.

Finally, the paper by Mitchell reviews results relevant to
the hydrological cycle from the five-layer atmospheric general
circulation model of the UK Meteorological Office, integrated
through three complete annual cycles. In general, there is good
agreement with the findings of Peixóto and Oort, and other
observational studies. There are, however, a few exceptions:
for example, the model predicts a more intense hydrological cycle
over the land than is actually observed and yields less extensive
deserts. Once these problems are overcome and the model is
adapted so that it can run with changing boundary conditions,
such as sea-surface temperature, then the causes of variability
in the global water budget may be sorted out experimentally and
our understanding of the cycle advanced considerably. In itself,
this paper represents a major step forward.

It is hoped that this book will fill what is perceived as
a gap in the literature and act as a text for students and
researchers from the various disciplines with an interest in the
global water cycle.

Our Commission and the various international organizations
associated with the symposium are most grateful to the co-editors
of this book, most particularly to Dr Alayne Street-Perrott, as
well as to Ann Barham who compiled the index, Mary Gollop who set
the texts, and Chris Argent who supervised the transformation of
the manuscripts to finished camera-ready copy.

Techniques of Measurement and Analysis:
I. Atmospheric Processes

INTRODUCTION TO TECHNIQUES OF MEASUREMENT AND ANALYSIS:
ATMOSPHERIC PROCESSES

J.T. Houghton

Rutherford Appleton Laboratory
Chilton, Didcot
Oxfordshire, England

The presence of water in solid, liquid or vapour form is probably the single most important parameter in determining the impact of climate on Man. Further, because of its large latent heat, the transport of water is closely linked with the transport of energy throughout the climate system. Therefore, any appreciation of, and understanding of, climate and climatic change must include an understanding of the various components of the global water budget and their variations. The contributions in this section examine in detail the sources, transports and sinks of atmospheric water vapour, with particular emphasis on new techniques of measurement and analysis.

It is only recently that hydrologists have begun to think in global terms, largely because it is only recently that meteorological information with global coverage has become available and that theoretical models of global extent have been developed. This global thinking is at an early stage. Greatly improved data and much better formulations of physical processes are required before further progress can be made.

Global hydrological studies need to be viewed in the context of the World Climate Research Programme (WCRP), a programme jointly organized by the World Meteorological Organization and the International Council of Scientific Unions, which is aimed at developing a physical understanding of the processes involved in climate and climatic change on a global scale and over periods of months to decades. These are the time scales which can be addressed by global circulation models and for which data are or can be expected to be available. Such time scales are those of importance to decision makers concerned with planning policy in the energy or agricultural fields.

The WRCP involves all parts of the climate system, namely the

A. Street-Perrott et al. (eds.), Variations in the Global Water Budget, 2–4.

atmosphere, land, oceans and ice. Although the ultimate aim may
well be the development of a grand, time-dependent model which
links all the components together, the programme has been planned
in stages. Three such stages have been identified by the
Programme's Joint Scientific Committee (JSC). The first aims at
establishing a foundation for the long-range prediction of atmospheric anomalies on time scales of several weeks to one or two
months, based on a deterministic computation of the evolution of
the atmosphere with fixed land- and ocean-surface conditions.
The second stage, known as TOGA (Tropical Oceans and Global
Atmosphere), considers atmospheric forcing of the tropical oceans,
the development of anomalies in tropical sea-surface conditions
and, in turn, the influence of such anomalies on the atmospheric
circulation. The third stage, which includes the WOCE (World
Ocean Circulation Experiment), aims at establishing the atmospheric and oceanic circulation regimes based on modelling of the
coupled dynamics of the global atmosphere and the global ocean,
without reference to a particular initial state.

A prime requirement for the first stage is a better theoretical understanding of the land surface and of the processes by
which it is coupled to the atmosphere. Among these processes, the
evaporation of water is especially important. For the second
stage, a much better description of the exchange of water vapour,
heat and momentum between the atmosphere and the ocean surface is
paramount. For the third stage, in addition to the requirements
for the first two, links between changes in the slowly-varying
components of the climate system, such as vegetation cover, and
the relevant physical processes, will need to be established in
the model.

An important means of assessing the accuracy of theoretical
models is through the development of atmospheric water-vapour
budgets which include all the important sources, sinks and transports. The paper by Alestalo is an example of such a budget on
a regional scale and that by Peixóto and Oort is of particular
significance because it tackles the global budget problem. A
valuable independent source of evidence on the atmospheric water-
vapour cycle is provided by measurements of the stable isotopes,
deuterium and oxygen-18, in precipitation; the paper by Sonntag
et al. summarizes recent progress in this field.

In addition to the development of much better theoretical
descriptions, considerable advances in measurement will be
necessary in order to provide all the different types of data
required by the WRCP. Satellite instruments are the only economic
means of providing many of the necessary data on a global basis.
Examples of satellite observations of atmospheric water vapour and
clouds are given in the papers by Gorodetsky and Syachinov, and
Taylor et al. Pathak's paper compares the results of water-vapour
measurements by satellite and conventional methods. These papers
illustrate the considerable advances in satellite measurement
techniques that have occured in recent years; observations of

temperature and humidity from infrared and microwave instruments,
of wind velocity from cloud motions, and of surface albedo, cloud
cover and radiation budgets from visible and infrared imaging
devices and radiometers, are all now acquired routinely by
various satellites, both polar-orbiting and geostationary. Other
possibilities exist, especially through the use of passive and
active microwave systems to measure surface wind and surface
stress over the oceans, precipitation (see, for example, the paper
by Ruprecht *et al.*), and ice cover and type (see Chiu's paper),
or through the application of radar altimetry to the study of
ocean currents. Further development in some of these areas is
urgently required, particularly in the measurement of precipitation and of sea surface temperature with very high accuracy. Data
analysis and information retrieval are areas that also need a lot
of attention, especially as the types of data available become
more varied and the quantities even larger than at present. These
developments will require, on the one hand, the skills of hydrologists in all areas of the subject and, on the other hand, will
have an enormous impact on the growth of global hydrology as a
discipline. For these reasons, the publication of these key
papers is very timely.

THE ATMOSPHERIC BRANCH OF THE HYDROLOGICAL CYCLE AND CLIMATE

José P. Peixóto

Geophysical Institute
University of Lisbon
Portugal

Abraham H. Oort
Geophysical Fluid Dynamics Laboratory/NOAA
Princeton University
Princeton, New Jersey 08540, USA

ABSTRACT. Based on daily observations from about 1000 rawinsonde stations, tables and global distributions of the various water vapour fields are presented for mean annual, winter and summer conditions covering the 10-year period, May 1963 through April 1973. The fields include horizontal maps of precipitable water, of total zonal, meridional and vertical transports by eddy and mean circulations, as well as meridional profiles and zonal mean cross-sections of these quantities.
   The connections between the atmospheric branch and the hydrology of the Earth's surface were studied with the aid of the divergence fields of mean total water vapour flux and through horizontal and vertical streamfunction analyses. The divergence maps agree quite well with maps of evaporation minus precipitation (E-P) obtained from classic, climatological surface data. Over the oceans, the divergence fields show a good correlation with the evaporation and surface salinity maps.
   From the zonal mean streamfunction analyses, the total vertical transports of water substance were inferred, and compared with the contributions by standing eddies and mean meridional circulations. The resulting vertical fluxes by transient eddies show the great importance of cumulus convection in the tropics for the atmospheric circulation. The main sources and sinks of atmospheric water vapour, as well as the dominant mean trajectories of water, are identified.
   In the light of various climatological and oceanographic

5

A. Street-Perrott et al. (eds.), Variations in the Global Water Budget, 5–65.
Copyright © 1983 by D. Reidel Publishing Company.

considerations, the results show that the study of the atmospheric
branch is essential to improve understanding of the Earth's water
balance on both regional and global scales.

1.   INTRODUCTION

The climatic system can be regarded as a global hydrosphere
consisting of various reservoirs (subsystems) interconnected by
the transfer of water in various phases. In decreasing order of
water amount held in storage, the five reservoirs of the hydrosphere are: the world oceans, the ice masses and snow deposits,
the terrestrial waters, the atmosphere and, finally, the biosphere. The residence time of water in the various reservoirs
varies from about 10 days for atmospheric water vapour to thou-
sands of years for the polar ice and the oceans.
    A loss or 'output' of water from the Earth's surface, through
evaporation and evapotranspiration from the oceans and continents,
is the 'input' of water for the atmospheric branch, whereas precipitation, the atmospheric 'output', may be regarded as a gain
for the terrestrial branch of the hydrological cycle. Water is
thus one of the crucial links between the various components of
the climatic system. Therefore, the atmospheric and terrestrial
branches of the hydrological cycle have to be taken as a whole,
bringing together the sister disciplines of meteorology and
hydrology.
    The fundamental role of the atmosphere and of its general
circulation as a forcing factor for the water cycle has long been
recognized by climatologists and hydrologists. However, quantitative studies of the gaseous hydrosphere and of its aerial runoff
have only been possible in recent decades, with the development of
an adequate network of aerological stations. The study of the
various fields which characterize the flow of water vapour in the
atmosphere is essential for improving comprehension of the global
water cycle and the interrelations between the terrestrial and
aerological branches. Traditionally, hydrologists have studied
the hydrological cycle in a piecemeal way, considering only the
Earth-bound branch. However, progress along this line has been
hindered seriously by the difficulty in obtaining reliable values
of evaporation, change of water storage and precipitation. Estimates of evaporation, based on the diffusion theory or on the
energy-budget approach, are uncertain and may differ widely.
Precipitation data over the oceans are still rather scarce and
difficult to interpret, especially island-based measurements.
    The extensive research on the general circulation of the
atmosphere, mainly by the MIT-school under the leadership of the
late Victor P. Starr, stimulated a number of investigations of the
behaviour of the various fields of water vapour in the atmosphere.
These studies are regarded now as essential to increase understanding of the dynamics of the general circulation of the atmo-

sphere, and to analyse all aspects of the energetics of the atmosphere and of the Earth. Furthermore, they proved to be a particularly relevant tool for the investigation of the water balance
of the climatic system on various time and space scales (Starr &
White 1955, Bannon & Steele 1960, Rasmusson 1971). This type of
research opens a new perspective in all domains of dynamic hydrology (Starr et al. 1958, Eagleson 1970).

The present work is an extension of the previous pole-to-pole
studies of water vapour based on data from 1958 during the International Geophysical Year (IGY) (Starr et al. 1969, Peixóto 1970,
Starr & Peixóto 1971, Peixóto 1972, Peixóto et al. 1976, 1978).
It extends the earlier work to a 10-year period (May 1963 through
April 1973) of homogeneous data obtained at more than 1000
stations of the global rawinsonde network. This study provides
a stable, long-term representation of the mean water vapour
fields in the atmosphere on a planetary scale.

The actual results, which are analysed and discussed in the
light of various meteorological and hydrological considerations,
show the importance of the aerological branch of the hydrological
cycle in dealing with global water problems.

This paper, together with an extensive study (based on the
same data set) of the angular momentum and energy cycles in the
global atmosphere (Oort & Peixóto 1982), should provide one of the
most complete summaries of the observed balance requirements for
the general circulation of the global atmosphere.

2.   THEORETICAL FRAMEWORK

2.1 Formulation of the Balance Requirements of Water Vapour in the Atmosphere

To a high degree of accuracy, the atmosphere may be assumed
to be in a state of hydrostatic equilibrium. This will allow the
use of pressure, p, as a vertical coordinate in the (λ,ϕ,p,t)-
coordinate system, where λ denotes longitude, ϕ latitude and t
time.

The amount of water vapour contained in a unit area column of
air which extends from the Earth's surface to the top of the atmosphere, at a given instant, is expressed by:

$$W(\lambda,\phi,t) = \int_0^{p_0} q\,(dp/g) \tag{1}$$

where q is the specific humidity, g the acceleration due to
gravity, and p_0 atmospheric pressure at the surface. The term W
is sometimes called precipitable water since it represents the
amount of liquid water that would result if all the water vapour
of the unit column of the atmosphere was condensed. It is usually
expressed in units of g cm^{-2} or cm.

At a given level of the atmosphere, we can define a vector

field $\vec{F}$ of the transport of water vapour in the atmosphere, such that:

$$\vec{F}(\lambda,\phi,p,t) = (1/g)q\vec{v} = (1/g)(qu)\vec{i} + (1/g)(qv)\vec{j} = F_\lambda\vec{i} + F_\phi\vec{j} \qquad (2)$$

where $\vec{i}$ and $\vec{j}$ are the zonal and meridional unit vectors, respectively, u the zonal wind component (positive if eastward) and v the meridional wind component (positive if northward). For a vertical column of the atmosphere with unit base, the total horizontal flux of water may be obtained by integrating the previous expression with respect to pressure defining a two-dimensional vector field $\vec{Q}(\lambda,\phi)$:

$$\vec{Q}(\lambda,\phi,t) = \int_o^{p_o}\vec{F}\,dp = \int_o^{p_o}q\vec{v}(dp/g) = Q_\lambda\vec{i} + Q_\phi\vec{j} \qquad (3)$$

In this expression, Q_λ and Q_ϕ are the zonal and meridional components of the total water vapour transport vector field, $\vec{Q}$, respectively. They are given by

$$Q_\lambda = \int_{p_o}^{p_o}qu(dp/g) = \hat{qu}(p_o/g)$$

$$Q_\phi = \int_o^{p_o}qv(dp/g) = \hat{qv}(p_o/g) \qquad (4)$$

where the vertical mean operator $(\,\hat{}\,)$ is defined by

$$(\,\hat{}\,) = \int_o^{p_o}(\ )dp/p_o$$

The vector $\vec{Q}$ represents the total instantaneous water vapour transport, and may be considered as the 'aerial runoff'. The term Q_ϕ represents the flux of water vapour across a unit strip of a latitudinal wall, ϕ = constant, and Q_λ the flux across a unit strip of a meridian plane, λ = constant. Equations (1) to (4) may be averaged over a time interval τ leading to the corresponding mean values $\bar{W}$, $\vec{F}$, $\vec{Q}$, $\bar{Q}_\lambda$ and $\bar{Q}_\phi$, where the bar denotes the time-average operator:

$$(\,\bar{}\,) = (1/\tau)\int_o^\tau(\ )dt$$

The balance requirement of water vapour in the atmosphere for a unit mass at a certain instant can now be expressed by a balance equation, making use of the continuity equation of the p-system. The resulting equation is:

$$(\partial q/\partial t) + \text{div } q\vec{v} + [\partial(q\omega)/\partial p] = s(q) \qquad (5)$$

where div is the horizontal divergence operator, ω the 'vertical velocity' in the p-system (counted positive if downward), and s(q) equals the rate of generation or destruction of water vapour within the unit mass due to phase changes. The main sources and sinks of water vapour in the atmosphere are due primarily to evaporation, condensation and, to some extent, diffusion from the surroundings. Thus, $s(q) = e - c$, where e and c are the rates of evaporation and of condensation per unit mass. As previously noted (Peixóto 1973), water transports in the solid and liquid phases, $q_c \vec{v}$ and $\vec{Q}_c$, by clouds are negligible when compared with the transport in the vapour phase, except possibly in cumulonimbus clouds in the tropics and over warm ocean currents.

After integration with respect to pressure from the Earth's surface to the top of the atmosphere, the time-mean balance equation assumes the final form:

$$(\overline{\partial W}/\partial t) + \mathrm{div}\ \overline{\vec{Q}} = \overline{E} - \overline{P} \tag{6}$$

This is a balance equation for the water vapour in the atmosphere. The excess of mean evaporation $\overline{E}$ over mean precipitation $\overline{P}$ at the Earth's surface is balanced by the local rate of change of water vapour storage $\overline{\partial W}/\partial t$, and by the inflow or outflow of water vapour, div $\overline{\vec{Q}}$.

The application of Equation (6) to a region of the Earth's surface of area A (e.g., a river-drainage basin or an interior sea) bounded by a closed vertical wall leads to another form of Equation (6):

$$\langle\overline{\partial W}/\partial t\rangle + \langle\mathrm{div}\ \overline{\vec{Q}}\rangle = \langle\overline{E} - \overline{P}\rangle, \tag{7}$$

where the brackets $\langle\rangle$ denote a space average over the area A. After using the Gauss' theorem, Equation (7) may be transformed into the following form, which is more useful for regional studies:

$$\langle\overline{\partial W}/\partial t\rangle + (1/A)\ \oint_\gamma (\overline{\vec{Q}}\cdot\vec{n}_\gamma)\,d\gamma = \langle\overline{E} - \overline{P}\rangle \tag{8}$$

where $\vec{n}_\gamma$ denotes the outward normal vector to the boundary, γ, namely the conceptual vertical wall.

Equations (7) and (8) may be regarded as equations of hydrology for the atmospheric branch of the hydrological cycle. The mean difference $\langle\overline{E} - \overline{P}\rangle$ within a given region is balanced by the change in water content in the atmosphere $\langle\overline{\partial W}/\partial t\rangle$ and the net

water vapour transfer through aerial runoff $\overline{\vec{Q}}$. Except in the case of severe storms for short intervals of time, the rate of change of precipitable water is very small when compared with the other terms. Thus, for sufficiently long periods of time, positive divergence is found where evaporation exceeds precipitation, whereas negative divergence (convergence) is found where precipitation is greater than evaporation.

For a given region, the classic equation of hydrology can be written in the following simplified form:

$$-\langle\overline{E} - \overline{P}\rangle = \langle\overline{R}_o\rangle + \langle\overline{\partial S}/\partial t\rangle \tag{9}$$

where $\langle\bar{R}_o\rangle$ is the average rate of runoff per unit area and $\langle\overline{\partial s/\partial t}\rangle$ is the rate of change of total surface and subterranean water storage. The mean difference $\langle\overline{E-P}\rangle$ is common to Equations (7) and (9) and establishes the connection between the terrestrial and the atmospheric branches of the hydrological cycle (Peixóto 1973). Elimination of $\langle\overline{E-P}\rangle$ between Equations (7) and (9) yields

$$\langle\bar{R}_o\rangle + \langle\overline{\partial S/\partial t}\rangle = - \langle\mathrm{div}\ \vec{\bar{Q}}\rangle - \langle\overline{\partial W/\partial t}\rangle \tag{10}$$

which links, in a simple manner, the two branches of the hydrological cycle.

If, besides the aerological data, $\langle\bar{R}_o\rangle$ and $\langle\bar{P}\rangle$ are known from stream flow and precipitation data, we can use the previous equations to estimate the average rate of change of ground water storage, $\langle\overline{\partial S/\partial t}\rangle$, and the average rate of evaporation $\langle\bar{E}\rangle$ for a catchment basin. Thus, using finite differences as is usual in hydrology, Equation (10) may be written:

$$\langle\Delta S/\Delta t\rangle = - \langle\Delta W/\Delta t\rangle - (1/A)\oint_\gamma (\vec{\bar{Q}}\cdot\vec{n}_\gamma)\mathrm{d}\gamma - \langle\bar{R}_o\rangle \tag{11}$$

Similarly from Equation (8), the areal average mean evaporation can be expressed as:

$$\langle\bar{E}\rangle = \langle\overline{\Delta W/\Delta t}\rangle + (1/A)\ \oint (\vec{\bar{Q}}\cdot\vec{n}_\gamma)\mathrm{d}\gamma + \langle\bar{P}\rangle \tag{12}$$

Furthermore, since over a long period of time such as a year storage changes in the land and in the atmosphere become negligible, it can be inferred from Equation (11) that the mean runoff R_o from the continents is exactly compensated by the aerial runoff from the oceans. Similarly, when the total climatic system is considered over a long period of time, the net transports and the changes in storage are zero, and we can conclude from Equation (12) that mean evaporation $\bar{E}$ counterbalances mean precipitation $\bar{P}$.

In order to define an average streamfunction, ψ_q, in the meridional-vertical plane, we take the zonal average of Equation (5), which leads to

$$[\partial\bar{q}/\partial t] + \{\partial[\overline{qv}]\ \cos\phi/(a\ \cos\phi\ \partial\phi)\} + (\partial/\partial p)[\overline{q\omega}] = [\overline{s(q)}]\tag{13}$$

where the [ ] operator is defined by $[\ ] = \int_0^{2\pi}()\mathrm{d}\lambda/2\pi$. A similar equation can be written for the condensed form, noting that $s(q_c) = - (e-c)$. As mentioned before, the local rate of change can be neglected for the long-term mean. Considering that $\overline{qv} + \overline{q_c v} \approx \overline{qv}$, the final balance equation becomes

$$\{\partial[\overline{qv}]\ \cos\phi/(a\ \cos\phi\ \partial\phi)\} + \partial[\overline{q\omega} + \overline{q_c\omega_c}]/\partial p = 0 \tag{14}$$

where ω_c is the vertical velocity of the condensed phase. Under these conditions, a streamfunction ψ_q can be defined by the equations:

$$\partial \psi_q / \partial p = (2\Pi a/g) \cos\phi \; [\overline{qv}] \tag{15a}$$

$$- \partial \psi_q / a \; \partial\phi = (2\Pi a/g) \cos\phi \; [\overline{q\omega} + \overline{q_c \omega_c}] \tag{15b}$$

satisfying the boundary condition, $\psi_q = 0$, at the top of the atmosphere. The integration of these equations leads to the field ψ_q which portrays the general features of the zonal-mean meridional flow of water substance in a vertical plane.

The ψ_q lines start at the sources of water vapour due to the excess of evaporation over precipitation at the Earth's surface, and terminate at those latitudes in which precipitation predominates thus allowing the average movement of water substance in the atmosphere to be followed in the meridional plane.

2.2 Formulation of the Modes of Water Vapour Transfer in the Atmosphere

In order to acquire an understanding of the various mechanisms responsible for the transport of water vapour by the general circulation of the atmosphere, it is convenient to expand the mean quantities [qu], [qv] and [qω] from the previous equations. It is then possible to analyse the various terms in the expansions by studying those scales of motion that contribute significantly to the total aerial runoff.

The mean quantities $\overline{qu}$, $\overline{qv}$ and $\overline{q\omega}$ may be expanded according to the scheme $\overline{qu} = \overline{q}\;\overline{u} + \overline{q'u'}$, $\overline{qv} = \overline{q}\;\overline{v} + \overline{q'v'}$ and $\overline{q\omega} = \overline{q}\;\overline{\omega} + \overline{q'\omega'}$, where the prime indicates a departure from the time mean. We then average spatially around a latitude circle to obtain a resolution in the 'space-time' domain, e.g.,

$$[\overline{qv}] = [\overline{q}][\overline{v}] + [\overline{q^*v^*}] + [\overline{q'v'}] \tag{16}$$

and similarly for [qu] and [qω], where the asterisks denote departures from the zonal mean. The term $[\overline{q}][\overline{v}]$ represents the transport of water vapour by the mean meridional circulation[1] The $[\overline{q^*v^*}]$ is associated with the mean standing eddies of the general circulation, such as the semi-permanent subtropical anticyclones and the lows prevailing in high latitudes. Finally, the term $[\overline{q'v'}]$ represents the mean transfer of moisture due to the transient perturbations which develop along the polar front and over the intertropical convergence zone (ITCZ).

When the horizontal fluxes are integrated vertically, the following expression is obtained for the total meridional transport resolved in terms of the various modes of transfer

$$[\overline{Q}_\phi] = (1/g) \int[\overline{q}][\overline{v}]dp + (1/g) \int[\overline{q^*v^*}]dp + (1/g) \int[\overline{q'v'}]dp$$

$$= [\overline{\tilde{Q}}_\phi] + [\overline{Q^*_\phi}] + [\overline{Q'_\phi}] \tag{17}$$

and a similar equation for the total mean zonal transport $[\overline{Q}_\lambda]$.

The mean vertical transport terms $[\overline{q\omega}]$ and $[\overline{q_c \omega_c}]$ are considered next owing to their importance in the overall water balance. The vertical transport of moisture constitutes a link between the Earth's surface, its boundary layer and the free atmosphere in providing water vapour and energy to the upper levels of the atmosphere. However, the evaluation of instantaneous vertical velocities, ω, is extremely difficult and subject to large uncertainties, which severely limit the practical use of the expansion of $[\overline{q\omega}]$. This is especially true in evaluating the dominant term, namely, the transient eddy transport $[\overline{q'\omega'}]$. Its importance stems from the fact that it describes the vertical eddy transfer of water vapour due to turbulent diffusion, cumulus convection and other mesoscale convective phenomena (Peixóto 1973). The term $[\overline{q_c \omega_c}]$ represents the vertical transport of water droplets or solid ice particles, which, in general, is downward when $[\overline{q\omega}]$ is upward. It represents the rate of precipitation at a given pressure level.

3.   RESULTS

3.1 Data and Analysis Technique

The basic data necessary to compute the various fields $\overline{u}$, $\overline{v}$, $\overline{\omega}$, $\overline{q}$, $\overline{W}$, $\vec{Q}$, $\vec{Q}'$ and div $\vec{Q}$ are the daily values of specific humidity q, and wind components u and v. The rawinsonde data for the 10-year period, May 1963 through April 1973, were processed at the Geophysical Fluid Dynamics Laboratory/NOAA. After many error checks and tests for consistency of the daily reports, monthly general circulation statistics were calculated at each station following the procedures described by Oort (1982) and earlier by Oort & Rasmusson (1971). Only those stations that reported more than 10 days of the maximum possible days in any month were used to compute monthly mean statistics. The statistics consist of the averages, variances and covariances of the wind components and of specific humidity at various isobaric levels, namely, surface/ 1000 mb, 950, 900, 850, 700, 500, 400 and 300 mb.

The network of rawinsonde stations covered the entire globe, from pole to pole. The horizontal distribution of the network used and of the number of years of observations for each station are shown in Figure 1 for the 850 mb level.

Values of the statistics at regular grid points were obtained from the values at the upper-air stations through an objective analysis scheme using the zonal average of all data in a latitudinal belt as a first approximation. The various two-dimensional fields were obtained using CRAM (Conditional Relaxation Analysis Method), an objective analysis technique described by Harris *et al.* (1966) and Rosen *et al.* (1979a).

The topography of the Earth's surface was taken into account in all computations so that only grid points above the mountains contributed to the mean zonal and mean vertical estimates. To

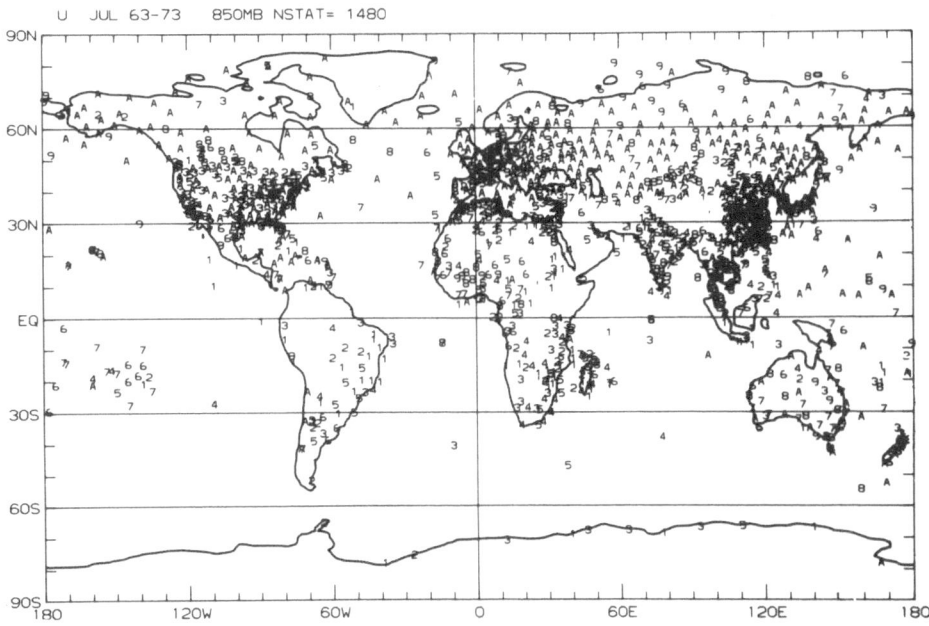

Fig. 1.   Location of rawinsonde stations included in the analyses
          at the 850 mb level for the 10-yr period 1963-73, showing
          the number of years of observations for each station -
          between 1 and 10 (denoted by A).

accomplish this, the topography was expressed in pressure coordinates, assuming that it was invariant throughout the year and
equal to the mean annual value.

   In addition to the mean annual fields for the entire 10-year
period, results were computed for two composite periods: December
through February (DJF) and June through August (JJA), for the
whole 10 years.

3.2 Mean Precipitable Water

   The spatial distributions of the mean precipitable water content[2], $\overline{W}$, on a planetary scale are represented in Figure 2.
With only a few exceptions, the analyses show a continuous decrease
of precipitable water from the equatorial regions, where it attains
the highest values, to the north and south poles. This distribution is understandable since the capacity of the atmosphere to
retain water vapour depends strongly on temperature.

   The departures from zonal symmetry are associated with the
physiography of the surface of the globe, and are apparent in both
hemispheres. As a general rule, the precipitable water is higher

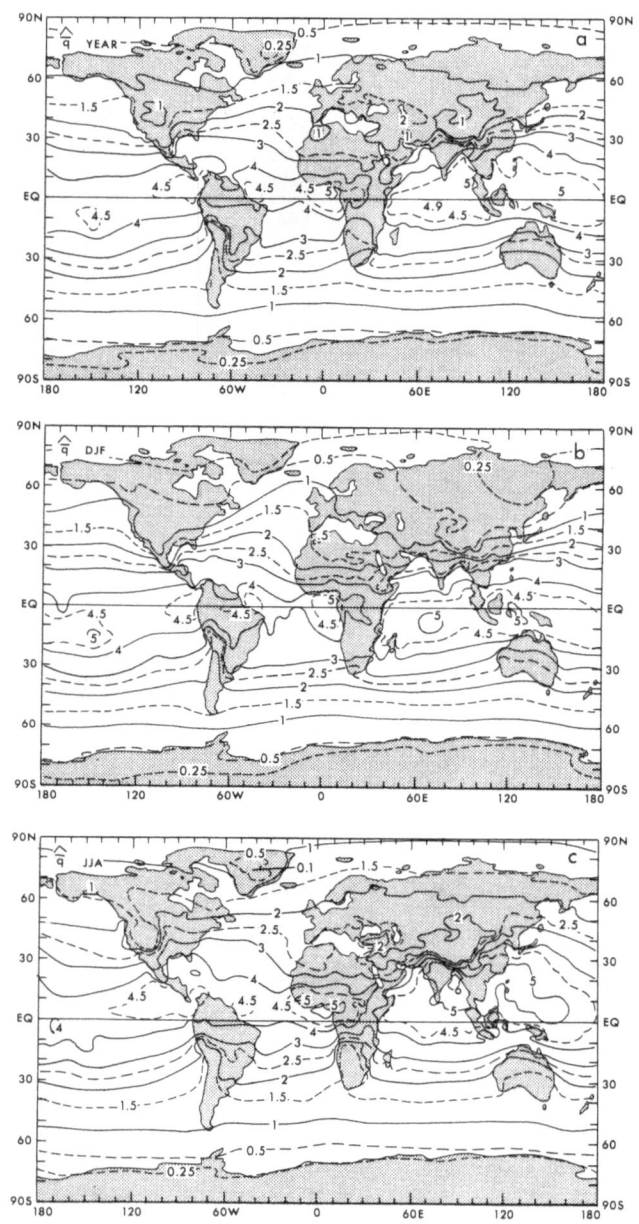

Fig. 2.   Planetary distributions of the vertically-averaged mean
         specific humidity $\hat{\overline{q}}$ (g kg^{-1}): _a, year; b, DJF season;
         c, JJA season. To convert to $\overline{W}$ (10 kg m^{-2}), multiply by
         p_o/g ($\simeq$ 1 over the oceans).

over the oceans than over the continents. The deflection of the
isolines near the western and eastern coasts of the continents is
reinforced by the topography and the presence of warm and cold
ocean currents. The distribution over the southern hemisphere is
practically zonal, since the ocean coverage exceeds by far that of
the continents. The lowest values of $\overline{W}$ occur, as expected, over
subpolar and polar regions (below 5 kg m^{-2}).

The precipitable water over the desert areas is considerably
smaller than the corresponding zonal average, mainly due to strong
subsidence. This effect is very pronounced in the eastern portions
of the large semi-permanent anticyclones of the subtropics of both
hemispheres. In addition, the effects of high terrain on the
precipitable water distribution are illustrated by relatively dry
areas (often with $\overline{W} < 10$ kg m^{-2}) over the major mountain ranges,
such as the Rockies, Himalayas, highlands of Ethiopia, and the
Andes. The effects of topography and the land-sea contrast in the
southern hemisphere are shown by the dipping of the 20 kg m^{-2}
isoline towards lower latitudes.

The seasonal variations tend to be most prominent over the
mid-latitude continents. This is particularly evident in the
northern hemisphere where the moisture content is greater in
summer than in winter at all latitudes, with the largest changes
occurring in the 20-30°N latitude belt. These changes are associated with monsoon circulations over India, southeastern Asia
and central Africa. On the other hand, seasonal variations over
the equatorial zone are not very pronounced. The influence of the
invasion of cold, dry air masses of continental origin is apparent
on the DJF map (Figure 2b) over Canada and Siberia.

In order to give the vertical structure of water vapour and
of its variability, the zonally-averaged values of the mean
specific humidity $[\overline{q}]$ and of its temporal standard deviation
$[\sigma(q)]$ were computed at various levels. They are shown in Figure
3 and in Tables A2 and A3 of the Appendix. Specific humidity
decreases rapidly with height, almost following an exponential
law; it also decreases with latitude. More than 50% of the water
vapour is concentrated below the 850 mb surface, while more than
90% is confined to the layer below 500 mb.

The variability of the vertical moisture distribution as
measured by the zonal average of the time standard deviation
$[\sigma(q)]$ may be inferred from Figure 3 and Table A3. The yearly and
seasonal cross-sections show a bimodal distribution with maxima in
the subtropics of both hemispheres below 700 mb. The intensity of
the maxima is slightly larger in summer, when the air masses are
more humid. In the equatorial region and poleward of 60°, the
variability is very small.

Using the grid point values of the $\overline{W}$ maps, the averaged
storage of water in the atmosphere $[\overline{W}]$ was evaluated, and is shown
in Figure 4(a). Actually, the values correspond to the vertical
integrals of the values presented in the previous cross-sections,
as also given in Table A2. These profiles give the gross distri-

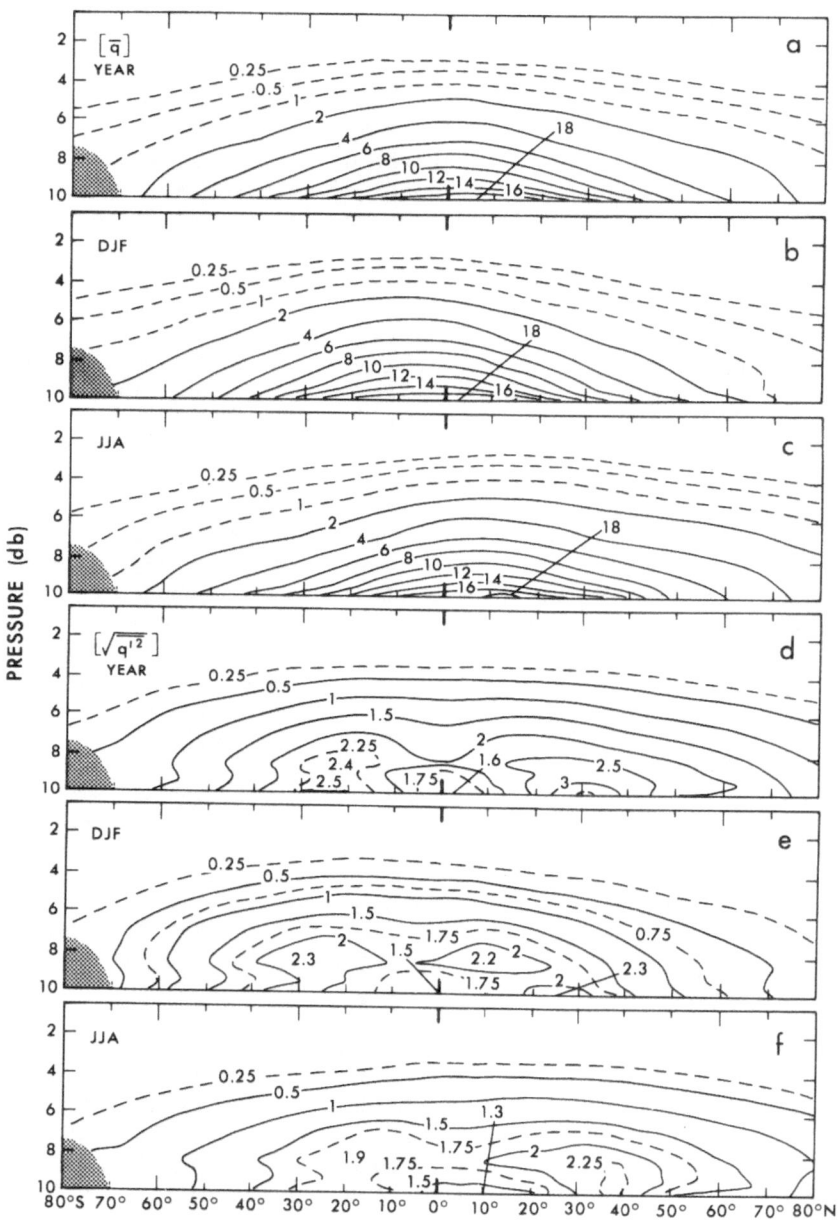

Fig. 3.   Vertical cross-sections of the specific humidity $[\bar{q}]$ and
           of its temporal standard deviation $[\sigma(q)]$ (g kg^{-1}):   a &
           d, year; b & e, DJF; c & f, JJA.

bution of water in the atmosphere. The seasonal profiles are
almost symmetrical with respect to the annual curve. They show
a maximum in the equatorial zone, with a slight seasonal migration
into the summer hemisphere, and also a monotonic decrease to polar
latitudes with the steepest gradient in the subtropics. As shown
in Table A2, the intra-annual variability in $[\overline{W}]$ is large except
over the equator.

The total water content of the atmosphere was evaluated by
using the expression $\{\overline{W}\} = 2\pi a^2 \int_{-(\pi/2)}^{(\pi/2)} [\overline{W}]\cos\phi \; d\phi$; the global value
obtained is of the order of 13.1×10^{15} kg. This value of the
water stored in the 'gaseous hydrosphere' agrees with previous
estimates provided by Starr *et al.* (1969), Peixóto (1972), Rosen
et al. (1979*b*). The seasonal change in the hemispheric water content is more pronounced in the northern hemisphere (JJA minus DJF
value is 3.0×10^{15} kg) than in the southern hemisphere (where
the comparable value is only 1.8×10^{15} kg).

The present value of $\{\overline{W}\}$, assuming a mean annual precipitation value of 1.0 m (Palmén & Newton 1969), gives a residence time
of water in the atmosphere of $\sim$ 9 days. This indicates that the
water vapour in the atmosphere is replenished about forty times
a year. As expected (and shown in Figure 4b), the variability,
estimated through the standard deviation, is largest for the year
since it includes the interseasonal variability, the maximum
variability occurring in the subtropics.

3.3 Zonal Transport of Water Vapour

The analyses of the field of total zonal transport of water
vapour $\overline{Q}_\lambda$ for yearly and seasonal conditions are presented in the
maps of Figure 5(a, b and c). The flow of water vapour reflects
the planetary behaviour of the general circulation in the lower
half of the atmosphere, since the specific humidity q acts as a
weighting factor for the wind field. Thus, the general pattern of
the $\overline{Q}_\lambda$ maps is consistent with the distribution of the mean zonal
flow, i.e., westerlies in mid-latitudes and easterlies in the
tropical belt.

The $\overline{Q}_\lambda$ field does not change substantially with season, except
for a northward displacement of the easterly belt in the JJA
analysis, and a less pronounced southward movement in the DJF map.
This intra-annual oscillation is accompanied by an intensification
of the $\overline{Q}_\lambda$ fields, mainly in the DJF season, in association with
the variations of the circulation around the subtropical anticyclones. The zero isoline coincides with the mean location of
the centres of the subtropical anticyclones in both hemispheres.

The zonal belt of easterly flow is interrupted over some continental areas, such as India, in the summer season associated with
the steady moist monsoon circulation. The westerly flow in mid-

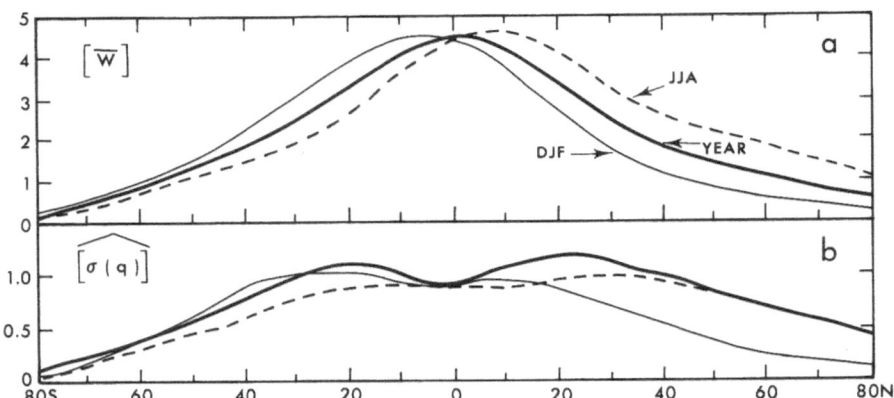

Fig. 4.   Meridional profiles of: a, the zonally-averaged precipitable water content of the atmosphere [W̄] (10 kg m⁻²);
b, the zonally- and vertically-averaged temporal standard

deviation of the specific humidity [σ̂(q)] (g kg⁻¹);
━━, year; ──, JDF; ---, JJA.

latitudes results from the circulations along the polar fronts and
along the polar borders of the subtropical anticyclones. In polar
regions, there is some evidence for weak westward flow, mainly
associated with the semi-permanent subpolar lows.
    The meridional profiles of $[\bar{Q}_\lambda]$ are presented in Figure 6.
These profiles summarize the main characteristics of the gross
zonal transport of moisture in the atmosphere. In each hemisphere,
the flow is from the west in mid to high latitudes, with a maximum
around 40° latitude. The southern hemisphere maxima exceed those
of the northern hemisphere. In tropical and equatorial latitudes,
the flux is westward with a bimodal distribution for the seasons,
the dominant maximum being in the winter hemisphere. The bimodal
configuration of the $[\bar{Q}_\lambda]$ is associated with the light and
variable winds over the equatorial trough. The intensity of the
westerly flow increases from winter to summer in the southern
hemisphere as the maximum moves to higher latitudes whereas, in
the northern hemisphere, the intensity decreases slightly from
winter to summer. The profile curves cross the zero line at 22°
latitude with a slight deviation poleward in the summer and
equatorward in the winter season.
    The vertical structure of the mean zonal flux is depicted in
the cross sections of [q̄u] given in Figure 7 and Table A4. They
show two well-developed centres of eastward flow in middle latitudes attaining their maximum values just above 850 mb. The
maximum in the southern hemisphere is larger than the corresponding
maximum in the northern hemisphere. The flow is still fairly large

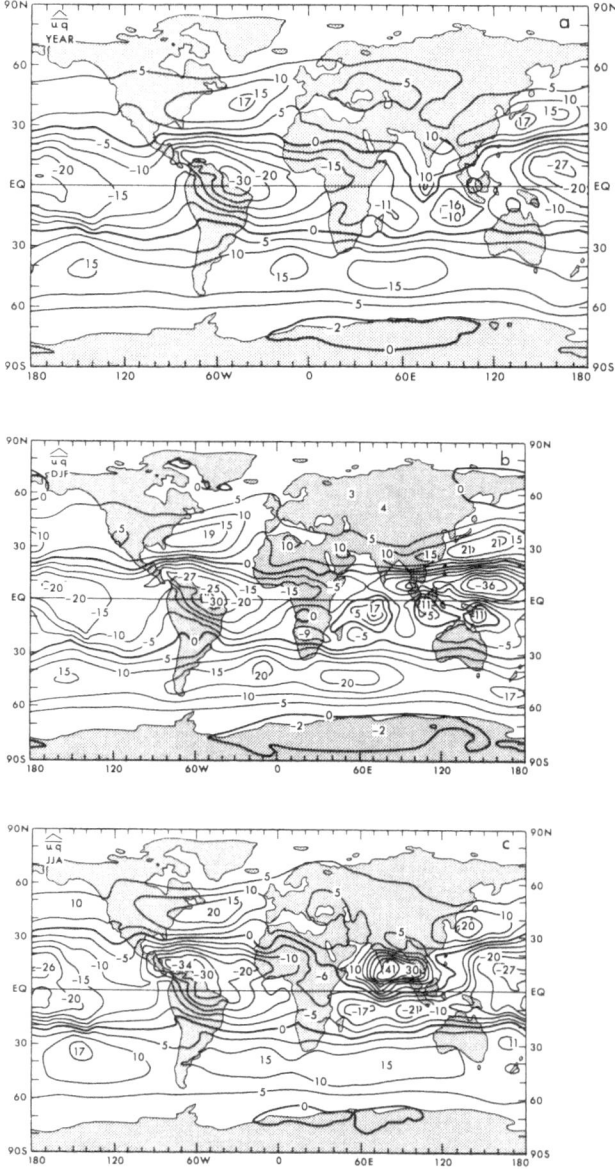

Fig. 5.   Planetary distributions of the vertically-averaged zonal
water vapour transport $\widehat{\overline{qu}}$ (g kg^{-1} m s^{-1}): a, year; b,
DJF; c, JJA. Negative values indicate westward flow.
To convert of $\overline{Q}_\lambda$ (10 kg m^{-1} s^{-1}), multiply by p_o/g
($\simeq$ 1 over the oceans).

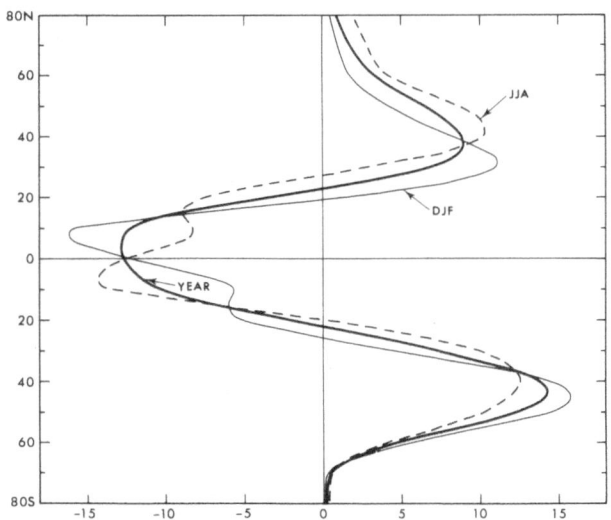

Fig. 6.   Meridional profiles of the mean-zonal water vapour transport $[\bar{Q}_\lambda]$ (10 kg m^{-1} s^{-1}): ▬▬, year; ——, DJF; - - -, JJA.

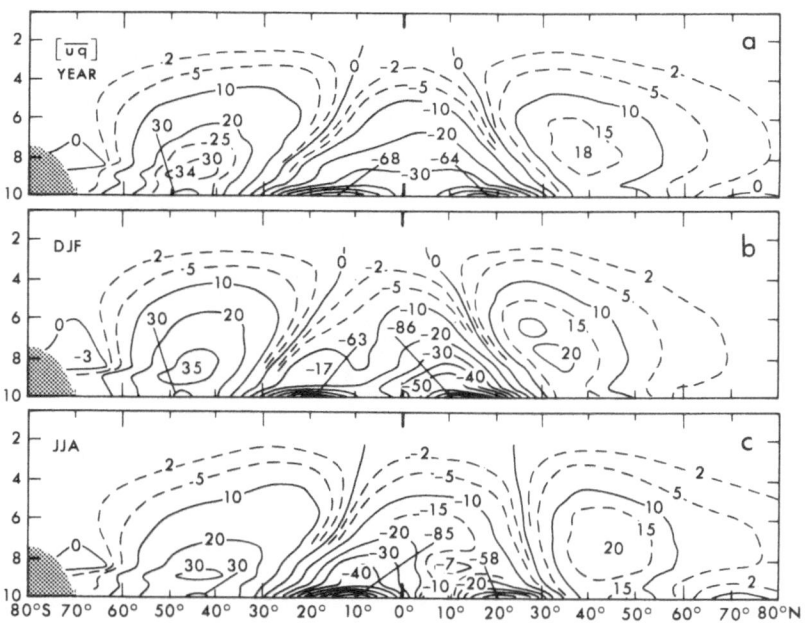

Fig. 7.   Vertical cross-sections of the mean-zonal water vapour
transport $[\overline{qu}]$ (g kg^{-1} m s^{-1}): a, year; b, DJF; c, JJA.

at 400 and 300 mb in both hemispheres, and should not be disre-
garded. Within the tropics, most of the westward flow occurs
below 800 mb with one main centre at the surface near 15° latitude
in each hemisphere. Maximum values are observed in the centre of
the winter hemisphere. Weak westward flow is also observed in
subpolar latitudes.

Inspection of the hemispheric and global mean values $\{\overline{Q}_\lambda\}$ in
Table A4 shows that the gaseous hydrosphere as a whole moves eastward, faster than the Earth, with relative values of 0.56, 0.83
and 0.49 m s^{-1} g kg^{-1} for the year, the DJF and JJA seasons,
respectively. However, the behaviour of the mean zonal fluxes for
the two hemispheres is very different. In the northern hemisphere,
the gaseous hydrosphere moves slower than the Earth (-0.54, -035
and -0.40 m s^{-1} g kg^{-1} for the years, DJF and JJA, respectively)
whereas it moves more rapidly in the southern hemisphere.

3.4 Meridional Transport of Water Vapour

The planetary distributions of the $\overline{Q}_\phi$ fields are presented in
Figure 8. The patterns reveal considerable detail and are not as
simple as those of the $\overline{Q}_\lambda$ fields. They reflect the main characteristics of the general circulation in the lower troposphere and
also show the effects of the inhomogeneity of the surface of the
globe, such as the land-sea contrast and variations of the topography. The $\overline{Q}_\phi$ values are two to three times smaller than the
corresponding $\overline{Q}_\lambda$ values. Nevertheless, the $\overline{Q}_\phi$ values are essential for the water balance over the Earth and play a very important
role in the energetics of the atmosphere. The most intense centres
of the $\overline{Q}_\phi$ fields are observed over the oceans and over the fringes
of the continents.

Although the meridional flux of moisture varies considerably
with the seasons over mid-latitude regions, it is predominantly
poleward throughout the year. This transport is mainly accom-
plished by baroclinic lows associated with the polar front and by
standing eddies, such as subpolar lows and subtropical anticyclones, together with their transient pulsations. The largest
variability during the year is associated with the movement and
changes in strength of the Hadley cells. The lower branches of
these cells are very effective in transporting moisture into the
ITCZ. During the JJA season, the monsoon effects are very intense
over India and Southeast Asia. These results agree with previous
climatological findings (Sellers 1965, Newton 1972, Viswanadham
et al. 1980).

Inspection of the $\overline{Q}_\phi$ maps shows a considerable cross-equatorial flow from the southern into the northern hemisphere during
the JJA season, and an opposite southward flux during the DJF
season. However, for the mean yearly conditions, there is a net
northward cross-equatorial flow of moisture. It explains the
excess of the mean annual precipitation over evaporation in the

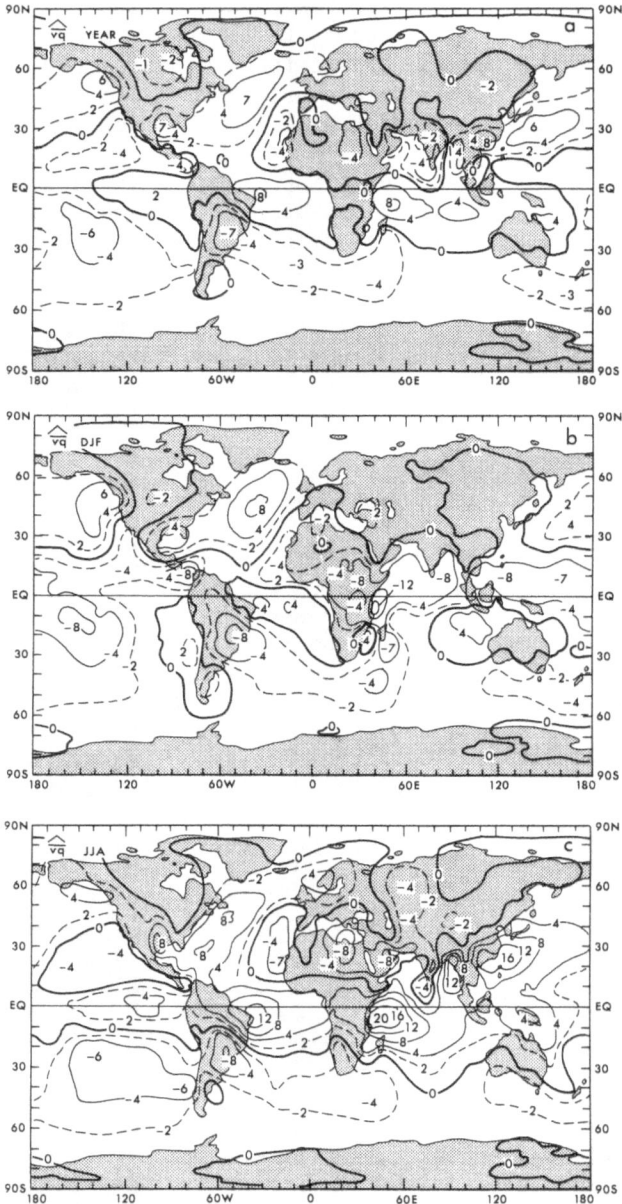

Fig. 8.   Planetary distributions of the vertically-averaged meri-
dional water vapour transport $\widehat{q\underline{v}}$ (g kg^{-1} m s^{-1}):   a, year;
b, DJF; c, JJA.   To convert to $\overline{Q}_\phi$ (10 kg m^{-1} s^{-1}), multiply
by p_o/g ($\simeq$ 1 over the oceans).

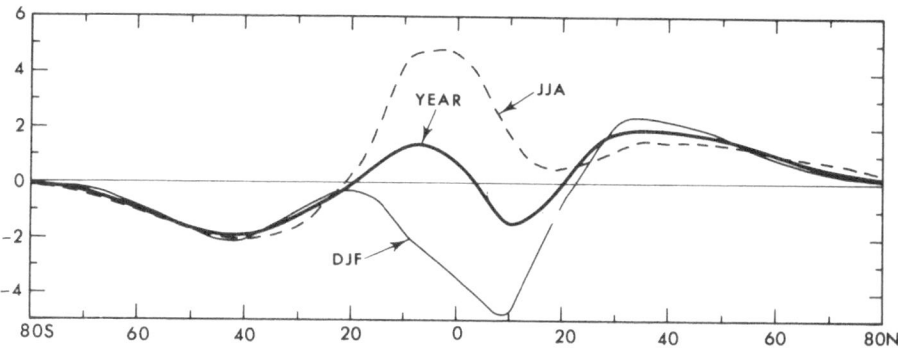

Fig. 9. Meridional profiles of the total meridional water vapour flux $[\bar{Q}_\phi]$ (10 kg m^{-1} s^{-1}) across latitudinal walls in the atmosphere: ▬, year; ——, DJF; ---, JJA. To convert to units of 10^8 kg s^{-1}, multiply by $4 \times \cos\phi$.

northern hemisphere (Palmén & Newton 1969). This flow feeds the ITCZ located, in the mean, to the north of the equator.

The $[\bar{Q}_\phi]$ profiles in Figure 9 synthesize the general behaviour of the Q_ϕ fields. The largest intra-annual fluctuations occur within the tropical regions. In mid-latitudes, the meridional transport is poleward in both hemispheres, with maxima near $40°$ latitude and with small seasonal variations.

In the tropical zone, the mean annual transport is positive south of the equator and negative to the north. There is a strong interaction between the hemispheres as shown by the profiles and the values in Table A5. The cross-equatorial flow presents a very marked seasonal variation both in intensity and sign. The flux into the northern hemisphere during JJA amounts to 18.8×10^8 kg s^{-1} and the outflow during DJF is about -13.6×10^8 kg s^{-1}. For the entire year, there is a net influx into the northern hemisphere of 3.2×10^8 kg s^{-1}. Thus, on an annual basis, the southern hemisphere supplies a considerable amount of water vapour to the northern. The cross-equatorial flow of water vapour implies (see Equation 8) an excess of precipitation over evaporation over the northern hemisphere of 39 mm yr^{-1} for the year as a whole and of 58 mm for the 3 months of the JJA season. The water vapour exported by the northern hemisphere during the DJF season corresponds to an excess of evaporation over precipitation in that hemisphere of about 42 mm per 3 months. These values are somewhat different from previous determinations using the aerological method (Peixóto *et al.* 1976) or classic climatological techniques (Palmén 1967, Budyko 1963), but are of the right order of magnitude in view of the uncertainties in estimating evaporation and precipitation.

The $[\overline{qv}]$ cross sections in Figure 10 and Table A5 show the

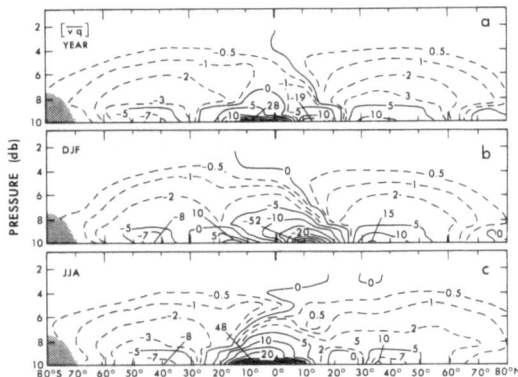

Fig. 10. Vertical cross-sections of the mean meridional water
         vapour transport $[\overline{qv}]$ (g kg^{-1} m s^{-1}): a, year; b, DJF;
         c, JJA.

vertical structure of the meridional transfer of water vapour and
its intra-annual variability. The largest values occur closer to
the surface than in the case of the zonal transport $[\overline{qu}]$.

3.5 Vertical Transport of Water Vapour

     The vertical transport of water vapour in the atmosphere plays
an essential role in the hydrological cycle, since it links the
terrestrial and aerological branches. However, it is very difficult to assess the instantaneous value of ω, or even of its local
time average $\overline{\omega}$, due to the present inadequacy of the network of
upper-air stations.
     The mean 'vertical' velocity was determined from the conti-
nuity equation by vertically integrating the horizontal divergence
from the Earth's surface upwards, assuming that $\overline{\omega}$ vanishes at the
ground,

$$\overline{\omega}(p) = \int_{p}^{P_o} \text{div } \overline{\vec{v}} \; dp. \tag{18}$$

The $\overline{u}$ and $\overline{v}$ fields were adjusted by subtracting the vertical mean
divergence value at each level. Thus the divergence for an entire
column of the atmosphere is equal to zero. The values of $\overline{\omega}$
obtained represent only the large-scale mean vertical motion.
Obviously, the $\overline{q\omega}$ flux does not include the transient eddy vertical
flux, nor the contributions from subgrid scale eddies. Neverthe-
less, the $\overline{q\omega}$ field at 850 mb, where the mean vertical flux of
water vapour attains high values, is given in Figure 11.
     Although the analyses are tentative, it is encouraging that
the planetary distributions appear to be consistent in time and

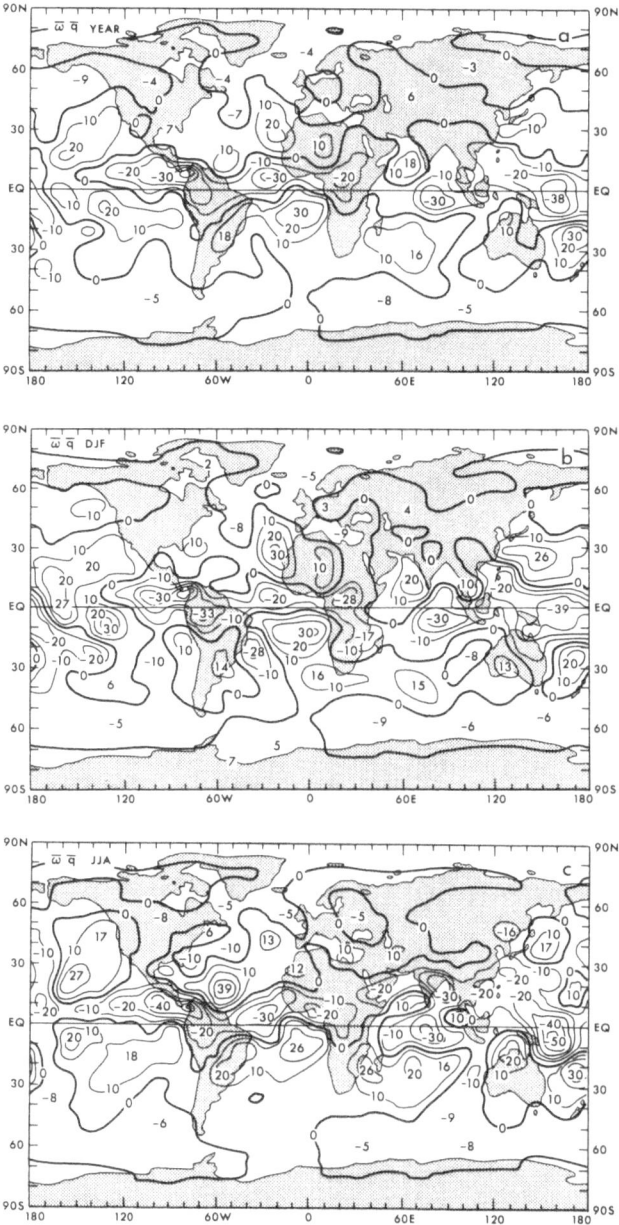

Fig. 11. Planetary distributions of the vertical water vapour
          transport by mean motions, q̄ω̄, at the 850 mb level
          (10^{-6} kg m^{-2} s^{-1}): a, year; b, DJF; c, JJA. Negative
          values indicate upward transport.

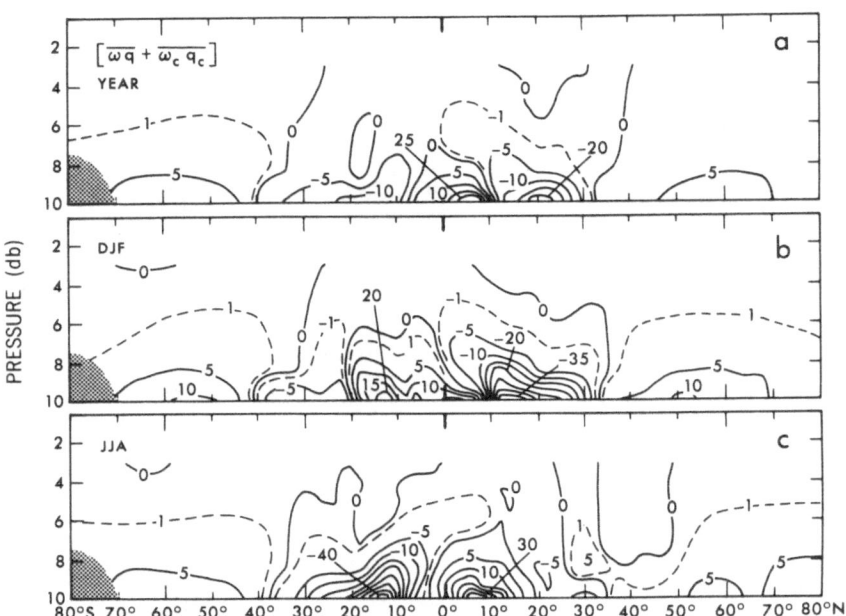

Fig. 12. Vertical cross-sections of the mean vertical water trans-
port $[\overline{q\omega} + \overline{q_c \omega_c}]$ (10^{-4} g kg^{-1} mb s^{-1}; $\simeq 10^{-6}$ kg m^{-2} s^{-1}):
a, year;   b, DJF; c, JJA.

space. For example, the flux is upward over the equatorial region
and at high latitudes, and downward in the subtropics, showing
clear seasonal shifts, generally toward the summer pole. The
belt of maximum upward transport over the equatorial region is
associated with the ascending branches of the Hadley cells, whereas the upward flux in middle and high latitudes must be due to the
semi-permanent lows. The centres of maximum downward flux occur
mainly in the eastern parts of the subtropical anticyclones over
the oceans due to prevailing subsidence.

Using the $[\overline{qv}]$ values and appropriate boundary conditions,
the ψ_q function was computed, from Equation (15a), and will be
discussed later. With this streamfunction, the total vertical
transport of water substance $[\overline{q\omega} + \overline{q_c \omega_c}]$ can be estimated with
the aid of Equation (15b). A similar technique was used previously
for angular momentum and energy (Starr *et al.* 1970, Oort & Peixóto
1982). In spite of the simplifications involved, the results, as
shown in Table A9 and Figure 12, seem to be consistent with the
general behaviour of the hydrological cycle.

Most of the vertical transport occurs in the layer below
700 mb with the highest values near the surface. The annual
distribution is almost symmetric with respect to the equator. The

subtropical regions of upward flux, where evaporation prevails at
the surface, are bordered by the equatorial region, and two mid-
latitude regions of downward flux, where precipitation exceeds
evaporation. In the winter hemisphere, there is an intensification of the pattern of vertical transport accompanied by a shift
toward the equator.

3.6 Mean Divergence of Water Vapour

Using the grid point values of $\overline{Q}_\lambda$ and $\overline{Q}_\phi$ the divergence of
water vapour in the atmosphere div $\vec{Q}$ was evaluated, applying the
approach and techniques already explained on several occasions
(e.g., Peixoto 1973). The analyses of the divergence fields for
mean yearly and seasonal conditions are represented in Figure 13.

The importance of the divergence field of the water vapour in
the atmosphere depends on its relationship with the mean difference
of evaporation and precipitation $(\overline{E-P})$, as shown earlier. Thus,
the divergence maps are of great interest for the study of the
planetary water balance. The regions of mean positive divergence
$(\overline{E-P} > 0)$ constitute the main sources of water vapour for the
atmosphere, whereas the centres of convergence $(\overline{E-P} < 0)$ are
regions of water vapour sinks.

The present analyses agree quite well with the pole-to-pole
analyses evaluated from IGY data (e.g., Starr *et al.* 1969, Peixóto
1972). Furthermore, the analyses also agree with Budyko's (1963)
maps of $\overline{E-P}$ as obtained by classic climatological methods. However, our JJA analysis shows an unexpected intense centre of
divergence just west of India that may be due to the sparseness of
data over the Indian Ocean and to the sensitivity of the divergence computations. Convergence generally prevails over the
equatorial and middle to high latitude zones, while divergence
predominates in the subtropics. The convergence and divergence
centres are, as a rule, more intense over the oceans than over
land. The equatorial convergence of water vapour is associated
with the ITCZ. Water vapour is carried towards the region of mean
rising motion by the lower branches of the Hadley cell, leading to
heavy precipitation. The belt of convergence consists of various
centres located over the headwaters and drainage basins of large
river systems, such as the Amazon in South America, the Ubangi,
Congo, Senegal and Blue Nile in Africa and the Indus, Ganges,
Mekong and Yangtze in Southeast Asia. The seasonal shifts of the
equatorial belt of convergence, to the north in the JJA season and
to the south in the DJF season, are related to the migration of
the rising branches of the Hadley cells.

The subtropical belts of divergence coincide largely with
the arid zones of the globe. They are associated with strong
evaporation over the oceans and with subsidence that prevails over
the large subtropical anticyclones. These regions constitute the
main sources of moisture for the entire atmosphere. Their lati-

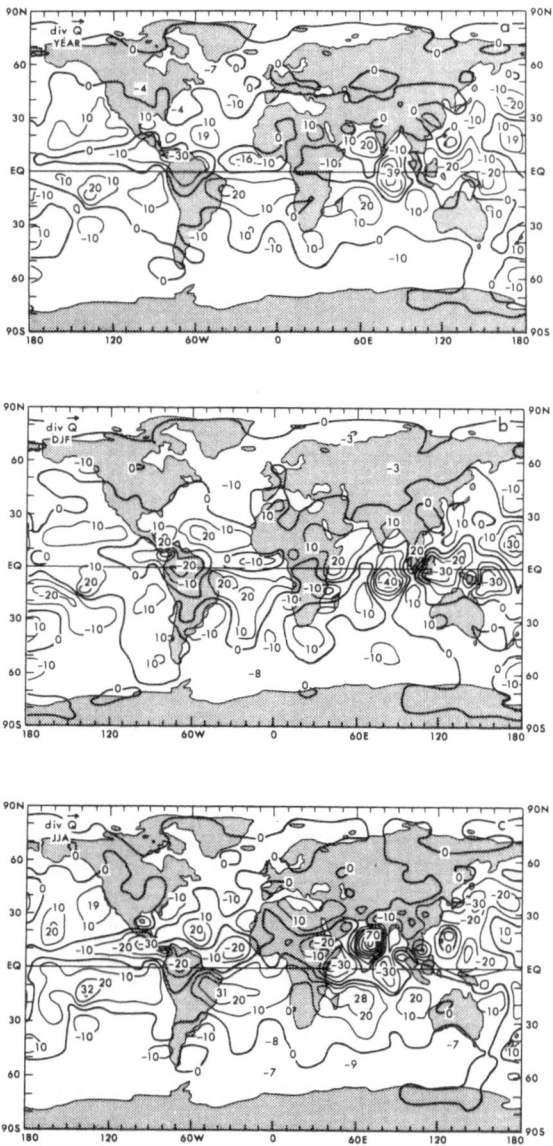

Fig. 13.   Planetary distributions of the horizontal divergence of
           the vertically-integrated total water vapour transport,

           div $\overline{\vec{Q}}$ (10 cm yr^{-1}):   a, year; b, DJF; c, JJA.   Negative
           values indicate convergence of moisture.

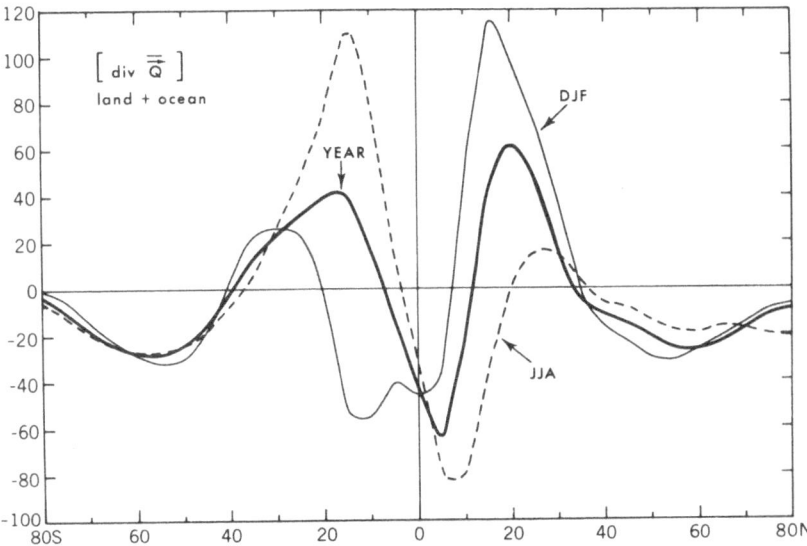

Fig. 14. Meridional profiles of the zonally-averaged divergence of
the water vapour transport [div $\vec{Q}$] (cm yr^{-1}): ━━, year;
─── , DJF; ---, JJA.

tudinal oscillation follows the yearly movement of the subtropical
anticyclones. The distribution of divergence is easy to understand over the oceans where there is always water available for
evaporation and the prevailing ocean currents will advect the
necessary fresh water to maintain equilibrium. The situation,
however, is more difficult to explain when divergence occurs over
land. In this case, surface and underground flows from less arid
regions must supply the water required to counterbalance the
observed excess of evaporation over precipitation (Starr & Peixóto
1958).

The middle to high latitude convergence in both hemispheres
is mainly associated with the transient baroclinic lows that
accompany the polar front. Over the polar regions, there are
some indications of a slight divergence, especially in the vicinity of the north pole.

The water balance equation (6), for a zone bounded by the
latitudes ϕ and $\phi + \Delta\phi$, can be transformed by applying the [ ]
operator, and can be reduced to

$$\Delta[\overline{Q}_\phi] \cos\phi/a \cos\phi \; \Delta\phi = [\overline{E-P}] \qquad (19)$$

Using this expression, the mean zonally-averaged divergence of the
total water vapour flux was calculated for the year and the DJF
and JJA seasons. The profiles are shown in Figure 14 and the
numerical estimates for 5°-latitude belts are given in Table I.

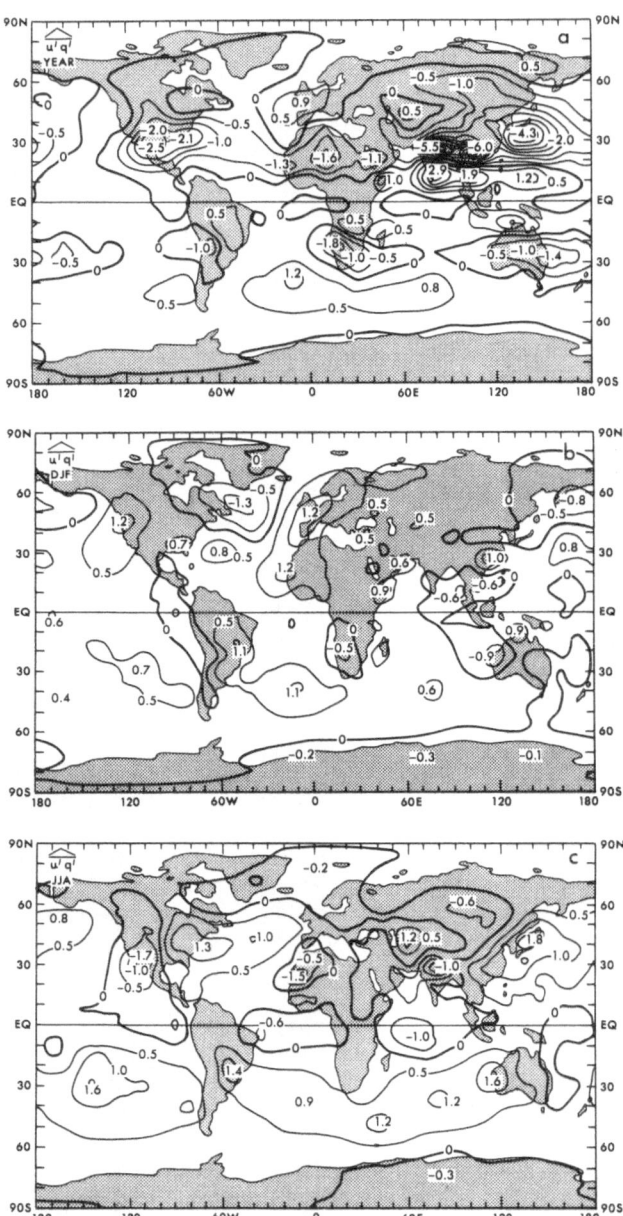

Fig. 15.   Planetary distributions of the vertically-averaged zonal
water vapour transport by transient eddies, $\widehat{q'u'}$ (g kg^{-1}
m s^{-1}): a, year; b, DJF; c, JJA. Negative values
indicate westward flow.

The profiles and tabulation synthesize the main features of the divergence maps already discussed. In all cases, they indicate a strong convergence with an excess of mean precipitation over mean evaporation, $[\overline{E-P}] < 0$, over the equatorial region due to convergence of the trade winds of both hemispheres. This convergence is responsible for the heavy precipitation observed over this region (see Figure 23). Other convergence zones with an excess of precipitation are found over middle and high latitudes. Strong divergence occurs in the subtropics. The divergence intensifies during the winter season. As can be seen from Figure 23, precipitation has a minimum whereas evaporation shows a maximum in each of the subtropical zones. It is worth noting that the values given in Table I for $\overline{E-P}$, obtained from the zonal mean cross-sections in Table A5, agree closely, as they should, with the values of the vertical transport of water substance at 1000 mb, as given in Table A9.

It is very instructive to superimpose the $[\overline{Q}_\phi]$ and $[\text{div } \overline{\vec{Q}}]$, profiles from Figures 9 and 14, since the latter profiles are the derivatives of the $[\overline{Q}_\phi]$ curves. They show a large export of water vapour from the zones of divergence and a strong import of moisture into the belts of convergence. The actual numerical values of $[\text{div } \overline{\vec{Q}}]$, as shown in Table I, are in fairly good quantitative agreement with previous findings (e.g., Starr et al. 1958, 1969, Peixóto 1972) obtained using a similar technique or using the classical climatological approach (Sellers 1965, Palmén & Newton 1969, Newton 1972, Jaeger 1976). The divergence profiles over the oceans (not shown here) do not differ much qualitatively from those for the entire globe, showing basically the same convergence and divergence zones.

4.   MODES OF WATER VAPOUR TRANSFER IN THE ATMOSPHERE

4.1   Zonal Flux of Water Vapour by Transient Eddies

The analyses of the field of total zonal eddy-transport of water vapour, in Figure 15, present a considerable degree of detail with centres of eastward (positive) zonal eddy flow alternating with centres of westward (negative) eddy transport in both hemispheres.

The distribution of the centres over the globe is associated with the land-sea contrast, and with the invasion of moist air masses of maritime origin into the continents. On the DJF map (Figure 15b), extensive regions of eastward flow are located over the west coasts of North America, Europe and North Africa, and over the east coast of South America. Simultaneously, the east coasts of Canada and of Asia are under the influence of westward flow, as are the west coasts of South America, South Africa and Australia.

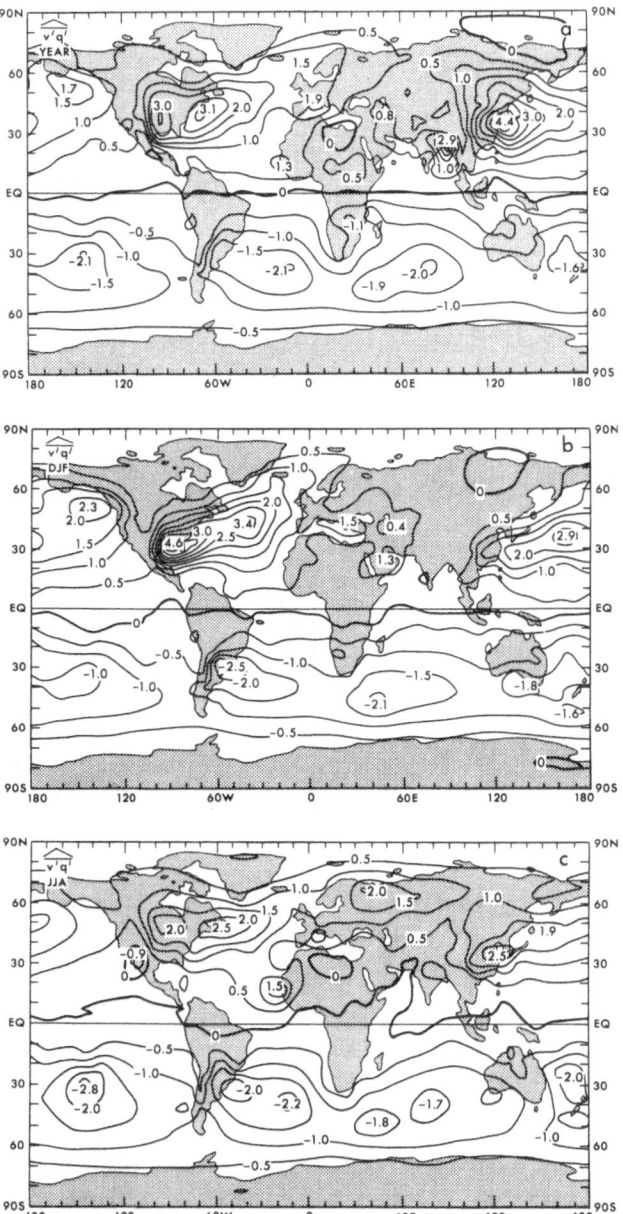

Fig. 16.   Planetary distributions of the vertically-averaged meridional water vapour transport by transient eddies, $\widehat{\overline{q'v'}}$
(g kg^{-1} m s^{-1}):   a, year;   b, DJF;   c, JJA.   Negative
values indicate southward flow.

During the JJA season (Figure 15c), the DJF conditions are
almost completely reversed. Centres of eastward flow are now
encountered over the east coasts of North America and Asia, and
eastward flow dominates in the southern hemisphere. Moist westward flows are found over the west coasts of North America,
southern Europe and North Africa. Over southern Asia and the
Indian Ocean, the seasonal variations are linked with the Asian
monsoon and perturbations along the ITCZ.

The year map (in Figure 15a) reveals the strong influence of
the continental character of the northern hemisphere. In this
hemisphere, the moist zonal eddy flow is predominantly westward,
with an extensive longitudinal belt of eastward flow at 10°N. In
the southern hemisphere, the flow is mostly eastward, except over
Australia, South Africa, South America and Antarctica where the
flow is westward.

We must point out that, since the transient eddy fluxes
evaluated for an annual period include the interseasonal variability, they are not equal to the average of the DJF and JJA
fluxes.

It may be noted that the maps of the zonal transient eddy
transport of water vapour $\overline{Q'_\lambda}$ show very little similarity with the
corresponding maps of the total zonal flux $\overline{Q}_\lambda$ (see Figure 5).
Indeed, they are much less orderly and their intensities are at
least one order of magnitude smaller than those for the total flux
(Peixoto *et al.* 1978).

4.2 Meridional Flux of Water Vapour by Transient Eddies

The spatial distributions of the vertically-integrated meridional transient eddy flux of water vapour are presented in
Figure 16; positive values indicate northward flow. In contrast
to the zonal eddy flow, the present patterns are much better
defined, with predominantly poleward flow in each hemisphere
throughout the year. Furthermore, the eddies are much more effective in transporting water vapour in the meridional than in the
zonal direction, when compared with the corresponding total tran-
sports. In fact, the configurations of the $\overline{Q'_\phi}$ and $\overline{Q}_\phi$ fields are
very similar in middle latitudes.

The analyses of the $\overline{Q'_\phi}$ fields show belts of strong poleward
flux in middle latitudes of both hemispheres. Near the poles and
in the tropics, the meridional eddy transports are small, almost
vanishing over the equator. The centres of poleward flow intensify and expand during the winter, reflecting the increased
activity along the polar front and its movement to the poles
during summer. The influence of the continents is evident in the
location of the maxima just east of the continents, especially in
the northern hemisphere. The maxima in the southern hemisphere
are less intense and more zonally uniform, so that the zonal mean
poleward fluxes are of the same magnitude in both hemispheres.

The vertical structure of the transient eddy meridional tran-

sport of water vapour can be inferred from the values of $[\overline{q'v'}]$
given in Table A6. The flow is directed polewards at all latitudes and levels reaching maximum intensity in middle latitudes
below 850 mb. The intensity is higher in winter than summer, with
a slight equatorward shift of the flow. Inspection of the $[\overline{q'v'}]$
cross-sections confirms that the vertical distribution of [qv]
(see Table A5) in middle latitudes is mainly determined by the
transient eddy flux. The same does not occur in the case of the
zonal eddy flux of moisture, as noted earlier.

4.3 Other Modes of Meridional Flux of Water Vapour

The contributions by standing eddies to the mean meridional
flux of water vapour are shown in Table A7, the vertically-integrated values $[\overline{Q^*_\phi}]$ being given at the foot of the table. The
contributions of the standing eddies to the total transport are
smaller and less coherent than those of the transient eddies. The
centres of northward flux in the northern hemisphere are more pronounced and subject to larger seasonal variations in position and
in intensity than the centres of southward flow in the southern
hemisphere. Their intensity increases during the summer in contrast to the behaviour of the transient eddy centres. The
standing eddy transport is associated with the quasi-stationary
features of the general circulation, particularly with the subtropical anticyclones and the semi-permanent lows at high
latitudes. The maximum observed in the northern hemisphere
between 20 and 30°N is reinforced during the JJA season by the
monsoon circulation over southern Asia.
The contributions by the mean meridional circulations[1] to
the total flux of water vapour are given in Table A8. The results
reveal a three-cell circulation structure (Starr *et al.* 1969) in
both hemispheres. The low-level branches of the Hadley cells
give the largest contributions to the total meridional flux of
water vapour over the tropics due to large values of $[\overline{v}]$ combined
with high values of $[\overline{q}]$. The upper branches of the cells do not
contribute much because of low humidities at these levels. In
contrast with the Hadley cells, the other cells are unimportant
compared with the other modes of transfer.
The seasonal variations in the Hadley cells are considerable.
The winter cell is most active and constitutes the dominant
mechanism for transferring water vapour into the summer hemisphere.
The consequences of the invasion into the northern hemisphere by
the Hadley cell of the southern hemisphere winter are particularly
decisive for the water balance of the globe. In fact, it is the
moisture supplied by the southern hemisphere that provides almost
all the water falling as precipitation during the summer monsoon
of the northern hemisphere.
The vertically-integrated values of the various modes of
meridional transport, shown at the foot of each table and in
Figure 17, synthesize many of the features already discussed; they

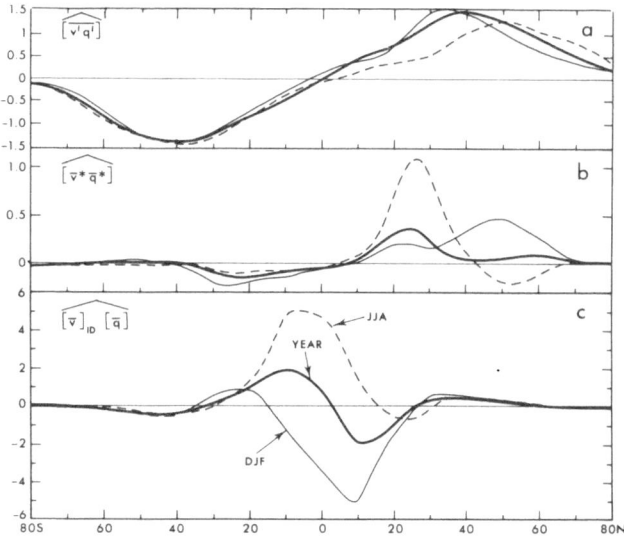

Fig. 17. Meridional profiles of the zonally- and vertically-
averaged values of northward water vapour transport
(g kg^{-1} m s^{-1}) by: a, transient eddiies [$\widehat{q'v'}$];
b, standing eddies [$\widehat{q*v*}$]; c, mean meridional circulations [$\widehat{q}$][$\widehat{v}$].

should be compared with the [$\overline{Q}_\phi$] profiles shown in Figure 9. From
the tabulated values of [$\overline{Q}'_\phi$], [$\overline{Q}^*_\phi$] and [$\tilde{Q}_\phi$], the divergence for
5°-latitude belts was evaluated using Equation (19) and presented
in Table I. The transient eddies, showing divergence in the
tropics, transport into middle and high latitudes part of the
moisture supplied by the Hadley circulation. The standing eddies
also contribute to divergence in low latitudes and convergence in
middle latitudes in both hemispheres. However, they are more
active in the northern hemisphere. Thus, the subtropics act as
the main sources of moisture for the entire globe.

4.4 Modes of Vertical Flux of Water Vapour

An expansion for [$\overline{q\omega}$] similar to Equation (16) can be derived.
Such terms as [$\overline{q*\omega*}$] and [$\overline{q}$] [$\overline{\omega}$] can be computed directly from our
data. Thus, since the total [$\overline{q\omega + q_c\omega_c}$] can be evaluated independently using the streamfunction approach, an idea of the contribu-
tion of the transient eddies, comprising the effects of tropical
disturbances and cumulus convection can be secured. However, this
contribution will also include the vertical transport in the liquid
phase.

Table I. Zonal mean divergence of the vertically-integrated atmospheric water vapour flux $\Delta[\overline{Q}_\phi]\cos\phi/a\ \cos\phi\Delta\phi(=\overline{E}-\overline{P})$ (cm yr^{-1}), by the various modes for the year and for the DJF and JJA seasons.

		-85°S	-80°	-75°	-70°	-65°	-60°	-55°	-50°	-45°	-40°	-35°	-30°	-25°	-20°	-15°	-10°	-5°	0°	5°
YEAR	TOTAL	-2.9	-3.8	-8.6	-18.1	-25.8	-28.4	-28.5	-25.8	-16.8	-0.7	14.0	23.7	30.9	39.3	40.4	16.7	-15.3	-43.9	-64.1
	TE	-5.4	-5.5	-9.1	-16.5	-21.5	-22.2	-20.1	-15.6	-11.3	-5.7	1.6	7.6	9.4	8.1	9.2	12.1	12.8	12.6	12.7
	SE	-1.1	-0.8	-0.3	0.1	0.3	0.4	0.4	-0.1	-0.1	-0.9	-3.1	-4.1	-2.2	0.8	2.0	1.7	0.8	0.9	2.0
	MMC	3.6	2.5	0.8	-1.7	-4.6	-6.6	-8.9	-10.2	-5.4	5.8	15.6	20.2	23.7	30.4	29.3	3.0	-29.0	-57.4	-78.9
DJF	TOTAL	-1.9	-2.2	-5.5	-14.2	-23.3	-29.2	-32.4	-30.4	-18.8	4.8	24.5	24.7	25.4	-2.0	-54.3	-55.5	-40.0	-46.4	-37.5
	TE	-4.6	-4.1	-6.6	-13.4	-19.2	-22.6	-22.6	-17.9	-12.7	-5.1	3.9	9.1	11.2	11.6	12.2	12.0	-10.5	9.7	7.0
	SE	-0.9	-0.6	-0.1	0.3	0.5	0.7	0.7	-0.4	-1.3	-3.0	-5.7	-4.4	-0.4	1.6	1.1	1.8	2.1	1.9	2.0
	MMC	3.6	2.5	1.3	-1.1	-4.6	-7.3	-10.6	-12.1	-4.7	12.9	26.2	20.1	14.5	-15.2	-67.7	-69.3	-52.6	-58.0	-46.5
JJA	TOTAL	-3.6	-5.8	-11.8	-20.0	-25.1	-26.5	-27.4	-25.5	-18.3	-6.2	6.7	27.4	48.8	76.0	111.8	79.7	13.7	-32.6	-79.4
	TE	-4.5	-6.1	-11.4	-19.0	-22.8	-21.5	-18.3	-14.5	-11.6	-7.4	-0.9	6.5	11.5	12.7	12.4	11.0	8.5	5.9	5.6
	SE	-1.0	-0.9	-0.6	-0.3	-0.1	-0.2	-0.2	-0.3	0.1	0.0	-1.6	-2.6	-1.1	0.3	0.2	0.2	1.0	2.1	3.6
	MMC	1.9	1.2	0.2	-0.7	-2.2	-4.9	-8.8	-10.7	-6.8	1.1	9.2	23.5	38.5	62.9	99.3	68.4	4.2	-40.6	-88.7

		10°N	15°	20°	25°	30°	35°	40°	45°	50°	55°	60°	65°	70°	75°	80°	85°N	SH	NH
YEAR	TOTAL	-19.3	43.2	61.8	48.8	14.7	-6.2	-11.5	-15.0	-21.1	-26.0	-27.1	-23.1	-16.4	-11.7	-9.8	-9.0	3.9	-3.9
	TE	9.9	6.9	9.5	11.6	7.7	0.9	-6.5	-15.1	-11.8	-17.9	-19.3	-18.2	-15.7	-13.7	-11.4	-9.9	0.0	0.0
	SE	5.4	8.2	4.9	-3.4	-5.5	-1.4	0.2	1.1	0.7	-1.2	-2.5	-1.7	-0.4	-0.5	-0.7	0.3	-0.3	0.3
	MMC	-34.6	28.1	47.4	40.6	15.6	-3.7	-3.4	-7.1	-8.8	-6.6	-2.4	1.0	2.4	2.0	1.6	4.2	-4.2	
DJF	TOTAL	48.6	116.2	96.1	73.0	35.9	-3.7	-16.6	-22.9	-30.5	-31.6	-26.1	-20.4	-14.3	-8.1	-7.4	-8.5	-16.6	16.6
	TE	5.6	9.7	15.8	16.0	5.6	-6.2	-12.9	-17.1	-17.9	-16.1	-13.5	-12.0	-11.0	-8.9	-8.4	-8.9	0.6	-0.6
	SE	3.8	5.3	2.1	-1.6	0.2	4.3	4.6	1.6	-3.9	-8.1	-8.8	-7.8	-4.2	-0.7	-0.3	-0.7	-0.4	0.4
	MMC	39.3	101.1	78.2	58.5	30.0	-1.8	-8.3	-7.4	-8.7	-7.5	-3.8	-0.6	0.9	1.4	1.3	1.1	-16.8	16.8
JJA	TOTAL	-80.4	-37.8	0.9	15.7	14.9	2.6	-6.3	-8.0	-13.8	-18.6	-18.9	-16.2	-17.4	-20.8	-20.3	-18.5	23.2	-23.2
	TE	6.7	5.3	2.9	3.4	7.6	10.7	7.6	1.9	-6.0	-14.4	-17.1	-16.6	-19.1	-22.0	-20.7	-18.8	-0.3	0.3
	SE	7.3	17.5	20.9	1.5	-20.5	-22.3	-14.0	-7.8	-1.7	3.6	5.2	3.5	1.1	-0.5	-0.8	-0.5	-0.2	0.2
	MMC	-94.5	-60.6	-22.9	10.8	27.8	14.3	-0.0	-2.1	-6.1	-7.8	-7.0	-3.2	0.6	1.7	1.1	0.8	23.7	-23.7

The standing eddy vertical flux of water vapour, $[\overline{q^*\overline{\omega}^*}]$ in Table A10, shows two main centres of upward transport near $22°N$ and $12°S$. Their intensity increases from winter to summer while their height changes from 850 to 700 mb. The standing eddy fluxes are more pronounced in the northern hemisphere. The mean vertical standing eddy transport is, by and large, upward over the entire globe and away from the Earth's surface leading to convergence and, thus, to condensation above the level of maximum transport.

The vertical transport of water vapour due to mean meridional circulations[1], $[\overline{q}]\,[\overline{\omega}]$ in Table A11, shows an hemispheric three-cell structure with a strong Hadley cell in the tropics, a weak indirect Ferrel cell in mid-latitudes and a very weak polar cell. The upward flux in the Hadley cells extends into the upper tropo-sphere with a maximum around 850 mb. A general downward flow prevails over the subtropics and lower mid-latitudes with highest values around 850 mb during the winter season.

The values of the 'transient' term, $[\overline{q'\omega'} + \overline{q_c\omega_c}]$ in Table A12, when compared with those of the other modes, dominate at the lowest levels where water vapour is much more abundant. Furthermore, its vertical distribution is very similar to that of the total tran-sport, although the eddy penetration appears to be deeper in the tropics. Thus, we conclude that the 'transient' modes constitute the principal mechanism in the total vertical transport. However, the transient eddies, *per se*, must always transport water vapour upward over the globe. Of course, the contribution of $[\overline{q_c\omega_c}]$ exceeds that of the transient effects in the equatorial region and in middle and high latitudes, where precipitation prevails. On the other hand, in the subtropics, the vertical transport must occur mainly in the vapour phase, where evaporation dominates and preci-pitation is sparse. The high penetration in the subtropics is related to deep trade-wind cumulus convection.

The profiles in Figure 18 summarize the relative contributions by the three modes and the total. They show that the vertical flux by the transient eddies, ranging in scale from organized convection to synoptic disturbances, must be at least as important for the total vertical transport as the mean meridional circulations. However, the latter oppose the transient eddy effects in the sub-tropics.

5.   IMPLICATIONS OF THE ATMOSPHERIC BRANCH FOR THE CLIMATIC SYSTEM

5.1 Aerial Runoff of Water

Our discussion of the role of the general circulation in the hydrological cycle commences with an analysis of the vertically-integrated atmospheric moisture flow in terms of the $\vec{Q}$ vector field, the aerial runoff.

By combining the $\overline{Q}_\lambda$ and the $\overline{Q}_\phi$ fields of Figures 5 and 8, the

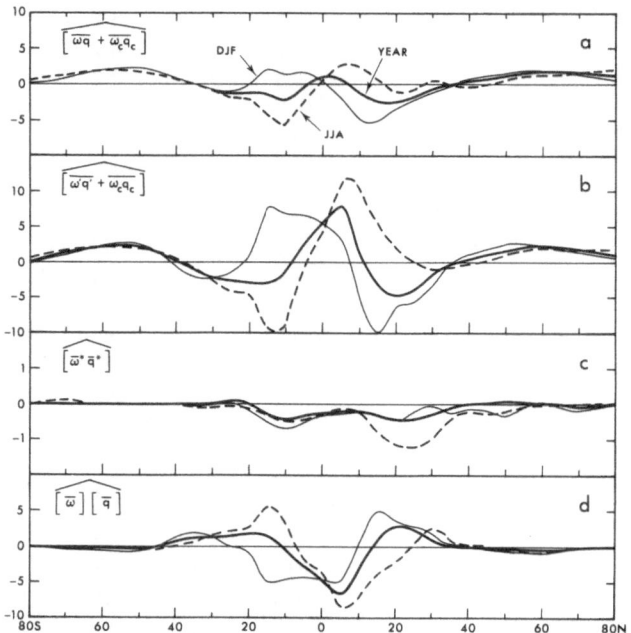

Fig. 18. Meridional profiles of the zonally- and vertically-
averaged values of vertical water vapour transport
(10^{-4} g kg^{-1} mb s^{-1}; $\simeq 10^{-6}$ kg m^{-2} s^{-1}); a, total
$[\overline{q\omega} + \overline{q_c \omega_c}]$; b, by transient eddies $[\overline{q'\omega'} + \overline{q_c \omega_c}]$;
c, by standing eddies $[\overline{q^*\omega^*}]$; d, by mean meridional
circulations $[\overline{q}][\overline{\omega}]$.

vector transport fields of water vapour in the atmosphere $\vec{Q}(\lambda,\phi)$
were obtained, as shown in Figure 19. In addition to the vectors,
streamlines were drawn. The maps are in good agreement with
similar maps based on data from other sources (Drozdov & Grigoreva
1963). In a hypothetical steady state, the streamlines show the
prevailing paths of water vapour in the atmosphere after its
release from the various source regions at the Earth's surface
(see also Rosen et al. 1979a). Figure 19 shows that the main
sources of water vapour for the atmosphere are localized over the
subtropical oceans. The $\vec{Q}$ maps provide a good indication of the
prevailing movements of the main moist air masses in the atmosphere and the sites of their formation. The $\vec{Q}$ maps also clearly
show that the water vapour necessary for precipitation over the
continents comes from the oceans. In steady state conditions, this
net inflow of moisture to the continents is compensated by the
runoff of the rivers into the oceans (see Equation 10).

a

STREAMLINE
VERTMN
DJF 63-73

b

STREAMLINE
VERTMN
JJA 63-73

c

STREAMLINE
VERTMN
YR 63-73

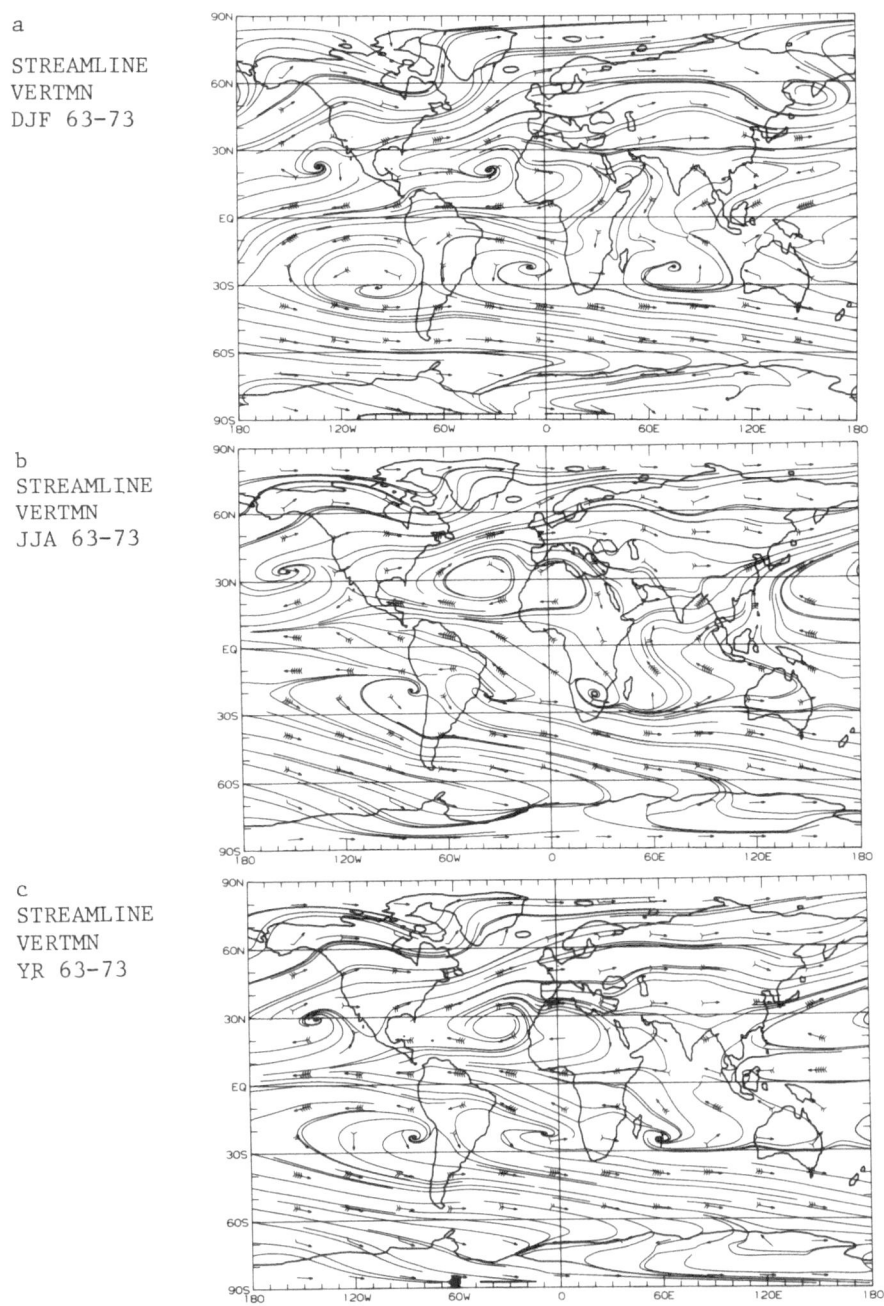

Fig. 19. Planetary distributions of the total aerial runoff, $\dot{Q}$,
and some corresponding streamlines; each barb represents
2 m s^{-1} g kg^{-1}: a, year; b, DJF; c, JJA.

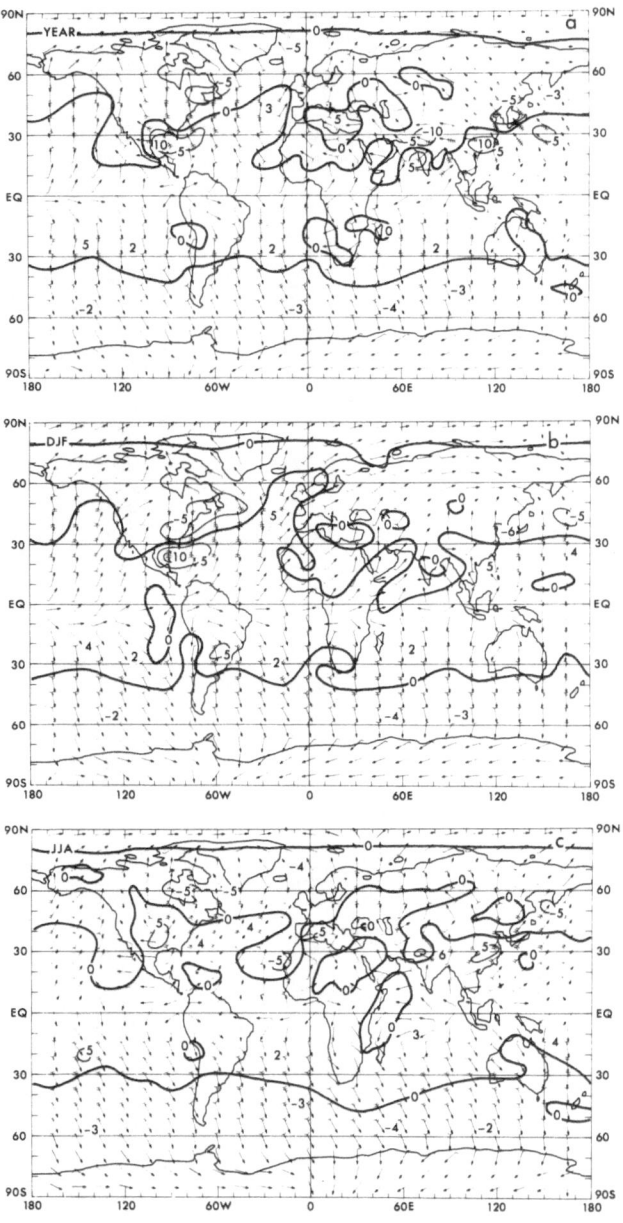

Fig. 20.   Planetary distributions of the water vapour flux diver-
           gence (10 cm yr^{-1}) and the aerial runoff by transient

           eddies, $\vec{Q}'$; each barb represents 0.5 m s^{-1} g kg^{-1}:
           a, year; b, DJF; c, JJA.

When the hydrosphere of the climatic system is considered *in toto*, there must be, on the average, a compensating meridional flux of water substance in the terrestrial hydrosphere opposite to that which is observed in the gaseous hydrosphere. Thus, balance considerations require a small net southward transport of water by ocean currents and rivers in mid-latitudes of the northern hemisphere, as well as across the equator in the subequatorial zone. At other latitudes, the net meridional flux of liquid water must be northward. Preliminary results (Hellerman 1981) indicate that there is a detectable meridional runoff by major rivers, although it is much smaller than the net contribution by ocean flows.

The eddy component of the aerial runoff $\vec{Q}'$ was obtained from the $\overline{Q}'_\lambda$ and $\overline{Q}'_\phi$ maps. As inspection of Figure 20 shows, the fields are now predominantly meridional. These fields are strongest in winter, when frontal activity is most intense. The $\vec{Q}'$ fields lead to marked divergence from the tropics as the superimposed isolines of divergence show. In middle and high latitudes, transient eddy convergence prevails.

The $\vec{Q}'$ fields illustrate clearly the importance of the eddy mechanisms for the achievement of regional water balances through the transport of moisture from the oceans deep into the mid-latitude continents. A seasonal reversal of prevailing conditions is apparent over the continents, mainly in the northern hemisphere. Of great interest is the intensification of the eddy divergence over the Gulf of Mexico during winter, showing the important role played by the eddies in the water balance of North America.

5.2 Interactions between the Atmospheric Branch and the Oceans

Further study of the div $\vec{Q}$ fields can shed light on certain other aspects of the global water cycle. Since evaporation exceeds precipitation in regions of divergence, there must be a correlation between the div $\vec{Q}$ field and the distribution of evaporation.

Using surface ship data from the 10-year period, May 1963 through April 1973, the evaporation over the oceans was evaluated using the expression

$$\overline{E} = \rho c_D |\vec{v}| (q_s - q_a) \tag{20}$$

where ρ is the density of air, c_D the coefficient of eddy diffusion (=0.0014), and $q_s - q_a$ the difference of the saturated specific humidity at the sea surface temperature (q_s) and the specific humidity in the free atmosphere (q_a). The corresponding maps presented in Figure 21 are preliminary in nature. They correlate fairly well with maps of the water vapour flux divergence. The highest values of evaporation, as well as the dominant divergence centres, occur over the subtropical oceans. The effects of warm and cold ocean currents, together with land-sea contrasts are very

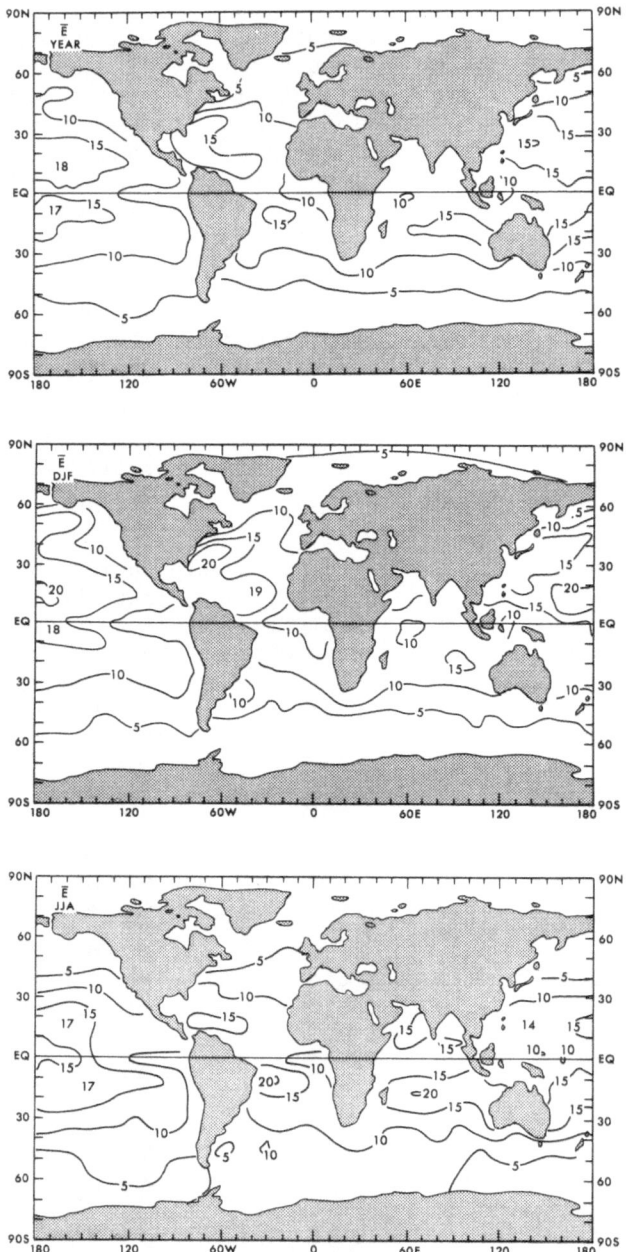

Fig. 21.   Distribution of the mean evaporation, $\bar{E}$, over the oceans
          (10 cm yr^{-1}):   a, year; b, DJF; c, JJA.

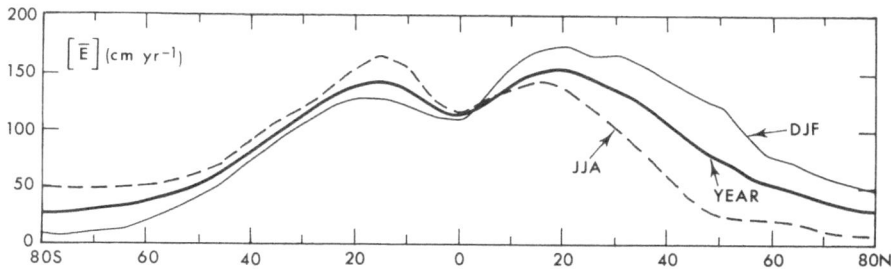

Fig. 22. Profiles of the zonal-mean evaporation [E̅] over the
          oceans (cm yr⁻¹): ▬▬, year; ──, DJF; ---, JJA.

important, as illustrated by the mid-latitude winter maxima of
evaporation, of ∿ 2 m yr⁻¹, just east of the major continents.
Over equatorial regions, where precipitation is abundant, evaporation is somewhat less intense due to lower sea surface
temperatures and weaker winds.

     The zonal mean profiles in Figure 22, obtained from our maps
of E̅, summarize the main aspects of the behaviour of evaporation
over the oceans. The annual profile agrees well with the corresponding one published by Baumgartner & Reichel (1975). The
profiles reveal that hemispheric evaporation rates during summer
tend to be lower than during winter due to variations in surface
wind intensity. Hemispheric and global averages are given in
Table II.

Table II. Precipitation rates according to Jaeger (1976)* and
           evaporation rates from the current study† (in italics)
           for the hemispheres and the globe. All evaporation
           rates refer to the oceans only.

	year (cm yr⁻¹)		DJF (cm per 3 months)		JJA (cm per 3 months)	
Northern hemisphere	97.8	*98*	20.2	*30*	30.2	*20*
Southern hemisphere	102.1	*109*	29.1	*24*	23.1	*31*
Globe	100.0*		24.7		26.6	
Globe (land only)	75.6		17.8		21.8	
Globe (oceans only)	109.9	*103*	27.5	*27*	28.6	*26*

* Jaeger multiplied the oceanic precipitation values by a factor
   of 1.062 to satisfy his model requirements for a global precipitation value of 100 cm yr⁻¹.

† Owing to limitations in our formulation and to the nature of
   the surface data used, the estimated uncertainty in our values
   is ∿ 30%.

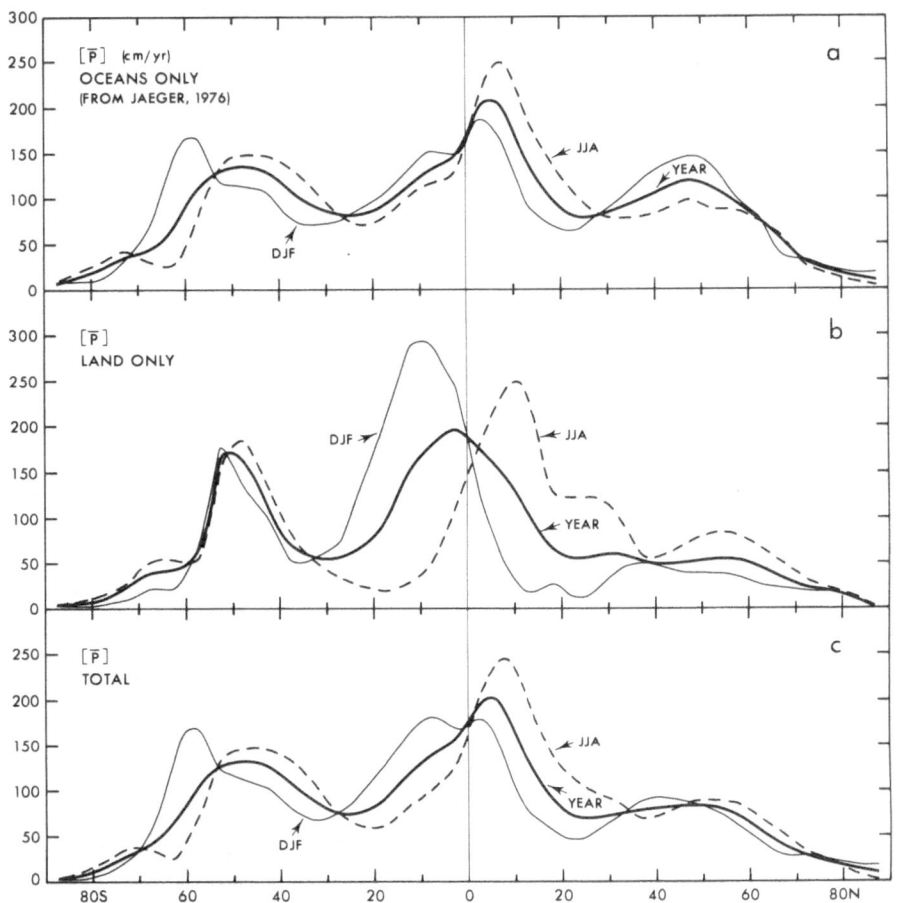

Fig. 23.   Profiles of the zonal-mean precipitation $[\bar{P}]$ (cm yr^{-1}),
(from Jaeger 1976):   a, over the oceans; b, over the land;
c, over the total land and ocean surface.

    For completeness, the zonal mean profiles of precipitation
rates, from Jaeger (1976), are given in Figure 23, the three parts
of which show sets of curves for the oceans only, land only, and
land plus oceans (total). The distributions of the precipitation
rates are far from symmetric with respect to the equator. Hemispheric and global averages, as computed by Jaeger (1976), are
also presented in Table II. The values show a slightly smaller
mean annual precipitation in the northern than in the southern
hemisphere, contrary to previous findings (Palmén & Newton 1969).
Further, there is more precipitation over the oceans than over the
continents, except in the tropics. Since the evaporation rates
also tend to be larger over the oceans (see, e.g., Sellers 1965,

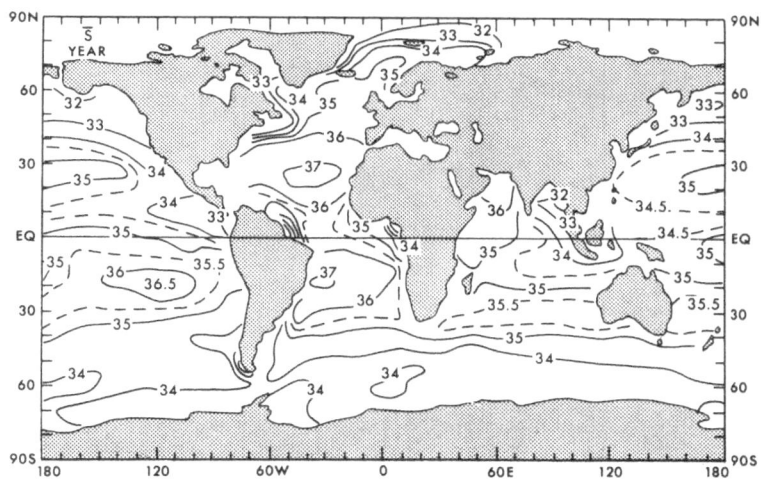

Fig. 24. Distribution of the ocean-surface salinity $\bar{s}_o$ (‰), for
          mean annual conditions (after Levitus 1982).

p. 85), a partial compensation will occur in the land-ocean differences for E-P. In fact, the net value for E-P has to be negative
over the continents and positive over the oceans to allow for
river and glacier runoff from the continents. In terms of the
atmospheric branch, this means a net inflow of water vapour to the
continents. Such interactions are essential to achieve and maintain the state of quasi-equilibrium of the climatic system.
Occasionally, however, imbalances may persist for an extended
period of time, leading to drought or flood conditions.
        As Figure 22 shows, the evaporation is very large over oceanic
regions of strong divergence, where an increase in salinity is to
be expected. However, in regions of convergence, the excess of
fresh water from rain and snow will dilute the ocean waters,
leading to lower salinity values. These relations were discussed
extensively by Peixóto (1959) and Lufkin (1959). To show these
connections, an annual salinity map, prepared by Levitus (1982),
is presented as Figure 24. The general configuration reveals a
high correlation with the mean annual divergence map of water
vapour (Figure 13a). The salinity is low over the equator,
where convergence of water vapour prevails, whereas it is high
in the subtropics, where divergence dominates. These facts are
confirmed by the zonal mean profiles in Figures 14 and 25(a),
which show that the divergence of water vapour in the atmosphere
over the oceans and the mean surface salinity are strongly
related.
        To analyse further the influence of the atmosphere on the
salinity distribution in the oceans, we consider the monthly rates

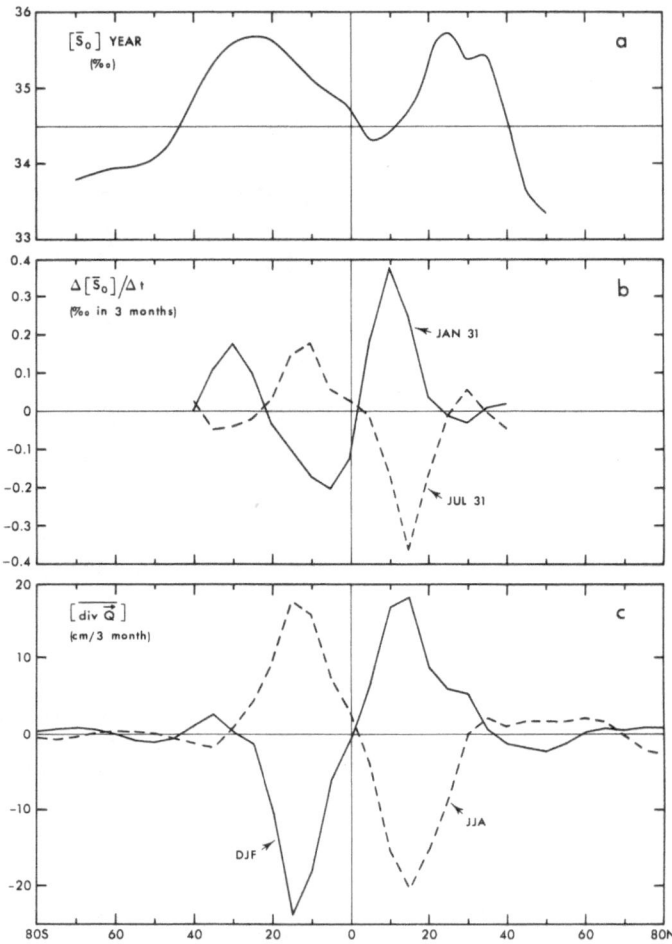

Fig. 25. Profiles of: a, the zonal-mean salinity (‰) at the
ocean surface; b, rate of change of zonal-mean salinity
(‰ per 3-month period); c, zonal-mean divergence of
total atmospheric water vapour flux [div $\vec{Q}$] (cm per
3-month period) for ocean and land, the annual mean profile of [div $\vec{Q}$] having been removed from the seasonal
profiles: ——, DJF; ---, JJA.

of change of zonally-averaged salinity $\Delta[S_o]/\Delta t$, centred on 31
January and 31 July (as illustrated in Figure 25b). The curves
are almost completely out of phase, showing that the rates of

change of salinity are practically reversed between January and
July. We suspect that these differences are due largely to
seasonal differences in precipitation and/or evaporation. If this
were the case, the $\Delta[S_o]/\Delta t$ profiles would be strongly correlated
with the real seasonal variability of the mean divergence profiles

[div $\vec{Q}$], after the mean annual signal was removed. With this in
mind, divergence profiles for DJF and JJA were constructed after
subtracting the corresponding yearly values. The filtered profiles are presented in Figure 25(c). The correspondence between
the two sets of curves in amplitude and phase is striking. In
fact, if 10 cm of rain were added to the ocean at a certain latitude in a 3-month period, and if it is assumed that this rain
would be mixed in a layer of ocean water 10 m thick with an
original salinity of 35‰, the variation of salinity in the 3-
month period would be approximately -0.4‰. This estimate seems
of the correct order of magnitude if the salinity change and water
vapour flux divergence curves in Figure 25(b, c) are compared.
Therefore, other processes, such as horizontal advection of
salinity by ocean currents and vertical mixing with the deeper
layers, may be less important on the seasonal time-scale, at least
in the tropics. Again, this analysis shows the importance of the
study of the aerological branch of the hydrological cycle for
oceanographic purposes.

6. FINAL COMMENTS

    The results presented in this study should prove useful in
investigating the global energetics of the atmosphere, especially
in the study of the energy cycle of the general circulation
(Peixóto 1965, Lorenz 1978, Oort & Peixóto 1982). We now present
some general comments.
    (i) The identification of the important sources of moisture
for the atmosphere, as shown in the div $\vec{Q}$ analyses (Figure 13) and
in the maps of $\vec{Q}$ (Figure 19), confirms that most of the water
vapour for precipitation over the continents is supplied by the
oceans. The air masses also receive some moisture over the continents due to evapotranspiration and evaporation from lakes, soil,
etc. However, this last moisture supply constitutes only a
fraction of the water that falls locally as precipitation over
land (Figure 23b). The present results confirm that the evapora-
tion-precipitation *in situ* theory, as a general rule, cannot be
accepted.
    (ii) The effects of the physiography are apparent. Precipi-
table water is much reduced over mountainous regions. Furthermore,
the water vapour transport, which occurs mainly in the lower troposphere, is clearly affected by the topography. Indeed, the
absence of large mountains along the Atlantic coast in Europe

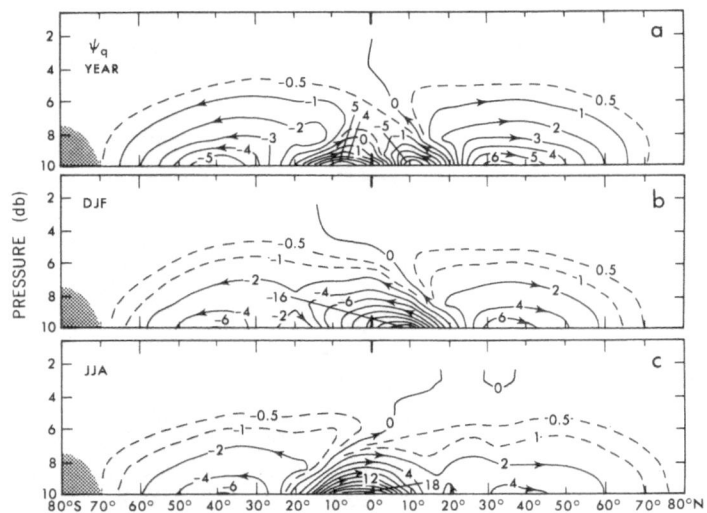

Fig. 26.   Streamlines of the mean meridional water vapour transport ψ_q (10^8 kg s^{-1}).

favours the deep penetration of moisture from the Atlantic Ocean
into the Eurasian continent and the Mediterranean region. On the
other hand, the existence of the Rocky Mts parallel to the west
coast of North America does not allow moisture from the Pacific
Ocean to penetrate deeply into the American continent. Most of
the moisture lost as precipitation over North America is supplied
by water vapour originating over the warm waters of the Gulf of
Mexico, with a deep northward intrusion of water vapour in all
seasons. The same applies to South America with respect to the
Andes. In fact, most of the moisture over South America comes
from the Atlantic Ocean, as is revealed by the $\vec{Q}$ maps in Figure 19.
     (iii) In order to get some idea of the three-dimensional
circulation of water in the atmosphere, a streamline representation of the mean meridional transport for yearly and seasonal
conditions is given (Figure 26). The streamlines were evaluated
from the [$\overline{qv}$] values using Equation (15a) and the boundary condition ψ_q = 0 at the top of the atmosphere.
     Inspection of the ψ_q cross-sections confirms that the water
vapour circulates in the lower troposphere, and that the maximum
transports occur in the planetary boundary layer. The streamlines
that begin at the Earth's surface are due to the existence of
sources of moisture (excess of evaporation over precipitation),
while those that end at the Earth's surface define the sink regions
for atmospheric water. The yearly ψ_q cross-section shows that

there is an export of water vapour from the zones between 12 and
35°N, and 10 and 35°S. In the northern hemisphere, a considerable
portion of this water vapour (to the right of the zero streamline)
is exported polewards throughout the lower half of the atmosphere,
while the remainder, confined to much lower levels, is directed to
the equatorial regions, mainly to feed the ITCZ. In the southern
hemisphere, a large part of the water vapour is exported to the
north in a shallow layer near the Earth's surface, some of it
crossing the equator and falling as precipitation in the ITCZ.
The other part is transported southwards at much higher levels, up
to 400 mb. Polewards of 40° latitude, in both hemispheres, the
streamlines terminate at the surface, indicating the existence of
an excess of precipitation over evaporation.

The seasonal cross-sections show simpler patterns of the
sources and sinks of moisture for the atmosphere. In the DJF
season, there is an export of moisture, produced over subtropical
regions, from the northern into the southern hemisphere. Most of
this moisture, circulating in the lowest layers, falls over the
southern subequatorial zone, while some of it is transported at
higher altitudes further to the south. From the moisture released
in this season, a small fraction (northward of the zero streamline)
flows to the north. In the JJA season, the main source of water
vapour is located in the southern hemisphere between 10 and 32°S,
invading practically the entire globe. However, most of the water
vapour is transported across the equator into the northern hemisphere.

(iv) Our results confirm the importance of the aerological branch
in the dynamics of the hydrological cycle, considered *in toto*. Of
course, solar radiation provides the bulk of the energy that sets
the water circulation, as a whole, in motion. Part of this solar
energy is redistributed within the climatic system by the aerological branch of the hydrological cycle when the phase transitions
of the water substance occur. If we add the role of water vapour
and clouds in the reflection, absorption and emission of both solar
and terrestrial radiation, we may say that the hydrological cycle
is one of the crucial factors in the dynamics of climate.

Thus we may add:

"If the clouds are full of rain, they empty themselves
upon the earth," and "all the rivers run into the sea,
yet the sea is not full; unto the place from which the
rivers come, to there and from there they return again."

Ecclesiastes 11:3, 1:7

ACKNOWLEDGEMENTS. We are grateful to Dr J. Smagorinsky for his
continuous support and interest in our work, to Professor R.E.
Newell, Mr D.G. Hahn and Dr N.-G. Lau for helpful discussions and
comments, and to Professor R.P. Pearce for his thorough review of
the manuscript. Thanks are also due to Mr M. Rosenstein for
handling the various computations, to Mr P.G. Tunison and his

group for drafting the figures, and to Ms J. Kennedy for typing
the manuscript. The visits of J.P. Peixoto to the Geophysical
Fluid Dynamics Laboratory were supported through NOAA Grant
04-7-022-44017.

NOTES

[1] Mean meridional velocities computed directly from the $\bar{v}$ fields
appear to be rather sensitive to the occurrence of spatial gaps in
the rawinsonde network, at least in middle and high latitudes.
Therefore, an indirect method was used to obtain the flux values
of water vapour by the mean meridional circulation, $[\bar{q}]$ $[\bar{v}]$ and
$[\bar{q}]$ $[\bar{\omega}]$, poleward of 20° latitude. The indirect method to calculate $[\bar{v}]$ is based on momentum balance (Kuo 1956, Newell *et al.*
1972: pp. 13, 251-252); for a detailed discussion of the actual
procedures used, see Oort and Peixóto (1982).

[2] Actually the map analyses were performed using vertical mean
values which do not differ much from the vertically-integrated
values over the oceans and low terrain, since the required correction factor equals $p_o g^{-1}$.

REFERENCES

Bannon, J.K. & Steele, L.P. 1960 Average water vapour content of
     air. *Met. Office Geophys. Memoir* No. 102, 38 pp.

Baumgartner, A. & Reichel, E. 1975 *The World Water Balance – Mean
     Annual Global, Continental and Maritime Precipitation,
     Evaporation and Runoff*, 179 pp. Amsterdam: Elsevier.

Budyko, M.I. 1963 *Atlas of the Heat Balance of the Earth*, 69 pp.
     (in Russian). Moscow: Glavnaia Geofiz. Observ.

Drozdov, O.A. & Grigoreva, A.S. 1963 *The Hydrological Cycle in the
     Atmosphere*. Leningrad: Gidromet. Izdatel. (English trans-
     lation: 1965 NTIS TT65-50119, 282 pp. Jerusalem: Israel
     Program for Scientific Translations.)

Eagleson, P.S. 1970 *Dynamic Hydrology*, 426 pp. New York: McGraw
     Hill.

Harris, R.G., Thomasell, A. Jr, & Welsh, J.G. 1966 Studies of
     techniques for the analysis and prediction of temperature in
     the ocean. Part III: Automated analysis and prediction.
     Interim Report, prepared by Travelers Research Center Inc.,
     for U.S. Naval Oceanographic Office, Contract #N62306-1675, 97 pp.

Hellerman, S. 1981 The net meridional flux of water by the oceans
    from evaporation and precipitation estimates, 13 pp. Prince-
    ton: GFDL/NOAA. [Unpublished.]

Jaeger, L. 1976 Monatskarten des Niederschlags für die ganze Erde.
    Ber. Dt. Wetterd., 18, No. 139, 38 pp.

Kuo, H.-L. 1956 Forced and free meridional circulation in the
    atmosphere. J. Met., 13, 561-568.

Levitus, S. 1982 Climatological Atlas of the World Ocean. NOAA
    Prof. Paper. Washington, D.C.: US Govt Printing Office.
    [In the press.]

Lorenz, E.N. 1978 Available energy and the maintenance of a moist
    circulation. Tellus, 30, 15-31.

Lufkin, D. 1959 Atmospheric water divergence and the water balance
    at the Earth's surface. General Circulation Project Sci.
    Rept No. 4, 44 pp. Cambridge, Mass.: MIT Press.

Newell, R.E., Kidson, J.W., Vincent, D.G. & Boer, G.J. 1972 The
    General Circulation of the Tropical Atmosphere, 258 pp.
    Cambridge, Mass: MIT Press.

Newton, C.W. 1972 Southern hemisphere general circulation in rela-
    tion to global energy and momentum requirements. In:
    Meteorology of the Southern Hemisphere, Meteorological Mono-
    graphs No. 13, pp. 215-246. Boston, Mass.: Am. Met. Soc.

Oort, A.H. 1982 Global atmospheric circulation statistics, 1958-
    1973. NOAA Prof. Paper. Washington, D.C.: US Govt Printing
    Office. [In the press.]

Oort, A.H. & Peixóto, J.P. 1982 Global angular momentum and energy
    balance requirements from observations. Adv. Geophys., 26,
    [In the press.]

Oort, A.H. & Rasmusson, E.M. 1971 Atmospheric circulation stati-
    stics. NOAA Prof. Paper No. 5, 323 pp. Washington, D.C.:
    US Govt Printing Office.

Palmén, E. 1967 Evaluation of atmospheric moisture transports for
    hydrological purposes, 63 pp. Geneva: WMO.

Palmén, E. & Newton, C.W. 1969 Atmospheric Circulations Systems,
    606 pp. New York: Academic Press.

Peixóto, J.P. 1958 Hemispheric humidity conditions during year 1950. General Circulation Project Sci. Rept No. 3, 142 pp. Cambridge, Mass.: MIT Press.

Peixóto, J.P. 1959 O campo da divergencia do transporte do vapor de agua na atmosfera. *Rev. Fac. Cien. Lisboa,* $\underline{B7}$, 25-56. [In Portugese.]

Peixóto, J.P. 1965 On the role of water vapour in the energetics of the general circulation of the atmosphere. *Portugalia Physica,* $\underline{4}$, 135-170.

Peixóto, J.P. 1970 Pole to pole divergence of water vapour. *Tellus,* $\underline{22}$, 17-25.

Peixóto, J.P. 1972 Pole-to-pole water balance for the IGY from aerological data. *Nord. Hydrol.* $\underline{3}$, 22-43.

Peixóto, J.P. 1973 Atmospheric vapour flux computations for hydrological purposes. *WMO Publ.* No. 357, 83 pp.

Peixóto, J.P., Rosen, R.D. & Salstein, D.A. 1978 Seasonal variability in the pole-to-pole modes of water vapour transport during the IGY. *Arch. Met. Geoph. Biokl.,* $\underline{A27}$, 233-255.

Peixóto, J.P., Rosen, R.D. & Wu, M.-F. 1976 Seasonal variability in the pole-to-pole water vapour balance during the IGY. *Nord. Hydrol.,* $\underline{7}$, 95-114.

Rasmusson, E.M. 1971 A study of the hydrology of eastern North America using atmospheric vapor flux data. *Mon. Weath. Rev.,* $\underline{99}$, 119-135.

Rosen, R.D., Salstein, D.A. & Peixóto, J.P. 1979a Streamfunction analysis of interannual variability in large-scale water vapor flux. *Mon. Weath. Rev.,* $\underline{107}$, 1682-1684.

Rosen, R.D., Salstein, D.A. & Peixóto, J.P. 1979b Variability in the annual fields of large-scale atmospheric water vapor transport. *Mon. Weath. Rev.,* $\underline{107}$, 26-37.

Sellers, W.D. 1965 *Physical Climatology,* 272 pp. Chicago: University of Chicago Press.

Starr, V.P. & Peixóto, J.P. 1958 On the global balance of water vapor and the hydrology of deserts. *Tellus,* $\underline{10}$, 189-194.

Starr, V.P. & Peixóto, J.P. 1971 Pole-to-pole eddy transport of water vapor in the atmosphere during the IGY. *Arch. Met. Geoph. Biokl.,* $\underline{A20}$, 85-114.

Starr, V.P. & White, R.M. 1955 Direct measurement of the hemi-
    spheric poleward flux of water vapor. *J. Mar. Res.*, 14,
    217-225.

Starr, V.P., Peixóto, J.P. & Livadas, G.C. 1958 On the meridional
    flux of water vapor in the Northern Hemisphere. *Geof. Pura e
    Appl.*, 39, 174-185.

Starr, V.P., Peixóto, J.P. & McKean, R.G. 1969 Pole-to-pole moisture
    conditions for the IGY. *Pure appl. Geophys.*, 75, 300-331.

Starr, V.P., Peixóto, J.P. & Sims, J.E. 1970 A method for the study
    of the zonal kinetic energy balance in the atmosphere. *Pure
    appl. Geophys.*, 80, 346-358.

Viswanadham, Y., Rao, N.J.M. & Nunes, G.S.S. 1980 Some studies on
    moisture conditions in the Southern Hemisphere. *Tellus*, 32,
    131-142.

APPENDIX

    In this appendix, zonally-averaged values of some selected
water vapour parameters are tabulated for the year and the Decem-
ber-February (DJF) and June-August (JJA) seasons. The data are
given for each 5°-latitude circle between 80°S and 80°N (except for
75°N and 75°S) at seven pressure levels between the surface/1000 mb
and 300 mb. In the last three columns of each table, areal
averages are also shown for the southern and northern hemisphere,
and the globe. In the last row, labelled 'MEAN', mass-weighted
vertically-averaged values are presented; these were computed using
the following formula for the 'MEAN' of an arbitrary parameter A:

$$[A] = [\int_{25\ \text{mb}}^{P_o} A\ dp/(p_o - 25\ \text{mb})]$$

Here, p_o is the surface pressure which equals a global-mean value
of 1012.5 mb at sea level, but is reduced over mountains according
to an assumed standard atmosphere profile. Since the q fields
could not be analysed reliably above the 300 mb level, the integration with respect to pressure was only performed between p_o and
250 mb. One should note that, in the zonal averages, only the grid
point values above the Earth's surface were included. To obtain
a vertically-integrated, rather than an averaged, value, the value
under 'MEAN' should be multiplied by the appropriate factor at the
corresponding latitude given in Table A1.

The quantities tabulated are:
Table A2: $[\bar{q}]$
Table A3: $[\sigma(q)]$
Table A4: $[\overline{qu}]$
Table A5: $[\overline{qv}] = [\overline{q'v'}] + [\overline{q*\bar{v}*}] + [\bar{q}][\bar{v}]_{ID}$
Table A6: $[\overline{q'v'}]$
Table A7: $[\overline{q*\bar{v}*}]$
Table A8: $[\bar{q}][\bar{v}]_{ID}$
Table A9: $[\overline{q\omega} + \overline{q_c\omega_c}]$
Table A10: $[\overline{q*\omega*}]$
Table A11: $[\bar{q}][\bar{\omega}]_{ID}$
Table A12: $[\overline{q'\omega'} + \overline{q_c\omega_c}]$

Table A1. Multiplication factor to convert the vertically-averaged
values (row 'MEAN') in Tables A2 through A12 into vertically-integrated values, while taking into account the
mean mass distribution over the globe (e.g., less mass
over the mountains). For simplicity, the sea level
pressure is assumed to be uniform over the globe. At
the equator, factor = 0.982.

latitude (°)	hemisphere		latitude (°)	hemisphere	
	southern	northern		southern	northern
80	0.748	0.974	40	0.997	0.937
75	0.777	0.971	35	0.992	0.930
70	0.903	0.961	30	0.985	0.944
65	0.991	0.953	25	0.980	0.968
60	1.000	0.962	20	0.981	0.976
55	1.000	0.966	15	0.977	0.982
50	0.999	0.956	10	0.985	0.985
45	0.998	0.952	5	0.980	0.982

Table A2. Vertical distribution of the zonally-averaged values (g kg^{-1}) of specific humidity $[\bar{q}]$ for the 10-yr period and for the DJF and JJA composite seasons. To obtain $[\bar{W}]$, the zonal mean precipitable water (10 kg m^{-2}), multiply the vertically-averaged values in the last 'MEAN' row by the corresponding values in Table A1.

year 1963 – 73

p(mb)	80S	70S	65S	60S	55S	50S	45S	40S	35S	30S	25S	20S	15S	10S	5S	EQ	5N	10N	15N	20N	25N	30N	35N	40N	45N	50N	55N	60N	65N	70N	80N	SH	NH	GLOBE
300	0.0	0.1	0.1	0.1	0.1	0.1	0.1	0.1	0.2	0.2	0.2	0.2	0.3	0.3	0.3	0.3	0.3	0.3	0.3	0.3	0.2	0.2	0.2	0.1	0.1	0.1	0.1	0.1	0.1	0.1	0.1	0.18	0.19	0.18
500	0.2	0.3	0.4	0.4	0.5	0.6	0.8	0.9	1.0	1.3	1.5	1.8	2.0	2.1	2.2	2.3	2.2	1.9	1.7	1.5	1.4	1.2	1.0	0.9	0.8	0.7	0.6	0.5	0.5	0.4	0.3	1.16	1.24	1.20
700	0.7	0.9	1.0	1.2	1.5	1.8	2.1	2.5	3.0	3.6	4.2	4.8	5.3	5.8	5.9	6.0	5.9	5.4	4.8	4.3	3.8	3.2	2.8	2.5	2.2	2.0	1.8	1.5	1.4	1.2	0.9	3.22	3.44	3.33
850	·····	1.1	1.4	1.9	2.4	3.0	3.6	4.2	5.0	5.9	6.9	8.0	9.0	9.9	10.4	10.7	10.5	9.9	9.0	8.1	7.0	6.0	5.2	4.5	3.9	3.4	3.0	2.6	2.2	1.9	1.3	6.16	6.28	6.22
900	·····	1.3	1.7	2.2	2.9	3.7	4.4	5.2	6.1	7.0	8.1	9.4	10.7	11.7	12.2	12.5	12.4	12.0	11.0	9.8	8.5	7.3	6.3	5.4	4.6	3.9	3.4	2.9	2.5	2.0	1.4	7.31	7.55	7.43
950	·····	1.7	2.2	2.9	3.4	4.2	5.0	6.0	7.0	8.1	9.4	10.9	12.3	13.4	14.1	14.5	14.6	14.2	13.1	11.8	10.3	8.7	7.5	6.3	5.2	4.4	3.8	3.2	2.7	2.2	1.6	8.39	8.96	8.67
1000	·····	2.2	2.7	3.3	4.1	5.1	6.4	7.7	9.2	10.8	12.4	13.9	15.6	16.6	17.2	17.6	17.7	17.8	17.9	16.9	15.6	13.9	12.2	10.4	8.6	6.8	5.7	4.8	4.0	3.3	2.5	10.34	12.09	11.10
MEAN	0.2	0.4	0.6	0.8	1.0	1.3	1.5	1.8	2.1	2.4	2.8	3.3	3.7	4.1	4.3	4.5	4.4	4.2	3.8	3.4	2.9	2.4	2.0	1.7	1.5	1.3	1.2	1.0	0.9	0.7	0.5	2.51	2.59	2.55

DJF 1963 – 73

p(mb)	80S	70S	65S	60S	55S	50S	45S	40S	35S	30S	25S	20S	15S	10S	5S	EQ	5N	10N	15N	20N	25N	30N	35N	40N	45N	50N	55N	60N	65N	70N	80N	SH	NH	GLOBE
300	0.0	0.1	0.1	0.2	0.2	0.3	0.3	0.3	0.3	0.3	0.3	0.3	0.3	0.3	0.3	0.3	0.3	0.2	0.2	0.2	0.2	0.1	0.1	0.1	0.1	0.1	0.1	0.0	0.0	0.0	0.0	0.21	0.15	0.18
500	0.2	0.3	0.4	0.4	0.5	0.7	0.8	1.0	1.2	1.4	1.7	2.0	2.2	2.3	2.3	2.2	1.9	1.5	1.2	1.0	0.9	0.8	0.7	0.6	0.5	0.4	0.3	0.3	0.3	0.2	0.2	1.41	0.89	1.15
700	0.7	1.0	1.2	1.4	1.8	2.2	2.6	3.2	3.8	4.6	5.3	5.7	5.9	6.0	5.9	5.9	5.3	4.5	3.7	3.1	2.6	2.2	1.8	1.5	1.3	1.1	0.9	0.8	0.7	0.6	0.6	3.78	2.56	3.17
850	·····	1.8	2.2	2.8	3.4	4.2	5.1	6.1	7.2	8.3	9.4	10.2	10.6	10.7	10.6	10.0	8.9	7.7	6.6	5.4	4.3	3.5	2.8	2.3	1.9	1.5	1.3	1.1	0.9	0.7	0.6	6.97	5.03	5.99
900	·····	2.1	2.7	3.4	4.3	5.1	6.2	7.3	8.3	9.5	10.8	11.8	12.4	12.5	12.5	12.1	11.1	9.8	8.8	8.0	6.7	5.4	4.4	3.6	2.8	2.5	1.7	1.4	1.1	0.6	0.6	8.16	6.22	7.19
950	·····	2.5	3.1	3.9	4.9	5.9	7.1	8.3	9.6	11.0	12.5	13.6	14.3	14.3	14.5	14.6	13.1	11.5	9.8	8.0	6.4	5.2	4.1	3.2	2.5	2.0	1.5	1.5	1.2	0.9	0.5	9.36	7.38	8.40
1000	·····	3.1	3.8	4.6	5.6	7.1	8.8	10.8	12.7	14.3	15.5	15.6	16.5	17.2	17.5	17.7	17.0	15.3	13.6	11.5	9.5	8.0	6.4	5.2	4.3	3.3	3.0	2.7	2.4	2.1	1.6	11.27	10.57	10.97
MEAN	0.3	0.6	0.8	1.0	1.2	1.5	1.8	2.2	2.6	3.0	3.5	3.9	4.2	4.5	4.5	4.4	3.7	3.2	2.7	2.2	1.7	1.4	1.1	0.9	0.8	0.6	0.5	0.5	0.4	0.3	0.3	2.86	2.06	2.46

JJA 1963 – 73

p(mb)	80S	70S	65S	60S	55S	50S	45S	40S	35S	30S	25S	20S	15S	10S	5S	EQ	5N	10N	15N	20N	25N	30N	35N	40N	45N	50N	55N	60N	65N	70N	80N	SH	NH	GLOBE
300	0.0	0.1	0.1	0.1	0.1	0.1	0.1	0.1	0.1	0.1	0.2	0.2	0.2	0.2	0.3	0.3	0.3	0.3	0.3	0.3	0.3	0.3	0.3	0.2	0.2	0.2	0.2	0.1	0.1	0.1	0.1	0.14	0.26	0.20
500	0.2	0.2	0.2	0.3	0.3	0.4	0.5	0.6	0.7	0.7	0.9	1.0	1.3	1.6	1.9	2.2	2.3	2.3	2.4	2.2	2.0	1.8	1.5	1.3	1.2	1.1	1.0	0.9	0.8	0.7	0.6	0.90	1.67	1.28
700	0.5	0.9	1.0	1.2	1.4	1.6	1.6	1.9	2.2	2.6	3.1	3.8	4.5	5.3	6.0	6.0	6.2	6.2	5.9	5.5	5.0	4.4	4.0	3.7	3.5	3.3	3.0	2.7	2.4	2.1	1.6	2.64	4.45	3.55
850	·····	1.1	1.6	2.1	2.6	3.5	4.0	4.7	5.0	4.7	5.5	6.6	7.8	8.9	9.5	10.9	11.0	12.7	12.2	11.4	9.7	7.9	7.1	6.6	6.0	5.5	5.1	4.6	4.1	3.5	2.6	5.36	7.84	6.60
900	·····	1.4	1.9	2.6	2.6	3.0	3.7	4.3	4.3	5.7	6.8	8.1	9.6	10.8	11.7	12.4	12.7	12.8	12.2	11.4	10.3	9.3	8.5	7.8	7.0	6.3	5.7	5.2	4.6	3.9	2.8	6.49	9.16	7.82
950	·····	1.5	2.1	2.9	3.6	4.3	5.0	5.8	6.6	6.6	7.8	9.3	11.0	12.4	13.5	14.3	14.8	14.5	13.7	12.7	11.3	10.2	10.2	9.0	7.9	6.8	6.3	5.8	5.1	4.3	3.0	7.42	10.79	9.06
1000	·····	2.0	2.8	3.7	4.6	5.6	6.7	7.9	9.2	9.2	10.7	12.3	14.2	15.7	16.6	17.2	17.8	18.3	18.0	17.3	16.2	15.1	13.7	11.5	9.1	7.5	6.7	6.1	5.4	4.6	3.3	9.34	13.70	11.24
MEAN	0.1	0.3	0.5	0.7	0.9	1.0	1.1	1.3	1.5	1.7	1.9	2.2	2.7	3.1	3.7	4.1	4.4	4.6	4.7	4.4	4.1	3.7	3.2	2.9	2.6	2.3	2.1	2.0	1.8	1.6	1.4	2.15	3.23	2.69

Table A3. Vertical distribution of the zonally-averaged values (g kg^{-1}) of the temporal standard deviation of specific humidity [σ(q)].).].

year 1963 – 73

p (MB)	80S	70S	65S	60S	55S	50S	45S	40S	35S	30S	25S	20S	15S	10S	5S	EQ	5N	10N	15N	20N	25N	30N	35N	40N	45N	50N	55N	60N	65N	70N	80N	SH	NH	GLOBE
300	0.0	0.0	0.0	0.0	0.1	0.1	0.1	0.1	0.1	0.1	0.1	0.1	0.1	0.1	0.1	0.1	0.1	0.1	0.1	0.1	0.1	0.1	0.1	0.1	0.1	0.1	0.1	0.1	0.1	0.1	0.1	0.10	0.11	0.11
500	0.2	0.2	0.3	0.4	0.5	0.6	0.6	0.7	0.8	0.9	0.9	1.0	1.0	1.0	0.9	0.9	0.9	1.0	1.0	1.0	0.9	0.9	0.9	0.8	0.7	0.6	0.5	0.4	0.4	0.3	0.2	0.67	0.75	0.71
700	0.4	0.4	0.6	0.7	0.9	1.0	1.2	1.4	1.6	1.8	2.0	2.1	2.1	2.0	1.8	1.7	1.8	2.0	2.1	2.1	2.0	1.9	1.9	1.8	1.6	1.4	1.2	1.1	0.9	0.7	0.7	1.52	1.67	1.59
850	0.6	0.7	0.9	1.1	1.3	1.4	1.6	1.9	2.1	2.3	2.4	2.6	2.7	2.6	2.4	2.1	2.0	2.4	2.6	2.7	2.7	2.6	2.4	2.2	2.1	1.7	1.5	1.4	1.1	0.9	0.9	1.89	2.25	2.07
900	0.7	0.7	0.9	0.9	1.1	1.3	1.5	1.8	2.0	2.2	2.3	2.5	2.3	1.9	1.8	1.7	1.8	2.0	2.3	2.5	2.7	2.7	2.6	2.4	2.2	1.8	1.5	1.3	1.1	0.9	0.9	1.74	2.20	1.97
950	0.7	0.7	0.9	0.9	1.1	1.3	1.5	1.8	2.0	2.2	2.3	2.4	2.2	1.9	1.7	1.6	1.7	2.0	2.3	2.5	2.7	3.0	3.0	2.8	2.5	2.0	1.3	1.1	1.0	0.8	0.9	1.76	2.31	2.03
1000	1.0	1.0	1.1	1.1	1.2	1.4	1.7	2.0	2.2	2.3	2.5	2.3	2.0	1.7	1.6	1.6	1.7	2.0	2.2	2.7	3.1	3.3	3.2	2.9	2.4	2.0	1.4	1.3	1.1	0.9	0.7	1.89	2.27	2.05
MEAN	0.2	0.3	0.4	0.5	0.6	0.7	0.8	0.9	1.0	1.1	1.1	1.2	1.1	1.0	1.0	0.9	1.0	1.1	1.2	1.2	1.2	1.1	1.0	1.0	0.9	0.8	0.7	0.6	0.5	0.4	0.3	0.83	0.96	0.90

DJF 1963 – 73

p (MB)	80S	70S	65S	60S	55S	50S	45S	40S	35S	30S	25S	20S	15S	10S	5S	EQ	5N	10N	15N	20N	25N	30N	35N	40N	45N	50N	55N	60N	65N	70N	80N	SH	NH	GLOBE
300	0.0	0.1	0.1	0.1	0.1	0.1	0.1	0.1	0.1	0.1	0.1	0.1	0.1	0.1	0.1	0.1	0.1	0.1	0.1	0.1	0.1	0.1	0.1	0.0	0.0	0.0	0.0	0.0	0.0	0.1	0.1	0.11	0.07	0.09
500	0.2	0.2	0.3	0.3	0.4	0.5	0.6	0.7	0.8	0.9	1.0	1.0	0.9	0.9	0.9	0.9	0.8	0.9	0.8	0.9	0.9	0.9	0.8	0.7	0.7	0.6	0.6	0.3	0.2	0.1	0.1	0.71	0.50	0.61
700	0.4	0.4	0.6	0.8	1.0	1.2	1.4	1.6	1.8	1.9	2.0	1.9	1.8	1.7	1.7	1.8	1.9	1.9	1.8	1.7	1.5	1.3	1.1	1.0	0.9	0.8	0.7	0.6	0.4	0.2	0.2	1.53	1.24	1.38
850	0.5	0.7	0.9	1.0	1.2	1.4	1.7	2.0	2.2	2.3	2.2	2.0	1.8	1.8	1.9	1.9	2.2	2.2	2.1	2.0	1.8	1.6	1.3	1.1	0.9	0.8	0.7	0.6	0.4	0.2	0.2	1.83	1.57	1.70
900	0.5	0.6	0.8	1.0	1.1	1.3	1.6	1.8	2.0	2.1	1.9	1.7	1.6	1.7	1.8	1.8	1.9	1.9	2.0	2.0	1.9	1.8	1.6	1.4	1.2	0.9	0.8	0.6	0.5	0.3	0.3	1.63	1.47	1.55
950	0.4	0.6	0.8	1.0	1.1	1.2	1.4	1.6	1.8	1.9	1.8	1.6	1.6	1.6	1.7	1.7	1.8	1.8	1.9	2.0	2.0	1.9	1.7	1.4	1.2	0.9	0.8	0.6	0.5	0.3	0.3	1.57	1.48	1.52
1000	0.5	0.7	0.9	1.0	1.1	1.3	1.5	1.7	1.9	2.0	1.8	1.6	1.6	1.5	1.6	1.7	1.8	1.8	1.9	2.1	2.3	2.2	1.9	1.6	1.3	1.1	0.9	0.8	0.5	0.3	0.3	1.63	1.59	1.61
MEAN	0.2	0.3	0.4	0.4	0.5	0.6	0.7	0.8	0.9	1.0	1.0	0.9	0.9	0.9	0.9	0.9	1.0	1.0	1.0	0.9	0.8	0.7	0.6	0.5	0.4	0.3	0.3	0.2	0.1	0.1	0.1	0.81	0.67	0.74

JJA 1963 – 73

p (MB)	80S	70S	65S	60S	55S	50S	45S	40S	35S	30S	25S	20S	15S	10S	5S	EQ	5N	10N	15N	20N	25N	30N	35N	40N	45N	50N	55N	60N	65N	70N	80N	SH	NH	GLOBE
300	0.0	0.1	0.1	0.1	0.1	0.1	0.1	0.1	0.1	0.1	0.1	0.1	0.1	0.1	0.1	0.1	0.1	0.1	0.1	0.1	0.2	0.2	0.1	0.1	0.1	0.1	0.1	0.1	0.1	0.1	0.1	0.07	0.12	0.10
500	0.1	0.2	0.5	0.6	0.7	0.8	0.9	1.1	1.2	1.4	1.6	1.7	1.8	1.8	1.8	1.6	1.6	1.6	1.5	1.7	1.8	1.8	1.8	1.7	1.6	1.5	1.4	1.2	0.8	0.5	0.3	0.50	0.78	0.64
700	0.4	0.5	0.6	0.7	0.9	1.1	1.2	1.4	1.6	1.8	2.0	1.9	1.9	1.9	1.9	1.9	2.0	2.1	2.3	2.3	2.3	2.3	2.1	2.1	1.5	1.4	1.2	1.1	1.0	0.8	0.8	1.27	1.56	1.42
850	0.6	0.7	0.8	0.9	1.1	1.2	1.4	1.6	1.8	1.8	1.9	1.9	1.8	1.6	1.6	1.6	1.7	1.8	2.1	2.2	2.3	2.3	2.3	2.1	1.7	1.5	1.3	1.2	1.0	0.9	0.6	1.57	1.98	1.78
900	0.7	0.7	0.9	1.1	1.2	1.4	1.5	1.6	1.8	1.8	1.7	1.6	1.5	1.6	1.7	1.6	1.7	1.8	1.9	2.0	2.1	2.2	2.2	2.3	1.9	1.6	1.4	1.0	0.9	0.8	0.7	1.45	1.83	1.64
950	0.6	0.6	0.8	0.9	1.1	1.2	1.4	1.5	1.7	1.7	1.6	1.6	1.6	1.5	1.6	1.7	1.7	1.8	1.8	2.0	2.0	2.1	2.3	2.3	2.1	1.6	1.5	1.4	0.8	0.6	0.6	1.41	1.77	1.58
1000	0.7	0.7	0.9	1.1	1.1	1.3	1.5	1.5	1.7	1.9	2.0	2.0	1.9	1.7	1.7	1.4	1.3	1.6	1.7	1.8	1.9	2.0	2.3	2.2	1.9	1.4	1.4	1.2	0.6	0.4	0.3	1.50	1.58	1.53
MEAN	0.2	0.3	0.4	0.5	0.6	0.7	0.8	0.8	0.9	1.0	1.0	1.0	0.9	0.9	0.9	0.9	0.9	0.9	1.0	1.0	1.0	1.0	1.0	0.9	0.9	0.8	0.7	0.6	0.4	0.3	0.2	0.68	0.86	0.77

Table A4. Vertical distribution of the zonally-averaged values (g kg^{-1} m s^{-1}) of the zonal transport of water vapour [$\overline{qu}$].

year 1963 – 73

p [mb]	80S	70S	65S	60S	55S	50S	45S	40S	35S	30S	25S	20S	15S	10S	5S	EQ	5N	10N	15N	20N	25N	30N	35N	40N	45N	50N	55N	60N	65N	70N	80N	SH	NH	GLOBE
300	0.2	0.3	0.5	0.9	1.2	1.6	2.1	2.9	3.5	3.7	3.1	2.0	0.9	-0.3	-1.0	-0.8	0.0	1.4	2.8	3.6	3.7	3.7	3.5	2.9	2.1	1.6	1.2	0.9	0.7	0.5	0.3	1.61	1.41	1.51
500	0.3	0.9	2.0	3.9	6.4	8.9	10.7	11.9	12.2	11.7	10.0	6.7	2.1	-2.9	-6.8	-9.4	-8.7	-6.5	-1.8	3.6	8.4	10.6	10.9	10.1	8.6	6.9	5.2	3.7	2.8	2.1	1.0	4.60	2.89	3.75
700	0.6	2.2	6.6	12.7	18.3	21.3	22.0	20.1	16.1	10.0	2.3	-5.2	-11.6	-16.3	-20.0	-21.5	-19.9	-11.3	-0.0	9.3	13.7	15.8	15.4	14.3	12.4	9.4	6.4	4.7	3.6	3.6	1.7	4.88	1.08	2.97
850	*****	-2.2	-1.0	5.5	16.3	26.6	31.1	30.1	23.0	11.0	-1.8	-14.6	-24.5	-29.2	-27.9	-26.2	-24.1	-14.0	-1.5	8.5	15.4	16.6	14.8	13.0	9.9	6.3	4.3	3.3	4.3	3.3	1.7	-0.55	-4.14	-2.35
900	*****	2.5	6.1	13.5	23.4	31.5	33.3	29.8	20.8	8.0	-4.5	-16.6	-26.1	-31.5	-32.3	-33.8	-35.1	-30.4	-18.4	-5.0	6.6	13.5	15.1	12.7	10.3	8.6	7.0	4.3	3.1	2.4	1.5	-0.49	-8.11	-4.29
950	*****	0.8	4.2	10.8	19.8	27.1	28.2	24.4	15.3	2.3	-10.4	-22.1	-30.5	-35.6	-36.1	-33.3	-34.0	-35.2	-34.2	-25.7	-11.3	1.5	10.5	12.6	10.3	8.6	5.7	3.2	2.2	1.2	1.1	-1.42	-11.32	-7.79
1000	*****	2.6	9.1	16.6	24.2	29.9	30.4	23.9	9.3	-13.0	-38.6	-58.7	-68.5	-61.8	-48.3	-34.7	-31.4	-47.9	-63.5	-60.8	-38.3	-12.1	7.8	14.5	13.9	13.4	6.8	2.9	1.0	-0.0	-0.0	-14.95	-22.69	-18.32
MEAN	0.2	0.3	1.7	4.6	8.7	12.3	14.0	13.8	11.5	7.7	3.1	-2.1	-7.0	-10.6	-12.0	-12.7	-12.8	-12.6	-9.3	-3.5	2.7	6.8	8.9	8.9	7.9	6.7	4.9	3.3	2.4	1.7	0.9	1.66	-0.54	0.56

DJF 1963 – 73

p [mb]	80S	70S	65S	60S	55S	50S	45S	40S	35S	30S	25S	20S	15S	10S	5S	EQ	5N	10N	15N	20N	25N	30N	35N	40N	45N	50N	55N	60N	65N	70N	80N	SH	NH	GLOBE
300	0.2	0.3	0.5	0.4	0.6	0.8	1.3	1.7	2.3	2.6	2.4	1.6	0.4	-0.6	-0.9	-0.9	-0.1	1.0	2.6	4.2	5.3	5.1	3.7	2.3	1.3	0.8	0.6	0.4	0.3	0.1	0.1	1.44	2.01	1.73
500	0.2	0.5	1.9	4.4	8.0	11.6	13.8	14.6	13.4	10.6	6.6	1.9	-1.3	-3.7	-6.5	-8.8	-7.0	-3.1	4.0	10.6	14.3	13.8	11.4	8.7	6.3	4.4	3.0	2.0	1.8	1.1	0.6	4.24	4.51	4.38
700	0.5	1.4	6.9	14.5	21.7	25.3	25.2	20.7	12.8	3.3	-5.0	-7.4	-6.8	-10.8	-17.4	-20.5	-17.3	-4.5	9.4	18.5	19.5	17.7	14.2	11.1	7.9	5.2	3.3	2.4	1.9	0.9	0.8	5.07	3.08	4.07
850	*****	-3.0	-1.5	6.1	18.0	29.1	32.9	29.6	18.4	2.6	-9.9	-16.9	-14.3	-9.0	-12.8	-23.0	-37.5	-46.2	-32.4	-10.9	6.8	16.9	20.3	16.8	11.8	8.6	5.4	3.1	1.9	1.9	0.6	2.18	-6.52	-2.18
900	*****	5.2	10.1	17.8	27.1	34.0	33.8	27.5	14.6	-1.0	-11.4	-16.5	-15.1	-12.8	-22.6	-36.4	-49.7	-55.3	-40.0	-17.0	3.2	16.1	19.2	15.8	10.6	6.7	3.7	2.4	2.2	2.0	0.5	1.79	-11.36	-4.76
950	*****	3.8	8.5	15.6	23.8	29.3	28.1	21.4	9.6	-4.9	-15.5	-21.0	-20.6	-20.9	-31.0	-41.2	-51.6	-56.1	-44.4	-24.1	-3.2	11.8	16.4	13.6	8.6	5.4	2.8	1.6	1.2	1.2	0.5	-2.86	-14.89	-8.73
1000	*****	1.4	6.5	14.4	23.4	29.8	29.6	20.0	-0.2	-28.1	-53.6	-62.5	-52.6	-30.2	-26.8	-34.8	-58.5	-86.1	-83.2	-59.6	-23.2	7.0	18.5	18.4	13.0	9.9	3.8	1.6	0.8	0.8	-0.1	-13.05	-28.56	-19.79
MEAN	0.1	0.1	0.6	1.0	1.8	2.5	3.2	3.8	3.8	2.0	0.3	-0.8	-2.0	-2.2	-2.2	-1.4	-0.8	0.3	2.4	4.2	5.0	5.4	4.0	6.1	4.3	2.7	1.8	1.2	1.0	0.8	0.4	2.01	-0.35	0.83

JJA 1963 – 73

p [mb]	80S	70S	65S	60S	55S	50S	45S	40S	35S	30S	25S	20S	15S	10S	5S	EQ	5N	10N	15N	20N	25N	30N	35N	40N	45N	50N	55N	60N	65N	70N	80N	SH	NH	GLOBE
300	0.2	0.3	0.5	0.4	0.4	0.6	1.0	1.7	2.8	4.1	4.2	3.4	1.9	0.3	-1.0	-2.0	-2.2	-2.0	-1.4	-0.8	0.3	2.0	3.5	3.8	3.2	2.5	1.8	1.2	1.0	0.8	0.5	1.56	0.61	1.08
500	0.5	1.0	2.8	4.2	5.8	7.3	9.0	10.5	11.9	12.1	10.2	5.0	-1.8	-6.6	-9.8	-10.6	-10.7	-8.8	-5.5	-0.2	5.4	10.0	12.2	11.4	9.3	6.7	4.7	3.5	2.0	2.0	1.0	4.56	0.57	2.57
700	0.9	1.8	5.1	9.7	14.0	16.8	18.3	18.4	17.1	14.2	8.5	-1.2	-12.4	-18.3	-19.3	-17.7	-18.2	-15.0	-9.1	-1.9	6.6	14.0	17.1	18.3	16.6	12.6	8.4	6.3	5.5	3.8	1.7	4.78	0.17	2.46
850	*****	-1.6	-1.2	3.9	13.0	22.1	27.3	28.9	25.7	17.6	5.2	-11.1	-29.4	-41.5	-37.9	-26.5	-12.9	-6.6	-10.2	-12.6	-7.6	3.3	13.5	17.9	18.2	16.9	13.2	8.4	6.2	5.5	4.1	-2.33	0.42	-0.95
900	*****	2.7	6.0	12.7	21.8	29.3	31.9	31.0	26.0	16.6	2.5	-15.8	-33.9	-45.0	-39.0	-27.4	-16.9	-12.3	-17.5	-15.9	-9.1	2.0	11.2	16.3	15.5	13.2	10.6	6.6	4.6	3.7	4.1	-1.36	-3.03	-2.20
950	*****	1.3	4.4	10.5	18.8	25.6	27.7	26.0	20.0	9.6	-4.3	-21.3	-37.1	-46.8	-40.1	-27.0	-17.3	-11.8	-21.1	-23.7	-14.2	-2.6	8.9	14.0	12.6	10.6	6.6	3.3	2.1	1.3	3.2	-4.66	-6.03	-5.33
1000	*****	2.4	7.5	14.1	21.4	27.8	30.4	28.7	19.9	2.2	-24.5	-54.5	-80.1	-84.9	-66.1	-38.0	-7.8	-8.6	-40.1	-57.1	-47.1	-22.4	4.0	13.9	15.0	15.1	8.6	3.9	0.4	0.4	-1.2	-16.34	-14.84	-15.69
MEAN	0.3	0.4	1.5	3.7	7.0	10.0	11.8	12.5	11.9	9.9	6.3	0.5	-6.9	-13.5	-14.3	-12.4	-9.2	-8.2	-9.0	-7.6	-3.2	2.8	7.9	10.1	10.1	8.9	6.4	4.2	2.9	2.0	1.37	1.37	-0.40	0.49

Table A5. Vertical distribution of the zonally-averaged values (g kg^{-1} m s^{-1}) of the meridional transport [1] of water vapour [qv].

year 1963 – 73

P (MB)	80S	70S	65S	60S	55S	50S	45S	40S	35S	30S	25S	20S	15S	10S	5S	EQ	5N	10N	15N	20N	25N	30N	35N	40N	45N	50N	55N	60N	65N	70N	80N	SH	NH	GLOBE
300	-0.1	-0.1	-0.1	-0.1	-0.1	-0.2	-0.2	-0.2	-0.2	-0.1	0.1	0.1	0.2	0.2	0.0	0.0	0.1	0.1	0.1	0.1	0.1	0.1	0.1	0.2	0.1	0.1	0.1	0.1	0.1	0.1	0.0	-0.10	0.11	0.01
500	-0.2	-0.3	-0.4	-0.5	-0.6	-0.8	-1.0	-1.1	-1.1	-1.0	-0.9	-0.9	-0.7	-0.5	-0.0	-0.7	-0.5	-0.0	0.6	0.7	0.8	1.0	1.1	1.2	1.1	1.0	0.9	0.7	0.5	0.4	0.1	-0.83	0.68	-0.07
700	-0.5	-0.8	-1.1	-1.5	-2.0	-2.3	-2.6	-2.7	-2.6	-2.3	-2.0	-1.6	-2.0	-1.2	-0.6	-1.0	-1.0	-0.8	0.5	1.1	1.9	2.3	2.7	2.7	2.6	2.5	2.3	1.9	1.5	1.1	0.4	-1.76	1.27	-0.24
850	*****	-1.2	-1.7	-2.3	-2.8	-3.0	-3.0	-2.8	-2.7	-2.5	-2.3	-1.9	-2.0	1.7	3.4	2.1	-1.3	-3.6	-0.1	2.7	3.6	4.0	3.6	3.2	3.1	2.6	2.9	2.5	1.9	1.4	0.7	-1.29	1.70	0.21
900	*****	-0.3	-1.6	-3.0	-4.5	-5.8	-6.7	-6.4	-4.5	-2.2	0.6	3.7	6.9	6.3	5.4	2.8	-1.6	-5.4	-6.4	-4.6	3.1	7.2	7.7	7.2	6.7	5.8	4.0	2.3	0.9	0.1	-0.04	1.20	0.58
950	*****	0.1	-1.2	-2.6	-4.1	-5.7	-7.0	-6.7	-4.8	-2.4	0.9	4.6	9.7	9.1	6.1	2.1	-4.1	-8.8	-9.1	-5.3	3.5	8.5	8.8	8.1	7.3	6.2	4.0	2.2	0.7	-0.2	0.1	0.51	0.46	0.49
1000	*****	0.7	-0.8	-2.1	-3.5	-5.2	-7.1	-7.6	-5.7	-3.1	1.4	6.2	23.1	28.3	23.9	21.0	8.6	-17.4	-18.6	-6.9	4.8	11.9	12.2	10.8	9.0	7.5	5.1	2.7	0.8	-0.2	-0.3	5.28	1.56	3.66
MEAN	-0.1	-0.3	-0.6	-1.0	-1.3	-1.7	-1.9	-1.7	-1.9	-1.7	-1.3	-0.7	-0.1	0.7	1.3	1.3	0.8	-0.3	-1.5	-1.0	0.0	1.3	2.0	2.0	1.8	1.7	1.4	1.0	0.7	0.4	0.2	-0.44	0.68	0.12

DJF 1963 – 73

P (MB)	80S	70S	65S	60S	55S	50S	45S	40S	35S	30S	25S	20S	15S	10S	5S	EQ	5N	10N	15N	20N	25N	30N	35N	40N	45N	50N	55N	60N	65N	70N	80N	SH	NH	GLOBE
300	-0.0	-0.0	-0.0	-0.0	-0.0	-0.1	-0.2	-0.2	-0.3	-0.2	0.2	0.3	0.3	0.3	0.2	0.3	0.3	0.5	0.6	0.9	0.8	0.7	0.7	0.9	0.8	0.5	0.7	0.5	0.4	0.3	0.0	-0.04	0.18	0.07
500	-0.2	-0.2	-0.3	-0.4	-0.6	-0.9	-1.1	-1.3	-1.3	-1.2	-0.9	-0.9	-0.9	-0.9	-0.8	-0.6	-0.1	0.5	0.6	0.6	0.8	0.9	1.1	1.1	0.8	0.7	0.7	0.5	0.4	0.3	0.0	-0.89	0.61	-0.14
700	-0.4	-0.4	-0.6	-1.1	-1.7	-2.3	-2.5	-2.7	-2.6	-2.3	-2.0	-1.6	-2.7	-3.4	-4.0	-4.7	-2.9	-0.7	0.6	1.6	2.1	2.7	3.2	3.1	2.9	2.6	2.2	1.6	1.2	0.7	0.4	-2.42	1.05	-0.67
850	*****	-0.9	-1.3	-2.0	-2.6	-3.0	-3.1	-3.0	-3.0	-2.9	-2.6	-2.0	-6.0	-8.6	-8.9	-11.0	-13.9	-13.9	-3.2	2.3	2.9	3.9	4.8	4.9	4.5	4.1	3.6	2.5	1.8	1.0	0.6	-4.30	-1.01	-2.65
900	*****	-0.2	-1.7	-3.2	-4.7	-7.6	-7.0	-3.6	0.2	1.7	4.2	0.6	-7.1	-9.9	-12.7	-15.8	-16.4	-13.5	-8.7	0.4	8.3	9.7	9.0	7.8	6.1	3.6	2.0	1.1	0.2	0.1	0.1	-3.58	-2.63	-3.10
950	*****	-0.1	-1.5	-2.9	-4.4	-6.2	-7.6	-7.1	-3.7	0.4	2.5	5.5	3.1	-4.3	-9.6	-14.1	-19.6	-22.9	-18.3	-10.1	0.6	10.0	11.3	10.1	8.4	6.5	3.7	2.0	1.0	0.1	0.1	-2.96	-4.12	-3.53
1000	*****	0.9	-0.4	-1.7	-3.2	-5.1	-7.4	-7.9	-4.6	0.1	3.3	6.9	12.8	4.4	-0.8	-3.8	-26.5	-51.8	-36.3	-15.2	0.0	14.0	15.1	13.3	10.6	8.5	4.8	2.4	1.2	0.4	0.0	-0.07	-9.48	-4.16
MEAN	-0.1	-0.2	-0.5	-0.8	-1.3	-1.7	-2.0	-2.1	-1.6	-1.0	-0.6	-0.7	-2.0	-2.6	-3.4	-4.7	-2.6	-0.6	0.9	2.2	2.5	2.1	1.8	1.3	0.9	0.6	0.3	0.2	-0.27	-1.34	-0.80

JJA 63-73

JJA 1963 – 73

P (MB)	80S	70S	65S	60S	55S	50S	45S	40S	35S	30S	25S	20S	15S	10S	5S	EQ	5N	10N	15N	20N	25N	30N	35N	40N	45N	50N	55N	60N	65N	70N	80N	SH	NH	GLOBE
300	-0.1	-0.1	-0.1	-0.1	-0.1	-0.1	-0.1	-0.1	-0.1	-0.0	-0.0	0.0	0.0	-0.1	-0.3	-0.2	-0.3	-0.1	0.0	0.0	0.1	0.1	0.1	0.2	0.2	0.2	0.1	0.1	0.1	0.1	0.1	-0.14	0.01	-0.07
500	-0.1	-0.3	-0.4	-0.5	-0.6	-0.8	-0.9	-1.1	-1.1	-1.0	-0.9	-0.9	-1.1	-1.4	-1.1	-0.7	-0.1	0.5	0.6	0.7	0.7	0.7	0.9	1.0	1.1	1.1	1.1	0.9	0.8	0.7	0.3	-0.89	0.61	-0.14
700	-0.5	-1.0	-1.4	-1.8	-2.2	-2.4	-2.7	-2.8	-2.8	-2.6	-2.0	-1.6	-1.3	0.5	2.6	3.0	2.0	0.4	0.4	1.4	2.0	1.8	1.8	1.9	2.1	2.0	1.9	2.0	1.8	0.7	1.64	0.23
850	*****	-1.1	-1.9	-2.7	-3.2	-3.3	-3.3	-3.1	-2.8	-2.6	-2.3	-1.8	1.2	10.6	14.3	13.7	9.9	5.6	3.0	3.3	5.4	5.3	4.2	3.3	2.6	2.2	2.3	2.0	1.9	1.8	0.7	-1.20	4.59	2.88
900	*****	-0.5	-1.4	-2.6	-4.0	-5.4	-6.4	-6.4	-5.5	-4.0	0.1	6.0	12.9	18.6	19.6	17.3	11.4	5.6	2.3	-0.5	0.5	3.4	6.5	6.2	6.0	5.1	3.7	2.2	1.1	0.5	3.49	4.56	4.02
950	*****	-0.4	-1.1	-2.0	-3.5	-5.2	-6.8	-7.0	-6.2	-4.7	-0.0	6.7	15.8	22.1	21.0	17.8	10.7	5.5	2.1	-1.6	-1.5	2.9	7.1	6.4	5.1	5.0	3.5	2.0	1.4	0.9	0.4	4.02	4.25	4.13
1000	*****	-0.8	-1.6	-2.3	-3.6	-5.4	-7.3	-8.1	-7.4	-5.8	0.8	10.0	32.9	48.3	44.9	45.4	42.9	16.2	1.5	-0.5	-1.4	3.8	10.0	9.0	7.1	5.0	3.5	2.3	1.3	0.5	0.3	10.18	11.28	10.66
MEAN	-0.1	-0.4	-0.7	-1.0	-1.3	-1.7	-1.9	-2.0	-1.9	-1.6	-0.8	0.2	2.0	4.2	4.8	4.7	3.6	1.8	0.7	0.6	1.0	1.4	1.7	1.6	1.5	1.4	1.3	1.1	0.9	0.8	0.4	0.35	1.55	0.95

Table A6. Vertical distribution of the zonally-averaged values (g kg^{-1} m s^{-1}) of the meridional transport of water vapour by transient eddies [$\overline{q'v'}$].

year 1963 – 73

p (HB)	80S	70S	65S	60S	55S	50S	45S	40S	35S	30S	25S	20S	15S	10S	5S	EQ	5N	10N	15N	20N	25N	30N	35N	40N	45N	50N	55N	60N	65N	70N	80N	SH	NH	GLOBE
300	-0.1	-0.1	-0.1	-0.1	-0.2	-0.2	-0.2	-0.2	-0.2	-0.1	-0.1	-0.0	-0.0	0.0	0.0	0.0	0.0	0.0	0.1	0.1	0.1	0.1	0.1	0.2	0.2	0.1	0.1	0.1	0.1	0.1	0.0	-0.10	0.08	-0.01
500	-0.2	-0.3	-0.4	-0.5	-0.6	-0.8	-1.0	-1.2	-1.2	-1.1	-1.0	-0.8	-0.6	-0.4	-0.1	0.1	0.3	0.4	0.5	0.5	0.7	0.9	1.1	1.2	1.1	1.0	0.9	0.7	0.5	0.4	0.1	-0.67	0.67	-0.00
700	-0.3	-0.7	-1.0	-1.5	-2.0	-2.4	-2.7	-2.9	-2.9	-2.6	-2.0	-1.5	-1.2	-0.8	-0.4	0.1	0.5	0.7	0.8	1.1	1.6	2.2	2.7	2.6	2.7	2.6	2.3	1.9	1.4	1.1	0.4	-1.65	1.53	-0.05
850	*****	-1.0	-1.6	-2.3	-2.8	-3.1	-3.0	-2.9	-2.7	-2.4	-1.9	-1.6	-1.4	-1.1	-0.6	0.6	1.3	1.5	1.6	2.3	2.9	3.5	3.6	3.6	3.1	2.9	2.5	2.1	1.7	1.3	0.6	-1.88	1.97	0.05
900	*****	-0.9	-1.5	-2.3	-3.0	-3.3	-3.2	-3.0	-2.6	-2.3	-2.0	-1.8	-1.6	-1.1	-0.4	0.8	1.3	1.3	1.8	2.4	3.0	3.4	3.7	3.6	3.1	2.7	2.1	1.7	1.3	0.8	0.3	-1.95	1.81	-0.08
950	*****	-0.6	-1.2	-1.8	-2.4	-2.8	-2.9	-2.6	-2.3	-2.0	-1.8	-1.5	-1.1	-0.6	-0.2	1.0	1.7	1.7	2.4	3.1	3.7	3.7	3.9	3.2	2.6	2.9	1.8	1.5	1.0	0.4	0.3	-1.84	1.97	0.02
1000	*****	-0.3	-1.0	-1.3	-1.7	-2.2	-2.6	-2.7	-2.4	-1.9	-1.5	-1.3	-0.9	-0.5	-0.0	1.3	2.1	3.0	5.3	4.0	5.3	5.1	3.7	3.7	2.9	2.2	1.4	0.8	0.4	-0.1		-1.52	2.37	0.17
MEAN	-0.1	-0.3	-0.6	-0.9	-1.1	-1.3	-1.4	-1.5	-1.4	-1.3	-1.0	-0.8	-0.7	-0.5	-0.2	0.0	0.2	0.5	0.6	0.8	1.0	1.3	1.5	1.5	1.4	1.3	1.1	0.9	0.7	0.5	0.2	-0.86	0.86	-0.01

DJF 1963 – 73

p (HB)	80S	70S	65S	60S	55S	50S	45S	40S	35S	30S	25S	20S	15S	10S	5S	EQ	5N	10N	15N	20N	25N	30N	35N	40N	45N	50N	55N	60N	65N	70N	80N	SH	NH	GLOBE
300	-0.1	-0.1	-0.1	-0.0	-0.0	-0.1	-0.1	-0.1	-0.2	-0.2	-0.1	-0.1	-0.0	-0.1	-0.1	0.1	0.1	0.1	0.1	0.0	0.1	0.1	0.1	0.1	0.1	0.2	0.2	0.2	0.2	0.1	0.0	-0.10	0.09	-0.01
500	-0.2	-0.2	-0.3	-0.4	-0.6	-0.7	-1.0	-1.2	-1.4	-1.3	-1.1	-0.9	-0.7	-0.4	-0.1	0.1	0.4	0.5	0.5	0.6	0.8	1.0	1.0	0.9	0.8	0.6	0.5	0.4	0.3	0.2	0.1	-0.75	0.60	-0.08
700	-0.3	-0.6	-1.1	-1.7	-2.3	-2.6	-2.7	-2.8	-2.7	-2.4	-1.9	-1.4	-1.0	-0.5	-0.0	0.3	0.7	0.9	1.1	1.5	2.0	2.8	3.0	2.6	2.3	1.9	1.6	1.2	0.9	0.7	0.3	-1.44	1.57	0.07
850	*****	-0.8	-1.3	-2.0	-2.6	-3.0	-3.0	-3.0	-2.7	-2.3	-1.8	-1.3	-0.9	-0.5	-0.2	0.1	0.5	0.8	0.8	1.6	2.6	3.4	3.8	3.8	3.1	2.4	1.9	1.5	1.3	0.9	0.5	-1.65	1.82	0.09
900	*****	-0.9	-1.6	-2.5	-3.1	-3.5	-3.4	-3.2	-2.9	-2.5	-2.0	-1.7	-1.4	-0.9	-0.4	0.1	0.4	0.4	0.8	1.4	2.5	3.4	3.7	3.6	3.0	2.2	1.6	1.2	1.0	0.6	0.3	-1.75	1.65	-0.05
950	*****	-1.0	-1.6	-2.2	-3.0	-3.1	-3.1	-2.9	-2.5	-2.0	-1.7	-1.3	-0.8	-0.4	-0.1	0.0	0.4	0.5	1.0	1.9	3.1	4.0	4.1	4.1	3.2	2.4	1.6	1.2	0.9	0.5	0.2	-1.57	1.85	0.10
1000	*****	-0.7	-1.1	-1.6	-2.2	-2.7	-2.6	-2.4	-1.9	-1.3	-1.2	-0.6	-0.3	-0.1	-0.1	0.4	0.7	1.0	1.7	3.5	5.3	4.9	4.6	4.1	2.7	1.7	1.1	0.8	0.5	0.1		-1.07	1.92	0.23
MEAN	-0.1	-0.3	-0.5	-0.7	-1.0	-1.2	-1.3	-1.4	-1.4	-1.2	-1.0	-0.7	-0.5	-0.3	-0.1	0.1	0.3	0.4	0.5	0.8	1.2	1.5	1.6	1.5	1.3	1.0	0.8	0.6	0.5	0.3	0.2	-0.78	0.81	0.01

JJA 1963 – 73

p (HB)	80S	70S	65S	60S	55S	50S	45S	40S	35S	30S	25S	20S	15S	10S	5S	EQ	5N	10N	15N	20N	25N	30N	35N	40N	45N	50N	55N	60N	65N	70N	80N	SH	NH	GLOBE
300	-0.1	-0.1	-0.1	-0.1	-0.1	-0.1	-0.1	-0.1	-0.1	-0.1	-0.1	-0.0	0.0	0.0	0.0	0.0	0.0	0.1	0.1	0.1	0.1	0.1	0.1	0.2	0.2	0.2	0.2	0.2	0.1	0.1	0.0	-0.09	0.08	-0.01
500	-0.1	-0.3	-0.4	-0.6	-0.8	-1.0	-1.1	-1.1	-1.0	-0.9	-0.8	-0.6	-0.5	-0.3	0.0	0.2	0.4	0.5	0.6	0.8	1.0	1.3	1.8	2.2	2.3	2.6	2.6	2.3	1.9	1.6	0.6	-0.66	0.60	-0.03
700	-0.4	-0.9	-1.3	-1.8	-2.5	-2.8	-3.0	-3.1	-2.9	-2.6	-2.2	-1.6	-1.1	-0.6	0.1	0.5	0.7	0.9	1.3	1.8	2.6	2.8	2.6	2.2	2.6	2.6	2.2	1.9	1.6	0.6	-1.73	1.12	-0.30
850	*****	-1.0	-1.8	-2.6	-3.4	-3.3	-3.1	-3.1	-3.0	-2.8	-2.2	-1.8	-1.3	-0.8	0.2	0.4	0.7	0.8	1.7	2.3	3.1	3.8	2.8	2.3	2.6	2.1	1.8	1.6	2.0	1.1	0.6	-1.98	1.34	-0.32
900	*****	-1.4	-2.2	-2.8	-3.1	-3.0	-3.0	-2.8	-2.5	-2.1	-1.6	-1.2	-0.7	-0.3	0.2	0.3	0.5	0.9	1.1	1.7	2.1	2.3	2.4	2.1	1.8	1.8	1.8	1.6	1.7	1.5	0.6	-1.84	1.14	-0.35
950	*****	-1.1	-1.6	-2.2	-2.6	-3.1	-3.1	-3.1	-3.0	-2.5	-2.0	-1.3	-0.8	-0.4	0.1	0.2	0.3	0.5	0.9	1.8	2.1	1.7	1.7	2.1	1.8	1.6	1.8	1.6	1.5	1.0	0.5	-1.90	1.03	-0.47
1000	*****	-1.2	-1.5	-1.7	-1.9	-2.0	-2.4	-2.7	-2.8	-2.6	-1.9	-1.2	-0.6	-0.3	-0.0	0.1	0.3	0.3	1.3	2.4	3.1	2.5	1.7	1.1	1.3	1.2	1.1	1.3	1.0	0.5	0.3	-1.55	0.89	-0.49
MEAN	-0.1	-0.4	-0.7	-0.9	-1.2	-1.3	-1.4	-1.5	-1.3	-1.1	-0.8	-0.6	-0.4	-0.3	0.0	0.1	0.3	0.4	0.5	1.2	1.3	1.3	2.4	1.2	1.3	1.2	1.1	0.9	0.8	0.3		-0.88	0.59	-0.14

year 1963 – 73   DJF 1963 – 73   JJA 1963 – 73

Table A7. Vertical distribution of the zonally-averaged values $(g\ kg^{-1}\ m\ s^{-1})$ of the meridional transport of water vapour by stationary eddies $[\overline{q^{*}v^{*}}]$.

year 1963 – 73

P (MB)	80S	70S	65S	60S	55S	50S	45S	40S	35S	30S	25S	20S	15S	10S	5S	EQ	5N	10N	15N	20N	25N	30N	35N	40N	45N	50N	55N	60N	65N	70N	80N	SH	NH	GLOBE
300	-0.0	-0.0	-0.0	0.0	0.0	0.0	0.0	0.0	0.0	0.0	0.0	0.1	0.1	0.0	-0.0	-0.0	0.0	0.0	0.1	0.1	0.1	0.1	0.0	0.0	0.0	0.0	0.0	0.0	0.0	0.0	-0.0	0.00	0.02	0.01
500	-0.0	-0.0	-0.0	0.1	0.1	0.1	0.1	-0.0	-0.0	0.0	0.0	0.1	0.1	-0.0	0.0	-0.0	0.1	0.1	0.2	0.2	0.4	0.3	0.0	0.1	0.0	0.0	0.0	0.0	0.1	0.0	-0.0	-0.00	0.05	0.02
700	-0.2	-0.1	-0.0	0.1	0.1	0.1	0.1	0.1	0.1	0.1	0.0	0.1	0.0	-0.1	0.0	-0.1	-0.0	0.1	0.6	0.2	0.4	0.3	0.1	0.1	0.0	0.0	0.1	0.1	0.1	0.0	-0.0	0.03	0.08	0.05
850	*****	-0.1	-0.0	0.0	-0.1	-0.0	0.2	-0.4	-0.5	-0.4	-0.8	-0.9	-0.6	-0.4	-0.0	0.0	-0.0	0.6	1.5	1.4	1.5	1.0	0.6	0.3	0.2	0.3	0.3	0.3	0.2	0.1	-0.1	-0.17	0.46	0.15
900	*****	-0.0	0.0	-0.0	-0.1	-0.1	-0.2	-0.8	-0.8	-0.8	-1.0	-1.1	-0.6	-0.4	-0.1	-0.1	-0.0	0.5	1.2	1.4	1.5	1.1	0.5	0.2	0.3	0.3	0.4	0.3	0.3	0.1	-0.0	-0.31	0.48	0.09
950	*****	0.0	0.0	-0.0	-0.1	-0.1	-0.3	-1.0	-1.0	-0.7	-1.4	-1.3	-0.8	-0.6	0.0	-0.2	0.1	0.6	1.4	1.3	1.4	1.2	0.7	0.3	0.3	0.4	0.5	0.4	0.4	0.1	0.0	-0.39	0.54	0.07
1000	*****	0.0	0.0	0.1	-0.1	-0.1	-0.5	-1.5	-1.8	-1.7	-2.2	-2.1	-1.2	-0.8	0.5	-0.5	0.5	1.6	2.1	2.1	1.4	1.5	0.8	0.3	0.5	0.5	0.8	1.0	0.9	0.3	0.0	-0.85	0.75	-0.16
MEAN	0.0	0.1	0.1	-0.1	-0.1	-0.1	-0.1	-0.2	-0.2	-0.1	-0.2	-0.1	-0.1	-0.1	-0.1	-0.1	0.0	0.0	0.2	0.4	0.5	0.3	0.1	0.1	0.1	0.1	0.1	0.1	0.1	0.0	0.0	-0.07	0.14	0.03

DJF 1963 – 73

P (MB)	80S	70S	65S	60S	55S	50S	45S	40S	35S	30S	25S	20S	15S	10S	5S	EQ	5N	10N	15N	20N	25N	30N	35N	40N	45N	50N	55N	60N	65N	70N	80N	SH	NH	GLOBE
300	0.0	0.0	-0.0	0.0	0.0	0.0	0.0	0.0	0.0	0.0	0.0	0.1	0.1	0.1	-0.0	-0.0	0.1	0.1	0.2	0.1	0.1	0.1	0.0	0.0	0.0	0.0	0.0	0.1	0.0	0.0	-0.0	0.01	0.03	0.02
500	-0.0	0.0	0.0	0.0	0.0	0.0	0.0	0.0	0.0	0.0	0.1	0.1	0.1	0.0	-0.0	-0.0	0.1	0.2	0.2	0.2	0.2	0.1	0.1	0.2	0.2	0.2	0.1	0.1	0.0	0.0	-0.0	0.03	0.12	0.07
700	-0.2	-0.1	-0.0	0.0	0.1	0.1	0.2	0.2	0.2	0.1	0.0	0.0	0.1	-0.0	-0.2	-0.2	-0.0	0.1	0.5	0.2	0.3	0.2	0.3	0.6	0.7	0.7	0.6	0.4	0.2	0.0	0.0	-0.04	0.26	0.11
850	*****	-0.1	0.0	0.0	0.0	0.0	-0.2	-0.4	-0.8	-0.5	-1.1	-0.9	-0.6	-0.3	-0.2	-0.0	-0.1	0.5	0.3	1.0	0.8	0.7	1.0	1.3	1.6	1.6	1.3	1.0	0.6	0.1	0.1	-0.35	0.68	0.17
900	*****	-0.0	0.1	0.0	-0.1	-0.1	-0.2	-0.5	-1.2	-0.8	-1.4	-1.1	-0.9	-0.6	-0.3	-0.1	-0.2	0.3	0.7	0.9	1.0	1.2	1.1	1.4	1.8	1.9	1.4	1.0	0.6	0.1	0.1	-0.51	0.67	0.08
950	0.0	0.1	0.1	0.1	-0.1	-0.2	-0.2	-0.5	-1.4	-1.0	-1.4	-1.3	-1.0	-0.5	-0.2	-0.1	-0.1	0.5	0.9	0.9	1.0	1.2	1.3	1.4	1.8	1.9	1.2	1.0	0.6	0.1	0.1	-0.60	0.74	0.05
1000	0.1	0.1	0.1	0.1	-0.1	-0.3	-1.2	-2.7	-2.7	-2.2	-2.8	-2.1	-1.1	-1.0	-1.1	-1.1	0.4	0.7	1.4	0.9	1.4	1.6	1.7	1.8	1.9	1.9	1.6	1.2	0.4	0.1	0.1	-1.31	0.92	-0.34
MEAN	-0.0	0.1	0.1	0.0	-0.0	-0.1	-0.2	-0.3	-0.3	-0.2	-0.3	-0.3	-0.2	-0.2	-0.1	-0.1	-0.0	0.0	0.2	0.3	0.3	0.3	0.3	0.4	0.5	0.6	0.5	0.3	0.2	0.0	-0.0	-0.13	0.23	0.05

JJA 1963 – 73

P (MB)	80S	70S	65S	60S	55S	50S	45S	40S	35S	30S	25S	20S	15S	10S	5S	EQ	5N	10N	15N	20N	25N	30N	35N	40N	45N	50N	55N	60N	65N	70N	80N	SH	NH	GLOBE
300	-0.0	-0.0	-0.0	0.0	-0.0	-0.0	-0.1	-0.0	-0.0	0.0	0.0	0.1	-0.0	-0.0	-0.0	0.0	-0.0	-0.0	-0.0	-0.0	-0.1	-0.1	-0.1	-0.1	-0.0	-0.0	-0.0	0.0	0.0	0.0	-0.0	-0.00	-0.04	-0.02
500	-0.0	-0.0	-0.0	0.0	-0.1	-0.1	-0.2	-0.2	-0.2	-0.0	-0.0	0.1	0.0	-0.0	-0.0	0.0	0.0	0.0	0.0	0.1	0.2	0.2	-0.0	-0.1	-0.1	-0.0	-0.1	0.0	0.0	0.1	-0.0	-0.07	0.01	-0.03
700	-0.1	-0.1	-0.0	0.1	0.0	0.1	0.1	0.1	0.0	0.1	0.1	0.5	0.1	0.1	0.1	0.0	0.1	0.5	1.1	1.0	1.0	0.6	-0.3	-0.4	-0.5	-0.3	0.0	0.1	0.1	0.1	0.0	0.02	0.21	0.11
850	*****	-0.0	0.0	0.0	0.1	-0.0	-0.3	-0.4	-0.4	-0.4	-0.7	-0.2	0.4	0.4	-0.0	-0.0	0.1	0.4	0.8	2.6	4.4	4.1	2.4	1.0	-0.5	-0.5	-0.1	-0.1	0.2	0.1	-0.1	-0.09	1.16	0.54
900	*****	-0.0	0.0	0.1	-0.1	-0.1	-0.3	-0.2	-0.5	-0.2	-0.8	-0.7	0.0	-0.3	-0.2	-0.1	0.0	0.2	0.9	3.1	4.8	4.1	2.3	1.2	-0.3	-0.4	-0.2	0.0	0.1	0.0	-0.0	-0.31	1.23	0.46
950	*****	-0.0	0.0	0.1	-0.1	-0.2	-0.2	-0.5	-0.8	-0.5	-0.6	-0.6	0.1	-0.2	-0.1	-0.1	0.1	0.5	1.1	2.9	4.1	4.0	2.4	1.3	-0.3	-0.4	-0.3	0.0	0.1	0.0	-0.0	-0.32	1.18	0.41
1000	*****	-0.1	0.0	0.0	-0.2	-0.3	-0.6	-1.4	-1.4	-1.1	-1.7	-0.8	0.9	-0.8	-1.3	-1.2	0.2	0.9	3.8	5.9	6.3	5.4	3.5	1.6	0.0	-0.3	0.0	0.3	0.2	0.0	0.0	-0.82	2.04	0.42
MEAN	-0.0	-0.0	-0.0	0.1	-0.0	-0.1	-0.1	-0.2	-0.2	-0.1	-0.3	-0.1	0.1	-0.1	-0.0	-0.0	0.1	0.3	0.8	1.2	1.1	0.6	0.2	-0.1	-0.2	-0.2	-0.1	0.0	0.1	0.1	-0.0	-0.08	0.31	0.12

Table A8. Vertical distribution of the zonally-averaged values (g kg^{-1} m s^{-1}) of the meridional transport of water vapour by mean meridional circulations[1] $[\bar{q}][\bar{v}]$.

year 1963 – 73

p (HB)	80S	70S	65S	60S	55S	50S	45S	40S	35S	30S	25S	20S	15S	10S	5S	EQ	5N	10N	15N	20N	25N	30N	35N	40N	45N	50N	55N	60N	65N	70N	80N	SH	NH	GLOBE
300	-0.0	-0.0	-0.0	0.0	0.0	0.0	0.0	-0.0	-0.0	-0.0	0.0	0.0	0.0	0.0	0.0	0.0	0.1	0.1	0.1	0.1	0.0	-0.0	-0.0	-0.0	-0.0	-0.0	-0.0	-0.0	0.0	0.0	0.0	-0.00	0.02	0.01
500	-0.0	-0.0	-0.0	0.0	0.1	0.1	0.1	0.1	0.1	0.0	-0.1	-0.1	-0.3	-0.6	-0.6	-0.6	-0.3	0.1	0.1	0.1	-0.0	-0.1	-0.1	-0.1	-0.1	-0.0	-0.0	-0.0	0.0	0.0	0.0	-0.15	-0.04	-0.09
700	-0.0	-0.0	-0.0	0.0	0.0	0.1	0.1	0.1	0.1	0.0	-0.1	-0.3	-0.5	-1.0	-1.4	-1.0	-0.3	0.0	0.0	0.0	-0.1	-0.1	-0.1	-0.1	-0.1	-0.1	-0.0	-0.0	0.0	0.0	0.0	-0.14	-0.34	-0.24
850	•••••	-0.0	0.0	0.0	0.1	0.1	0.1	0.1	0.1	0.2	0.1	0.1	-0.3	2.9	4.0	2.1	-1.8	-4.8	-2.1	-0.2	0.1	0.1	0.0	-0.1	-0.1	-0.1	-0.0	-0.0	0.0	0.0	0.0	0.76	-0.73	0.01
900	•••••	0.6	-0.1	-0.7	-1.5	-2.5	-3.4	-3.3	-1.7	0.5	3.3	6.4	9.1	7.7	6.2	2.9	-2.0	-6.2	-8.2	-7.5	-0.8	3.0	3.7	3.4	3.4	2.8	1.5	0.2	-0.7	-0.2	-0.2	2.22	-1.08	0.58
950	•••••	0.7	-0.1	-0.8	-1.7	-2.9	-4.0	-3.8	-2.0	0.6	3.9	7.4	12.0	10.7	6.9	2.4	-4.5	-9.9	-11.4	-9.0	-1.0	3.6	4.4	4.0	3.9	3.1	1.6	0.2	-0.8	-0.2	-0.2	2.74	-2.05	0.40
1000	•••••	1.0	-0.1	-1.0	-2.1	-3.5	-5.0	-4.9	-2.6	0.8	5.1	9.4	25.8	31.0	25.9	22.7	9.1	-19.2	-22.3	-12.0	-1.4	5.1	6.1	5.4	5.0	4.1	2.1	0.2	-0.9	-0.3	-0.3	7.65	-1.56	3.65
MEAN	0.1	0.1	-0.0	-0.1	-0.2	-0.4	-0.5	-0.5	-0.2	0.1	0.5	0.9	1.5	1.9	1.6	0.8	-0.5	-2.0	-1.8	-1.1	-0.1	0.4	0.5	0.4	0.4	0.3	0.2	0.0	-0.1	-0.0	-0.0	0.49	-0.31	0.09

DJF 1963 – 73

p (HB)	80S	70S	65S	60S	55S	50S	45S	40S	35S	30S	25S	20S	15S	10S	5S	EQ	5N	10N	15N	20N	25N	30N	35N	40N	45N	50N	55N	60N	65N	70N	80N	SH	NH	GLOBE
300	-0.0	-0.0	-0.0	0.0	0.0	0.0	0.0	-0.0	-0.0	-0.0	0.1	0.1	0.1	0.2	0.2	0.3	0.2	0.1	0.1	0.1	0.0	-0.0	-0.0	-0.0	-0.0	-0.0	-0.0	-0.0	0.0	0.0	0.0	0.05	0.06	0.06
500	-0.0	-0.0	-0.0	0.0	0.1	0.1	0.1	0.1	0.0	-0.1	-0.3	-0.5	-0.7	-0.7	-0.4	-0.1	-0.0	0.1	0.1	0.0	-0.1	-0.1	-0.1	-0.1	-0.1	-0.0	-0.0	-0.0	0.0	0.0	0.0	-0.17	-0.10	-0.14
700	-0.0	-0.0	-0.0	0.0	0.0	0.1	0.1	0.1	0.0	-0.1	-0.3	-1.6	-3.0	-4.8	-3.5	-1.6	-0.6	-0.1	0.1	0.0	-0.3	-0.2	-0.1	-0.1	-0.0	-0.0	-0.0	-0.0	0.0	0.0	0.0	-0.94	-0.77	-0.86
850	•••••	-0.0	0.0	0.0	0.1	0.1	0.1	0.1	0.1	-0.1	-1.6	-3.9	-7.8	-8.6	-11.1	-14.1	-14.3	-4.5	-0.3	-0.5	-0.2	-0.1	-0.1	-0.0	-0.0	-0.0	-0.0	0.0	0.0	0.0	0.0	-2.30	-3.50	-2.90
900	•••••	0.8	-0.0	-0.7	-1.6	-2.9	-4.0	-3.4	-0.5	3.1	4.6	6.7	2.4	-6.1	-14.5	-12.7	-16.0	-10.9	-2.9	4.0	3.0	4.9	4.9	4.0	3.0	2.0	0.7	-0.2	-0.5	-0.2	-0.2	-1.32	-4.96	-3.13
950	•••••	0.9	-0.0	-0.8	-1.8	-3.3	-4.5	-3.9	-0.6	3.5	5.3	8.3	4.8	-9.5	-16.6	-14.0	-19.3	-12.8	-3.4	4.8	3.5	5.8	5.8	4.6	3.5	2.3	0.8	-0.2	-0.5	-0.2	-0.2	-0.79	-6.71	-3.68
1000	•••••	1.1	-0.0	-1.0	-2.2	-3.8	-5.4	-4.9	-0.8	4.7	6.9	9.7	15.4	-2.6	-26.5	-52.6	-37.7	-17.8	-4.9	7.1	8.5	6.9	5.3	3.9	1.3	-0.3	-0.6	-0.1	2.31	-12.32	-4.05	2.31	-12.32	-4.05
MEAN	0.1	0.1	-0.0	-0.1	-0.2	-0.4	-0.5	-0.6	-0.1	0.5	0.6	0.9	0.0	-1.5	-2.4	-3.4	-4.5	-3.3	-1.6	0.5	0.6	0.6	0.6	0.4	0.2	0.1	-0.0	-0.0	-0.0	-0.0	-0.0	-0.43	-1.32	-0.87

JJA 1963 – 73

p (HB)	80S	70S	65S	60S	55S	50S	45S	40S	35S	30S	25S	20S	15S	10S	5S	EQ	5N	10N	15N	20N	25N	30N	35N	40N	45N	50N	55N	60N	65N	70N	80N	SH	NH	GLOBE
300	-0.0	-0.0	-0.0	0.0	0.0	0.0	0.0	-0.0	-0.0	-0.0	0.1	0.1	0.1	0.2	0.2	0.3	0.2	0.1	-0.0	-0.2	0.1	0.1	0.1	0.0	-0.0	-0.0	-0.0	-0.0	0.0	0.0	0.0	-0.05	-0.03	-0.04
500	-0.0	-0.0	-0.0	0.0	0.0	0.1	0.1	0.0	0.0	0.0	0.1	0.1	0.0	-0.1	-0.2	-0.7	-0.6	-0.2	0.3	0.2	0.1	0.1	0.1	-0.0	-0.1	-0.1	-0.1	-0.0	0.0	0.0	0.0	-0.15	0.00	-0.07
700	-0.0	-0.0	-0.0	0.0	0.0	0.1	0.1	0.1	0.1	0.2	0.2	0.0	1.2	2.8	3.0	1.8	0.2	-0.1	-0.1	-0.0	-0.1	-0.0	-0.1	-0.1	-0.1	-0.0	-0.0	-0.0	0.0	0.0	0.0	0.51	0.31	0.41
850	•••••	-0.0	0.0	0.0	0.0	0.1	0.1	0.0	0.0	0.2	0.3	0.2	2.5	11.3	14.7	13.7	9.7	5.9	1.6	-0.1	0.0	0.2	0.1	-0.1	-0.2	-0.1	-0.0	-0.0	0.0	0.0	0.0	3.23	2.09	2.66
900	•••••	0.3	0.0	-0.3	-1.1	-2.2	-3.2	-3.2	-2.5	-1.1	2.9	8.3	14.6	19.7	20.2	17.4	11.2	5.0	0.9	-4.3	-5.2	-1.8	2.6	2.9	3.5	3.0	1.9	0.6	-0.3	-0.6	-0.1	5.64	2.18	3.92
950	•••••	0.3	0.0	-0.4	-1.2	-2.5	-3.7	-3.7	-2.9	-1.3	3.3	9.5	17.7	23.2	21.6	18.0	10.5	4.8	0.7	-5.2	-6.4	-2.2	3.1	3.4	3.9	3.3	2.1	0.7	-0.4	-0.6	-0.1	6.25	2.04	4.20
1000	•••••	0.4	0.0	-0.5	-1.5	-3.2	-4.8	-5.0	-4.0	-1.8	4.5	12.6	34.6	50.0	46.3	43.9	15.1	-2.5	-6.6	-8.1	-2.9	4.1	4.3	4.6	3.6	2.3	0.7	-0.4	-0.7	-0.2	12.54	8.35	10.72
MEAN	0.0	0.0	-0.0	-0.1	-0.2	-0.3	-0.5	-0.5	-0.4	-0.1	0.5	1.2	2.7	4.7	5.0	4.8	3.6	1.6	0.2	-0.6	-0.7	-0.2	0.3	0.3	0.4	0.3	0.2	0.1	-0.0	-0.1	-0.0	1.31	0.65	0.98

Table A9. Vertical distribution of the zonally-averaged values (10^{-4} g kg^{-1} mb s^{-1}) of the total vertical transport of water [$\overline{q\omega} + \overline{q_c\omega_c}$]. To convert into units of g cm^{-2} yr^{-1}, multiply values by 3.16. In principle, the global averages in the last column should vanish. The non-zero values are a measure of the computational error in the calculations of the total vertical transport using the streamfunction approach.

year 1963 – 73

P (MB)	80S	70S	65S	60S	55S	50S	45S	40S	35S	30S	25S	20S	15S	10S	5S	EQ	5N	10N	15N	20N	25N	30N	35N	40N	45N	50N	55N	60N	65N	70N	80N	SH	NH	GLOBE
300	0.0	0.0	0.0	0.0	0.0	0.0	0.0	0.1	0.1	0.1	-0.0	-0.1	-0.1	-0.0	-0.0	0.0	0.1	0.1	0.0	0.0	0.1	0.0	-0.0	-0.0	0.0	0.0	0.0	0.0	0.0	0.0	0.0	-0.00	0.00	-0.00
500	0.3	0.3	0.4	0.4	0.6	0.7	0.6	0.5	0.3	0.1	-0.2	-0.4	-0.5	-0.6	-0.8	-1.2	-0.7	-0.1	0.2	-0.1	-0.1	-0.4	0.1	0.5	0.5	0.6	0.6	0.6	0.5	0.6	0.4	0.04	-0.05	-0.00
700	1.1	1.7	1.9	2.2	2.3	2.2	1.8	1.4	0.4	-0.5	-1.0	-0.2	0.1	-2.7	-1.1	-0.9	-2.6	-3.1	-2.4	-0.8	-1.1	-0.4	0.5	1.5	1.6	2.4	2.5	2.6	2.4	2.0	2.0	0.27	-0.30	-0.01
850	*****	3.7	4.4	4.6	4.1	3.4	2.5	1.7	0.4	-1.0	-1.6	0.5	-2.5	-10.3	-1.2	3.9	2.9	-6.3	-9.1	-3.5	-2.2	-0.6	1.4	2.1	2.7	3.7	4.8	4.7	4.4	3.6	3.6	0.03	-0.16	-0.06
900	*****	4.7	6.0	6.3	5.9	5.0	3.4	1.0	-1.3	-3.3	-4.0	-2.5	-4.2	-10.0	0.9	7.0	7.4	-5.2	-8.8	-8.9	-7.1	-0.9	0.6	2.0	2.6	4.2	5.6	6.5	5.7	4.7	3.5	-0.13	0.01	-0.06
950	*****	5.5	7.4	8.0	7.7	6.9	4.5	0.3	-3.1	-5.8	-6.7	-6.5	-7.2	-8.7	4.9	11.0	12.7	-3.7	-9.8	-15.7	-12.4	-2.0	1.7	2.6	3.3	5.9	7.2	8.3	6.7	4.9	3.4	-0.20	0.11	-0.05
1000	*****	6.0	8.4	9.2	9.0	8.5	5.7	-0.0	-4.6	-8.0	-9.0	-12.4	-14.9	-7.5	7.2	13.0	24.6	4.3	-13.8	-22.2	-16.3	-3.4	2.7	3.5	4.6	7.0	8.0	9.3	7.1	5.2	3.1	-0.76	1.13	0.06
MEAN	0.4	1.2	1.8	2.0	1.9	1.8	1.3	0.6	-0.2	-0.9	-1.2	-2.6	0.0	1.2	1.1	-1.6	-2.5	-2.2	-1.7	-0.3	-0.0	0.5	1.0	1.2	1.6	1.9	1.5	1.2	1.5	1.7	1.5	0.03	-0.05	-0.01

DJF 1963 – 73

P (MB)	80S	70S	65S	60S	55S	50S	45S	40S	35S	30S	25S	20S	15S	10S	5S	EQ	5N	10N	15N	20N	25N	30N	35N	40N	45N	50N	55N	60N	65N	70N	80N	SH	NH	GLOBE
300	0.0	0.0	0.0	0.0	0.0	0.0	0.1	0.1	0.1	0.1	-0.0	-0.1	0.2	-0.1	-0.1	0.0	0.1	0.1	0.0	0.1	0.1	0.1	-0.1	-0.0	0.0	0.0	0.0	0.0	0.0	0.0	0.0	0.05	0.05	-0.00
500	0.2	0.2	0.3	0.4	0.6	0.8	0.7	0.6	0.2	0.1	-0.4	-0.7	-0.6	-0.2	-0.8	-1.2	-0.6	-0.1	0.2	0.1	0.1	0.5	-0.4	0.3	0.5	0.5	0.5	0.5	0.4	0.3	0.3	0.3	-0.01	-0.00
700	0.5	0.9	1.4	2.4	2.9	2.7	2.1	1.6	0.7	-0.1	-1.5	-1.0	0.0	1.6	-2.1	1.7	-1.4	-0.7	-1.0	-0.1	-0.1	1.1	1.9	1.9	2.1	2.2	2.3	2.4	2.1	1.9	1.2	0.87	-0.89	-0.01
850	*****	3.5	4.8	5.1	4.2	2.8	2.1	0.8	-0.3	-2.5	-3.5	3.5	8.0	2.0	5.4	1.7	-6.3	-18.1	-16.6	-5.2	-2.3	-2.4	-0.3	1.7	3.1	4.6	4.9	5.0	4.4	3.9	2.1	2.94	-3.05	-0.06
900	*****	5.2	6.6	7.0	6.2	5.0	3.9	0.5	-2.5	-2.8	-3.6	2.4	13.2	6.0	8.4	4.8	-4.2	-19.6	-20.2	-11.5	-10.0	-5.2	0.4	2.7	4.1	6.7	7.0	6.4	5.2	4.3	1.9	3.91	-4.13	-0.10
950	*****	6.8	8.3	9.0	8.4	7.0	5.0	-1.0	-5.9	-5.7	-5.0	0.5	17.8	10.0	13.7	9.5	-0.6	-19.3	-26.7	-20.0	-18.3	-8.2	1.5	3.9	5.4	9.1	8.7	7.7	6.0	4.6	1.8	4.96	-5.42	-0.10
1000	*****	7.6	9.4	10.3	10.2	9.8	6.5	-1.7	-8.6	-8.4	-6.0	-3.8	18.3	15.2	15.3	14.1	-17.6	-38.9	-28.5	-25.0	-11.2	2.6	5.2	7.0	10.6	9.8	8.5	7.7	6.0	4.7	1.7	5.28	-6.26	0.26
MEAN	0.3	0.8	1.5	2.1	2.3	2.2	1.5	0.6	-0.4	-0.7	-1.2	0.1	2.6	1.2	2.2	0.7	-1.9	-5.0	-5.2	-3.0	-2.2	-1.2	0.1	0.8	1.3	1.8	1.9	1.8	1.5	1.3	0.7	0.99	-1.01	-0.01

JJA 1963 – 73

P (MB)	80S	70S	65S	60S	55S	50S	45S	40S	35S	30S	25S	20S	15S	10S	5S	EQ	5N	10N	15N	20N	25N	30N	35N	40N	45N	50N	55N	60N	65N	70N	80N	SH	NH	GLOBE
300	0.1	0.0	0.0	0.0	0.0	0.0	0.0	0.0	0.0	0.1	-0.1	-0.1	0.0	-0.2	-0.1	-0.0	-0.2	-0.2	-0.1	-0.1	0.0	0.1	-0.1	-0.0	0.0	0.0	0.0	0.1	0.1	0.1	0.0	0.05	-0.05	-0.00
500	0.6	0.3	0.2	0.3	0.4	0.6	0.5	0.5	0.2	0.0	0.0	0.0	0.5	0.0	-0.8	-1.0	-1.4	-0.4	0.0	0.0	0.5	0.5	-0.4	-0.6	-0.1	0.2	0.6	0.6	0.8	0.8	0.8	0.11	-0.11	-0.00
700	1.5	2.0	1.9	2.0	2.0	1.9	1.5	0.4	-0.1	-0.7	-1.0	-0.2	-0.7	-3.9	-4.2	-1.1	-0.0	0.7	-1.2	-2.0	-0.2	1.8	-0.7	-0.8	-0.2	0.2	1.7	2.1	2.9	3.2	3.2	-0.28	0.24	-0.02
850	*****	4.6	5.0	4.6	3.8	3.0	2.6	1.7	0.3	-1.6	-2.0	-1.8	-11.6	-19.6	-8.5	3.8	8.7	6.7	1.4	-5.8	-1.2	4.9	1.1	0.2	0.3	0.9	2.7	3.0	3.2	5.0	5.9	-2.61	2.46	-0.07
900	*****	5.4	6.2	6.1	5.5	4.7	3.5	1.6	-0.4	-3.8	-6.6	-6.9	-18.0	-23.3	-7.4	7.1	14.9	10.2	3.8	-4.4	-2.7	1.7	0.8	0.6	0.6	2.1	4.3	4.4	3.8	5.4	6.4	-3.84	3.83	-0.02
950	*****	6.0	7.2	7.5	7.4	6.7	4.9	1.0	-1.1	-6.1	-11.6	-13.0	-26.1	-26.0	-4.2	10.6	21.4	12.7	6.5	-2.3	-4.1	-2.7	0.5	1.0	1.1	3.4	5.6	4.5	5.6	6.8	-5.00	5.33	0.04
1000	*****	6.6	8.0	8.5	8.7	8.3	6.1	1.7	-1.8	-8.3	-16.0	-20.9	-38.9	-28.1	-2.5	8.1	30.7	26.4	9.7	-1.4	-4.8	-6.3	0.3	1.8	2.3	4.4	6.1	6.3	5.6	7.0	6.8	-6.40	8.25	-0.03
MEAN	0.6	1.3	1.8	1.8	1.8	1.7	1.4	0.9	0.0	-1.1	-1.7	-1.8	-4.2	-5.6	-2.3	1.0	3.0	2.5	0.6	-1.3	-0.6	0.7	-0.1	-0.2	0.1	0.5	1.3	1.4	1.3	1.8	2.1	-0.95	0.84	-0.01

Table A10. Vertical distribution of the zonally-averaged values (10^{-4} g kg^{-1} mb s^{-1}) of the vertical transport of water vapour by stationary eddies [q*ω*].

year 1963 – 73 (summary columns)

P (MB)	SH	NH	GLOBE
300	-0.02	-0.02	-0.02
500	-0.20	-0.19	-0.19
700	-0.23	-0.45	-0.34
850	-0.11	-0.29	-0.20
900	-0.11	-0.19	-0.15
950	-0.08	-0.10	-0.09
1000	-0.10	-0.07	-0.09
MEAN	-0.11	-0.17	-0.14

DJF 1963 – 73 (summary columns)

P (MB)	SH	NH	GLOBE
300	-0.03	-0.01	-0.02
500	-0.20	-0.18	-0.19
700	-0.40	-0.51	-0.46
850	-0.33	-0.46	-0.39
900	-0.25	-0.36	-0.30
950	-0.18	-0.23	-0.20
1000	-0.16	-0.11	-0.14
MEAN	-0.17	-0.20	-0.19

JJA 1963 – 73 (summary columns)

P (MB)	SH	NH	GLOBE
300	-0.04	-0.09	-0.06
500	-0.32	-0.61	-0.46
700	-0.24	-1.07	-0.66
850	-0.08	-0.74	-0.41
900	-0.14	-0.53	-0.33
950	-0.11	-0.31	-0.21
1000	-0.10	-0.12	-0.11
MEAN	-0.13	-0.45	-0.29

Table A11. Vertical distribution of the zonally-averaged values (10^{-4} g kg^{-1} mb s^{-1}) of the vertical transport of water vapour by mean meridional circulations(1) $[\bar{q}][\bar{\omega}]$.

year 1963 – 73

p (MB)	80S	70S	65S	60S	55S	50S	45S	40S	35S	30S	25S	20S	15S	10S	5S	EQ	5N	10N	15N	20N	25N	30N	35N	40N	45N	50N	55N	60N	65N	70N	80N	SH	NH	GLOBE
300	-0.0	-0.0	-0.0	-0.0	-0.0	-0.0	0.0	0.0	0.1	0.1	0.1	0.1	0.1	0.1	-0.3	-0.4	-0.5	-0.1	0.3	0.2	0.2	0.2	0.1	-0.0	-0.0	-0.0	-0.0	0.0	0.0	0.0	0.0	-0.00	-0.01	-0.01
500	-0.0	-0.1	-0.2	-0.2	-0.2	-0.2	-0.1	0.4	0.6	0.8	0.6	0.5	0.7	0.9	-2.1	-3.5	-4.5	-0.7	1.6	1.8	1.7	0.7	-0.0	-0.0	0.1	0.0	0.1	0.5	0.1	0.0	0.0	-0.02	-0.19	-0.11
700	0.0	-0.3	-0.6	-0.6	-0.7	-0.8	-0.2	1.2	2.0	2.5	2.1	2.4	3.8	1.7	-5.3	-8.8	-13.5	-4.1	3.2	5.6	5.0	2.0	-0.2	0.1	0.2	0.3	-1.4	-2.2	-2.0	0.2	0.8	0.27	-0.85	-0.29
850	*****	-0.6	-1.1	-1.3	-1.4	-1.8	-0.4	2.6	4.1	4.1	4.9	4.3	6.6	6.1	-2.7	-8.1	-9.3	-17.9	-8.9	0.6	10.0	9.2	3.7	-0.2	0.4	0.3	-1.3	-1.9	-1.7	-0.7	0.1	0.66	-1.29	-0.32
900	*****	-0.6	-1.0	-1.2	-1.4	-1.8	-0.4	2.6	4.1	4.7	4.1	7.1	6.3	-4.3	-7.7	-7.5	-16.5	-9.7	0.2	9.9	8.8	3.6	-0.2	0.5	0.2	-1.3	-1.9	-1.7	-0.7	0.1	0.7	0.66	-1.23	-0.28
950	*****	-0.3	-0.6	-0.8	-0.9	-1.1	-0.2	1.7	2.6	2.9	2.4	5.6	6.1	-2.9	-4.9	-4.3	-13.6	-8.9	1.7	6.9	5.3	2.2	-0.1	0.3	-0.1	-0.8	-1.0	-0.9	-0.3	-0.0	0.4	0.68	-1.16	-0.22
1000	*****	-0.1	-0.2	-0.2	-0.2	-0.3	-0.0	0.4	0.7	0.7	0.6	1.7	2.2	-0.7	-1.0	-0.8	-4.4	-3.1	0.9	1.9	1.3	0.6	-0.0	0.1	-0.1	-0.2	-0.2	-0.2	-0.1	-0.0	0.1	0.28	-0.49	-0.06
MEAN	0.0	-0.2	-0.3	-0.4	-0.4	-0.5	-0.1	0.8	1.3	1.5	1.3	1.8	2.1	0.0	-2.9	-4.0	-6.9	-2.8	1.2	3.3	2.9	1.2	-0.1	0.1	0.1	-0.4	-0.6	-0.6	-0.2	0.0	0.2	0.18	-0.45	-0.13

DJF 1963 – 73

p (MB)	80S	70S	65S	60S	55S	50S	45S	40S	35S	30S	25S	20S	15S	10S	5S	EQ	5N	10N	15N	20N	25N	30N	35N	40N	45N	50N	55N	60N	65N	70N	80N	SH	NH	GLOBE
300	-0.0	-0.1	-0.1	-0.2	-0.2	-0.2	-0.0	0.1	0.1	0.2	0.2	0.7	0.3	0.2	-0.5	-0.2	0.5	0.7	0.3	0.2	0.2	0.2	0.1	-0.0	-0.1	-0.1	-0.0	-0.0	0.0	0.0	0.0	-0.19	0.12	-0.04
500	-0.1	-0.1	-0.2	-0.2	-0.3	-0.4	-0.0	0.4	1.2	0.8	0.2	1.8	3.8	-1.7	-4.4	-3.5	-1.7	2.5	3.8	1.8	1.8	0.1	-0.1	-0.3	-0.2	-0.9	-0.6	-0.2	0.1	0.0	0.0	-1.36	0.60	-0.38
700	-0.2	-0.2	-0.6	-0.8	-0.9	-1.1	-0.0	2.3	3.6	2.5	3.6	5.5	6.2	-9.3	-8.5	-8.7	3.7	10.1	5.5	6.2	3.7	-0.1	-0.3	-0.9	-0.6	-0.2	0.1	0.5	0.1	0.7		-2.54	1.05	-0.74
850	*****	-0.5	-1.2	-1.5	-1.8	-2.2	-0.2	4.4	7.1	5.0	0.8	2.0	-11.1	-8.2	-9.4	-11.9	-15.9	-2.0	10.7	10.6	12.7	6.5	-0.2	-0.5	-0.5	-1.6	-1.7	-1.0	-0.3	0.1	0.7	-2.22	0.82	-0.69
900	*****	-0.5	-1.2	-1.5	-1.8	-2.2	-0.2	4.4	6.9	4.7	0.7	3.5	-8.0	-9.7	-9.2	-10.6	-16.0	-4.7	9.3	11.0	12.5	6.5	-0.2	-0.4	-0.5	-1.5	-1.5	-0.8	-0.4	-0.1	0.0	-1.53	0.54	-0.50
950	*****	-0.3	-0.7	-0.9	-1.1	-1.4	-0.1	2.8	4.3	2.9	0.4	3.5	-3.2	-6.4	-4.7	-7.7	-14.7	-4.0	8.8	7.7	7.5	4.0	-0.2	-0.2	-0.4	-0.9	-0.8	-0.4	-0.1	-0.0	0.3	-0.68	0.00	-0.35
1000	*****	-0.1	-0.2	-0.2	-0.3	-0.3	-0.0	0.7	1.1	0.7	0.1	1.2	-0.3	-1.7	-0.8	-2.1	-5.2	-1.1	3.2	2.2	1.9	1.1	-0.0	-0.1	-0.2	-0.3	-0.2	-0.1	-0.0	-0.0	0.0	-0.05	-0.14	-0.09
MEAN	-0.0	-0.1	-0.4	-0.5	-0.6	-0.7	-0.0	1.4	2.2	1.5	0.2	1.8	2.0	-5.1	-4.4	-4.3	-4.7	-5.2	0.8	5.0	3.8	2.2	-0.0	-0.1	-0.3	-0.5	-0.5	-0.3	-0.1	0.0	0.2	-1.14	0.48	-0.33

JJA 1963 – 73

p (MB)	80S	70S	65S	60S	55S	50S	45S	40S	35S	30S	25S	20S	15S	10S	5S	EQ	5N	10N	15N	20N	25N	30N	35N	40N	45N	50N	55N	60N	65N	70N	80N	SH	NH	GLOBE
300	-0.0	-0.0	-0.0	-0.0	-0.0	-0.0	0.0	0.0	0.1	0.1	0.1	0.1	0.5	0.6	0.1	-0.3	-0.8	-0.8	-0.5	-0.2	-0.0	0.2	0.1	-0.0	0.0	0.0	0.0	0.0	0.2	0.0	0.0	0.12	-0.19	-0.04
500	-0.0	-0.1	-0.1	-0.2	-0.2	-0.2	-0.1	0.2	0.4	0.9	1.1	0.9	3.7	3.8	-0.2	-3.0	-6.6	-5.3	-2.9	-1.4	0.1	1.8	0.6	-0.0	0.1	-0.2	-0.3	-0.3	-0.1	0.1	0.1	0.74	-1.42	-0.34
700	-0.1	-0.2	-0.3	-0.5	-0.7	-0.9	-0.3	0.6	1.1	2.4	3.7	4.0	11.5	8.9	-2.8	-8.2	-17.3	-13.2	-7.4	-3.7	1.0	5.0	1.3	0.2	0.3	-0.8	-1.1	-1.3	-0.6	0.1	0.2	2.00	-3.66	-0.85
850	*****	-0.3	-0.5	-0.9	-1.6	-2.1	-0.7	1.4	2.4	4.9	8.2	9.2	15.4	4.7	-7.4	-7.4	-19.8	-16.1	-9.6	-6.5	1.8	9.2	2.4	0.9	0.5	-1.4	-2.0	-2.3	-1.1	0.3	0.3	2.62	-4.32	-0.86
900	*****	-0.3	-0.5	-0.9	-1.6	-2.1	-0.7	1.4	2.4	4.8	8.1	9.6	14.0	1.1	-7.5	-4.7	-17.2	-15.1	-7.8	-5.7	1.6	8.6	2.3	0.9	0.4	-1.3	-1.7	-2.0	-0.9	0.2	0.2	2.26	-3.73	-0.73
950	*****	-0.2	-0.3	-0.6	-1.0	-1.3	-0.4	0.9	1.6	2.9	4.9	6.9	10.9	0.2	-4.7	-0.7	-12.6	-13.9	-5.3	-2.8	1.0	5.3	1.4	0.5	0.0	-0.8	-1.0	-1.0	-0.5	0.0	0.1	1.69	-2.84	-0.52
1000	*****	-0.0	-0.1	-0.1	-0.3	-0.3	-0.1	0.3	0.4	0.8	1.2	2.1	3.5	0.0	-1.0	0.6	-3.6	-5.1	-1.4	-0.4	0.3	1.2	0.3	0.1	-0.0	-0.2	-0.1	-0.2	-0.1	-0.0	0.0	0.54	-0.94	-0.10
MEAN	-0.0	-0.1	-0.1	-0.3	-0.5	-0.6	-0.2	0.4	0.7	1.5	2.4	2.7	5.9	3.3	-1.8	-3.3	-8.4	-7.0	-3.8	-2.1	0.5	2.8	0.8	0.2	0.1	-0.4	-0.6	-0.7	-0.3	0.1	0.1	1.03	-1.80	-0.38

Table A12. Vertical distribution of the zonally-averaged values (10^{-4} g kg^{-1} mb s^{-1}) of the vertical 'transient eddy' transport $[\overline{q'\omega'}] + [\overline{q_c\,\omega_c}]$.

year 1963 – 73

P(MB)	80S	70S	65S	60S	55S	50S	45S	40S	35S	30S	25S	20S	15S	10S	5S	EQ	5N	10N	15N	20N	25N	30N	35N	40N	45N	50N	55N	60N	65N	70N	80N	SH	NH	GLOBE
300	0.0	0.0	0.0	0.1	0.0	0.1	0.1	0.1	-0.0	-0.0	-0.0	-0.1	-0.1	-0.1	0.3	0.4	0.4	0.1	-0.2	-0.1	-0.1	-0.1	-0.0	-0.0	0.0	0.1	0.1	0.1	0.1	0.0	0.0	0.02	0.03	0.03
500	0.4	0.4	0.5	0.6	0.7	0.7	0.1	0.1	-0.0	-0.4	-0.7	-0.7	-0.8	-0.5	2.2	3.2	3.6	0.4	-1.2	-1.2	-1.6	-0.6	-0.2	0.2	0.5	0.6	0.9	0.9	0.7	0.6	0.4	0.26	0.34	0.30
700	1.1	2.1	2.4	2.8	3.0	3.1	2.0	0.1	-0.6	-1.7	-3.1	-2.5	-2.9	-3.3	5.0	8.5	11.5	1.5	-4.6	-5.3	-5.1	-1.5	-0.4	0.5	1.2	2.2	3.2	3.7	3.0	2.5	1.6	0.23	1.01	0.62
850	*****	4.4	5.4	5.7	5.2	2.9	2.9	-0.8	-3.5	-5.8	-6.2	-7.9	-6.9	-6.9	7.0	13.4	20.9	2.7	-9.5	-12.4	-10.3	-2.9	0.4	1.1	1.9	4.1	5.7	6.7	5.6	4.5	2.9	-0.52	1.42	0.45
900	*****	5.3	6.3	7.5	7.2	6.8	3.8	-1.5	-5.1	-8.2	-9.6	-11.9	-8.9	-5.2	8.6	14.5	24.0	4.5	-8.9	-18.1	-15.3	-4.1	1.0	1.6	2.5	5.5	7.5	8.2	6.6	4.9	2.9	-0.68	1.43	0.37
950	*****	5.9	6.9	8.0	8.1	8.0	4.7	-1.3	-5.5	-8.6	-9.2	-14.0	-11.4	-5.4	9.8	15.3	26.1	5.0	-11.4	-22.0	-17.2	-4.0	2.0	2.0	2.3	6.7	8.1	9.2	7.2	5.2	3.0	-0.80	1.37	0.26
1000	*****	5.9	7.2	8.7	8.2	8.8	5.7	-0.4	-8.7	-9.5	-14.0	-16.8	-14.6	-6.4	8.3	14.0	28.9	7.4	-14.6	-23.8	-17.3	-3.9	2.8	3.4	4.7	7.2	8.2	9.5	7.2	5.3	3.0	-0.93	1.68	0.20
MEAN	0.4	1.3	2.1	2.5	2.2	2.3	1.4	0.9	-0.2	-1.4	-2.3	-2.7	-3.0	-2.1	3.2	5.5	8.3	1.3	-3.5	-5.0	-4.2	-1.2	0.1	0.4	0.9	1.6	2.2	2.5	2.0	1.6	1.0	-0.05	0.57	0.26

DJF 1963 – 73

P(MB)	80S	70S	65S	60S	55S	50S	45S	40S	35S	30S	25S	20S	15S	10S	5S	EQ	5N	10N	15N	20N	25N	30N	35N	40N	45N	50N	55N	60N	65N	70N	80N	SH	NH	GLOBE
300	0.0	0.0	0.0	0.1	0.0	0.0	0.1	0.0	-0.1	-0.0	-0.0	-0.2	-0.3	-0.5	0.7	0.5	0.5	0.2	-0.3	-0.2	-0.2	-0.2	0.1	0.0	0.0	0.0	0.0	0.0	0.0	0.1	0.0	0.17	-0.06	0.05
500	0.3	0.3	0.4	0.6	0.9	1.2	0.8	-0.0	-0.6	-0.9	-0.9	-1.5	-2.7	-3.7	5.1	3.2	0.9	0.1	-1.6	-1.4	-1.5	-1.0	0.1	0.5	0.1	1.0	0.8	0.6	0.5	0.6	0.1	1.57	-0.44	0.56
700	1.2	2.0	3.1	3.1	3.7	3.7	2.1	-0.7	-3.0	-2.8	-2.2	-4.5	-6.1	-8.0	11.5	7.9	2.8	0.6	-6.0	-4.1	-4.5	-3.7	0.6	1.5	2.3	2.3	3.4	2.9	2.4	2.2	0.7	3.82	-1.43	1.18
850	*****	2.8	4.6	6.1	6.7	6.3	3.1	-2.3	-5.9	-4.9	-6.1	-14.4	-26.7	-11.7	15.1	13.9	9.9	1.4	-14.4	-14.1	-14.4	-9.3	0.7	2.3	4.1	3.7	6.9	5.9	5.2	4.4	1.4	5.49	-3.41	1.03
900	*****	3.9	6.3	8.0	8.7	8.3	4.1	-3.8	-7.1	-4.0	-6.0	-21.3	-29.2	-14.9	16.7	16.7	16.1	1.5	-15.5	-21.3	-22.3	-11.4	1.3	3.2	5.3	5.3	8.8	7.2	6.0	4.9	1.4	5.68	-4.30	0.71
950	*****	4.6	7.5	9.2	10.0	9.6	5.1	-3.7	-9.9	-5.2	-2.5	-26.9	-35.3	-16.6	18.5	14.4	26.7	1.5	-14.0	-26.9	-25.6	-14.4	2.2	4.2	6.3	9.4	9.8	8.1	6.5	5.1	1.5	5.82	-5.20	0.45
1000	*****	4.6	7.8	9.7	10.0	10.5	6.5	-2.3	-8.9	-6.0	-4.7	-30.4	-41.9	-17.4	16.2	14.0	19.3	1.7	-16.4	-30.4	-26.7	-12.1	2.9	5.2	7.3	11.0	10.0	8.5	6.4	4.9	1.7	5.49	-6.01	0.49
MEAN	0.3	0.9	1.8	2.5	2.8	2.8	2.1	-1.5	-2.5	-0.8	-2.1	-6.0	-10.0	-5.6	6.9	5.7	3.5	0.5	-8.3	-6.0	-5.7	-3.4	0.4	1.0	1.6	2.7	2.5	2.1	1.7	1.5	0.5	2.30	-1.29	0.51

JJA 1963 – 73

P(MB)	80S	70S	65S	60S	55S	50S	45S	40S	35S	30S	25S	20S	15S	10S	5S	EQ	5N	10N	15N	20N	25N	30N	35N	40N	45N	50N	55N	60N	65N	70N	80N	SH	NH	GLOBE
300	0.1	-0.0	0.0	-0.0	0.0	0.1	0.1	0.1	0.0	-0.0	-0.0	-0.3	-0.3	-0.3	0.1	0.3	0.6	0.7	0.5	0.4	0.2	0.1	-0.1	-0.1	0.0	0.1	0.1	0.1	0.1	0.1	0.0	0.23	-0.03	0.10
500	0.6	0.4	0.3	0.4	0.6	0.6	0.8	0.4	-0.1	-0.0	-0.4	-0.7	-0.4	-4.8	0.0	2.5	5.7	5.9	4.3	2.7	1.0	-0.5	-0.7	-0.3	0.1	0.6	0.9	1.0	0.8	0.8	0.8	1.91	-0.31	0.80
700	1.7	2.4	2.4	2.7	2.9	2.2	2.2	0.8	-0.7	-3.4	-4.8	-4.2	-11.6	-11.7	-0.6	7.8	17.6	14.7	8.5	4.5	1.8	-1.1	-1.1	-1.5	0.1	1.5	2.8	3.4	2.6	2.8	3.1	4.97	-2.04	1.49
850	*****	5.1	5.4	5.4	5.1	3.3	3.3	0.4	-1.9	-6.5	-10.8	-11.4	-26.5	-23.3	-0.8	11.4	28.4	22.9	11.7	1.8	2.1	-1.7	-0.1	-0.2	0.5	2.7	4.7	5.5	4.2	4.8	5.7	7.52	-5.15	1.20
900	*****	5.9	6.7	6.9	7.1	4.3	4.3	0.3	-2.4	-8.4	-15.1	-16.6	-31.5	-23.6	0.4	11.9	32.0	25.2	12.0	2.1	-2.6	-5.0	-0.5	0.3	0.8	3.6	6.1	6.6	4.8	5.3	6.2	8.09	-5.96	1.04
950	*****	6.3	7.5	8.0	8.3	5.3	5.3	0.8	-2.5	-9.0	-16.8	-13.9	-36.4	-25.4	0.7	11.4	33.8	26.2	12.0	1.5	-4.0	-7.0	-0.2	0.9	1.4	4.3	6.4	6.8	5.0	5.6	6.7	8.47	-6.58	0.76
1000	*****	6.7	8.1	8.9	8.7	6.2	6.2	1.5	-2.2	-9.0	-17.2	-22.9	-42.2	-27.7	-1.3	7.8	34.3	26.2	11.2	-0.5	-4.6	-7.3	0.2	1.8	2.4	4.6	6.2	6.5	4.8	5.6	7.0	9.31	-6.84	0.18
MEAN	0.6	0.4	0.3	0.4	0.6	0.9	1.6	0.4	-0.7	-2.5	-4.1	-4.4	-9.7	-8.3	-0.1	4.5	11.5	9.8	5.2	1.8	0.1	-1.2	-0.5	0.2	1.1	1.9	1.9	2.1	1.6	1.8	2.0	-1.74	3.08	0.66

THE ATMOSPHERIC WATER VAPOUR BUDGET OVER EUROPE

Mikko Alestalo

Department of Meteorology
University of Helsinki
Finland

ABSTRACT. Vertically-integrated annual and monthly mean divergences of horizontal water vapour flux ($\nabla \cdot \underset{\sim}{Q}$) and corresponding rates of change of water vapour storage ($\partial W/\partial t$) have been computed using aerological observations for Europe. Long-term averages for precipitation P, evapotranspiration E and runoff R are used together with the atmospheric data to elucidate the mean annual and monthly water vapour budget for Europe.

Relatively good agreement is obtained between the annual mean values of $\nabla \cdot \underset{\sim}{Q}$ and E-P. However, E-P reaches a maximum around May, whereas ($\partial W/\partial t$)+$\nabla \cdot \underset{\sim}{Q}$ is at its highest value two months later. As this time lag occurs in all the study areas, it seems to be of a systematic nature.

1. INTRODUCTION

Two aspects at least of the Earth's hydrological cycle are of fundamental importance. First, it is the driving mechanism responsible for the maintenance of fresh water supplies that are essential for life. Secondly, the latent heat energy bound up in atmospheric water vapour is a significant element in the energy budget of the atmosphere.

Inherent difficulties in measuring the components of the hydrological cycle have made their detailed analyses hard to carry out and, sometimes, even their areal averages may exhibit large uncertainties. The most serious sources of error in conventional climatological studies are the problems involved in measuring precipitation and evapotranspiration, in estimating the spatial distribution of runoff and in acquiring representative data on changes in soil moisture content. If atmospheric wind and humidity

A. Street-Perrott et al. (eds.), Variations in the Global Water Budget, 67–79.

observations are used to estimate the net water vapour flux over a specific area, errors in wind data often weaken the reliability of flux divergence computations, especially for smaller areas.

This paper presents a study of the annual and interannual variations of the atmospheric net flux of water vapour over Europe using aerological observations between 1969 and 1977. Quantitative estimates of the vertically-integrated divergence of the total water vapour flux have been compared with long-term mean estimates of the difference between evapotranspiration and precipitation over particular areas. This study extends to a much larger area of Europe than similar studies of less extensive European areas, e.g., southern England (Hutchings 1957), an 1800 × 2200 km^2 rectangular area in eastern Europe (Drozdov & Grigoreva 1963), southern Sweden (Nyberg 1965), the Baltic Sea (Palmén & Söderman 1966) and northwest Europe (Alestalo & Savijärvi 1977, Baese & Liebing 1977).

2. EVALUATION OF COMPONENTS OF THE HYDROLOGICAL CYCLE

The atmospheric part of the cycle is governed by the equation (Peixoto 1973):

$$\partial W/\partial t + \nabla \cdot Q = E-P \qquad (1)$$

where W is the vertically-integrated specific humidity in an atmospheric column of unit horizontal cross-sectional area extending from the surface of the Earth to the top of the atmosphere, Q represents the vertically-integrated horizontal flux of specific humidity in that column, E is evapotranspiration and P precipitation, both measured at ground level. In Equation (1), contributions from water in liquid and solid phases are neglected, as usual. Nevertheless, it is recognized that the net transport of water due to clouds in mountainous areas near coastlines may be important (Ninomiya 1964, Korzun 1974).

The terrestrial branch of the cycle is governed by an equation analogous to Equation (1):

$$\partial S/\partial t + R = P-E \qquad (2)$$

where S is the moisture content of a soil-ocean-cryosphere column integrated to the depth where there is no significant temporal variation of S within the relevant time scales, and R is runoff (including surface, subsurface and underground flow); P-E serves as a link between the atmospheric and terrestrial branches of the hydrological cycle.

In contrast to changes in the atmospheric storage term W, $\partial S/\partial t$ generally does not vanish for a time span of one year. However, for mean multi-annual conditions, the convergence of the atmospheric water vapour flux over a given region must be compensated, on average, by the terrestrial runoff from that region:

$$\nabla \cdot \bar{Q} = \bar{P} - \bar{E} = \bar{R} \qquad (3)$$

3. HYDROLOGICAL CYCLE DATA

The atmospheric quantities ($\partial W/\partial t$, $\nabla \cdot Q$) in Equation 1 were calculated from radiosonde observations from the 3-year period 1974-1976, and the 8-year period July 1969-June 1977. In the former data set (SMHI data), described by Alestalo and Holopainen (1980), observations from standard levels below 150 mb as well as from the surface and the 950, 900, 800 and 600 mb levels were used. The radiosonde observations for the 8-year data set (DWD data - initially compiled by Deutscher Wetterdienst 1969-77) include only standard level data (up to 70 mb). Owing to the paucity of 1000 mb observations, a special extrapolation procedure was used to estimate flux divergence at 1000 mb (Alestalo 1981a). As the vertical representation is very crude in the DWD data, the main emphasis in the results derived from these data is on the annual and interannual variations rather than on quantitative estimates of $\nabla \cdot Q$.

In order to compare the computed values of mean annual net atmospheric water vapour transport over the study areas with the long-term averages of P, E and R, the latter were extracted from: Möller (1951), Budyko (1963), L'vovich (1973), Korzun (1974), Baumgartner and Reichel (1975) and Jaeger (1976).

4. METHOD OF COMPUTING FLUX DIVERGENCE

The boundaries of the areas for which flux divergences were computed (Figure 1) were defined by the locations of the aerological stations contributing data to the study; about 90 (SMHI data) and 60 (DWD data) stations were included in the largest study area A. The elementary computational entity used was a triangle, with the results being averaged to give values for the main polygons. The main criterion applied in the construction of the polygons was the availability of original observations (see § 3).

In computing the vertically-integrated divergence, the flux components are assumed to vary linearly between different stations and different pressure levels. Specific humidity is assumed to be zero at 300 mb. The same mass balance correction technique as in Alestalo (1981b) is used, i.e., at each level, the total divergence is obtained as

$$\nabla \cdot \overline{q\underline{v}}\Big|_{\text{corrected}} = \nabla \cdot \overline{q\underline{v}}\Big|_{\text{computed}} - \overline{q}\widehat{\nabla \cdot \underline{v}} \qquad (4)$$

where q is specific humidity, $\underline{v}$ is the horizontal wind vector and $\widehat{\nabla \cdot \underline{v}}$ represents the computed vertical average of $\nabla \cdot \underline{v}$. The mass balance is required for the whole atmospheric column, i.e., surface to 150 mb for SMHI data and 1000 to 100 mb for DWD data. This correction, which is of utmost importance in the case of the total energy budget, is also important for water budget computations, at least with the present data (see Alestalo 1981b); for example, its magnitude for region A (Figure 1), amounted to 87% of the uncorrected annual mean value of $\nabla \cdot Q$.

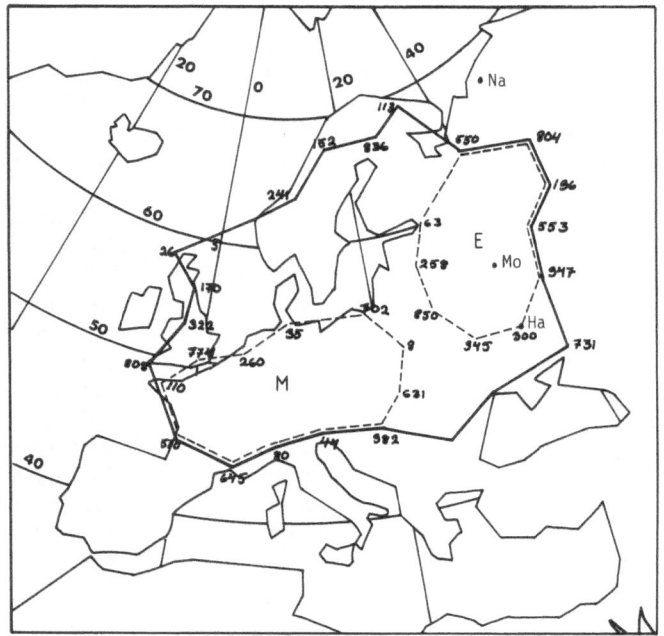

Fig. 1. Regions for which flux divergence was computed. Region A
         is delineated by a solid line and its subareas, M and E,
         with dashed lines. Numbers at the corners of the poly-
         gons refer to the WMO station index of aerological
         stations; Na, Narjan Mar; Mo, Moscow; Ha, Harkov.

5.   THE WATER BALANCE OF EUROPE

5.1   Intercomparison of the Estimates of P, E and R

     In order to secure a measure of the general reliability of the
annual means of P, E and R as explicitly given in some recent comprehensive climatological studies, their values for the European
region (an area of about 10×10^6 km^2) are shown in Table I. The
smallest scatter is in the different estimates of annual mean runoff (about 300 ± 20 mm yr^{-1}), as would be expected on the basis of
the rather straightforward measuring technique. The largest scatter
is associated with $\bar{P}$ and is due mainly to the different correction
procedures employed.

5.2   The Annual Mean Water Budget

     Table II gives the annual mean values of P, E and R, for the
study areas delineated in Figure 1, that were extracted from various

Table I. Long-term means of precipitation (P), evapotranspiration
          (E) and runoff (R) in the European region $(10 \times 10^6 \ km^2)$.
          Units: mm yr^{-1}.

Author	P	E	R
L'vovich (1973)	734	415	319
Korzun (1974)	789	470	306
Baumgartner & Reichel (1975)	657	375	282

isopleth maps, together with computed values of net atmospheric
water vapour transport. In this study, the values of $\bar{P}$, $\bar{E}$ and $\bar{R}$
include errors arising from reading by eye the grid values from the
maps, which sometimes have quite coarse resolution. It is assumed
that these errors are not larger than 10 mm yr^{-1}.

Table II. Computed annual mean values of $-\nabla \cdot Q$ and estimated pre-
           cipitation (P), evapotranspiration (E) and runoff (R)
           for the study areas A and M shown in Figure 1. Units:
           mm yr^{-1}.

Source	$-\nabla \cdot \bar{Q}$	$\bar{P}$	$\bar{E}$	$\bar{P} - \bar{E}$	$\bar{R}$
Region A					
Present study (3-year data set)	235				
Budyko (1963)		690	460	230	
Baumgartner & Reichel (1975)		715	440	275	
Jaeger (1976)		720			
Region M					
Present study (3-year data set)	420				
Möller (1951)		845			
Budyko (1963)		885	510	375	
Korzun (1974)		955	600	355	360
Baumgartner & Reichel (1975)		875	465	410	
Jaeger (1976)		880			

     From Table II, it can be seen that, in region A ($6.5 \times 10^6 \ km^2$),
$\bar{P}$ and $\bar{E}$ are about 710 and 450 mm yr^{-1}, respectively, which yields
about 260 mm yr^{-1} for $\bar{P} - \bar{E}$ or $\bar{R}$. The 3-year (1974-76) data set gave
a value for $-\nabla \cdot \bar{Q}$ of 235 mm yr^{-1} $\pm$ 15 mm yr^{-1}, the uncertainty being
due to random errors (see § 6). Accordingly, the terrestrial and
atmospheric approaches converge to the same solution with $\bar{P} - \bar{E}$ and
$-\nabla \cdot \bar{Q}$ being around 250 mm yr^{-1} with an uncertainty of $\pm$ 20 mm yr^{-1}
for region A. Of course, owing to interannual variability, a time
span of three years is not necessarily representative of long-term

conditions. According to the annual summaries of Deutscher Wetterdienst (1969-1977), precipitation in region A was slightly below
normal between 1974 and 1976.

The uncertainty of the estimates of $\bar{P}-\bar{E}$ and $-\nabla\cdot\bar{Q}$ for the
smaller study area M (1.6×10^6 km^2) can be expected to be larger
than for region A. The range of the $\bar{P}-\bar{E}$ values in Table II is
55 mm yr^{-1} so that it is inferred that $\bar{P}-\bar{E}$ (or $\bar{R}$) for region M is
about 390 ± 30 mm yr^{-1}, which is in good agreement with the
computed $-\nabla\cdot\bar{Q}$ value of 420 mm yr^{-1}.

To sum up, in areas with a dense network of aerological
stations, estimates of the annual mean value of $-\nabla\cdot\bar{Q}$ are as
reliable as those of $\bar{P}-\bar{E}$ or $\bar{R}$ for areas larger than about
1.5×10^6 km^2. A similar result for North America was obtained
by Rasmusson (1968). Just as the mean annual runoff from a
drainage basin gives the boundary condition to be satisfied in
evaluating $\bar{P}$ and $\bar{E}$, so may $-\nabla\cdot\bar{Q}$ values be used for irregular
areas or when runoff measurements are not available or are
difficult to apply (e.g., over deserts and oceans).

5.3 The Seasonal Variation of the Water Budget

The preceding analysis can be extended to a mean monthly
basis provided $\partial W/\partial t$ is taken into consideration in accordance
with Equation (1). This term is quite small in magnitude, however, varying sinusoidally with an amplitude of about 5 mm per
month, with the maximum and minimum in April and September,
respectively.

The long-term annual trend of mean monthly values of $\bar{P}$ and $\bar{E}$,
taken from Budyko (1963), Jaeger (1976) and Korzun (1974), is
shown in Figure 2. The USSR compilation, which is used only for
region M, gives the annual cycle in the form of histograms
(showing the percentage of each month's contribution to the annual
total) for some locations in the region (fourteen for P and four
for E); accordingly, the estimated mean monthly values of $\bar{P}$ and $\bar{E}$
have a relatively large uncertainty (apart from possible inherent
errors in the analyses) which is assumed to be about 5 mm per
month.

In Figure 3, the annual trend of $\partial W/\partial t + \nabla\cdot\bar{Q}$ based on the
8-year DWD data set and of the long-term mean values of $\bar{E}-\bar{P}$ are
shown for regions A and M. For most of the year, as can be seen
from Figure 3, $\bar{P} > \bar{E}$ (or $\partial W/\partial t + \nabla\cdot\bar{Q} < 0$), with $\bar{E}$ (divergence)
dominating only in the late spring to early summer period.

The only discrepancy between $\bar{E}-\bar{P}$ and $\partial W/\partial t + \nabla\cdot\bar{Q}$ is the phase
difference between their maxima, which is especially evident for
region A where $\bar{E}-\bar{P}$ has its peak in May while $\partial W/\partial t + \nabla\cdot\bar{Q}$ reaches
its maximum in July. To find out whether this discrepancy is of
a systematic nature, the degree of natural year-to-year variability of the monthly values of $\nabla\cdot\bar{Q}$ and the random errors must be
considered.

If the random error in the multi-annual mean value of $\nabla\cdot\bar{Q}$ is

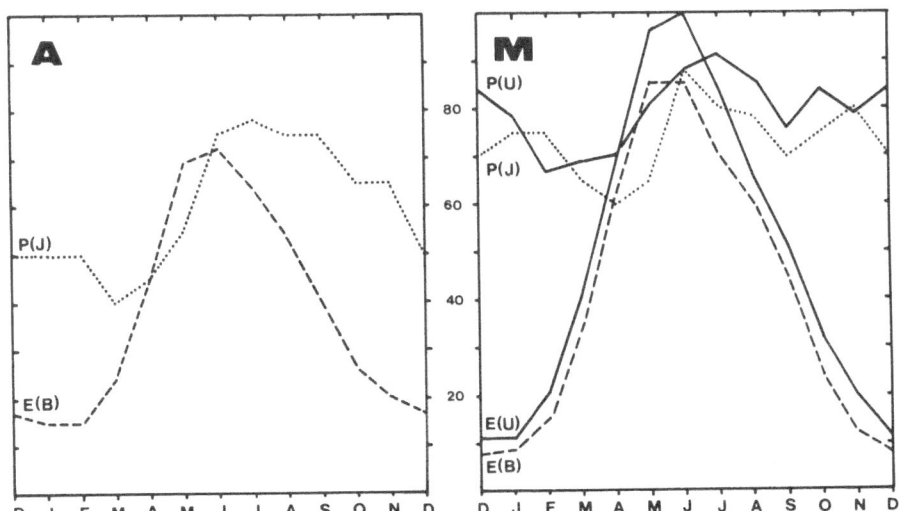

Fig. 2.   Annual cycles of evapotranspiration (E) and precipitation
(P) for regions A (left) and M (right) in mm per month;
···, from Jaeger (1976) (J); ---, from Budyko (1963) (B);
———. from Korzun (1974) (U).

15 mm yr^{-1}, the corresponding error in the monthly mean $\nabla \cdot \bar{Q}$ is
about 4 mm per month (i.e., $(12 \times 4^2)^{-\frac{1}{2}} \simeq 15$). For comparison,
Rasmusson (1968) reported corresponding errors of 5 and 2.5 mm per
month for winter and summer, respectively, using a 5-year data set
for an area of 6.4×10^6 km^2 in North America. The standard deviations for the individual monthly values of $\nabla \cdot Q$, shown in Figure 3,
include the contributions of natural variability and random errors,
and are about 10 and 30 mm per month for regions A and M, respectively. Because the random error due to the method of estimating
grid values of the monthly $\overline{E-P}$ values is obviously not much larger
than 5 mm per month (see § 5.2), it follows from the preceding
discussion that the discrepancy between the $\overline{E-P}$ and $\partial W/\partial t + \nabla \cdot Q$
curves is systematic in region A. Although there is a similar
tendency in region M, corresponding differences also appear for
other months so that these may only be due to computational noise.
        In order to ascertain whether the systematic discrepancy for
region A is due to some inherent systematic errors in the computation method or in the data employed, some possible error sources
are discussed below.
        In the case of the DWD data, an extrapolation of divergences
below 850 mb was necessary (see § 3). Using the SMHI data set
with its good vertical resolution, the same annual cycle was
observed as with the DWD data at least on a qualitative basis:

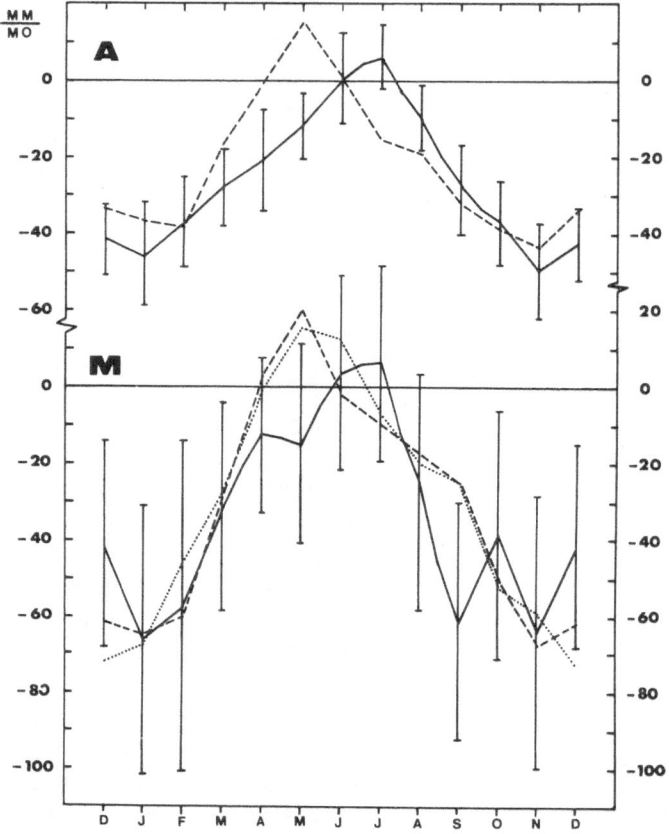

Fig. 3. Annual cycles of the divergence of atmospheric water
         vapour flux for July 1969 to June 1977 and the long-term
         mean values of E-P (mm per month): ——, $\partial W/\partial t + \nabla \cdot \underline{Q}$;
         ---, E-P obtained from Budyko (1963) and Jaeger (1976);
         •••, E-P obtained from Korzun (1974). The results are
         given for regions A (top) and M (bottom) (see Figure 1).
         Standard deviations of the individual monthly mean $\nabla \cdot \underline{Q}$
         values (N=8) are shown by vertical lines for each month.

the values for winter (Dec.-Feb.), spring, summer and autumn were,
respectively, -35, -10, 7 and -33 mm per month based on the SMHI
data and -41, -20, -1 and -38 mm per month based on the DWD data.
Furthermore, the mass balance correction (a necessary part of the
the computations to secure annual values) merely lowers the $\nabla \cdot \underline{Q}$
values rather evenly throughout the year, and causes no shift in
the occurrence of the maxima or minima.
         To isolate possible errors arising from the use of stations

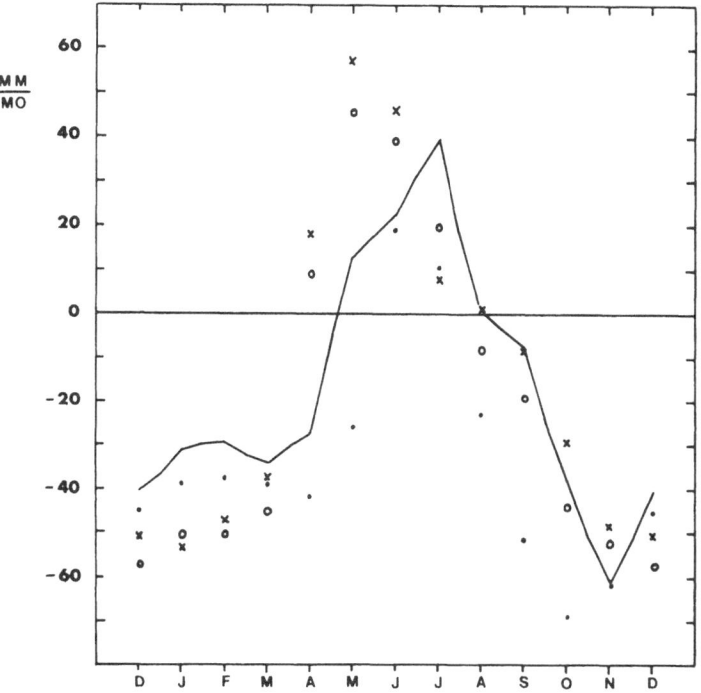

Fig. 4. Annual cycles of $\partial W/\partial t + \nabla \cdot \bar{Q}$ for region E and of the long-
term mean value of E-P in mm per month for the three loca-
tions: •, Narjan Mar; O, Moscow; ×, Harkov. See Figure 1,
for locations.

near coasts or mountains (where local effects are most likely in
the wind observations) or from the use of many different types of
radiosonde in central Europe, a third study area (E) covering
1.4×10^6 km^2, was devised using only USSR stations (see Figure
1). Figure 4 presents the annual cycle of $\partial W/\partial t + \nabla \cdot \bar{Q}$ for this
area based on the DWD data set together with corresponding $\bar{E}$-$\bar{P}$
values provided by Korzun (1974) for three locations, also shown
in Figure 1 (Narjan Mar to the north, Moscow in the centre and
Harkov on the southern boundary of region E). The availability
of these station data only made it fruitless to estimate areal
averages of monthly values. From Figure 4, it can be seen that
the same discrepancy as with the largest study area A is also
present in region E.

Possible systematic errors in calculation of the $\bar{E}$-$\bar{P}$ cycles
may arise from changes in soil moisture content and in potential
evapotranspiration, both of which parameters are needed for
computing the annual cycle of $\bar{E}$ (see Equation 2). Owing to the
lack of objective ways of measuring these parameters, empirical
relationships are widely used (Budyko 1956, Korzun 1974). The
reliability of the annual cycle of $\bar{E}$ and, consequently, of $\bar{E}$-$\bar{P}$ may

therefore be suspect. In his computation of the annual cycle of $\bar{E}$
on the basis of atmospheric data and observed precipitation,
Rasmusson (1968) found that the equivalent estimate of Budyko (1963)
for the study area in question led Rasmusson's curve by 2 to 4
weeks. In the current study, this phase difference is as large
as two months, but is only present during the spring and summer.

5.4 Interannual Variability of the Water Budget

The observed monthly deviations of areal precipitation from
long-term mean values taken from monthly summaries of the DWD data
were compared with the monthly values of $\nabla \cdot \bar{Q}$ from July 1969 to June
1977. The annual cycle was first removed from this 96-month time
series in order to obtain the monthly deviations from average
conditions. A similar study by Rasmusson (1968) showed correlations around -0.9 between areal monthly precipitation and $\nabla \cdot Q$,
using a 5-year data set.

By classifying the precipitation data (for region M only) into
three categories (namely, monthly values of P less than 90%, 90-110%
or more than 110% of normal) and the computed $\nabla \cdot \bar{Q}$ values similarly
into three categories (monthly $\nabla \cdot \bar{Q}$ value below, within or above the
interval ± 20 mm per month around the 8-year mean value), the frequencies of each of the nine possible events were examined. The
permutations are shown in Figure 5 for four temporal periods:
winter, summer, spring and autumn combined, and the whole year.

The negative correlation found by Rasmusson (1968) is as
expected in the light of Equation (1). This outcome is also evident,
on a qualitative basis, in all the combinations in Figure 5 as the
largest frequencies are found near the diagonal. However, the
scatter is quite large and the close relationship between monthly
values of P and $\nabla \cdot Q$, found by Rasmusson (1968), is not evident.

6. ERROR ANALYSIS OF THE ANNUAL $\nabla \cdot Q$ VALUES

The random error estimate (15 mm yr^{-1}) of the mean annual value
of $\nabla \cdot Q$ (235 mm yr^{-1}), given in § 5.2, is based on the following considerations. For random errors in $\nabla \cdot \bar{Q}$ arising from assumed errors
in measuring q and $\underline{v}$, the error computation procedures (Rasmusson
1977, Alestalo 1981$\bar{b}$) lead generally to quite small errors, about
10 mm yr^{-1} for the one-year mean value for region A.

The sensitivity of the computed $\nabla \cdot \bar{Q}$ values to modifications in
the plan of the polygon (i.e., when different sets of stations were
used), to different choices for the upper boundary of the layer
where the mass balance was required (200, 150, 100 or 70 mb) or to
the use of standard level and surface data only rather than of all
available information was investigated. This resulted in a total
range of about 30 mm yr^{-1} of $\nabla \cdot \bar{Q}$ for region A. Thus the error in
the multi-annual mean value of $\nabla \cdot \bar{Q}$ can be inferred to be about
15 mm yr^{-1}.

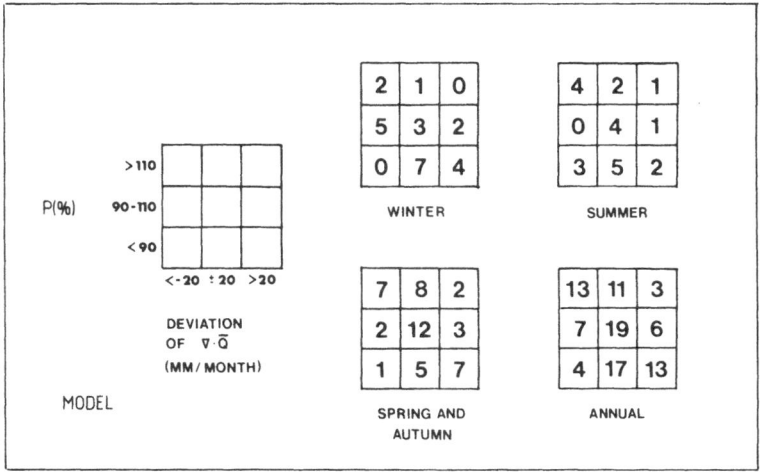

Fig. 5. Frequency distribution of simultaneous pairs of monthly
         values of $\nabla \cdot \bar{Q}$ and P for July 1969 to June 1977 for
         region M (see Figure 1). P values (from Deutscher
         Wetterdienst 1969-1977) are classified into three
         categories: monthly P value less than 90%, 90-110% or
         more than 110% of the norm. Computed $\nabla \cdot \bar{Q}$ values are
         similarly classified into three categories: monthly $\nabla \cdot \bar{Q}$
         value below, within or above the interval 20 mm per month
         around the 8-year mean value.

7. SUMMARY

    The vertically-integrated divergence of the total atmospheric
flux of water vapour ($\nabla \cdot \bar{Q}$) for areas larger than about 1.5 × 10^6 km^2
provides information as reliable as river runoff or $\bar{E}$-$\bar{P}$ measurements
when the water budget on a mean annual basis is determined over a
particular area. For the main study area, consisting of most of
Europe, the uncertainty of the long-term mean value of these quantities is about ± 20 mm.
    The annual cycles of $\partial \bar{W}/\partial t + \nabla \cdot \bar{Q}$ and $\bar{E}$-$\bar{P}$ show a systematic dis-
crepancy in all three areas studied in that $\bar{E}$-$\bar{P}$ has its maximum in
May and $\partial \bar{W}/\partial t + \nabla \cdot \bar{Q}$ in July. Further study is needed to solve this
discrepancy.
    The mass balance requirement is found to be an essential part
of the flux divergence computations because the wind observations
used in this study lead to a non-zero net divergence of air. Associated with this spurious mass divergence, the uncorrected long-
term mean value of $\nabla \cdot \bar{Q}$ deviated about 100 mm yr^{-1} from the corresponding corrected value in the main study region.

ACKNOWLEDGEMENT. The author is greatly indebted to Professor E.
Holopainen for his guidance during the present work as well as to

Professor E. Palmen and Professor J. Virta for valuable discussions. Professor P. Speth provided the aerological observations (DWD data set) for the flux and flux divergence computations. Mrs P. Saarikivi prepared the figures.

REFERENCES

Alestalo, M. 1981*a* On the annual course of the energy balance for the earth-atmosphere system. In: *Proc. 10th Geophys. Symp., Helsinki 1981*, pp. 113-121.

Alestalo, M. 1981*b* The energy budget of the earth-atmosphere system in Europe. *Tellus*, <u>33</u>, 360-377.

Alestalo, M. & Holopainen, E. 1980 Atmospheric energy fluxes over Europe. *Tellus*, <u>32</u>, 500-510.

Alestalo, M. & Savijärvi, H. 1977 Experiences with the use of the aerological method in evaporation studies in northwestern Europe. *Nord. Hydrol.*, <u>8</u>, 47-56.

Baese, K. & Liebing, H. 1977 An investigation of the atmospheric heat and moisture balance in the Baltic Sea region. Part III: Preliminary results of a determination of evaporation minus precipitation (April, May, June 1976) in the frame of the International Hydrological Programme. *Met. Rdsch.*, <u>30</u>, 185-192.

Baumgartner, A. & Reichel, E. 1975 *The World Water Balance - Mean Annual Global, Continental and Maritime Precipitation, Evaporation and Runoff*, 179 pp. Amsterdam: Elsevier.

Budyko, M.I. 1956 *The Heat Balance of the Earth's Surface*. Leningrad. English translation 1958, 258 pp. Washington, D.C.: U.S. Weather Bureau.

Budyko, M.I. (ed.) 1963 *Atlas of the Heat Balance of the Earth*, 69 pp. (in Russian). Moscow: Glavnaia Geofiz. Obs.

Deutscher Wetterdienst 1969-1977 *Die Grosswetterlagen Europas*. Nos. 22-29. Offenbach: Deutscher Wetterdienst.

Drozdov, O.A. & Grigoreva, A.S. 1963 *The Hydrological Cycle in the Atmosphere*. Leningrad: Gidromet. Izdatel. (English translation: 1965 NTIS TT65-50119, 282 pp. Jerusalem: Israel Program for Scientific Translations.)

Hutchings, J.W. 1957 Water-vapour flux and flux-divergence over southern England: summer 1954. *Q. Jl Roy. met. Soc.*, <u>83</u>, 30-48.

Jaeger, L. 1976 Monatskarten des Niederschlags für die ganze Erde. *Ber. Dt. Wetterd.*, 18, No. 139, 38 pp.

Korzun, V.I. (ed.) 1974 *World Water Balance and Water Resources of the Earth*. Report of the USSR Committee for the IHD, 663 pp. Leningrad: Gidromet. Izdatel. (English translation 1977-78: *Studies and Reports in Hydrology* No. 25. Paris: UNESCO.)

L'vovich, M.I. 1973 The global water balance. *EOS, Wash.*, 54, 28-42.

Möller, F. 1951 Vierteljahreskarten des Niederschlags für die ganze Erde. *Petermanns Geogr. Mitt.*, 95, 1-7.

Ninomiya, K. 1964 Water-substance budget over the Japan Sea and the Japan Islands during the period of heavy snow storm. *J. met. Soc. Japan (II)*, 42, 317-329.

Nyberg, A. 1965 A computation of the evaporation in southern Sweden during 1957. *Tellus*, 17, 473-483.

Palmén, E. & Söderman, D. 1966 Computation of the evaporation from the Baltic Sea from the flux of the water vapour in the atmosphere. *Geophysica*, 8, 261-279.

Peixóto, J.P. 1973 Atmospheric vapour flux computations for hydrological purposes. *WMO Publ*. No. 357, 83 pp.

Rasmusson, E.M. 1968 Atmospheric water vapor transport and the water balance of North America: II. Large-scale water balance investigations. *Mon. Weath. Rev.*, 96, 720-734.

Rasmusson, E.M. 1977 Hydrological application of atmospheric vapor-flux analyses. *Operational Hydrology Rept* No. 11 (*WMO Publ*. No. 476), 50 pp.

REMOTE SENSING OF ATMOSPHERIC WATER CONTENT FROM SATELLITES

A.K. Gorodetsky and V.I. Syachinov

Space Research Institute
USSR Academy of Sciences
Moscow, USSR

ABSTRACT. A satellite-borne technique to determine atmospheric water vapour amount from reflected solar radiation in the near infrared is described. The use of microwave and infrared techniques are essential to the investigation of clouds, and provide estimates of liquid water content. The means whereby the different water phases (vapour, liquid, ice) can be distinguished using spectrophotometric and infrared radiometer observations are described. Results obtained from experiments on Cosmos-series satellites are presented.

Remote sensing methods for determining atmospheric water vapour content using infrared (6-20 μm) and microwave (1.35 cm) radiation are subject to some physical limitations in their application. Therefore, an additional method has been proposed based on measurements of solar radiation reflected from the atmosphere in a water vapour absorption band around 0.95 μm.

Gorodetsky *et al.* (1974) investigated the use of the atmospheric transmission function

$$P(w,\theta,\zeta) = [I_1(w,\theta,\zeta)/I_2(\theta,\zeta)](S_2/S_1) \qquad (1)$$

where $I_1(w,\theta,\zeta)$ is the spectral radiance in a passband at 0.95 μm within the water vapour absorption band, $I_2(\theta,\zeta)$ is the spectral radiance in a reference channel at 1.03 μm which is essentially free of absorption, S_1 and S_2 are the extraterrestrial solar intensities at these wavelengths, and θ and ζ are the zenith angles of the incident solar beam and the observed reflected radiation, respectively. (An alternative pair of wavelengths are 0.72 and

81

A. Street-Perrott et al. (eds.), Variations in the Global Water Budget, 81–87.
Copyright © 1983 by D. Reidel Publishing Company.

0.74 μm). The mass of water vapour, w, in the inclined incident
and reflected ray paths can be determined by using the absorption
function

$$A(w) = 1-P(w,\theta,\zeta) \tag{2}$$

where $A(w) = c(w)^{\frac{1}{2}}$.

The water vapour content of an atmospheric vertical column w_f
is related to w by

$$w_f = w/(\sec\theta + \sec\zeta) \tag{3}$$

The coefficient c in Equation (2) is much more dependent on
spectral resolution than on the atmospheric aerosol thickness or
scattering function owing to the small difference in wavelength
between the two channels. Nevertheless, the validity of Equation
(2) has been confirmed under atmospheric conditions with multiple
scattering. The spectral half-widths of the interference filters
used in the experiment by Gorodetsky et al. (1974) were 0.011 and
0.018 μm in the 0.95 and 1.03 μm channels, respectively.

The results of the 'Cosmos-arrow' experiments (Gorodetsky
et al. 1974) confirmed that, while absorption in the 0.72 μm range
was not strong enough for water vapour determination, the 0.95 μm
range was satisfactory. A comparison of determinations of water
vapour content by this photometric method (w_f) and from radiosonde
data (w_z) (Figure 1) gives a correlation coefficient between w_f
and w_z of 0.9 for values of w_z up to 4.2 g cm^{-2}, with a standard
deviation of 0.28 g cm^{-2}.

In studies of water resources, cloudiness can be regarded as
the water amount in the vapour, liquid and ice phases. The point
of interest is to find links between macrophysical characteristics
(water content, cloud state or amount of precipitable water) and
those parameters which are accessible to measurement by remote
sensing. Possible linking parameters are: cloud thickness, water
content defined by the number density of droplets and their mean
dimensions, and mean temperature. The following parameters can be
determined by remote sensing techniques with varying degress of
confidence: temperature and height of the cloud top (Gorodetsky
et al. 1971, 1974), liquid water content (Basharinov et al. 1969,
Akvilonova et al. 1973), the dominant phase component (Gorodetsky
et al. 1975) and optical depth (Gorodetsky et al.1980, Rozenberg
et al. 1974).

Complex infrared and microwave experiments are essential for
these investigations of clouds. Direct comparison of microwave
brightness temperature T_B (at 0.8 cm) and infrared radiative
temperature T_r at 10-12 μm, by the Cosmos 243 and 384 experiments,
demonstrated the possibility of determining the phase composition
of clouds (Gorodetsky et al. 1975). The cloud-related parameters
measured in these satellite experiments were liquid water content
W, determined from the 0.8 cm band observations (Basharinov et al.

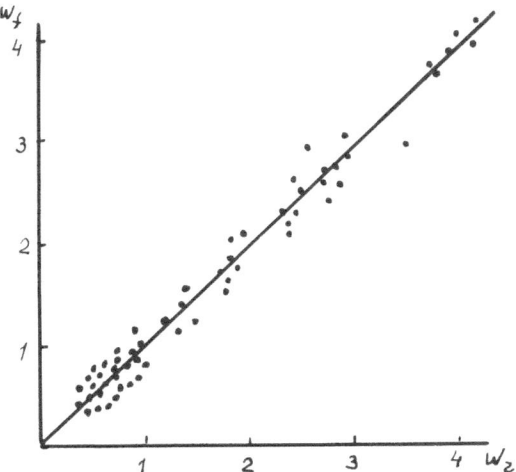

Fig. 1.   The regression between water vapour content from photo-
          metric data w_f and from radiosonde data w_z.

1969, Akvilonova *et al.* 1973), the difference between the cloud top
radiative temperature T_r as observed in the 10-12 µm band and the
mean radiative temperature T in cloud-free conditions ($\Delta T_r = T_r - T$),
and the correlation coefficient between T_B and T_r (r_{TT}).
    Figure 2 and Table I show the results obtained (mean values of
W, ΔT_r and r_{TT}) for three categories of typical clouds all of which
had horizontal scale lengths of >500 km. Where $r_{TT} \simeq 0$, ice was
the dominant phase in the upper parts of a cloud; when r_{TT} varied
between -0.6 and -0.8, the liquid phase predominated. Thus the
technique enables one to distinguish between super-cooled water
and ice clouds.

Table I. Liquid water content and phase components.

Category	ΔT_r (°C)	W (kg m^{-2})	r_{TT}	Predominant phase
I	25	0.05	-0.2 to 0.0	liquid water
		0.03	0	ice
II	10 - 30	0.05 - 0.2	-0.8 to -0.6	liquid water
			0	mixed phases
III	15 - 60	0.2 - 0.6	-0.8 to -0.6	liquid water
			0	ice in cloud top, liquid water at cloud base

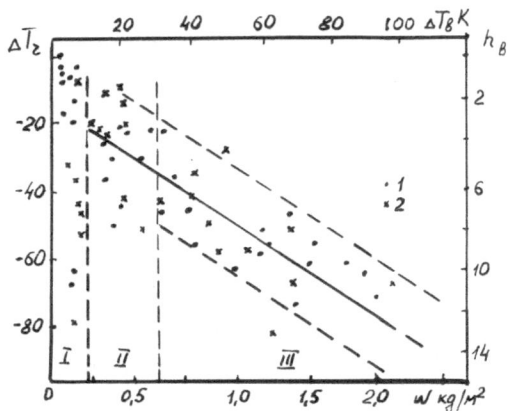

Fig. 2.   Predominant phases in clouds: W, liquid water content
          (kg m^{-2}); ΔT_r and ΔT_B, variations of the brightness
          temperature (K) at 10-12 μm and 0.8 cm from their clear-
          sky values.

     A distinction between the water vapour and ice content of ice
clouds can be achieved from complex measurements including spectrophotometric observations in the water vapour absorption band
(0.94 μm) and the ice band (1.03 μm), and observations with a
radiometer in the 10-12 μm band (Gorodetsky et $al.$ 1974). Variations of the ratio (P) between reflected radiances I_1 (at 0.94 μm)
and I_2 (at 1.03 μm) are related to the water vapour-to-ice content
ratio in the cloud. From Figure 3, which shows the variation of P
as a function of radiative temperature (T_r) of the cloud upper-
boundary, it can be seen that $0.6 < P < 1.6$ if $T_r < 273$ K. P increases
as absorption in the ice band (1.03 μm) increases and consequently
I_2 decreases, indicating growth of ice content of the cloud.
     For measurements over the sea, which has a low albedo, the
intensities I_1 and I_2 depend on multiple scattering and absorption
principally in the boundaries of the cloud. However, if water
vapour content is determined by microwave measurements, the
resulting value is a measure of the total quantity of water vapour
in, above and below the cloud.
     Given this combination of techniques, it can be seen that each
complements the other.
     Spatial and temporal variations in cloud composition depend on
cloud type. One of the results of a study of optical properties of
clouds, using an airborne back-scattering nephelometer and photo-
meter, was the demonstration of internal inhomogeneity of some
cloud types (Gorodetsky et $al.$ 1980). For example, space inhomogeneities within nimbostratus clouds varied typically from 60-80 m
to 200-300 m.
     The probability distribution of the scattering coefficients

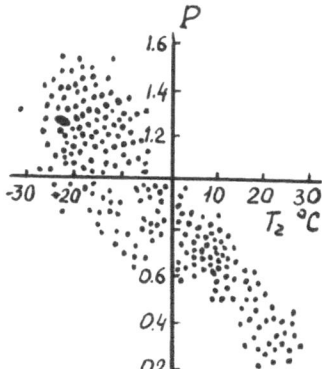

Fig. 3. Dependance of the ratio P between reflected radiances at
         0.94 and 1.03 μm on the temperature of the cloud upper
         boundary temperature T_r (°C).

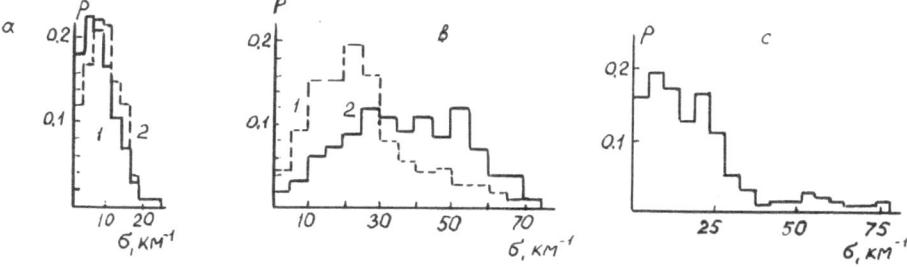

Fig. 4. Probability distribution of scattering coefficient (km^{-1})
         for different cloud types: (a) at a distance of 200 km
         from a frontal zone with rainfall: 1, nimbostratus;
         2, altostratus; (b) near the frontal zone: 1, opaque
         nimbostratus; 2, stratus; (c) stratocumulus.

$\rho(\sigma)$ varies markedly with cloud type. As shown in Figure 4, stratus
clouds have the most uniform distribution of $\rho(\sigma)$ with σ in the
range 25-55 km^{-1}, implying a high degree of uniformity of their
internal structure. Distributions of $\rho(\sigma)$ for stratocumulus and
nimbostratus clouds are very asymmetrical. For stratocumulus, this
is because of the influence of high values of σ; for nimbostratus,
the asymmetry is caused by low values of σ (<2.5 km^{-1}). Small
values of σ are also typical of altostratus.
      Spectral radiance measurements of the upper cloud boundary
make it possible to estimate cloud optical depth τ (Rozenberg 1972,

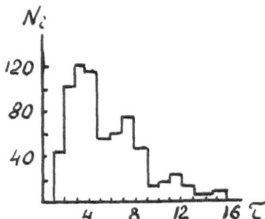

Fig. 5. Histogram of cloud optical depth τ, from *Cosmos-320*
spectrophotometric data.

Rozenberg *et al.* 1974). Figure 5 shows a histogram of values of τ
from *Cosmos 320* spectrophotometric measurements at 0.738 μm
(Gorodetsky *et al.* 1980). The values of τ were determined from an
expression for the brightness coefficient R (Rozenberg 1972,
Rozenberg *et al.* 1974):

$$R = h-[(1-A) \ g(\mu) \ g(\mu_o)]/[1 + (1-A)(\tau/\ell)] \tag{4}$$

where g is the functional angular dependence for scattering, h is
the brightness coefficient for an infinitely thick non-absorbing
layer, ℓ is a parameter describing cloud microstructure which has
a value in the range 8 to 10, A is the surface albedo, μ_o and μ are
the cosines of the solar zenith angle and observation angles,
respectively. Equation (4) can only be applied in the case of small
specific absorption coefficients (i.e., $\beta \simeq 10^{-5}$). As β is found
to be in in the range 10^{-3} to 10^{-2} at wavelengths around 0.74 μm,
brightness coefficients R(λ) observed at wavelengths with different
values of the volume absorption coefficient α(λ) can be extrapolated
to the value R_o, corresponding to α(λ) = 0 (Rozenberg *et al.* 1974).
By doing so, the cloud optical depth would be increased in comparison to the values of τ in Figure 5.
    Gorodetsky *et al.* (1980) evaluated mean cloud thickness $\bar{t}$
assuming an equal probability of occurence of different cloud types
(Ns-As, St, Sc, As). On the basis of the data in Figures 4 and 5,
the mean value of the scattering coefficient $\bar{\sigma}$ for these cloud types
is 18.7 km^{-1} and the mean optical depth $\bar{\tau}$ is 5.4 km. Consequently,
the mean cloud thickness $\bar{t} = \bar{\tau}/\bar{\sigma} = 0.29$ km. This value is in
agreement with measurements made from aircraft (Feigelson *et al.*
1981).
    These methods of monitoring cloud properties make it possible
to evaluate the mean total water content in clouds over a given
locality. For example, taking the following mean values for the
following cloud parameters: cloud thickness $\bar{t}$ = 0.29 km, temperature T = 273 K, mean specific water concentration 0.2 g m^{-3}
(Feigelson *et al.* 1981), it can be shown that the clouds over an
area of 1 km^2 contain 0.6×10^5 kg of liquid water and up to
14.5×10^5 kg of saturated water vapour.

These results appear to justify the continued development of a range of remote sensing techniques and the accumulation of statistical data on correlations between the various optical parameters of clouds.

REFERENCES

Akvilonova, A.B., Basharinov, A.E., Gorodetsky, A.K., Gurvitch, A.S., Kzylova, M.S., Kutuza, B.G., Matveyev, D.T. & Ozlov, A.P. 1973 The cloud parameters investigation from the satellite *Cosmos-384*. *Izv. Akad. Nauk SSSR, Physica Atmosphery i Oceana*, 9, 187-189.

Basharinov, A.E., Gurvitch, A.S. & Egorov, S.T. 1969 The determination of geophysical parameters from the measurements of the Earth's microwave emission on the satellite *Cosmos-243*. *Dokl. Akad. Nauk SSSR*, 188, 1273-1276.

Feigelson, E.M. *et al*. 1981 Radiation in the cloud atmosphere. *Gidrometeoizdat, Leningrad*, pp. 9-21.

Gorodetsky, A.K., Malkevitch, M.S. & Syachinov, V.I. 1971 The determination of cloud altitude from the radiation measurements on satellite *Cosmos-320*. *Dokl. Akad. Nauk SSSR*, 200, 588-590.

Gorodetsky, A.K., Kasatkin, A.M., Malkevitch, M.S., Rozenberg, G.V., Syachinov, V.I. & Faraponova, G.P. 1974 Scientific program, science apparatus complex and main results of *Cosmos* arrow experiment: Optical atmosphere investigations. *Science, Moscow*, 178-186.

Gorodetsky, A.K., Dombkovskaya, E.P., Manuylova, N.I., Matveev, D.T. & Orlov, A.P. 1975 Determining phase characteristics of cloudiness from results of measurements with the aid of artificial satellites. *Met. Hydrol.*, No. 5, pp. 66-71.

Gorodetsky, A.K., Kuznetsov, I.V., Lystsev, V.E. & Syachinov, V.I. 1980 The light scattering coefficients of layered clouds. *Izv. Akad. Nauk SSSR, Physica Atmosphery i Oceana*, 16, 705-711.

Rozenberg, G.V. 1972 On spectroscopical sounding of clouds. *Izv. Akad. Nauk SSSR, Physica Atmosphery i Oceana*, 8, 355-368.

Rozenberg, G.V., Malkevitch, M.S., Malkova, V.S. & Syachinov, V.I. 1974 Determination of cloud optical characteristics from the reflected solar radiation measurements made by *Cosmos-320*. *Izv. Akad. Nauk SSSR, Physica Atmosphery i Oceana*, 10, 14-24.

COMPARISON OF WATER VAPOUR DATA FROM MONEX-79 AND THE *TIROS-N*
SATELLITE

P.N. Pathak

Meteorology Division
Space Applications Centre
Ahmedabad-380053
India

ABSTRACT. During MONEX-79 (from May to July 1979), water vapour
observations were made, remotely, from the *Tiros-N* satellite and,
directly, from drop-sondes and up-sondes, respectively released
from aircraft and ships. From these data, and dew-point temperatures acquired from the sondes, values for water vapour were
calculated and compared for three separate pressure levels, the
spatial separation between sub-satellite and sub-sonde positions
being limited to $1°$ in latitude and longitude, and the temporal
separation between measurements that were compared being restricted
to not more than 5 hours. The analysis demonstrated an excellent
agreement between the satellite-derived and directly-measured data,
with correlation coefficients of + 0.98 in both cases.

1. INTRODUCTION

     During the ICSU/WMO Monsoon Experiment (MONEX) between May
and July 1979 extensive in-situ measurements of various meteorological parameters were made using a variety of measuring platforms.
Water vapour data acquired directly from aircraft-launched drop-
sondes and ship-launched up-sondes have been compared with like
data derived from *Tiros-N* observations.

2. DATA AND ANALYSIS

     The *Tiros-N* data for the period May to July 1979 were obtained
as 'finished-product' data tapes from the NOAA Environmental Data

A. Street-Perrott et al. (eds.), Variations in the Global Water Budget, 89–92.
Copyright © 1983 by D. Reidel Publishing Company.

and Information Service, USA. These data give water vapour values
in terms of amount (in mm) of water above the dew point for three
broad atmospheric layers, namely 1000-700, 700-500 and 500-300 mb.

From the dew-point temperature data, acquired from both the
drop-sondes and up-sondes at different pressure levels, the
corresponding water vapour density was obtained using standard
meteorological tables (Berry *et al.* 1945). To convert pressure
levels to corresponding heights (in km), the data on the 'Standard
Indian Atmosphere' (Ananthasayanam & Narasimha 1979) were used.
From these basic data, like values of water (mm) above the dew
point for the three atmospheric layers were calculated.

To facilitate the comparison of the sonde data with those
from *Tiros-N*, the spatial separation between ship/aircraft
position and the reported position of *Tiros* was limited to $1°$ in
latitude and $1°$ in longitude. The average time difference between
the drop-sonde/up-sonde measurement and the *Tiros* pass was about
4 to 5 hours. The data compared are daily spot values, no spatial
or temporal averaging having been essayed.

The geographical locations of the aircraft and ships responsible for launching the sondes, from which water vapour data were
acquired and subsequently compared with *Tiros-N* data, are shown in
Figure 1. As can be seen, these positions are mostly limited to
the Arabian Sea region.

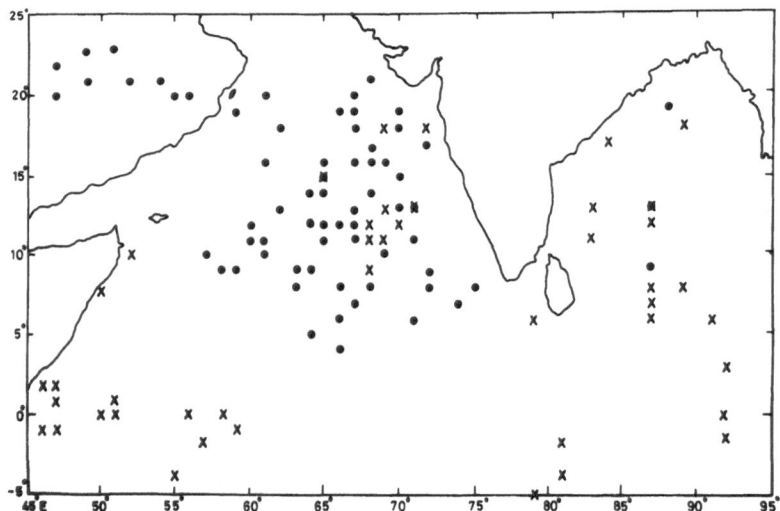

Fig. 1. Geographical locations of MONEX-79 aircraft (●) and ships
         (×) launching sondes yielding water vapour data that were
         compared with like data from *Tiros-N*.

The scatter plots of the drop-sonde and up-sonde water vapour data for the three atmospheric layers, with respect to the corresponding *Tiros-N* data, are given in Figure 2. The excellent agreement between the in-situ and *Tiros* water vapour data can be clearly seen, the correlation coefficients in both cases (a, drop-sonde; b, up-sonde) being +0.98. The least-squares-fit lines, which agree very closely with the 45° lines of perfect fit, have the following equations:

$$W_s = 0.91 \, W_a + 0.91 \text{ mm} \tag{1}$$

$$W_s = 1.01 \, W_d - 0.3 \text{ mm} \tag{2}$$

where W_s = water vapour (mm) measured by *Tiros-N* satellite

W_a = water vapour (mm) measured by up-sonde (ships), and

W_d = water vapour (mm) measured by drop-sonde (aircraft).

In these cases, the RMS errors are about 2 to 3 mm.

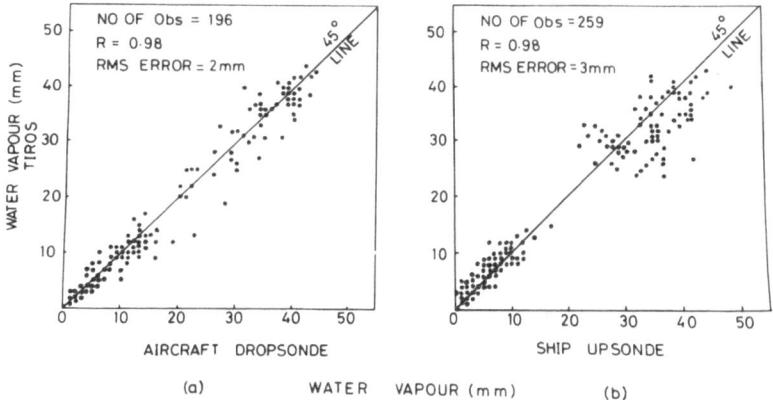

Fig. 2.   Scatter plots of *Tiros-N* water vapour data compared with (a) aircraft drop-sonde data and (b) ships' up-sonde data. R = correlation coefficient.

CONCLUSION

The analysis shows that the *Tiros-N* water vapour compare extremely well with nearby in-situ data, even on a day-to-day basis.

ACKNOWLEDGEMENT. The author is grateful to Dr T.A. Hariharan, Head of the Meteorology Division, for his encouragement and interest in the present work. Thanks are due to Dr Pranav S. Desai for useful discussion and to Shri B.S. Gohil for his help in the computations.

REFERENCES

Ananthasayanam, M.R. & Narasimha, R. 1979 *Standard Atmosphere for Aerospace Applications in India.* Bangalore: Indian Institute of Science (Dept. of Aeronautical Engineering) Report No. 79,F.M 15. [Unpublished.]

Berry, F.A., Bollay, E. & Beers, N.R. (ed.) 1945 *Handbook of Meteorology,* 70 pp. New York: McGraw Hill.

ATMOSPHERIC WATER DISTRIBUTIONS DETERMINED BY THE *SEASAT* MULTI-CHANNEL MICROWAVE RADIOMETER

P.K. Taylor, T.H. Guymer

Institute of Oceanographic Sciences
Wormley, Surrey
United Kingdom

K.B. Katsaros

University of Washington
Seattle
Washington, USA

R.G. Lipes

Jet Propulsion Laboratory
Pasadena
California, USA

ABSTRACT. The *Seasat* Scanning Multichannel Microwave Radiometer (SMMR) provided determinations of total atmospheric water vapour q_v, liquid water q_ℓ and rain rate. Radiosonde data from the Joint Air Sea Interaction Experiment, JASIN, show that variations in q_v and q_ℓ reflect changes of structure in the first 500 mb of the atmosphere. The SMMR measured q_v at least as accurately as the JASIN radiosondes. SMMR q_ℓ and rain rate values were in qualitative agreement with JASIN observations. Quantitative evaluation was hampered by lack of independent measurements; for those comparisons which could be made, agreement was good.

Examples are given of the use of SMMR data in the detection of atmospheric fronts, the mapping of areas of widespread rainfall, and the detection of a mesoscale thunderstorm area. The SMMR is shown to be a valuable research tool contributing to the analysis of the JASIN data.

A. Street-Perrott et al. (eds.), *Variations in the Global Water Budget, 93–106.*

1.   INTRODUCTION

     Seasat, the first oceanographic satellite using microwave
sensors, was launched on 28 June 1978 and failed on 10 October
1978.   The satellite (Born *et al.* 1979) carried five instruments
including a Scanning Multichannel Microwave Radiometer (Njoku *et
al.* 1980); the SMMR was similar to that currently in operation on
Nimbus 7.   The SMMR measured microwave radiation from the Earth at
five frequencies between 6.6 and 37 GHz for both vertical and
horizontal polarizations (Table I).   This allowed estimation of
sea surface temperature and wind speed as well as atmospheric
water vapour, liquid water and rain rate (Wilheit 1978, Wilheit
et al. 1980).   This paper considers the last three variables which
represent major components of the atmospheric water budget.

Table I.   Characteristics of the *Seasat* SMMR channels.   The
           retrieval grid is referred to a 3 dB limit.

Frequency (GHz)	Wavelength (cm)	Retrieval Grid (km)	Principal Predicted Quantity
6.6	4.6	150 × 150	Sea surface temperature
10.7	2.8	85 × 85	Wind speed
18.0	1.7	54 × 54	Precipitation rate (18H)
			Water vapour
			Liquid water
21.0	1.4	54 × 54	Water vapour
			Liquid water
37.0	0.8	27 × 27	Precipitation rate (37H)
			Windspeed in clear air
			Water vapour
			Liquid water

     Passive microwave radiometry has been used previously to
measure atmospheric water from satellites such as *Nimbus 5* and *6*.
Both the *Nimbus 5* Microwave Spectrometer, NEMS (Staelin *et al.*
1976), and the Scanning Microwave Spectrometer, SCAMS (Rozenkranz
et al. 1978) used a 22 GHz water vapour channel and 31 GHz window
channel for atmospheric water vapour determinations.   Electrically
Scanning Microwave Radiometers, ESMR, carried on *Nimbus 5* (Wilheit
et al. 1977) and *Nimbus 6* (Rodgers *et al.* 1979) were designed to
map rainfall areas.   The advantages of the SMMR over these previous
instruments are the incorporation of additional channels to allow
better discrimination between the effects of liquid water, water
vapour and sea state, and improved spatial resolution (Table I).
     Seasat was an experimental satellite and the performance of
the sensors required validation by comparison with *in situ* measurements.   The Gulf of Alaska Experiment (GOASEX) was designed to

validate the SMMR and the results for atmospheric water vapour are
summarized by Katsaros *et al.* (1981). The Joint Air-Sea Interaction experiment (JASIN), a large, independently-organized field
experiment (Royal Society 1979) took place in the North Atlantic
Ocean from July to September 1978. Comparisons of *Seasat* data and
JASIN measurements were reported in the *Seasat*-JASIN (1980) and
SMMR-Mini III (1980) reports and by Katsaros *et al.* (1981).

The aim of this paper is to demonstrate the value of SMMR-
derived water vapour distributions in analysing atmospheric structure by presenting examples from the JASIN experiment. SMMR data
are used to detect mesoscale features and the position of atmospheric fronts, and to map rainfall regions. However, first it is
necessary to consider how two SMMR-measured quantities, the total
atmospheric water vapour content q_v and total liquid water content
q_1, relate to the vertical structure of the atmosphere and to consider the accuracy of SMMR measurements of these quantities.

2. RADIOSONDE AND SMMR DETERMINATIONS OF ATMOSPHERIC WATER

2.1 JASIN Radiosonde Observations

During JASIN, radiosondes were released from three ships which
formed an equilateral triangle of *ca.* 200 km side centred at about
$59\frac{1}{2}°$N, $12\frac{1}{2}°$W in the North Rockall Trough region of the North
Atlantic. The ships were on station between 23 July and 8 August
and 22 August to 4 September 1978. During certain periods, a
fourth radiosonde ship was positioned at the triangle's centre.
Radiosondes were released from each ship at intervals of, at most,
six hours and, during daylight hours on chosen days, about hourly
intervals. VIZ 1220 and 1223 radiosondes were used with factory-
calibrated 'premium' sensors of US National Weather Service type
sampling every 0.8 s. The resulting data set provides a uniquely-
detailed description of the troposphere at a mid-latitude open
ocean site during a summer time period.

Figure 1 shows the mean water vapour content q_v, and its standard deviation σ_v, for 5 mb atmospheric layers, using data from
about 600 radiosonde flights throughout the JASIN period. As is
well known, the decrease of temperature with height results in a
rapid decrease of q_v; some 80% of the total occurs in the range
$p* = 0$ to $p* = 250$ mb (where $p* = p_o - p_i$, with p_o the surface
pressure and p_i the pressure at the level of interest). In contrast, σ_v is a maximum not at the surface, but in the $p*$ range 120
to 230 mb. This is due to the relative uniformity of the surface
boundary layer and the large fluctuations in q_v for those layers
which sometimes are within the boundary layer and at other times
are within the drier air above. The standard deviation remains
comparable to the surface value to about $p* = 550$ mb, and 80% of
the variation is spread over the $p*$ range from 0 to 400 mb. Thus,
despite the maximum of q_v at the surface, the radiosondes show that
variations of q_v may reflect changes in at least the first 500 mb
of the atmosphere.

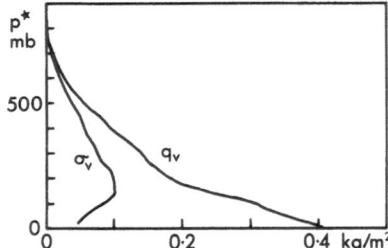

Fig. 1. Mean atmospheric water vapour content q_v during JASIN
         (June to September 1978) for 5 mb layers and the standard
         deviation σ_v of individual radiosonde observations for
         each layer: p*, pressure (mb).

2.2 Comparison of SMMR and Radiosonde Measurements of Water Vapour

      A comparison of SMMR and radiosonde determinations of q_v
at the *Seasat*-JASIN Workshop was presented by Katsaros *et al.*
(1980) and Taylor *et al.* (1981). Comparisons for JASIN ascents
and tropical soundings are shown in Figure 2. For JASIN, the
mean difference, SMMR-minus-radiosonde, was 1.2 ± 1.6 kg m^{-2}
which compares very favourably with the estimated accuracy of the
radiosondes of ± 1.7 kg m^{-2}.

      In the tropics, radiosonde errors would be expected to be
greater; after elimination of values contaminated by land effects
or, in one case, heavy rain, the mean difference was 4.9 ± 3.7
kg m^{-2}. Changes in the SMMR evaluation procedures following the
Seasat-JASIN Workshop have removed this bias (SMMR-Mini III 1980),
and it appears that the SMMR can measure total atmospheric water
vapour at least as accurately as radiosondes.

2.3 Radiosonde determinations of liquid water

      Radiosondes do not measure liquid water and the only direct
measurements available for JASIN were obtained by aircraft equipped
with Johnson-Williams type instruments (Slingo *et al.* 1982). As
these measurements were limited in space and time, an attempt has
been made to infer liquid water content from the radiosonde flights
so as to provide comparisons with *Seasat* estimates. Slingo *et al.*
1981 show that, for the marine stratocumulus layer (which frequently occurred during JASIN), the liquid water content was very
close to that calculated by adiabatic ascent of a parcel of air
from the cloud base. The level at which JASIN radiosonde balloon
was observed to enter cloud was noted and, hence, for stratocumulus,
the specific liquid water content Q_ℓ is given by

$$Q_\ell = Q_{CB} - Q_i$$

where Q_{CB} and Q_i represent the specific humidity at cloud base and

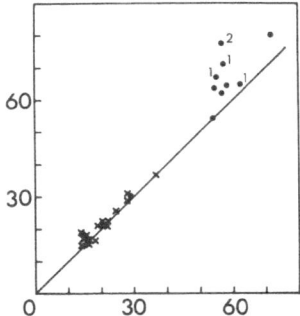

Fig. 2.   Comparison of integrated atmospheric water vapour q_v (kg
          m^{-2}) as measured by the SMMR (ordinate) and by radiosondes
          (abscissa), from the 1980 *Seasat*-JASIN Workshop (Taylor *et
          al.* 1981):   ×, JASIN values; •, tropical soundings;
          1, tropical soundings for which SMMR values may be too
          high due to the island of Guam being within the field of
          view; 2, SMMR values possibly affected by heavy rain.
          Processing changes since the Workshop have removed the
          tropical data bias.

at the level of interest. The total liquid water content is
obtained by integrating between the cloud base and cloud top. For
the JASIN radiosondes, the mean relative humidity at cloud entry
was 98 ± 3%, a fall to ≤ 95% being chosen arbitrarily as marking
cloud top. This gave good agreement with the available aircraft
measurements. Upper cloud layers were assumed if the relative
humidity rose above 98%, extending until values of < 95% were
encountered. All clouds were assumed to have stratocumulus liquid
water distributions, though the comparisons with SMMR measurements
are presented as classified by cloud type.
     The resulting radiosonde-derived liquid water data are pre-
sented in Figure 3; the continuous line shows the number of ascents
reporting 'cloud' at each level from a total of about 600 ascents
of which all but 50 reached 500 mb. The maximum occurred just
below the maximum variation of q_v (Figure 1), that is below the
mean boundary layer top. The dashed line shows the mean liquid
water content for each 5 mb layer where the mean is taken over only
those ascents reporting liquid water at that level. Values
increased through the boundary layer and then remained high as
long as 'clouds' were present. Although this is probably caused
by the adiabatic assumption, it illustrates that all layers to
500 mb could make a significant contribution to liquid water content. The dotted line shows the number of ascents reporting cloud
where only the lowest cloud layer, that which the balloon was
observed to enter, is included. The liquid water content (not
shown) was similar to that for all layers, except that the cutoff
was at about $p^* = 300$ mb.

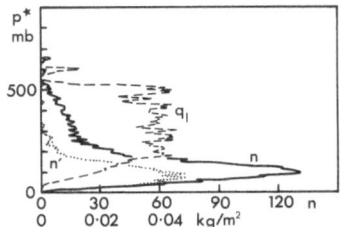

Fig. 3.   Number of radiosonde ascents n for which cloud is inferred
          for 5 mb layers of the atmosphere: ——, multiple cloud
          layers; •••, first cloud layer only; ---, mean liquid
          water content q_ℓ (kg m^{-2}) for cases reporting cloud in
          each layer; p*, pressure (mb).

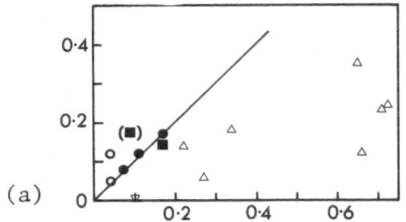 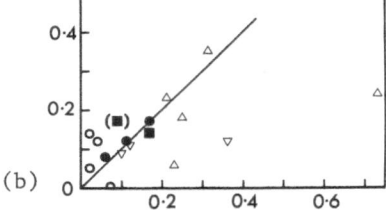

(a)   (b)

Fig. 4.   Comparison of SMMR (ordinate) with aircraft and radiosonde
          (abscissa) estimates of total liquid water q_ℓ (kg m^{-2}):
          (a) where radiosonde values include all cloud layers;
          (b) radiosonde values including the first cloud layer
          only: ■, aircraft measurements - mean of several profiles
          (Slingo *et al.* 1982); (■), derived from a single profile
          (S. Nicholls, personal communication); remaining points
          represent radiosonde values with the following standard
          deviations (kg m^{-2}): ○, < 0.05; •, < 0.05 for days when
          a relatively uniform stratocumulus cloud covered the area;
          ∇, < 0.1; Δ, < 0.5.

2.4 Comparison of SMMR and Radiosonde Values of Liquid Water

     Katsaros *et al.* (1981) present comparisons of two different
algorithms used in evaluating SMMR data. For JASIN, the agreement
was within about 0.06 kg m^{-2} for q_ℓ < 1 kg m^{-2}. For higher q_ℓ
values and in the tropics, agreement was less good with the Wentz
algorithm (GOASEX I 1979) giving higher values that that of Wilheit
and Chang (1979). Under certain circumstances, both algorithms
exhibited negative values suggesting a bias (SMMR-Mini III 1980),
but this was not evident in the Wentz determination for the JASIN
area, used here.
     A comparison of SMMR q_ℓ measurements and *in situ* estimates is
shown in Figure 4. The SMMR values represent the mean liquid water

content for all passes on a given day for the area 58.5°N to 61°N
and 9°W to 16°W. The radiosonde values represent the mean for all
flights on a given day for all the ships, different symbols indicating the magnitude of the standard deviation between the three
ships; for both aircraft measurements, there was a uniform stratocumulus cover. Figure 4(a) shows that for $q_\ell < 0.2$ kg m^{-2} the
standard deviation of the radiosonde values was small and the
agreement with SMMR was good, particularly for the stratocumulus
cases. At higher q_ℓ values, the radiosondes showed high variability and tended to overestimate compared to the SMMR measurements. This would be expected, both because the adiabatic assumption is less likely to hold for deeper cloud layers, and because
of the assumption that a higher cloud layer always occurred when
over 98% relative humidity was observed. Figure 4(b) shows that,
if only the first, observed, cloud layer is considered, the standard deviation of many radiosonde values is decreased and more
cases are in agreement with the SMMR. On low q_ℓ days with
scattered small cumulus clouds, radiosondes tend to produce underestimates, an understandable tendency since such clouds are poorly
sampled by radiosondes.

In summary, where liquid water measurements by radiosondes
should be reliable, agreement with SMMR data is good; in other
cases, the differences between the SMMR and radiosonde measurements are physically comprehensible. The comparison further
emphasizes the difficulty of obtaining liquid water data by means
other than microwave radiometry.

2.5   SMMR Measurements of Rain Rate

SMMR rain rate determinations have been discussed by Katsaros
et al. (1981) and in the SMMR-Mini III (1980) report. For JASIN,
the Wentz and Wilheit algorithms (§ 2.4) showed a scatter of about
0.5 mm h^{-1} in rain rate. For the tropics, the scatter was much
greater. The Wilheit algorithm predicted rain for more cases than
did the Wentz. For JASIN, the SMMR data predicted rain on a qualitative basis when precipitation was 'widespread', i.e., reported
by several JASIN ships. Light scattered showers, with horizontal
dimensions very much less than the SMMR resolution, were not
detected. Lack of surface 'truth' has prevented a qualitative
comparison, except in a single instance in which SMMR (Wentz
algorithm) and ship-measured rain rates were in reasonable agreement (see § 3.2).

3.   EXAMPLES OF SMMR ATMOSPHERIC WATER DISTRIBUTIONS FOR THE
JASIN AREA

3.1   The Detection of Atmospheric Fronts

Taylor *et al.* (1981) have shown that the position of atmospheric fronts could be detected by the SMMR on occasions where

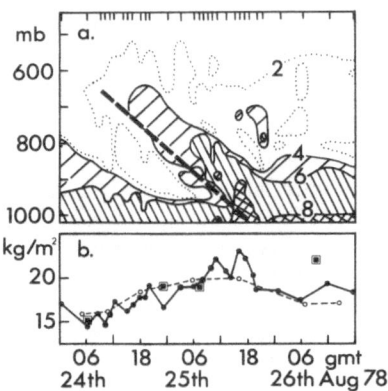

Fig. 5.   (a) Time-height atmospheric cross-section showing the
          variation of specific humidity (g kg^{-1}) with the passage
          of a warm front. The section is based on 31 radiosonde
          flights (at the times denoted by scale ticks) from the NW
          corner of the JASIN triangle: ---, approximate region of
          frontal surface.
          (b) Values of integrated water vapour q_v (kg m^{-2}) (after
          Taylor et al. 1981) from: ——, radiosondes; □ & ▣ SMMR;
          ---, SMMR q_v distribution at 0730 GMT 24 August 1978.

lack of distinct cloud structures made detection by visible or
infrared imagery difficult. A time-height atmospheric cross-
section (Figure 5a) shows the variation of specific humidity with
the passage of a warm front; the values of q_v for the 31 radiosonde flights used are shown in Figure 5(b). The warm moist air
above the front results in an increase of q_v to a maximum at about
the time of frontal passage. There is then a sharp fall to the
warm air boundary-layer value. SMMR values showed good agreement
with the radiosondes except on one occasion (26 August) when strong
gradients of q_v existed near the JASIN area and within the SMMR
resolution cell.
     Figure 6 shows the SMMR-derived q_v distribution at 0730 GMT
on 24 August 1978 (near the start of the period shown in Figure 5)
superimposed on the $NOAA$ 5 VHRR infrared image for 1138 GMT.
Values of q_v, measured by the JASIN radiosonde ship triangle, are
in good agreement with the SMMR data. The JASIN radiosonde observations indicated little change in the frontal structure with time.
Hence, the horizontal distribution in Figure 6 can be transformed
into a time variation of q_v using the appropriate propagation
speed (2 m s^{-1} toward 005°): the result (see Figure 5b) agrees
well with the radiosonde values.
     Thus the surface frontal position has been marked on Figure 6
using the SMMR q_v distribution and the frontal movement has been

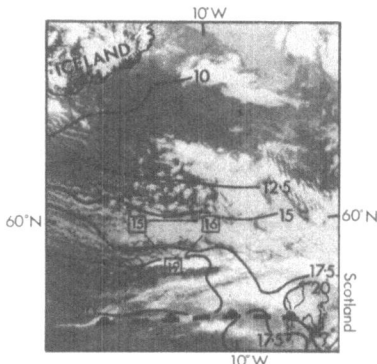

Fig. 6.   SMMR distribution of q_v (kg m^{-2}) at 0730 GMT 24 August
           1978 superimposed on *NOAA 5* Infrared image for 1138 GMT:
           □, radiosonde q_v values.

calculated from succeeding SMMR passes. This was not possible
using visible and infrared imagery owing to the lack of any distinct cloud band associated with the front. Nor was the frontal
position detectable in the standard synoptic surface data and the
front was shown as discontinuous in the UK Meteorological Office
analysis. Frontal detection by the SMMR will be a considerable
aid in interpreting JASIN radiosonde data.

3.2   Detection of Rainfall Areas

       During the evening of 30 August 1978, an occluding frontal
system passed over the JASIN area. The SMMR q_v and q_ℓ distributions at 2311 GMT are shown in Figure 7. The warm air sector
ahead of the cold front had $q_v > 35$ kg m^{-2} compared to a minimum
in the cold air to the east of the frontal system of < 15 kg m^{-2}.
Liquid water content was generally < 0.2 kg m^{-2}, except in the
occlusion point region where a large area had $q_\ell > 0.6$ kg m^{-2} with
a maximum of > 1.0 kg m^{-2}. The results in this figure may be
interpreted in terms of the ascent of warm moist air in the
occlusion region with condensation followed by precipitation.
       SMMR rainfall rates are shown in Figure 8 superimposed on the
NOAA 5 VHRR IR image from 2030 GMT. From SMMR data on 30 and
31 August, the rain area moved at 12 m s^{-1} towards 110°. Using
these values, rain would be predicted at the NE corner of the JASIN
triangle between 1600 and 2300 GMT with a mean SMMR rain rate of
0.4 mm h^{-1}. Measurements on HMS *Hecla* at that position showed that
rain occurred between 1600 and 2400 GMT with a mean rate of
0.7 mm h^{-1}. Considering the difficulty of making accurate ship-
board measurements of rainfall, the difference in magnitude is not
significant.
       Although quantitative rainfall measurements were not available

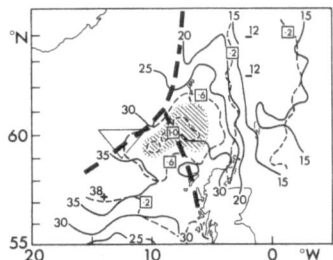

Fig. 7.   SMMR values of q_v (kg m^{-2}) and q_ℓ (kg m^{-2}) at 2311 GMT
30 August 1978: ———, q_v; ---, q_ℓ (values in boxes);
hatched area, region with over 0.4 mm h^{-1} rain rate.

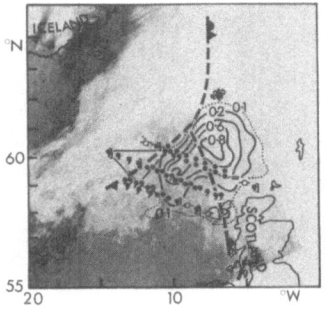

Fig. 8.   SMMR rain rate values (mm h^{-1}) at 2311 GMT 30 August 1978
superimposed on *NOAA 5* Infrared image for 2030 GMT,
showing WMO present-weather codes reported from the three
JASIN ships assuming that the rainfall area propagated at
12 m s^{-1} toward 110°.

from the other JASIN ships, Figure 8 shows the 'present-weather
types' (reported in standard meteorological code) plotted in a
frame of reference moving with the SMMR-identified rain area.
Observations of rain generally lie within the 0.2 mm h^{-1} SMMR
contour; SMMR values between 0.1 and 0.2 mm h^{-1} correspond to
drizzle. However, the distinctions between moderate and slight,
and between intermittent and continuous rain do not correspond
with SMMR rain rate values. This example shows that the SMMR
rainfall rates can be used to map areas of widespread rain over
the ocean.

3.3 Mesoscale Features

The SMMR can resolve atmospheric structures having length
scales of $\geqslant$ 100 km. A thunderstorm area was detected by several

Seasat sensors on 4 August 1978, and its detection has been discussed by Guymer *et al.* (1981). Figure 9 shows the synoptic
situation at 2100 GMT. A near-stationary cold front lay to the
southwest, while a cold occlusion moved southwestward through the
JASIN area at about 5 m s^{-1}. Radiosonde ascents showed that,
between these fronts, boundary layer convection was prevented from
reaching above about 2 km height by a strong temperature inversion which marked the warm front surface of the occlusion. The
air above this inversion was potentially unstable and could have
supported deep convection extending from 2 to 6 km if lifted by a
small amount. The SMMR-derived q_v and q_ℓ distributions are shown
in Figure 10. There was a region of high q_v between the fronts
with low values in the cold air sectors. A maximum in q_ℓ occurred
at the cold front presumably indicating ascent of the warm air
over the frontal surface. Maxima in q_v and q_ℓ occurred to the
southeast of the JASIN area at a position marked by deep convective
cloud in the *NOAA 5* VHRR and *Seasat* Visible and Infrared Radiometer imagery. Assuming that this feature was steered by the mid-
tropospheric flow (Newton & Katz 1958), it would have passed over
the southern part of the JASIN triangle at about 2330 GMT: three
of the JASIN ships in that region reported thunderstorm activity
between 2100 GMT and midnight.

   Thus a mesoscale region of mid-level thunderstorm activity
was detected in the SMMR data as a maximum in q_v and q_ℓ values.
Despite the deep convection, the ships reported little precipitation at the surface during the period and the SMMR indicated a
zero rain rate over the area.

4.   SUMMARY

   For a mid-latitude open-ocean area, variations in integrated
water vapour q_v and liquid water q_ℓ reflect changes in the
structure of the lower 500 mb of the atmosphere. SMMR measurements of q_v are at least as accurate as radiosonde values. SMMR
values of q_ℓ agree with radiosonde estimates for uniform marine
stratocumulus cloud cover. For other cases, the SMMR estimates
appear reasonable, but lack independent confirmation. Similarly,
the SMMR estimates for rain rate appear qualitatively correct and
quantitatively reasonable for cases of widespread rainfall, as
was demonstrated for the ascent of the warm air at an occlusion.
However, good independent measurements are lacking.

   SMMR-derived q_v distributions may be used to position
synoptic-scale fronts even in cases where this is not possible
from visible or infrared imagery. A mesoscale region of mid-level
thunderstorm activity was detected in the SMMR data as maxima in
q_v and q_ℓ.

   The examples which have been shown represent the first stages
in the use of SMMR measurements in the analysis of the JASIN data.
The JASIN radiosondes, JASIN surface observations and SMMR measure-

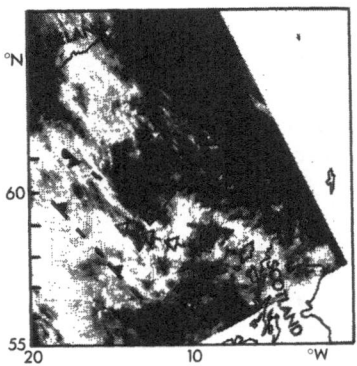

Fig. 9. Synoptic chart for 2100 GMT 4 August 1978 superimposed on *NOAA 5* Infrared image for 2010 GMT: →, near surface flow; ⇒, mid-level flow.

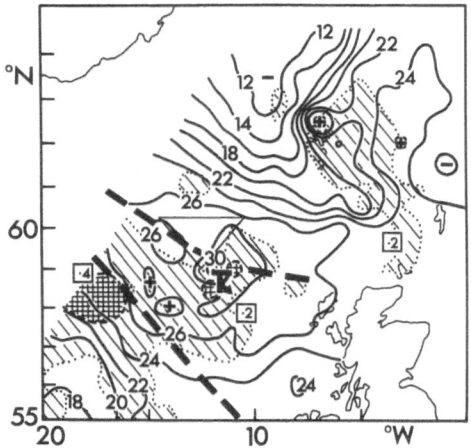

Fig. 10. SMMR distribution of q_v (kg m^{-2}) and q_ℓ (kg m^{-2}) for 2134 GMT 4 August 1978: ——, q_v; hatched areas, q_ℓ (value in boxes); thunderstorm symbol marks the region of deep convective activity. Note: The maximum in the region of 62°N, 7°W probably represents an erroneous SMMR value due to the presence of land (Faeroe Islands).

ments are complementary data sets with which it is hoped to improve understanding of the mid-latitude atmosphere over the ocean. Despite its short life, the *Seasat* SMMR has graduated from sensor validation to an actively-used research tool.

ACKNOWLEDGEMENTS. Katsaros appreciates support by NOAA/NASA
contract NA78 SAC 04102. Figures 2 and 5 are reprinted by permission from *Nature, Lond.*, 294, 737-739: Copyright © 1981
Macmillan Journals Limited. The *NOAA 5* imagery used in Figures 6,
8 and 9 is by courtesy of the University of Dundee.

REFERENCES

Born, G.H., Dunne, J.A. & Lame, D.B. 1979 *Seasat* mission overview.
    Science, N.Y., 204, 1405-1406.

GOASEX I 1979 *Seasat* Gulf of Alaska Workshop Report, JPL-622-101-
    NASA, Vol. 1. Pasadena: Jet Propulsion Laboratory.

Guymer, T.H., Businger, J.A., Jones, W.L. & Stewart, R.H. 1981
    Anomalous wind estimates from the *Seasat* Scatterometer.
    Nature, Lond., 294, 735-737.

Katsaros, K.B., Taylor, P.K., Alishouse, J.C. &·Lipes, R.G. 1981
    Quality of *Seasat* SMMR (Scanning Multi-Channel Microwave
    Radiometer) atmospheric water determinations. In: *Oceano-
    graphy from Space*, pp. 691-706. Plenum Publ. Corp.

Newton, C.W. & Katz, S. 1958 Movement of large convective rain-
    storms in relation to winds aloft. *Bull. Am. met. Soc.*, 39,
    129-136.

Njoku, E.G., Stacey, J.M. & Barath, F.T. 1980 The *Seasat* Scanning
    Multichannel Microwave Radiometer (SMMR): instrument descrip-
    tion and performance. *IEEE J. oceanic Eng.*, OE-5, 100-115.

Rodgers, E., Siddalingaiah, H., Chang, A.T.C. & Wilheit, T.T. 1979
    A statistical technique for determining rainfall over land
    employing *Nimbus 6* ESMR measurements. *J. appl. Met.*, 18,
    978-991.

Royal Society 1979 *Air-Sea Interaction Project: Summary of the
    1978 Field Experiment (JASIN 1978)*, 139 pp. London: The
    Royal Society.

Rozenkranz, P.W., Staelin, D.H. & Grody, N.C. 1978 Typhoon June
    (1975) viewed by a Scanning Microwave Spectrometer. *J. geo-
    phys. Res.*, 83, 1857-1868.

Seasat-JASIN 1980 *Seasat*-JASIN Workshop Rept Vol. 1: Findings and
    Conclusions, JPL 80-62. Pasadena: Jet Propulsion Laboratory.

SMMR-Mini III 1980 *Seasat* SMMR-Mini Workshop III, JPL-622-224-NASA.
    Pasadena: Jet Propulsion Laboratory.

Slingo, A., Nicholls, S. & Schmetz, J. 1982 Aircraft observations
    of marine stratocumulus during JASIN. *Q. Jl R. met. Soc.*
    [In the press.]

Staelin, D.H., Kunzi, K.F., Pettyjohn, R.L., Poon, R.K.L., Wilcox,
    R.W. & Waters, J.W. 1976 Remote sensing of atmospheric water
    vapour and liquid water with the *Nimbus 5* microwave spectro-
    meter. *J. appl. Met.*, 15, 1204-1214.

Taylor, P.K., Katsaros, K.B. & Lipes, R.G. 1981 *Seasat* Multichannel
    Microwave Radiometer atmospheric water determinations and the
    detection of synoptic fronts. *Nature, Lond.*, 294, 737-739.

Wilheit, T.T., Chang, A.T.C., Rao, M.S.V., Rodgers, E.B. & Theon,
    J.S. 1977 A satellite technique for quantitatively mapping
    rainfall rates over the oceans. *J. appl. Met.*, 16, 551-560.

Wilheit, T.T. 1978 A review of applications of microwave radiometry
    to oceanography. *Bound.-Layer Met.*, 13, 277-293.

Wilheit, T.T. & Chang, A.T.C. 1979 An algorithm for retrieval of
    ocean surface and atmospheric parameters from the observations
    of the Scanning Multichannel Microwave Radiometer (SMMR).
    NASA Tech. Mem. 80277. Greenbelt: NASA Goddard Space Flight
    Center.

Wilheit, T.T., Chang, A.T.C. & Milman, A.S. 1980 Atmospheric
    corrections to passive microwave observations of the ocean.
    Bound.-Layer Met., 18, 65-77.

VARIATIONS OF DEUTERIUM AND OXYGEN-18 IN CONTINENTAL PRECIPITATION
AND GROUNDWATER, AND THEIR CAUSES

C. Sonntag, K.O. Münnich, H. Jacob

Institute of Environmental Physics
University of Heidelberg
Federal Republic of Germany

K. Rozanski

Institute of Nuclear Physics and Techniques
University of Mining and Metallurgy, Crakow
Poland

ABSTRACT. Spatial and temporal variations in the deuterium- and
^{18}O-content of precipitation are caused by isotope fractionation
resulting from evaporation and condensation processes during
circulation of atmospheric water vapour. During these phase
transitions, the isotope-labelled water molecules are preferentially transferred into the liquid phase. Water loss by moist
adiabatic cooling of air masses therefore leads to progressive
isotopic depletion, which is described by a Rayleigh condensation
formula. Application of this formula to the moist adiabatic
ascent of discrete air masses yields exponential vertical profiles
of the moisture and its D- and ^{18}O-content assuming a constant
relative humidity and a mean temperature lapse rate of $-6°C$ km^{-1}.
These isotope profiles agree fairly well with observed profiles in
tropospheric water vapour. The spatial variation of D and ^{18}O in
precipitation and groundwater across Europe and North America can
also be described by a simple Rayleigh condensation model, which
links the local isotope data with the parameters of the vertically-
integrated water vapour flux. This model uses as input data
monthly means of local temperature, relative humidity, precipitation and evapotranspiration. The model also yields an estimate of
the variation of the isotope content of local precipitation with
temperature. A one-year record of D and ^{18}O in daily mean values
of water vapour at Heidelberg shows the seasonal variation already

107

A. Street-Perrott et al. (eds.), Variations in the Global Water Budget, 107–124.

known from monthly mean precipitation data and successfully
simulated by the model. Superimposed on this seasonal pattern are
strong short-term variations with a dominant periodicity of 22
days. This periodicity seems to be related to the phase velocity
of long waves in the free atmosphere, which steer the water vapour
transport in the lower troposphere.

1.   INTRODUCTION

     In natural waters, the heavy stable isotopes, deuterium and
oxygen-18, exist almost exclusively in the form of the isotopi-
cally-labelled water molecules HDO and $H_2{}^{18}O$. The mean concentration of these heavy molecules, expressed as an isotope ratio R
(Craig & Gordon 1965), is given by

$$\bar{R}_D = (HDO)/(H_2O) = 3.2 \times 10^{-4}; \quad \bar{R}_{(18O)} = (H_2{}^{18}O)/(H_2O) = 2.0 \times 10^{-3}.$$

     Since oceanic water forms 97% of the total water inventory
and the observed variations of D and ^{18}O within the water cycle
are relatively small, the values quoted represent the D- and ^{18}O-
content (abbreviated to 'heavy isotope content') of Standard Mean
Ocean Water (SMOW), to which the deviations δD and $\delta^{18}O$ of the
isotope content of water samples (in parts per 1000:‰)

$$\delta D \text{ (or } \delta^{18}O) = [(R_{sample} - R_{SMOW})/R_{SMOW}]10^3 \tag{1}$$

The observed variations of D and ^{18}O in natural waters are intimately related to isotopic fractionation during the evaporation
and condensation of water, when the heavy water molecules HDO and
$H_2{}^{18}O$ preferentially remain in, or pass into, the liquid phase.
This isotope fractionation is commonly described by the 'fractionation factor' α, which is the ratio of the isotope content of the
liquid (or solid) and of the vapour phase, such that $\alpha = R_L/R_V > 1$.
In systems at thermodynamic equilibrium, the fractionation factor
α_e corresponds to the ratio of the saturation vapour pressure of
normal water H_2O to that of 'heavy' water (HDO or $H_2{}^{18}O$), α_e
decreasing with temperature (Majoube 1971) as shown in Table I.
In the case of non-equilibrium, e.g., during evaporation from an
open water surface into an unsaturated atmosphere (relative
humidity h below 100%), slight differences in the transfer of the
normal and heavy water molecules through the viscous boundary
layer at the water surface cause additional fractionation. This
kinetic fractionation (α_k) is controlled by the molecular diffusion
coefficients of the various water-molecule species in air and the
moisture deficit (1-h) of the atmosphere (Merlivat & Coantic 1975).
Acting (during evaporation) in the same direction as the equilibrium fractionation, the total isotope fractionation can be
expressed as $\alpha = \alpha_k \alpha_e > 1$. Variations in D and ^{18}O, from isotope
fractionation during evaporation and condensation, occur almost

Table I. Temperature dependence of the equilibrium isotope fractionation $\alpha_e - 1$ for deuterium and oxygen-18; α_e = equilibrium fractionation factor.

	temperature (°C)				
	-10	0	10	20	30
$(\alpha_e - 1)_{HDO} \times 10^3$	129.6	112.6	97.9	85.2	74.2
$(\alpha_e - 1)_{H_2{}^{18}O} \times 10^3$	12.9	11.7	10.7	9.8	9.0
$(\alpha_e - 1)_{HDO}/(\alpha_e - 1)_{H_2{}^{18}O}$	10.0	9.6	9.1	8.7	8.2

exclusively in the atmospheric branch of the water cycle, including the Earth's surface. Once water has penetrated deep into the ground, its isotopic composition remains practically unchanged during subsurface movement and storage. Thus, D and ^{18}O can act as tracers of circulation of atmospheric water vapour.

The IAEA/WMO Global Precipitation Network has been in operation since 1961, providing monthly mean data on D, ^{18}O and tritium in local precipitation (IAEA 1969-79). In the early 1960s, several theoretical studies considered the extent to which information on the global water vapour transport might be obtained from this data collection (Dansgaard 1964, Eriksson 1965, Gat 1980). However, in the following decade, hydrometeorological interest in these data was rather low, since models of water vapour circulation (Holloway & Manabe 1971, Peixóto 1973, Manabe & Holloway 1975) established during this period did not even consider the isotope data from the network. One reason for this may have been the lack of isotope data on tropospheric water vapour (i.e., D and ^{18}O vertical profiles within the moist layer at various locations and times) needed to test isotope models. Another reason may have been the time-scale problem which arises if the isotope data for monthly mean precipitation, as supplied by the network, are used. Since numerical-model estimates of water vapour circulation are normally carried out in time steps of a few hours, it is questionable whether data on a monthly basis would be sufficiently detailed. On the other hand, even detailed isotope data for individual precipitation events can only be expected to provide information on the circulation of water vapour in special circumstances which may not be typical of the overall period being studied.

Increasing interest in D and ^{18}O, as indicators of climatic changes in the past (variations of δD and $\delta^{18}O$ in palaeo-waters, plant material, speleothems, etc.), has recently drawn fresh attention to the problem of how the isotope signature of local precipitation is linked to the parameters of the atmospheric water vapour transport.

2.   D AND ^{18}O IN THE CIRCULATION OF WATER VAPOUR

2.1 Vertical Distribution of D and ^{18}O in Tropospheric Water
     Vapour

    Consider first the subtropical ocean which is the main source
for atmospheric water vapour. As mentioned above, water vapour
released by the ocean is isotopically depleted in comparison with
oceanic water. The latter's isotope content is given by
$R_o = \alpha^{-1}R_{SMOW}$. Condensation of water vapour to form clouds during
the moist adiabatic ascent of discrete air masses leads to further
isotopic depletion of the vapour, even without moisture loss by
rainfall, since the heavy molecules HDO and $H_2{}^{18}$O are preferentially converted into water droplets (or ice crystals). The
partition of the isotopes in this two-phase system is obtained
from the combination of the moisture mass and the isotope balance
$R_o = R_v(1-L) + R_\ell L = R_v(1-L) + \alpha_e R_v L$, where L is the cloud-water
fraction of the total specific humidity, R_v and R_ℓ are the isotope
ratios of the vapour and the liquid phase, and α_e is the fractionation factor at the cloud temperature. For the isotope content of
a cloudy marine air mass, which has not yet undergone any moisture
loss by precipitation, this balance equation yields

$$R_v = [1 + (\alpha_e-1)L]^{-1}R_o = [\alpha(1+(\alpha_e-1)L)]^{-1}R_{SMOW} \qquad (2)$$

Although a water content of up to 4 g m^{-3} (and thus high L values)
have been observed in cumulo-nimbus clouds, the mean cloud-water
fraction L is only about 0.1 or less (Liljequist & Cehak 1979).
Then $R_v \simeq R_o$, i.e., $R_v \simeq \alpha^{-1}R_{SMOW}$. During further moist adiabatic
cooling in the course of vertical and/or horizontal displacement
of the air mass towards cooler regions, selective loss of the heavy
water molecules by precipitation leads to increasing isotopic
depletion of the residual vapour and thus of the total air-mass
moisture. The same happens to the precipitation derived from this
residual moisture. The isotope variation as a consequence of this
batch-like process can be described by the Rayleigh condensation
formula $R/R_o = F^{\alpha_e-1}$ which relates the isotope content R of the
vapour (or condensate derived from it) to the fraction F of the
original moisture remaining in the air mass at the same time
(Craig & Gordon 1965, Dansgaard 1964).
    For simplicity, advective water-vapour transport only has
been assumed, though turbulent mixing does play an important role
in global vapour circulation, particularly in the vertical and
meridional directions. The vertical eddy diffusion coefficients
D are $\sim 10^5$ cm^2 s^{-1}, whereas the horizontal ones are about
10^{10} cm^2 s^{-1} (Eriksson 1965).
    A prediction of the vertical distribution of D and ^{18}O in the
tropospheric moist layer can be essayed, assuming a mean moist
adiabatic temperature lapse rate of $-6°$C km^{-1} and a constant mean
relative humidity of 60% (i.e., h = 0.6) at all tropospheric levels

(Möller 1973). Then the specific humidity $q(z) = h(M/V)(p_s(T)/p_o)$ (g H_2O m^{-3}), where p_o = 1013 mb (atmospheric pressure at sea level), M = 18 (molecular weight of water) and V = 22.4 × 10^{-3} m^3 (the molecular volume), shows an exponential decrease with altitude z, since the vapour saturation pressure $p_s(T)$ [see Magnus formula (Möller 1973)] can be approximated by function $p_s(T)$ = 6.1 [exp T/T_o)] (mb) where the characteristic temperature T ≃ +15°C. If T_g denotes the air temperature at ground level, the vertical temperature profile can be expressed by the temperature lapse rate, $\Delta T/\Delta z$ ≃ -6°C km^{-1}, to be $T(z) = T_g + Z(\Delta T/\Delta z)$. This leads to the analytical expression for the specific humidity

$$q(z) = 4.8h[\exp(T_g/T_o)][\exp(-z/\bar{z})] \quad (g\ H_2O\ m^{-3}) \quad\quad (3)$$

where $\bar{z} = T_o(\Delta z/\Delta T)$ = 15/6 = 2.5 km which is the scale height of the atmospheric water vapour distribution, and the specific humidity near the ground at vapour saturation is q_g = 4.8 [exp(T_g/T_o]. The moisture decrease with altitude is expressed by the factor exp($-z/\bar{z}$). In the case of the moist adiabatic ascent of an otherwise discrete air mass, continuously losing the condensate by 'rainout' (one-phase system), the Rayleigh batch condensation formula, with $F = \exp(-z/\bar{z})$, becomes

$$R(z)/R_g = F^{\alpha_e - 1} = \exp[-(\alpha_e - 1)(z/\bar{z})] \quad\quad (4)$$

i.e., the isotope content of tropospheric water vapour is also expected to show an exponential decrease with altitude. The variation in the δ-values is thus

$$\delta(z) = (1 + \delta_g)\{\exp[-(\alpha_e - 1)(z/\bar{z})]-1\} \quad\quad (5)$$

Taking average values for (α_e-1) of 0.1 for D and 0.013 for ^{18}O, mean gradients of $\Delta\delta D/\Delta z$ = -40 ‰ km^{-1} and $\Delta\delta^{18}O/\Delta z$ = -5 ‰ km^{-1} are obtained for the vertical decrease of δD and δ^{18}O. In the case of turbulent vertical vapour transport, only half this decrease would be expected (Eriksson 1965). The experimental data for D in tropospheric water vapour (Figure 1), from Heidelberg (Germany), Scotts Bluff (Nebraska, USA) and the Pacific Ocean (off Santa Barbara, California), seem to support the advection model of vapour transport, whereas the δD-profile from Death Valley (California) tends towards the turbulent type as could be expected from the strong thermal convection in desert areas. Excluding the Death Valley data, these isotope profiles, which show an even higher gradient than that obtained from our simple advection model, have been interpreted by a 'multibox' cloud model (Rozanski & Sonntag 1982) taking into account the isotopic exchange between falling rain drops and ambient moisture (vapour with cloud droplets). This exchange acts like a continuous preferential advection of heavy water molecules from upper cloud levels causing a steepening of the isotope gradient. However, it is not possible at present to

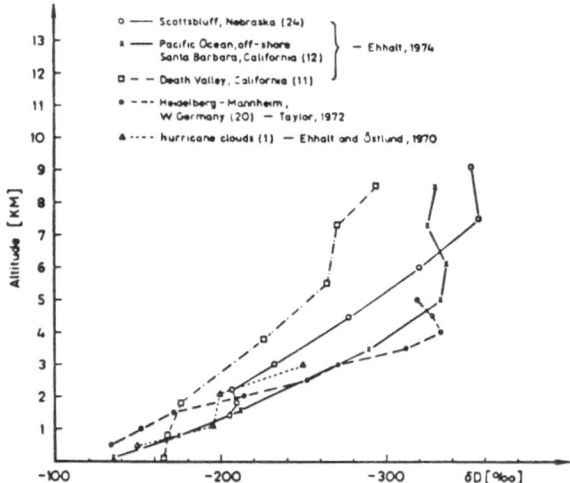

Fig. 1.   Vertical distribution of δD in tropospheric water vapour
          (from Rozanski & Sonntag 1982).

explain how these steep isotope profiles are maintained against
vertical eddy mixing during cloudless weather conditions.

    Isotope measurements of local water vapour and precipitation
at ground level approach isotopic equilibrium between vapour and
rain in most cases, i.e., $R_{prec} = \alpha_e R_{vap}$. However, from the
preceding discussion, the precipitation is expected to be in
isotopic equilibrium with the vapour at the cloud base, with an
isotope content significantly lower than near ground level. The
condensate derived from cloud-base vapour should be lower, therefore, in isotope content than liquid water in equilibrium with
average ground-level vapour. However, this difference is reduced
by the increase in the equilibrium-fractionation factor with
decreasing temperatures (see Table I) and, moreover, by molecular
exchange between the falling rain drops and atmospheric water
vapour, leading to an isotopic adjustment.

    Horizontal transport of water vapour from regions of high dew
point towards those with a lower one implies a decrease of precipitable water in the direction of flow, i.e., the moisture loss by
precipitation exceeds the supply by evaporation.

2.2   Isotopic Variations in Precipitation over the Oceans

    If a subtropical marine air mass is transported across the
ocean towards high latitudes, vapour loss by precipitation leads
to isotopic depletion of its residual moisture, a loss partly
compensated by the input of isotopically-heavier water vapour by
evaporation from the ocean surface crossed. Moreover, vapour

exchange through the ocean surface has to be considered. The
isotope content of vapour released from the ocean is defined by
the isotope ratio of ocean water and the total fractionation on
evaporation from the water surface $R_o = \alpha^{-1}R_{SMOW}$. Thus, it is not
related to the isotope content of the residual moisture of the air
mass, and the spatial variation of D and ^{18}O in atmospheric vapour
and precipitation cannot be simulated by the simple Rayleigh condensation model described above. The isotopic development of an air
mass passing over the ocean has to be estimated numerically and is
expected to be smaller than that predicted by the Rayleigh model.

2.3 Isotopic Variations in Precipitation over Land Masses

In contrast to the complex situation over the ocean, the
spatial variation of isotopes in continental precipitation and
groundwater can be satisfactorily described by a simple Rayleigh
condensation formula (see § 3). Over the continents, the isotope
content of water vapour (which is fed back into the atmosphere by
local evapotranspiration) is derived from local soil moisture and
thus from local rainfall, i.e., it is determined by the mean
isotopic composition of the local atmospheric moisture. The water
vapour released by evaporation from bare soils and by evaporation
of intercepted water is isotopically depleted in comparison to the
soil water (kinetic-fractionation factor α). However, the vapour
released by plant transpiration has the same isotope content as
the soil water. This relationship, although surprising at first
glance, becomes clear if the water transport from the soil via
plants into the atmosphere is considered in more detail. Soil
water, transported through plant roots and stems into the leaves,
is entirely converted into vapour at the leaf surface, i.e., the
amount of soil water passing through a plant is equal to the transpiration, and the same must be true for the isotope flow. In
view of isotope fractionation during evaporation from the leaf
water, this continuous process forces the leaf water to become
isotopically enriched with respect to the soil water, to the extent
that $R_{leaf} = \alpha R_{vap} = \alpha R_{soil}$. The isotopical enrichment of leaf
water has been verified experimentally (Gonfiantini $et\ al.$ 1968,
Förstel 1978).

3. A SIMPLIFIED ADVECTION MODEL OF THE VARIATIONS OF D AND ^{18}O IN PRECIPITATION AND GROUNDWATER

The present-day spatial variation of deuterium in precipitation across North America and in European groundwaters is shown in
Figures 2 and 3 in the form of δD-contour maps. The distribution
pattern for $\delta^{18}O$ is very similar to that for δD, since δD and $\delta^{18}O$
are linearly related (Figure 3b) by $\delta D = 8\delta^{18}O + 10\ ‰$, the so-
called 'meteoric water line' (Craig & Gordon 1965). The slope of
this line is mainly due to the ratio of the equilibrium fractionation of D and ^{18}O, respectively (see Table I). The 'deuterium

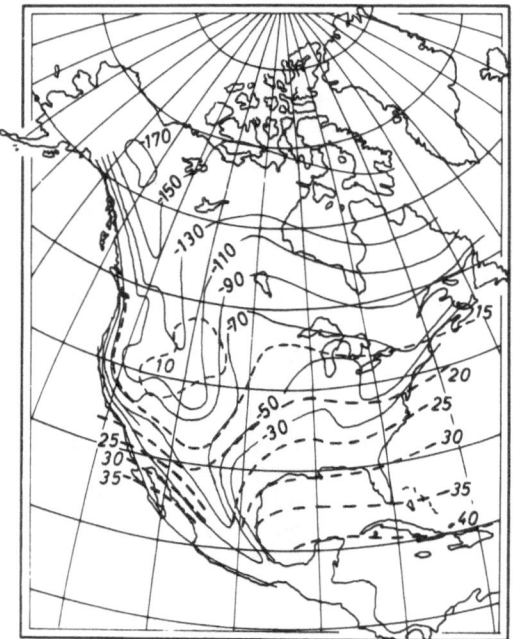

Fig. 2. Isolines of δD (from Yapp & Epstein 1977) and of precipi-
        table water W (mm) (from Reitan 1960) for North America:
        ——, δD-contours; ---, W-lines.

excess', d = $\delta D - 8\delta^{18}O$, is caused by kinetic fractionation during
evaporation over the ocean; it depends on the moisture deficit
(1-h) of the marine atmosphere, d = +10 ‰ corresponding to h = 78%
(Sonntag *et al.* 1978). In this context, it is interesting to note
that a value of d ≃ 5 ‰ has been found in Saharan groundwater
(Sonntag *et al.* 1978) and in Antarctic ice core sections, dated at
about 18 000 yr BP, from which a higher humidity (h ≃ 85%) of the
marine atmosphere during the last ice age has been inferred
(Merlivat & Jouzel 1979).

     From the above reasoning, it can be deduced that the direc-
tion of the vapour flow of the *precipitating* moist air masses is
perpendicular to the isotope (or precipitable water) contour lines,
from higher to lower δ-values. Roughly speaking, North America
receives rain from the moist air masses of the Gulf of Mexico
since moisture inflow from the Pacific Ocean is drastically
reduced by high rainfall over the Rockies, while Europe mainly
receives moisture from the Gulf Stream Drift (Figure 3a).

     The observed isotope pattern can be described by a simple
Rayleigh model linking the local isotope data to the parameters of
the horizontal water vapour flux. In this model, the vertically-
integrated vapour flux, $\bar{Q} = \int_0^\infty q(z)u(z)dz$ (where u(z) is the
horizontal wind vector), is expressed by the precipitable water

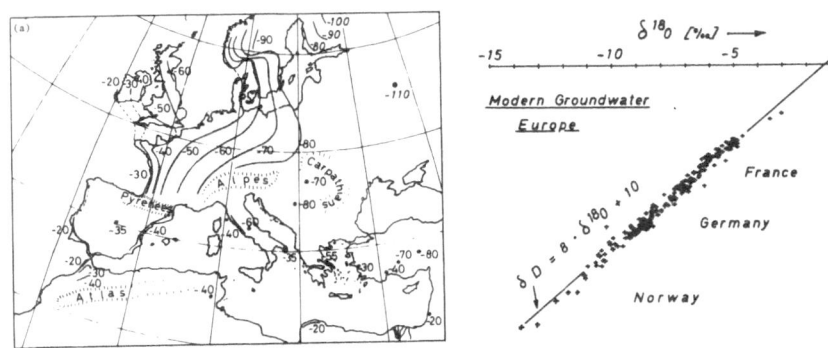

Fig. 3.   (a)   Spatial distribution of δD in present-day European
groundwaters. The data points (about 400) uniquely fall
into distinct isozones, the resolution across the isolines
(excluding the influence of altitude effects) may be not
much more than σ ≃ 50 km or about ±2 ‰ in δD.
(b)   Diagram comparing δD with δ^{18}O for present-day
European groundwaters. The spread around the regression
line is predominantly due to the existing analytical
precision of ±1 ‰ for deuterium.

$W = \int_{o}^{\infty} q(z)dz$, such that $\bar{Q} = \bar{u}W$, where the moisture advection
velocity u is the mean wind velocity weighted by the humidity q(z).
Assuming an exponential decrease of specific humidity q with
altitude as well as of its isotope content R, the precipitable
water W and its bulk isotope content R are determined exclusively
by the ground-level values of air temperature T_g, relative
humidity h and the isotope content of the water vapour R_g,
respectively. Thus:

$$W = hq_s(T_g)\bar{z} \tag{6}$$

and $R = R_g/\alpha$, where α is the mean isotope fractionation factor.
Making the assumption of time-independent water storage in the
atmosphere, $\partial W/\partial t = 0$, which is believed to hold for the monthly
means, then the continuity equations of the horizontal water-
vapour flux (Peixóto: this volume) and of the isotope flux read

$$\text{div}\bar{Q} = \bar{u}\, \text{grad}W = E-P \tag{7}$$

$$\text{div}(R\bar{Q}) = \bar{u}\, \text{grad}(RW) = \alpha_e R(E-P) \tag{8}$$

where E is the local evapotranspiration and P is the precipitation.
These equations are based on assumptions which may prove to be
over-simplified. The critical points are:
(i)   Eddy flow is neglected although it is known to play a dominant
role in meridional vapour flow. (This problem is discussed below.);
(ii)   At any time and location, an exponential decrease of

specific humidity and of the isotope content of precipitation with
altitude is assumed, as well as isotopic equilibrium between local
rainfall and water vapour at ground level (see above);
(iii) A mean vapour-advection velocity independent of location
also implies a mean wind profile independent of location;
(iv) All vapour-flow parameters are considered to be mean values
(monthly, seasonal or annual means), while the precipitation P
only represents mean values during periods of precipitation;
(v) The isotope content of the vapour produced by evaporation is
assumed to be identical to that released by plant transpiration,
which has been found to transfer soil moisture into the atmosphere
without isotope fractionation. However, water released from bare
soils yields vapour of lower isotope content owing to isotopic
fractionation during evaporation. The isotope feedback $\alpha_e RE$,
therefore, is overestimated in our model, particularly during
periods of reduced transpiration in winter. However, in the winter
of temperate zones (with which this study is concerned), P greatly
exceeds E, so that the isotope flux divergence $\alpha_e R(E-P)$ is nevertheless a reasonable estimate. In warm seasons, E and P are of
similar magnitude (i.e., E-P is a small number), but then the
transpiration plays the dominant role in E.

The argument can be simplified by considering just the
component Q_n of the total horizontal water-vapour flux $\bar{Q}$ in the
direction of gradW (the component of the advection velocity $\bar{u}$ in
this direction is denoted by u_n). Elaboration of the differential
operations in Equation (8) and utilizing Equation (7) yields

$$dR/R = (\alpha_e-1)(dQ_n/Q_n) = (\alpha_e-1)(dW/W) \tag{9a}$$

$$d(\ln R) = (\alpha_e-1)d(\ln Q_n) = (\alpha_e-1)d(\ln W) \tag{9b}$$

Integration of this differential equation leads to a Rayleigh condensation formula in terms of either the horizontal vapour flux
or the precipitable water

$$R/R_o = (Q_n/Q_{n,o})^{(\alpha_e-1)} = (W/W_o)^{(\alpha_e-1)} = F^{(\alpha_e-1)} \tag{9c}$$

where $Q_{n,o}$, W_o and R_o are the initial data at the vapour source
area, somewhere over the subtropical ocean.

The initial isotope content over the ocean is $R_o = \alpha^{-1}R_{SMOW}$,
and Equation (9c) in δ-notation reads

$$\delta_{vap} = \alpha^{-1}(F^{(\alpha_e-1)}-1) \tag{10}$$

or

$$\delta_{prec} = (\alpha_e/\alpha)(F^{(\alpha_e-1)}-1) \simeq F^{(\alpha_e-1)}-1 \tag{10a}$$

This simple vapour-advection model can be applied to the isotope
pattern for North America and Europe. As expected from the model,
the contours of δD for the mean annual precipitation of North
America (Yapp & Epstein 1977) are similar (see Figure 2) to the

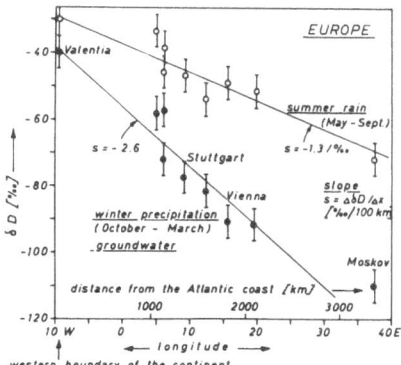

Fig. 4.   Zonal variation of δD in present-day European ground-
          waters and in winter and summer precipitation (data from
          IAEA 1969-1979).

contours of the mean annual precipitable water (Reitan 1960). The
change of δD from one isoline to the other can be described by the
relative change of W using Equation (9a) and $(\alpha_e-1) \approx 0.1$ for
deuterium; thus

$$\Delta\delta D/(1+\delta D) = (\alpha_e-1)(\Delta W/W) \tag{11}$$

In Europe, the δD-pattern for present-day groundwater (Sonntag et
al. 1978) follows the contours of mean W during the cold season
between October and March (not shown explicitly in Figure 3a).
The zonal variations of δD in groundwater and in winter- and
summer-precipitation presented in Figure 4 show that groundwater
is largely derived from winter precipitation, while summer rain
displays a much smaller decrease of δD with distance inland.
Groundwater recharge by winter rains is also supported by regional
water balances. Figure 4 suggests a significant seasonal variation in the pattern of precipitation for heavy stable isotopes, as
is also expected from the vapour-flow model. High relative
vapour-flux divergence (E-P)/Q_n in winter (P > E) implies strong
inland gradients of δD and δ^{18}O, whereas small relative divergence
in summer (P≈E) leads to small gradients; this seasonal variation
in the isotope pattern has been simulated with a numerical model
(Rozanski et al. 1982). Here, an attempt is made to interpret the
observed west-to-east decrease of δD in European groundwaters by
means of hydrometeorological data. According to Equations (9a
and b), the relative vapour-flux divergence equals the gradient
of the δ-values in the direction of flow dδ/dn divided by the
isotope fractionation (α_e-1):

$$(dQ_n/dn)/Q_n = (E-P)/Q_n = (d\delta/dn)/(\alpha_e-1) \tag{12}$$

P and E are known from precipitation records and from local (or regional) water balances, whereas the local horizontal vapour flow is not well known. Making use of Equation (9c), the isotope data yield the local flux divergence as a fraction of the original horizontal flow $Q_{n,o}$, thus:

$$(E-P)/Q_{n,o} = (d\delta/dn)/(\alpha_e - 1)[(1+\delta)^{1/(\alpha_e - 1) - 1}] \qquad (13)$$

The initial vapour flux $Q_{n,o}$ may be obtained from a moisture balance for the marine atmosphere as the water vapour flows from the oceanic area into the continental atmosphere. Figure 5 shows the zonal variation of the relative vapour-flux divergence esti-mated from hydrometeorological data together with the same parameter obtained from isotope data, according to the zonal rain-out model (Equation 13). The remaining vapour fraction $F(n) = W(n)/W_o = (1 + \delta)^{1/(\alpha_e - 1)}$ is also represented in this figure. E and P represent mean values of the local evapotranspiration and precipitation over the period October to March (Rozanski et al. 1982). For $Q_{n,o}$, published vapour-flow data (Korzun 1974) were used; W(n) was estimated from local mean temperature and relative humidity data (CLINO 1962) using the water-vapour scale height z = 2.5 km (see above).

As can be seen from Figure 5, the data points obtained from the hydrometeorological and isotope data agree up to about 15°E. The increasing discrepancy further east might be explained by an increasing contribution from the meridional vapour-flow component which supplies moisture from the Mediterranean area for winter precipitation in eastern Europe. The departure of the hydro-meteorological data points from the isotope data points is in the expected direction in this case. The additional moisture supply leads to higher precipitation and thus to a higher $(P-E)/Q_{n,o}$ ratio than predicted by the one-dimensional zonal rain-out model. Moreover, the higher original isotope content of meridionally-transported vapour yields a departure of $(d\delta/dn)(1 + \delta)^{1/(\alpha_e - 1) - 1}$ of opposite sign. Therefore, in Figure 5, both sides of Equation (13) depart from the predicted straight line in opposite direc-tions if the observed data are inserted. The interpretation of the discrepancy in the relative vapour-flow divergence is sup-ported by map data for the zonal and meridional horizontal vapour flow (Korzun 1974). The increasing discrepancy between the F data $(Q_n/Q_{n,o})$ across eastern Europe is also consistent with this interpretation in so far as concerns the F values obtained from the isotope data. However, an additional (meridional) moisture supply suggests more precipitable water than that estimated using average local winter temperatures. This discrepancy might be due to assumption (iv) (see above) for the vapour advection model. It can be explained by assuming that local temperatures on days of precipitation are considerably higher than the mean winter tempera-tures used in the estimate of W(n) for eastern Europe. In connection with this interpretation of the isotope variation in

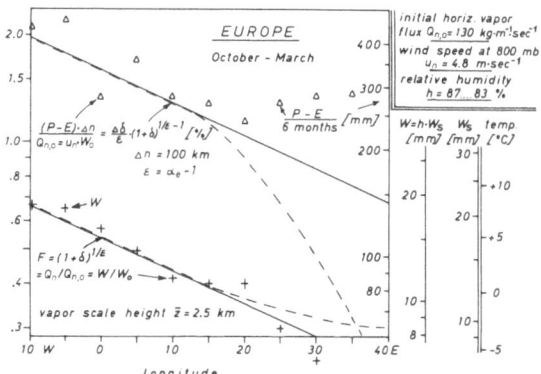

Fig. 5. Zonal variation of the relative vapour-flux divergence
(in % per 100 km flow distance) and of the residual
vapour fraction of the air masses (left-hand logarithmic
vertical scale) over Europe during the winter season:
—— (upper), prediction of the zonal rain-out model
(right-hand side of Equation 13) compared with the values
derived from observed hydrometeorological data (Δ points
= $(P-E)\Delta n/Q_{n,o}$) and from isotope data: broken line =
$\Delta\delta/\alpha_e - 1)[(1+\delta)^{1/(\alpha_e-1)-1}]$, respectively; +, vapour fraction $F = W/W_o$; --- (lower), $F = (1+\delta)^{1/(\alpha_e-1)}$. Right-
hand logarithmic vertical scales represent the winter
infiltration P-E (mm) over the six months from October to
March, precipitable water (mm) and the local mean winter
temperature ($^\circ$C).

European waters, an interesting climatological phenomenon emerged.
Western and central Europe, where this model works, correspond to
Köppen's C_f-climate zone (Lamb 1972), whereas eastern Europe,
beyond 15°E longitude, falls into the D_f-zone, and the deuterium
content of the winter precipitation falls below $\delta D = -80$ ‰. In
the latter zone, the moisture supply is low because of the preceding precipitation across the C_f-zone. In other words, in order
to allow for the precipitation actually observed, water vapour of
different origin is needed, in this case, meridionally-transported
vapour. In the case of North America, this δD margin again
approximately coincides with the margin between the C_f-zone in the
south and the D_f-zone in the north. As with eastern Europe, the
zonal vapour component comes into play, but in the reverse
direction.

The simple vapour advection model also yields an estimate of
the variation of D and ^{18}O in precipitation with the local
temperature, assuming the ocean temperature is unchanged. Equation
(9a, b) can be written as $d\delta/dT \simeq (\alpha_e-1)[(dW/dT)/W]$ from which

$$d\delta/dT \simeq (\alpha_e - 1)/T_o \tag{14}$$

is obtained assuming an exponential temperature dependence of the local precipitable water $W(T) = W_o \exp(T/T_o), T_o \simeq +15°C$ (see above). With $(\alpha_e - 1) = 0.1$ for deuterium, a temperature gradient of $\delta D/dT \simeq 6.7 ‰ °C^{-1}$ is obtained, which is fairly similar to the empirical relation between the annual mean values of δD (and $\delta^{18}O$) in precipitation and of the local temperature (Dansgaard 1964). This temperature aspect is important, if D and ^{18}O are to be used as palaeotemperature indicators.

4.   DAILY VARIATION OF D AND ^{18}O IN ATMOSPHERIC WATER VAPOUR AT HEIDELBERG

The Rayleigh condensation model of the spatial variation of D and ^{18}O in precipitation and groundwater, which yields the isotopic composition of the residual moisture content of discrete air masses, suggests the existence of strong short-term isotopic variations in local atmospheric water vapour due to the varying influence of different air masses on actual weather conditions and to the contribution of regional evapotranspiration to local water vapour. To some extent, these variations are reflected by the isotope data for individual rainfall events. A systematic study of D and ^{18}O in atmospheric water vapour in relation to local weather conditions was tried for the first time a few years ago (Hübner *et al.* 1979). Encouraged by this study, the present authors established complete time series of D and ^{18}O in atmospheric water vapour at Heidelberg (daily means) between June 1980 and April 1981 (Figure 6).

The monthly means of the isotope data show the seasonal variation previously observed in monthly precipitation data. Correlations of the short-term variations with specific humidity and wind direction are poor. This suggests that classification of the isotope data according to the various air masses controlling local weather will not be as straightforward as might have been expected. Therefore, it is intended to compare a Rayleigh model prediction of D and ^{18}O in the atmospheric moisture at Heidelberg (based on daily meteorological data) with the observed isotope record. To do this, air-mass trajectories over the Atlantic Ocean and Europe (estimates of which can be obtained from daily weather charts) are required. A frequency analysis of the δD series (after subtraction of the seasonal variation) has been encouraging. As can be seen from the power spectrum in Figure 6, the δD time series shows, in addition to the seasonal variation, a predominant periodicity of about 22 days, seemingly caused by the phase velocity of the planetary long waves at the 500 mb level. The troughs and ridges of these waves evidently control the water vapour transport within the moist layer below 500 mb.

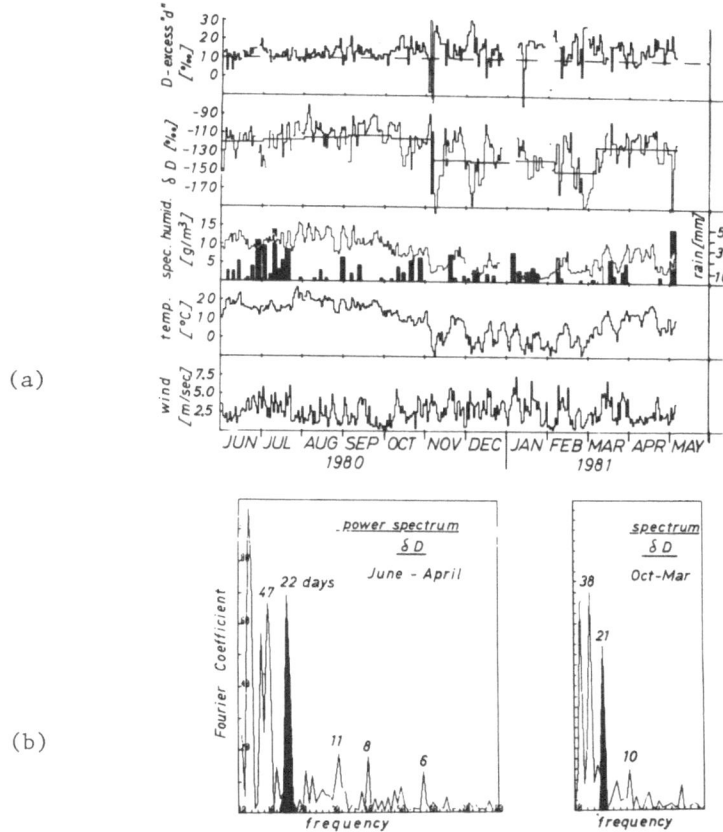

Fig. 6.   (a) Time series of D and ^{18}O (δ^{18}O implicitly presented
by the deuterium excess, d = δD-8δ^{18}O) in daily mean
water vapour at Heidelberg together with relevant hydrometeorological data;
(b) Power spectrum of the δD series after removal of the
seasonal component.

5.   CONCLUDING REMARKS

A synoptic study of the daily variations in D and ^{18}O in
atmospheric water vapour to be undertaken at selected locations in
western and central Europe, e.g., Paris, Hannover, Heidelberg,
München and Cracow, can be expected to yield considerably more
information on the relation between these isotopes and the water-
vapour circulation than has been available so far. Although such
studies are unlikely to bring new insights on present-day water
vapour circulation (since the available hydrometeorological and

aerological data are far superior in this respect), a better understanding of the extent to which the spatial variation of D and ^{18}O in precipitation reflects the actual vapour circulation is urgently needed for palaeoclimatic research using these isotopes. The relationship between horizontal vapour-flux divergence and local isotope data obtained by the simple vapour-advection model is a step in this direction. The isotope time series for atmospheric water vapour should be of particular relevance to climatic research based on isotopes in tree rings. In addition to the water-vapour record, a time series of D and ^{18}O in leaf water (to which the isotope content of lignin is in turn related) has been collected. Owing to fast molecular exchange between atmospheric moisture and leaf water (equilibrium being achieved within a few hours), the isotope variations in leaf water closely follow those in atmospheric vapour. Consequently, the isotope content of tree rings depends on the isotopic composition of the local soil water - which contains both summer and winter precipitation, and of the atmospheric water vapour during the growing season. These parameters depend on local temperatures, so that the temperature dependence of the isotopes in tree rings is fairly complex, even with the assumption that global vapour circulation patterns have remained constant during different climatic periods.

Finally, as the effect of turbulent mixing on horizontal vapour flow is omitted from the isotope model here presented, it is relevant to comment on this type of mass transport in the context of the model. The meridional water-vapour flux is known to be controlled almost exclusively by stationary and by transient eddies (Peixóto 1973). In this case, Equation (9a, b) would read (Eriksson 1965):

$$d(\ell nR) = [(\alpha_e)^{-\frac{1}{2}}-1]d(\ell nW) \approx \frac{1}{2}(\alpha_e-1)d(\ell nW), \text{ rather than}$$

$$d(\ell nR) = (\alpha_e-1)d(\ell nW)$$

However, the observed meridional variation of δD across North America (Figure 2) seems to support the advection rather than the turbulent flow model. It seems that this apparent discrepancy is due to the extremely large mixing length $\ell = 2D/|v_x|$ (D=horizontal eddy coefficient, $|v_x|$ = average wind component w.r.t. in direction x). Inserting reasonable values for D and $|v_x|$ yields a value of ℓ of $\sim$ 1000-2000 km (pressure cell diameter). Within this distance, therefore, air masses undergo isotope variations according to the advective Rayleigh depletion model as discussed in the present paper.

ACKNOWLEDGEMENT. This investigation was supported by the Heidelberg Akademie der Wissenschaften.

REFERENCES

CLINO 1962 Climatological normals for climat and climat ship
    stations for the period 1931-1960. *WMO Rept* No. 117 (TP 52).

Craig, H. & Gordon, L.I. 1965 Deuterium and oxygen-18 variations
    in the ocean and the marine atmosphere. In: *Stable Isotopes
    in Oceanographic Studies and Paleotemperatures* (ed.
    E. Tongiorgi). Spoleto: Conferences in Nuclear Geology.

Dansgaard, W. 1964 Stable isotopes in precipitation. *Tellus,* 16,
    436-468.

Eriksson, E. 1965 Deuterium and oxygen-18 in precipitation and
    other natural waters. Some theoretical considerations.
    Tellus, 17, 498-512.

Förstel, H. 1978 The enrichment of ^{18}O in leaf water under natural
    conditions. *Rad. Environ. Biophys.,* 15, 323-344.

Gat, J. 1980 The isotopes of hydrogen and oxygen in precipitation.
    In: *Handbook of Environmental Isotope Geochemistry,* Vol. 1:
    The Terrestrial Environment (ed. P. Fritz & J.C. Fontes), A,
    pp. 21-47. Amsterdam, Oxford, New York: Elsevier.

Gonfiantini, R., Gratziu, S. & Tongiorgi, E. 1968 Oxygen isotope
    composition of water in leaves. *IAEA Proc. Ser.,* 108,
    405-410.

Holloway, J.L., Jr & Manabe, S. 1971 Simulation of climate by a
    global general circulation model. *Mon. Weath. Rev.,* 99,
    335-370.

Hübner, H., Kowski, P., Hermichen, W.D., Richter, W. & Schütze, H.
    1978 Regional and temporal variations of deuterium in preci-
    pitation and atmospheric moisture of Central Europe. In:
    Isotope Hydrology 1978, Vol. I, pp. 289-310. Vienna: IAEA.

IAEA 1969, 1970, 1971, 1973, 1975, 1979 Environmental Isotope
    Data: World survey of isotope concentrations in precipita-
    tion. *IAEA Tech. Rept Ser.* Nos 69, 117, 129, 147, 165, 192.
    Vienna: IAEA.

Korzun, V.I. (ed.) 1974 *World Water Balance and Water Resources of
    the Earth.* Report of the USSR Committee for the IHD, 663 pp.
    Leningrad: Gidromet. Izdatel. (English translation 1977-78:
    Studies and Reports in Hydrology, No. 25. Paris: UNESCO.)

Lamb, H.H. 1972 *Climate: Present, Past and Future,* Vol. I.
    London: Methuen.

Liljequist, G.H. & Cehak, K. 1979 *Allgemeine Meteorologie*.
Braunschweig, Wiesbaden: Vieweg.

Majoube, M. 1971 Fractionnement en oxygène-18 et en deutérium
entre l'eau et sa vapeur. *J. chim. Phys.*, 10, 1423.

Manabe, S. & Holloway, J.L., Jr 1975 The seasonal variation of the
hydrologic cycle as simulated by a global model of the
atmosphere. *J. geophys. Res.*, 80, 1617-1649.

Merlivat, L. & Coantic, M. 1975 Study of mass transfer at the
air-water interface by an isotope method. *J. geophys. Res.*,
80, 3455.

Merlivat, L. & Jouzel, J. 1979 Global climatic interpretation of
the deuterium-oxygen 18 relationship for precipitation.
J. geophys. Res., 84, 5029-5033.

Möller, F. 1973 *Einfuhrüng in die Meteorologie*, Vol. 1. Mannheim,
Wien, Zürich: Bibliographisches Institut (Hochschultaschen-
bücher, Vol. 276).

Peixóto, J.P. 1973 Atmospheric vapour flux computations for
hydrological purposes. *WMO Publ.* No. 357, 83 pp.

Reitan, C.H. 1960 Distribution of precipitable water vapour over
the continental United States. *Bull. Am. met. Soc.*, 41,
79-87.

Rozanski, K. & Sonntag, C. 1982 Vertical distribution of deuterium
in atmospheric water vapour. *Tellus*. [In the press.]

Rozanski, K., Sonntag, C. & Münnich, K.O. 1982 Factors controlling
stable isotope composition of modern European precipitation.
Tellus. [In the press.]

Sonntag, C., Klitzsch, E., Löhnert, E.P., El-Shazly, E.M.,
Münnich, K.O., Junghans, Ch., Thorweihe, U., Weistroffer, K.
& Swailem, F.M. 1978 Palaeoclimatic information from
deuterium and oxygen-18 in carbon-14-dated North Saharan
groundwaters. In: *Isotope Hydrology 1978*, Vol. II, pp. 569-
581. Vienna: IAEA.

Yapp, C.J. & Epstein, S. 1977 Climatic implications of D/H ratios
of meteoric water over North America (9.500-22.000 B.P.) as
inferred from ancient wood cellulose C-H hydrogen. *Earth
planet. Sci. Lett.*, 34, 333-350.

Techniques of Measurement and Analysis:
II. Surface Processes

INTRODUCTION TO TECHNIQUES OF MEASUREMENT AND ANALYSIS:
SURFACE PROCESSES

M.A. Beran   F.Alayne Street-Perrott

Institute of Hydrology   School of Geography
Wallingford   University of Oxford
Oxfordshire, England   England

Land-surface processes stand in a very special relation to
the study of the global water budget. Quite apart from their role
(which varies in importance with the time and space scale of
study) within the overall water cycle, they have a major impact
on Man's activities. Apart from the obvious harm caused by
extremes - unhappily little studied in the water-budget context -
even the average pattern of availability of water in its solid,
liquid and gaseous phases limits human settlement, agriculture
and industry, leading thereby to the expenditure of energy and
resources (as well as ingenuity) in endeavours to circumvent these
natural constraints.

The practical benefits of a good understanding of surface
processes are particularly apparent in a number of vital areas.
First, in agronomy, the aim is to optimize biomass production,
subject to the constraints set by soil moisture supply, the
evaporative demand of the atmosphere and plant physiology.
Secondly, in the domain of water supply, the amount and temporal
availability of water need to be matched to the pattern of demand,
through management of surface and subsurface sources. Thirdly,
the fields of navigation, power production and effluent disposal
require a thorough comprehension of surface processes.

In the scientific study of land-surface processes, one ought,
in principle, to be able to include the near-surface pathways of
water as elements in an expanded global circulation model (GCM).
Reference is made elsewhere (in the section on modelling and
prediction) to some of the problems of achieving this explicitly.
Němec in his paper underlines these difficulties by referring to
the different traditions of climatic and hydrological modellers;
in particular, the tendency of the former to regard runoff as a
closing term, i.e., a sink for rainfall excess after soil moisture

A. Street-Perrott et al. (eds.), Variations in the Global Water Budget, 126–128.
Copyright © 1983 by D. Reidel Publishing Company.

deficit has been satisfied, is irreconcilable with the approach usually adopted in hydrology.

Of course, hydrologists are often guilty of finding convenient sinks to effect closure of their models. Often, the evaporation term is used in this way. Shuttleworth provides a systematic review of the efforts that have been made to estimate evaporation from more fundamental climatic measurements.

In our ability to assess evaporation - potential or actual - at a global or regional scale, this element of the near-surface water budget lags considerably behind the other elements. Detailed maps, such as those provided by Jaeger in his paper on precipitation patterns, are simply not possible for evaporation at present. Even for rainfall, which is a comparatively well measured and understood variable, the lack of data over the oceans necessitated some gross assumptions about global evaporation, based on a consideration of energy inputs from the Sun. This contribution from atmospheric physics to the resolution of the debate about a phenomenon as Earth-bound as evapotranspiration reminds us that scientific advance may well come through unifying principles derived from ostensibly well-separated academic disciplines.

A practical example of interdisciplinary cooperation is provided by Ruprecht et al. Considerable progress is currently being made in the estimation of rainfall rates over the oceans using satellite-borne microwave radiometers. Encouraging agreement was found between rainfall estimates derived from Nimbus 5 observations and data collected by ship-mounted radars during the GARP Atlantic Tropical Experiment (GATE).

The papers highlighted above should be read in conjunction with those on modelling methods which also address the near-ground phase of the hydrological cycle and, likewise, deal with the status quo. The vast literature on climatic variability, however, has alerted theoretical hydrologists as well as practising water-resource managers to the likelihood that future conditions will differ markedly from those experienced in the past. This, in turn, has aroused interest in secular and long-term climatic fluctuations and their possible impact on the amounts, temporal distribution and pathways of water within the hydrological cycle. As yet, there are few places in the world where the data record is sufficient to allow the consequences of climatic variability to be studied directly. Nevertheless, Liebscher's analysis of long series of Rhine river flows is a contribution in this direction.

In many mountain areas, the best historical evidence for variations in water balance is provided by the fluctuations of alpine glaciers. Unhappily, the response of glacier termini to changes in accumulation and ablation (mass loss by melting, evaporation or calving) is complicated by glacier dynamics. Two papers in this section discuss new methods of determining the past and present 'state of health' of valley glaciers. In areas where

long series of mass-balance measurements are available, it is
possible to evaluate the temporal and spatial variability of net
accumulation using multivariate statistical techniques. Reynaud
summarizes the correlations between climatic parameters and the
net mass balance of glaciers in the Alps, and demonstrates the
regional nature of mass-balance variations. In contrast,
Bhandari et al. employ new geophysical dating techniques to
investigate the dynamics of glaciers in the Himalayas, where
records of mass balance are scarce. They also discuss the implications of oxygen-isotope and dust measurements carried out on ice
cores. As a result of isotopic fractionation during the evaporation, transport and precipitation of water vapour (Sonntag et al.:
this volume), the isotopic composition of glacier ice varies
across the Himalayas. The results of Bhandari et al. are supported
by recent Chinese and Japanese data which have revealed significant
differences between the stable isotope contents of Himalayan
summer and winter precipitation. This raises the interesting
possibility that ice-core data may eventually be of assistance in
determining secular changes in atmospheric moisture transport over
glacierized regions.

   One aspect of climate prediction which is receiving
considerable attention is the effect of rising levels of atmospheric CO_2 on the global water balance. Miller examines the
water budget of Arctic and boreal forest ecosystems in which the
storage of water, in the biomass, organic layer and mineral soil,
is a particularly important element. In a novel approach, the
water contents of the various ecosystem components were estimated
from the extensive data on carbon contents collected during the
International Biological Programme. Miller suggests that a CO_2-
induced climatic warming would result, paradoxically, in an
increase in water storage in the Arctic permafrost, due to
greater insulation of the ground by moisture-loving mosses. This
example illustrates the complex non-linear interactions within
the biosphere and cryosphere that govern the response of the
surface branch of the hydrological cycle to climatic change.

MONTHLY AND AREAL PATTERNS OF MEAN GLOBAL PRECIPITATION

Lutz Jaeger

Department of Meteorology
Freiburg University
Federal Republic of Germany

ABSTRACT.   After summarizing the history of the cartographic
representation of global and continental-scale precipitation, and
demonstrating in so doing the gradual refinements used in deducing
annual and seasonal variations of rainfall over large areas, a
presentation is made of a gridding technique for evaluating precipitation over the continents and the oceans based on the sub-
division of the Earth's surface into nearly 2600 sectors and data
obtained prior to 1965. The results of this analysis, comprising
the record of the monthly meridional distribution of rainfall,
and maps of mean precipitation - variously ordered to show the
monthly and annual variations in respect to latitudinal zones as
well as land and sea surfaces, are presented and described.

1.   REPRESENTATION OF GLOBAL PRECIPITATION DISTRIBUTION AND WATER BALANCE

A Scotsman, E. Loomis, was the first to investigate rainfall
or precipitation on a global scale. His map (Loomis 1882) was
based on observations of rainfall depth over continental regions
and ship-borne observations of rainfall frequencies over the ocean.
Supan's (1898) map was the basis of further investigations into
global precipitation mapping and also in computing of global water
balances.
The latter map, redrawn for ease of comparison to the Mercator-
Samson cylindrical projection (Figure 1), shows rainfall estimates
for the Atlantic and southern Indian Oceans though not for the
Pacific, Caribbean, Baltic and western Mediterranean sea areas nor
for the polar regions.
Herbertson (1908), Jefferson (1926) and Ekhart (1930) designed

A. Street-Perrott et al. (eds.), Variations in the Global Water Budget, 129–140.
Copyright © 1983 by D. Reidel Publishing Company.

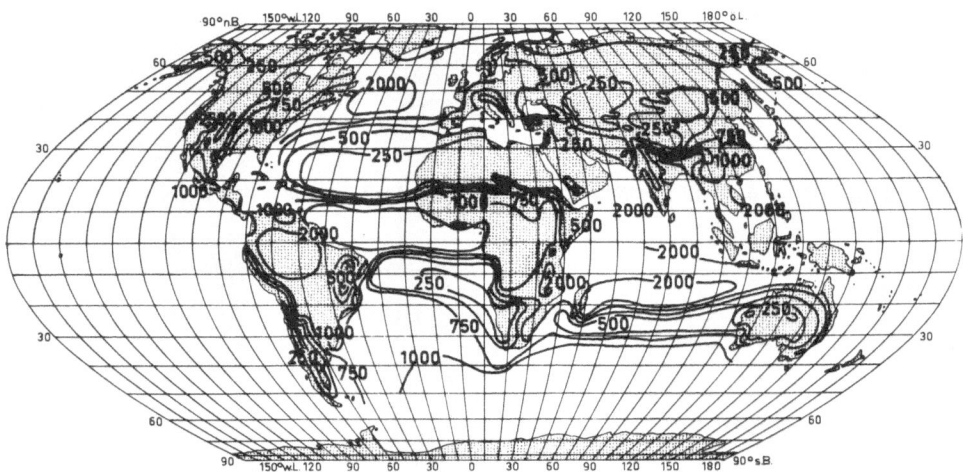

Fig. 1.   Mean annual precipitation (mm) over the globe, after
          Supan (1898).

maps of global precipitation which utilized the growing number of
maritime rainfall observations from ships and oceanic islands.
Ekhart's work was notable as being the first to use a standard
recording period – in his case, 1911-1920. Schott (1926, 1935)
paid special attention to the oceanic regions and his views influenced Meinardus (1934) whose map is one of the most important
in the German literature. Möller's (1951) maps of the seasonal
precipitation distribution are also important as they are often
used in verification of global precipitation modelling. A list of
of important global precipitation maps is given in Table I.
        The history of attempts to balance the elements of the hydro-
logical cycle predates the history of cartographic representation
of these elements. Of course, the appearance of global rainfall
maps stimulated studies of the global balance especially as map
originators (e.g., Schott and Meinardus) often added balances.
A survey of the development of terms used in water balance compilations was presented by Baumgartner and Reichel (1971); Table II
is an updating of their table of terms. Using Murray's (1887)
continental estimates, Brückner (1905) was the first to evaluate
all the terms which he separated into continental and oceanic
fractions (Table II). Brückner discussed extensively the problem
of marine evaporation which he determined from data from points
close to the sea and large lakes, a 5% reduction being made to allow
for the effect of brine. He estimated the error in continental
precipitation to be 20% and in runoff to be 50%.
        An entirely new estimate was given by Brooks and Hunt (1930)
based upon a study of the zonal distribution of annual precipitation. Their technique was to merge rainfall maps for every country

Table I. Global precipitation maps

Author	Year of publication	Spatial cover	Period
Loomis	1882	Globe	
Supan	1898	Globe	
Herbertson	1908	Continents	
Schott	1926	Atlantic Ocean	
Jefferson	1926	Continents	
Ekhart	1930	Globe	
Schott	1932	Indopacific Ocean	
Meinardus	1934	Globe	
McDonald	1938		Seasonal
Jacobs	1951	Oceans	Annual & Seasonal
Albrecht	1951	Indopacific Ocean	Annual & Monthly
Möller	1951	Globe	Seasonal
Drozdov	1953	Globe	
Albrecht	1960	Oceans	
L'vovich & Ovtchinnikov	1964	Continents	
Geiger	1965	Globe	
Jacobs	1968	Oceans	Seasonal
Baumgartner & Reichel	1971	Globe	
Jaeger	1975	Globe	Annual & Monthly

into continental maps (though data for the Antarctic continent were
almost totally lacking). The maps were divided into 1° squares in
order to compute zonal and global averages over the land area. For
the ocean areas, Brooks and Hunt extracted the annual rainfall from
Schott's (1926) map for the Atlantic and from Hann and Süring's
(1936) version for the Pacific and Indian oceans. Their results
were presented as tables of mean monthly meridional distribution
over land, ocean and whole Earth as well as isopleths showing the
annual variation of zonal rainfall over the globe (Figure 2).

Meinardus (1934) capitalized on these studies to produce a
fresh evaluation of 5°-meridional rainfall means and their annual
variations. Finally, mention must be made of Möller's (1951)
calculation of seasonal water balances from maps of 3-monthly
rainfall means.

2. EVALUATION OF PRECIPITATION OVER THE CONTINENTS

In order to calculate monthly, seasonal and annual variations
of precipitation over the whole globe, both hemispheres and various

Table II. Values of global water balance terms, after Baumgartner and Reichel (1971). Values of P_L, E_L, D_L, P_S, E_S, $\Sigma P=\Sigma E$ in 10^3 km^3, and of $P_E=E_E$ in mm. Areas: continents, 148.9×10^6 km^2; oceans, 361.1×10^6 km^2; globe, 510×10^6 km^2.

Author	Year	P_L	E_L	D_L	P_S	E_S	$\Sigma P=\Sigma E$	$P_E=E_E$
Johnson	?			56				
Black	1864/81				368			
Reclus	1883			28				
Woeikof	1886	78	61	17				
Murray	1887	122	97	25				
Bezdek	1904	121						
Brückner	1905	122	97	25	359	384	481	940
Fritzsche	1906	112	81	31	353	384	465	910
Lütgens	1911					506		
Schmidt	1915	112	81	31	242	273	354	690
v. Kerner	1919				363			
Wüst	1922	112	75	37	267	304	379	743
Kaminsky	1925	81	51	30	307	337	388	760
Ekhart	1930						495	970
Brooks & Hunt	1930	99			398		497	975
Cherubim	1931	112	75	37	334	371	446	880
Meinardus	1934	99	62	37	412	449	511	1002
Halbfass	1934	100	52	48	410	458	510	1000
Wüst	1936	99	62	37	297	334	396	780
Mosby	1936					383		
Wundt	1938	99	62	37	346	383	445	880
L'vovich	1945	107	71	36	412	448	519	1020
Albrecht	1949						393	770
Möller	1951	99	62	37	324	361	423	832
Reichel	1952	100	70	30	315	345	415	810
Wüst	1954	100	73	27	324	351	424	830
Budyko	1955	100	66	34/38	370	408	470/474	930
Albrecht	1960	100	67	33	378	411	478	940
Budyko	1963	107	61	46/48	404	452	512	1000
Marcinek	1964			36				
Mira Atlas	1964	108	72	36	412	448	520	1020
Nace	1968	100	69	31	319	350	419	820
Kessler	1969	100	60	40	410	450	510	1000
L'vovich	1969	109	72	37	411	448	520	1020
Mather	1970	106	69	37	382	419	488	955
Budyko	1970	107	64	43	412	455	519	1020
Baumgartner and Reichel's evaluations								
from Jacobs	1951		65			382	447	875
from Privett	1960		65			428	493	965
from Albrecht	1960	100	67	33	385	418	485	950
from Knoch	1961	100			396		496	970
Baumgartner	1970	100	65	35	383	418	483	950
Jaeger	1975	113			397		510	1000

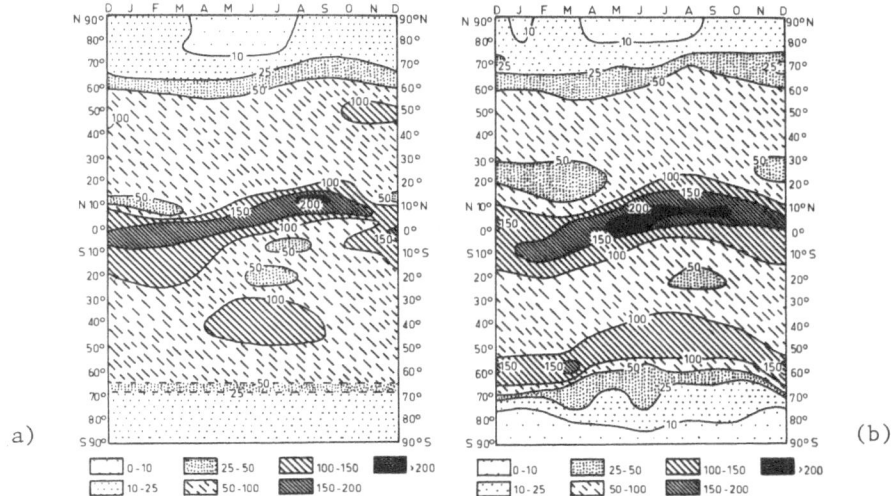

a)   (b)

Fig. 2.   Mean annual precipitation (mm) over the globe, after
           (a) Brooks and Hunt (1930); (b) Jaeger (1976).

meridional zones, a gridding technique was used on data spanning
1931 to 1960 over the continents, and 1955 to 1965 over the oceans.
The steps for assembling the data and adjusting them for cartographic representation are outlined by the block diagram in
Figure 3. Having divided the Earth's surface into 2592 sectors,
each containing a grid point at its centre, an assessment was made
of the proportion of land and sea in each sector, so that the total
area of land or sea within the meridional zones could be calculated.

For the continental regions, the grid point values were
obtained as eye estimates from isopleth maps prepared from up-to-
date climatic atlases containing annual and monthly rainfall
values, supplemented by other data sets. Although it was
initially intended to use data for the standard period 1931-1960,
this did not prove possible for all regions. It appears that the
attainment of a homogeneous global climatic data set is still a
distant goal.

3.   GRID POINT VALUES OF PRECIPITATION OVER THE OCEANS

Möller's (1951) method for estimating rainfall frequencies was
adopted to provide ocean precipitation data. Monthly percentage
frequencies were extracted from the mapped isolines of the US
Marine Climatic Atlas (US Naval Weather Service 1955-1965) and
interpolated to the grid points (see § 2). After re-expressing

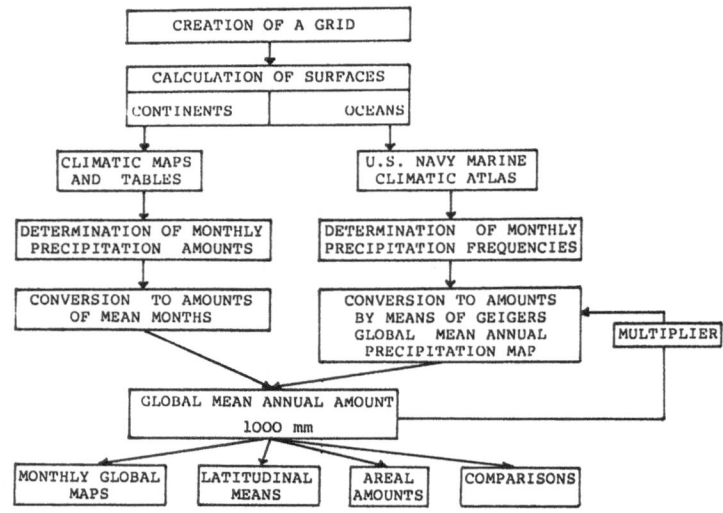

Fig. 3.   Block diagram of data processing procedure.

the monthly frequencies as annual percentages, the values were
scaled to rainfall depth units using Geiger's (1965) precipitation
map shown in Figure 4 to yield monthly precipitation means.

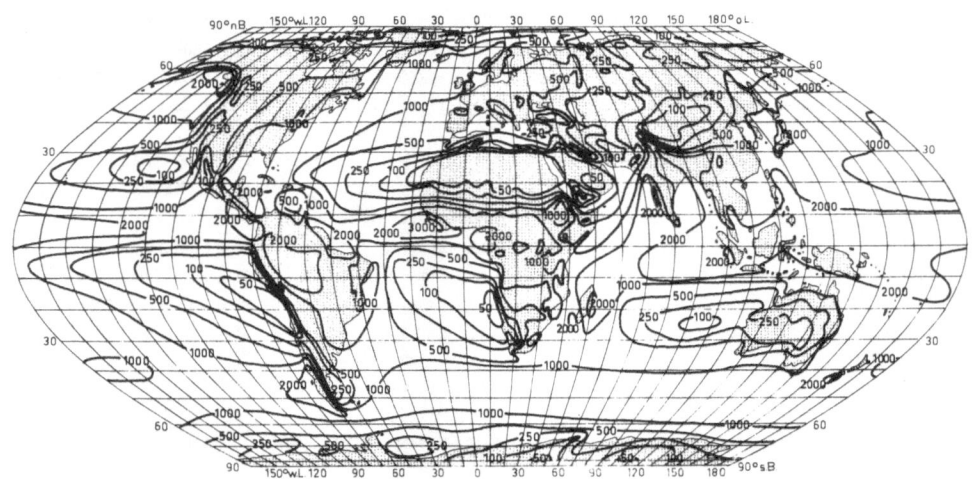

Fig. 4.   Mean annual precipitation (mm) over the globe, after
           Geiger (1965).

4.   DESIGN OF MAPS OF MEAN MONTHLY GLOBAL PRECIPITATION

    Having gridded the map with monthly precipitation values, it
was, in principle, possible to produce maps, tables and diagrams
of the types already presented. However, the sum of the monthly
values so produced yielded an annual global precipitation of 966 mm.
This value appeared low by comparison with other authorities, e.g.,
Kessler (1968) who estimated global evaporation (on average equal
to global precipitation) to be 1000 mm. Other approaches based on
the global heat balance and on satellite observations suggested
that 1000 mm was a lower limit, and Kessler (1982), in a critical
examination of recent research, considered the value to lie between
1000 and 1200 mm.
    It was decided, therefore, to use the value of 1000 mm; this
was attained by adjusting the marine precipitation values (which
together with the polar regions are the most suspect parts of the
calculation) by a factor of 1.062. Although the estimate of precipitation in the polar regions remains unsatisfactory, it is a
relatively dry region so the overall effect on the global budget
is slight.

5.   MONTHLY AND SEASONAL VARIATIONS OF PRECIPITATION

    Figure 5 shows two examples from the final output: the monthly
global rainfall for January and July. Zonal monthly means calculated from the grid point values are presented in Table III, which
also shows annual means and values for each hemisphere and the
whole globe. This indicates that the northern hemisphere received
4% less rainfall than the southern hemisphere, no doubt owing to
the greater land area north of the equator. The seasonal distribution of rainfall in the two hemispheres is out of phase, while
the absolute maximum and minimum are both to be found in the
northern hemisphere (102.3 mm in August, 60.6 mm in March).
    The overall global pattern does not show such a clear cyclicity
although the positions of the maximum and minimum are tied to the
summer and winter positions of the northern hemisphere. This same
tendency for the variation of global precipitation within the year
to be tied more to the northern than the southern hemisphere's
behaviour has been observed in other climatic variables such as
surface temperature and ozone concentration (Kessler 1968). However, the within-year variation of global evaporation, net
radiation and surface pressure are found to be forced by their
variations in the southern hemisphere.
    The zonal information of Table III is also represented in
Figure 2 where it may be contrasted with the earlier estimates by
Brooks and Hunt (1930). Three features stand out: (a) The
improved data situation for high southern latitudes reveals a well-

Table III. Meridional distribution of precipitation depth by 5°-latitude zones (mm)

latitude (°)	area (km²)	J	F	M	A	M	J	J	A	S	O	N	D	year
90 – 85 N	978 908	6	19	18	1	2	3	3	4	15	16	31	19	137
85 – 80 N	2 928 681	8	18	18	4	4	6	7	10	20	19	29	19	162
80 – 75 N	4 854 424	19	17	15	17	19	14	16	17	20	18	22	20	214
75 – 70 N	6 740 454	24	23	15	13	14	15	24	31	28	26	26	26	265
70 – 65 N	8 571 586	22	23	19	24	26	35	46	53	45	37	29	24	383
65 – 60 N	10 333 297	36	35	31	33	36	55	60	65	60	51	41	38	541
60 – 55 N	12 011 886	53	52	44	46	50	65	75	74	76	69	69	56	729
55 – 50 N	13 594 589	67	64	61	56	59	70	75	72	68	67	69	66	794
50 – 45 N	15 069 680	72	69	63	61	59	73	74	69	65	67	77	75	830
45 – 40 N	16 426 533	76	76	68	67	65	67	74	62	64	66	74	74	820
40 – 35 N	17 655 661	79	78	77	68	68	57	57	54	56	58	67	76	784
35 – 30 N	18 748 731	63	65	60	60	57	57	57	72	65	55	56	60	761
30 – 25 N	19 698 557	48	49	43	51	59	66	80	85	70	52	44	45	707
25 – 20 N	20 499 076	37	34	32	33	56	78	83	101	85	55	39	42	694
20 – 15 N	21 145 318	49	42	37	42	61	80	100	129	135	101	72	58	946
15 – 10 N	21 633 367	59	50	47	58	101	109	111	188	184	137	106	81	1333
10 – 5 N	21 960 325	109	88	91	120	186	156	166	200	202	192	165	143	1918
5 – 0 N	22 124 278	149	144	159	197	216	217	205	158	151	168	154	152	2013
0 – 5 S	22 124 278	141	142	157	166	149	203	162	101	103	110	126	134	1542
5 – 10 S	21 960 325	154	162	156	148	111	113	100	82	88	102	121	146	1447
10 – 15 S	21 633 367	145	151	133	106	81	94	83	72	74	82	103	122	1213
15 – 20 S	21 145 318	115	115	99	77	61	74	70	48	51	60	76	98	908
20 – 25 S	20 499 076	93	91	76	64	61	56	56	48	48	53	61	77	773
25 – 30 S	19 698 557	65	66	67	65	64	51	51	60	57	55	54	57	735
30 – 35 S	18 748 731	57	61	60	69	78	63	62	84	73	71	61	55	840
35 – 40 S	17 655 661	60	65	67	81	97	84	87	111	95	89	77	66	1030
40 – 45 S	16 426 533	86	85	90	102	118	110	112	121	113	105	96	92	1255
45 – 50 S	15 069 680	93	96	95	107	118	123	124	122	116	107	100	96	1295
50 – 55 S	13 594 589	98	100	100	106	117	123	122	117	114	107	103	98	1295
55 – 60 S	12 011 886	127	138	161	60	58	118	62	76	51	65	114	156	1116
60 – 65 S	10 333 297	126	119	144	25	17	48	21	21	22	24	43	121	685
65 – 70 S	8 571 586	53	37	46	14	25	15	27	29	28	33	21	52	412
70 – 75 S	6 740 454	20	21	27	12	24	36	27	29	25	30	18	19	300
75 – 80 S	4 854 424	8	8	12	8	14	35	16	17	14	17	9	6	155
80 – 85 S	2 928 681	4	5	7	6	8	20	9	9	8	9	5	3	86
85 – 90 S	978 908	5	6	5	5	5	11	5	5	5	5	5	5	61
90 – 0 N	254 975 352	66.7	62.7	60.6	67.3	82.6	99.4	99.9	102.3	98.7	87.2	78.2	72.8	978.4
0 – 90 S	254 975 352	98.0	99.8	99.6	87.1	82.8	78.6	76.3	76.1	72.7	75.3	81.4	93.3	1021.0
90 – 90	509 950 680	82.3	81.2	80.1	77.2	82.7	89.0	88.1	89.2	85.7	81.2	79.8	83.1	1000

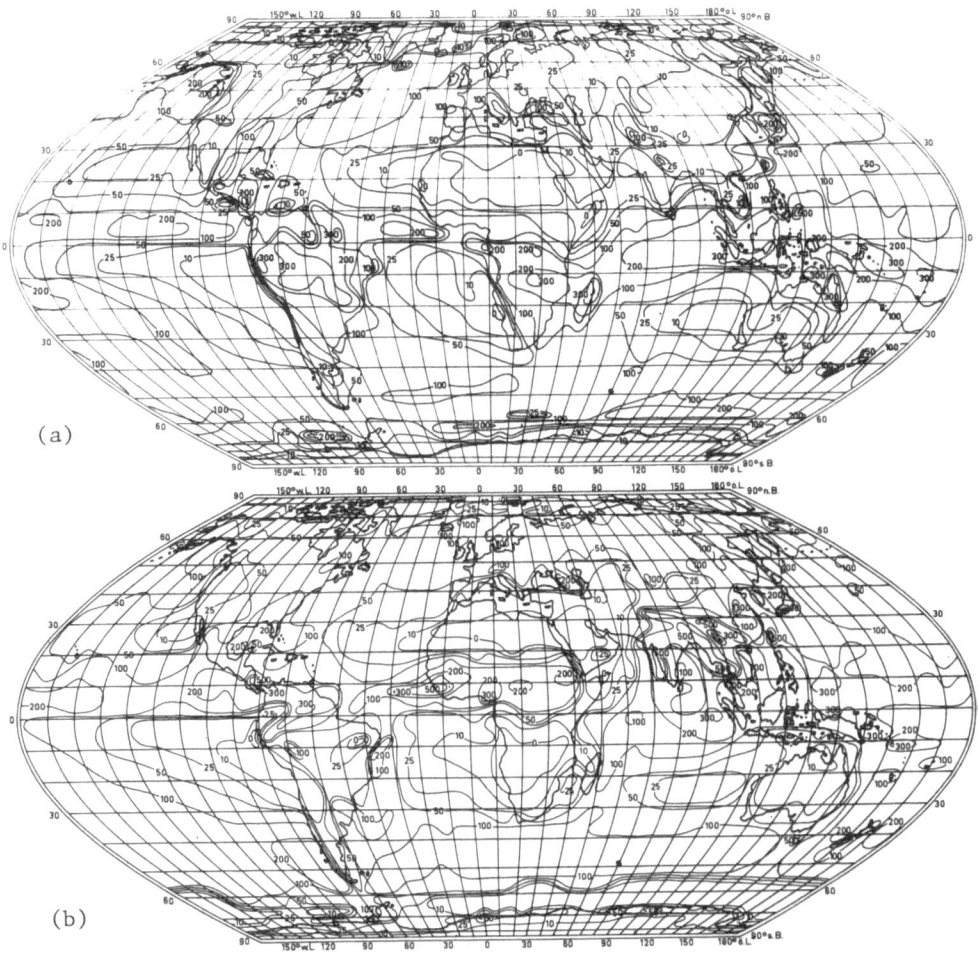

(a)

(b)

Fig. 5.   Mean monthly precipitation (mm) over the globe:
          (a) January; (b) July.

defined isopleth pattern;   (b) The maximum in the southern hemisphere westerlies, which appeared only in (northern) summer months,
is seen to be present in all the months of the year, its position
varying with the Sun's altitude;   (c) The rainfall in the intertropical convergence zone is found to be higher than found in the
earlier evaluation.

     Aggregate zonal information is shown in Figure 6 for January,
July and the whole year. Each period shows the characteristic
3-peak pattern; the zones of westerlies in both hemispheres and
the equatorial zone. The maximum in the southern hemisphere
westerlies is more pronounced than in the northern hemisphere

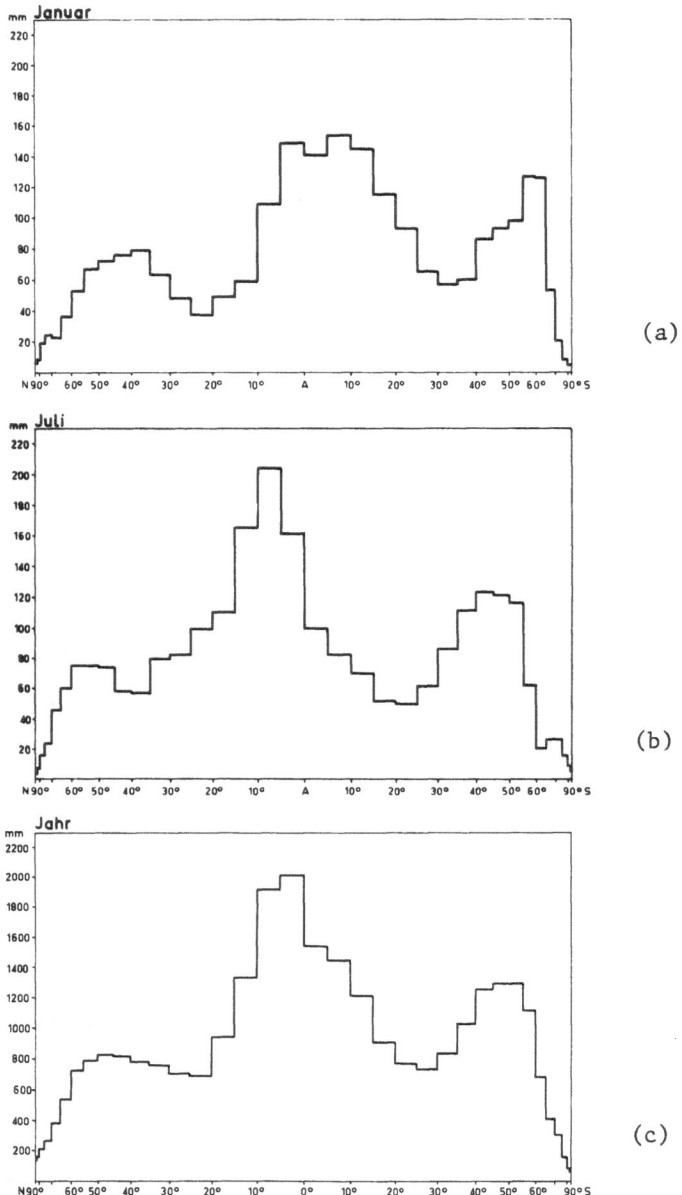

(a)

(b)

(c)

Fig. 6.   Variation of mean precipitation (mm) with latitude:
           (a) January; (b) July; (c) year.

during all three periods due to the asymmetric global distribution
of land and sea.
    Figure 7 gives some insights into the distribution of rainfall

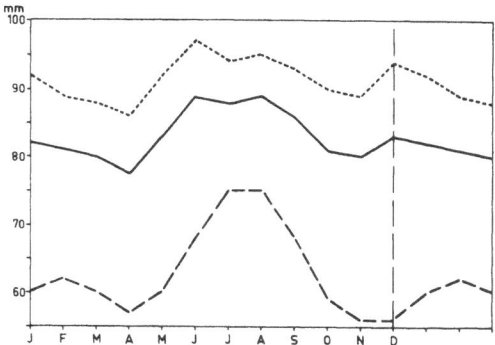

Fig. 7.   Variation of mean annual precipitation (mm) over:
·····, the oceans; ---, the continents; ——, the globe.

between land and sea. The within-year variability of rainfall is
more marked over land than over the sea, although the latter does
receive the larger overall quantity and so forces the global
pattern. However, the temporal pattern of continental rainfall is
very similar with some tendency to lag slightly.

REFERENCES

Baumgartner, A. & Reichel, E. 1971 Preliminary results of new
    investigations of world's water balance. In: *Proc. Symp.
    World Water Balance, Reading, 1970*, pp. 67-78. Gentbrugge:
    Int. Ass. Sci. Hydrol. (IAHS Publ. 92).

Brooks, C.E.P. & Hunt, T.M. 1930 The zonal distribution of rain-
    fall over the earth. *Mem. R. met. Soc.*, 3, 139-159.

Brückner, E. 1905 Die Bilanz des Kreislaufes des Wassers auf der
    Erde. *Geogr.Z.*, 11, 436-445.

Ekhart, E. 1930 Eine neue Regenkarte der Erde. *Petermanns Geogr.
    Mitt.*, 76, 57-64.

Geiger, R. 1965 *World Maps 1:30 M: The Earth's Atmosphere Nr. 5,
    Mean Annual Precipitation*. Darmstadt: Justus Perthes.

Hann, J. & Süring, R. 1936 *Lehrbuch der Meteorologie* (5th Ed.),
    480 pp. Leipzig.

Herbertson, A.J. 1908 *The Distribution of Rainfall over the Land*,
    60 pp. London.

Jaeger, L. 1976 Monatskarten des Niederschlags für die ganze Erde. *Ber.Dt.Wetterd.*, 18, No. 139, 38 pp.

Jefferson, M. 1926 A new map of world rainfall. *Geogr.Rev.*, 16, 285-290.

Kessler, A. 1968 Globalbilanzen von Klimaelementen. *Ber.Inst. Meteorol.Klimatolog.Techn.Hochschule Hannover* No. 3, 141 pp.

Kessler, A. 1982 Heat balance climatology. In: *World Survey of Climatology* (ed. H. Landsberg), Vol. 1. [In the press.]

Loomis, E. 1882 Contributions to meteorology, being results derived from an examination of the observations of the United States Signal Service, and from other sources: Seventeenth paper. *Am.J.Sci.*, 24, 1.22.

Meinardus, W. 1934 Eine neue Niederschlagskarte der Erde. *Petermanns Geogr. Mitt.*, 80, 1-4.

Möller, D. 1951 Vierteljahreskarten des Niederschlags für die ganze Erde. *Petermanns Geogr.Mitt.*, 95, 1-7.

Murray, J. 1887 On the total annual rainfall on the land of the globe, and the relation of rainfall to the annual discharge of rivers. *Scott.Geog.Mag.*, 3, 65.

Schott, G. 1926 *Geographie des Atlantischen Ozeans*, 368 pp. Hamburg.

Schott, G. 1932 *Geographie des Indischen und Stillen Ozeans*, 413 pp. Hamburg.

Supan, A. 1898 Die jährlichen Niederschlagsmengen auf den Meeren. *Petermanns Geogr.Mitt.*, 44, 179-182.

US Naval Weather Service 1955-65 *US Marine Climatic Atlas of the World*, Vols 1-6. Washington D.C.: US Government Printing Office.

COMPARISON OF RAINFALL RATES DERIVED FROM RADAR AND *NIMBUS 5* MICROWAVE OBSERVATIONS IN THE TROPICAL ATLANTIC

Eberhard Ruprecht, Wolfgang Medrow and Ehrhard Raschke

Institüt für Geophysik und Meteorologie
Universität zu Köln
Federal Republic of Germany

ABSTRACT. Rainfall rates over the tropical Atlantic (GATE-B oceanic area) on five days have been calculated from thermal radiation data received from the Earth-atmosphere system using a microwave radiometer on board the *Nimbus 5* satellite. These rainfall rates were compared with rainfall rates calculated from radar data for the same days. Reasons for the differences revealed are discussed.

1. INTRODUCTION

The lack of sufficient data is mainly responsible for short-comings in our knowledge of the global hydrological cycle. This is particularly true for the vast oceanic regions. When the first meteorological satellites were launched 20 years ago, the expectation was great that their observations would fill these gaps. Indirect methods were developed using visible and infra-red (IR) data to determine the rainfall over the oceans; see, for example, the review of Martin and Scherer (1973). In 1972, *Nimbus 5* was launched with a passive microwave radiometer on board which made it possible to measure precipitation directly. The Electrically Scanning Microwave Radiometer (ESMR) measures the thermal radiation of the Earth-atmosphere system at 19.35 GHz. Buettner (1963) was the first to suggest that such observations could be a direct measure of the rainfall rate. The theory for this method was later derived with various degrees of approximation by Wilheit (1972), Kidder (1976) and Jung (1980). Because of the very small and nearly constant microwave (MW) emission of the ocean surface compared to the large and variable emission of the land surface the method can only be applied over oceans.

141

A. Street-Perrott et al. (eds.), Variations in the Global Water Budget, 141–146.
Copyright © 1983 by D. Reidel Publishing Company.

Two factors militate against a straightforward application of this method:

(a) Owing to the lack of ground truth data (i.e., precipitation measurements at ground level over the oceans), a test of the calculated rainfall rates is often not possible over large areas. With the radar rainfall data provided by the GARP Atlantic Tropical Experiment (GATE), this data set made such tests possible;

(b) Sampling problems raise large difficulties. In the main rain producing area (the tropics), rainfall processes are organized in scales which can hardly be resolved by the MW radiometers. Even within cloud clusters which generate significant large extended signals in the visible and IR images, the precipitation areas cover only a small region (Ruprecht & Gray 1976, Houze & Cheng 1977).

In this study, the calculated rainfall rates were compared with selected rainfall data from radar observations in the B-scale area ($5°-15°N$, $20°-27°W$) of the GARP Atlantic Tropical Experiment (GATE). The information from the visible and IR images from the Synchronous Meteorological Satellite (*SMS-1*) was used to determine the precipitation in the field of view of the ESMR onboard *Nimbus 5*. A few case studies are given to illustrate the method.

2. DATA

Five days (31 August, 2, 4, 5, 7 September) were chosen from Phase III of GATE (30 August–19 September 1974). During these days, the sub-satellite path of *Nimbus 5* was directly over or very close to the GATE B-area. The microwave radiances are available once a day at noon with a spatial resolution of 25 × 25 km near nadir.

The rainfall data, derived from radar observations (Hudlow & Patterson 1979), are given for each hour with a spatial resolution of 28 × 28 km. All 148 data fields of the so-called 'Master Array'[1] (of ship-mounted radars) were used.

The *Nimbus 5* data were rearranged in a grid system with a spatial resolution similar to the radar rainfall data. For the 3 days, 31 August, 2 and 7 September, the resolution was 28 × 28 km. For the other two days, when the sub-satellite path did not directly cross the 'Master Array', the observations of the data fields were averaged to give a resolution of 28 × 56 km.

3. THEORETICAL ASPECTS

In the spectral range of the microwave region (here, 19.35 GHz), the following statements are true:

(a) Clouds producing precipitation are not transparent, whereas the cloud-free atmosphere and non-precipitating clouds are both nearly 90% transparent;

(b)   Over land, emittance is approximately 0.9 depending on soil
      moisture content;
(c)   The calm ocean emittance is about 0.4, though increasing to
      0.5 for a rough ocean surface (winds greater than 10 m s^{-1}).
      Thus, when looking through a clear atmosphere over land, the
brightness temperature (T_{IR}) as seen by a microwave radiometer
onboard a satellite is about 0.9 times the surface temperature
or about 260 to 270 K. Over the oceans, the brightness temperature
will be of the order of 0.5 times 300 K, i.e., 150 K. Rain clouds
consequently appear as 'warm islands' over a cold ocean in a microwave image.
      The equation of radiative transfer for the microwave region
is iteratively solved (Jung 1980) with the following assumptions
and parameters:
(i)   The vertical profiles of temperatures and humidities are
       given by the Phase III GATE B-scale mean. A small correc-
       tion, as given by Gray (1977), was applied to the temperature
       profiles in the rain area. Constant mixing ratios are
       assumed for CO_2 and O_2;
(ii)   As the surface winds over the GATE area did not exceed
       7 m s^{-1} during the observation period, a smooth ocean sur-
       face and undisturbed Fresnel-reflection have been assumed;
(iii) The rain clouds are modelled with 2 layers: a lower, pure
       rain cloud layer with a Marshall-Palmer rain drop distri-
       bution and an upper layer with a thickness of 1 km and
       containing either cloud drops with a liquid water content
       of 0.1 g m^{-3} or a Marshall-Palmer ice particle distribution
       for an equivalent rain rate of 5 mm h^{-1}. The height of the
       cloud tops, i.e., the top of the upper layer, is determined
       from the radiation temperature in the IR imagery.

4.   CORRECTION METHOD

      The veracity of the derived rainfall rates was tested first
with averaged data. The rainfall rates were calculated for each
data field of the 'Master Array' and then averaged. These mean
rainfall rates for each of the 5 days were then compared with the
mean radar rainfall (Figure 1). The curve depicts the theoreti-
cally-derived relation between microwave brightness temperatures
and rainfall rates for cloud tops above 5 km.
      It is obvious that the rainfall rates will be underestimated
if the theoretical curve is applied. That is also true for all
individual data fields as shown by the error bars (standard deviation). The results in Figure 1 can also be interpreted in another
way: namely, that the observed brightness temperatures are too low
for the given rainfall rates. We believe that the main reason for
this underestimate is due to the fact that the microwave data with
a spatial resolution of 25 × 25 km cannot resolve the individual
rain clouds or cells, but give an average over a region with rain
which included large portions without rain.

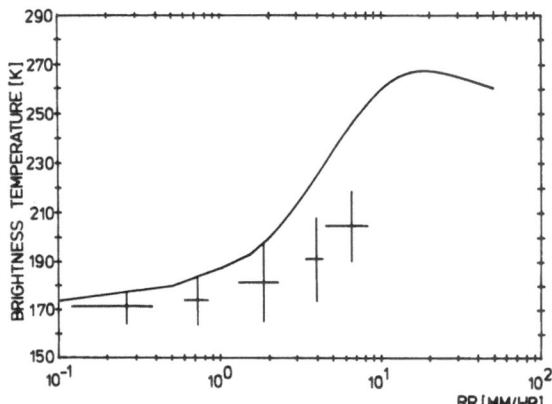

Fig. 1. Mean brightness temperature compared with mean radar rainfall rates (RR) averaged over all data fields with mean
cloud top heights greater than 5.5 km. The bars show the
standard deviation; continuous line represents the theore-
tically-derived curve.

It is assumed that the observed brightness temperature T_{RG}
consists of two components: the brightness temperature of the rain
region, T_{RR} and of the non-rain region, T_{RK}. It is further assumed
that T_{RG} is an area-weighted mean of these two components. Thus,

$$T_{RG} = f\ T_{RR} + (1 - f)\ T_{RK} \tag{1}$$

where f = horizontal fractional area of the rain clouds. From
Equation (1), T_{RR} can be calculated with knowledge of f.
    Determination of f is based on knowledge of the nature and
distribution of clouds in the tropics and, especially, on the
observations during GATE. The following observational evidence
can be used:
(a) Clouds with tops below 3.5 km produce no rain (see, for
     example, Riehl 1979);
(b) During Phase III of GATE, only 0.2% of the total rain was
     observed in areas with echo top heights below 5.5 km (Cheng
     & Houze 1979);
(c) In the presence of deep convective clouds in the GATE B-area,
     rain occurred only beneath those clouds with top heights above
     8.8 km ($T_{IR} < 247$ K) (Stout *et al.* 1979);
(d) Nearly 50% of the region with deep convective clouds was with-
     out rain (Cheng & Rodenhuis 1977);
(e) From the radar observations, the fractional coverage of each
     data field with rain rates greater than 0.5 mm h^{-1} is given.
From conditions (a) to (e), the following assumptions can be
inferred:
1. f_1 = fractional area of all clouds with top heights > 3.5 km
     ($T_{IR} < 279$ K);

2. f_2 = fractional area of all clouds with top heights > 5.5 km
     (T_{IR} < 268 K);
3. f_3 = fractional area of all clouds with top heights > 8.8 km
     (T_{IR} < 247 K), though if f_3 < 20%, we let $f = f_1$ whereas, if
     f_3 > 80%, we let $f = \frac{1}{2}f_3$.
Then T_{RR}, the brightness temperature required to give the rainfall
rates for raining clouds, is such that

$$T_{RR} = [T_{RG} - (1 - f) \, T_{RK}]/f \tag{2}$$

T_{RK} is assumed constant for a non-raining atmosphere (and T_{RK} =
160 K).

5.   RESULTS

For each data field, the observed brightness temperature was
corrected by applying criteria 1 to 3. The heights of the cloud
tops were determined from the IR observations from *SMS 1* with a
spatial resolution of 7.5 × 3.5 km. The rainfall rates, computed
with the corrected temperatures for each data field, are shown in
Table I together with the radar-derived rainfall given as total
water output over the 'Master Array'.

Table I. Total water output over all data fields (10^3 kg s^{-1})

Date	Time (GMT)	Ground Resolution (km)	Radar-derived Rainfall	Satellite-derived Rainfall	
				Uncorrected	Corrected
31 Aug.	1243	28 × 28	5.2	2.0	4.3
2 Sept.	1300	28 × 28	34.4	14.2	20.9
4 Sept.	1316	28 × 56	14.6	7.2	12.9
5 Sept.	1231	28 × 56	37.5	27.2	42.1
7 Sept.	1247	28 × 28	5.3	2.2	3.1

It can be seen that the best agreement is with the values of
total water output calculated after the application of criterion 3
(Table I, last column), though the satellite-derived rainfall rate
is still an underestimate in four out of five cases.
The frequency distributions (not shown) give a hint of the
cause of the underestimate in calculated rainfall rate; namely,
that the frequency of both intense and low rainfall is too small.
The results have shown that it is in principle possible to
overcome some of the sampling problems by applying a simple
correction method to the brightness temperatures. The method will
be further tested with more GATE data and, if data become available,
for other regions.

NOTE
(1)   'Master Array' (Hudlow & Patterson 1979) is a circular area
      of *ca.* 400 km diameter centred at 8°30'N, 23°30'W.

REFERENCES

Buettner, K. 1963 Regenortung von Wettersatelliten mit Hilfe von
    Zentimeterwellen. *Naturwissenschaften,* 50, 591-592.

Cheng, C.-P. & Houze, R.A.Jr 1979 The distribution of convective
    and mesoscale precipitation in GATE radar echo patterns. *Mon.
    Weath. Rev.,* 107, 1370-1381.

Cheng, N. & Rodenhuis, D. 1977 An intercomparison of satellite
    images and radar rainfall rates. *Proc. 11th Tech. Conf. on
    Hurricanes and Tropical Meteorology, 13-16 December 1977,
    Miami Beach, Florida.*

Gray, W.M. 1977 Tropospheric mean state and variability. *US GATE
    Central Program Workshop Rept,* pp. 199-213.

Houze, R.A. & Cheng, C.-P. 1977 Radar characteristics of tropical
    convection observed during GATE: Mean properties and trends
    over the summer season. *Mon. Weath. Rev.,* 105, 964-980.

Hudlow, M.D. & Patterson, V.L. 1979 GATE radar rainfall atlas.
    NOAA Spec. Rept, 155 pp. Boulder: NOAA (Center for Environ-
    mental Assessment Services).

Jung, H.J. 1980 The determination of rainfall rates from satellite
    measurements of the thermal microwave emission. *Beitr. Phys.
    Atm.,* 53, 366-388.

Kidder, S.Q. 1976 Tropical oceanic precipitation frequency from
    Nimbus 5 microwave data. *Colorado State Univ. (Fort Collins)
    Atm. Sci. Paper* No. 248, 50 pp.

Martin, D.W. & Scherer, W.D. 1973 Review of satellite rainfall
    estimation methods. *Bull. met. Soc.,* 54, 661-674.

Riehl, H. 1979 *Climate and Weather in the Tropics,* 611 pp. London,
    New York, San Fransisco: Academic Press.

Ruprecht, E. & Gray, W.M. 1976 Analysis of satellite-observed
    tropical cloud clusters. II Thermal, moisture and precipita-
    tion fields. *Tellus,* 28, 414-426.

Stout, J.E., Martin, D.W. & Sikdar, D.N. 1979 Estimating GATE rain-
    fall with geosynchronous satellite images. *Mon. Weath. Rev.,*
    107, 585-598.

Wilheit, T.T. 1972 The electrically scanning microwave radiometer
    (ESMR) experiment. *Nimbus 5 User's Guide,* pp. 55-105. NASA
    Goddard Space Flight Center.

EVAPORATION MODELS IN THE GLOBAL WATER BUDGET

W.J. Shuttleworth

Institute of Hydrology
Wallingford
Oxon, UK

ABSTRACT. Despite the key role of the evaporation process in the global water budget, it lags behind other elements such as precipitation, runoff and the state of water bodies, in terms of our ability to monitor its variation in time and over the surface of the globe. Attempts to introduce evaporation into the water budget as an independent term depends on our ability to model the process. This paper presents a classification scheme for such models which highlights their suitability for different tasks. Some forecasts and speculations regarding the response of evaporation and energy partition in the atmosphere, to specified changes in climatic inputs and land use, are also made.

1. INTRODUCTION

Evaporation is, with precipitation, one of the two principal elements in the global water budget. Changes in the global hydrological cycle will occur if there are changes in the magnitude and distribution of the evaporative sources of water vapour. For example, since two-thirds of the Earth is covered with liquid water, variations in sea surface temperature in time and space exert a vital influence on the global water budget. However, our ability to describe and measure evaporation directly is in its infancy. Evaporation, on the global, continental or regional scale, must be deduced indirectly from the other terms in the water budget. This paper looks forward to the prospects for an independent assessment of evaporation by surveying the properties of current evaporation models. Even in the present state of the art, it is possible to infer changes in evaporation which can be induced by other major changes in climate and/or land-surface characteristics.

A. Street-Perrott et al. (eds.), Variations in the Global Water Budget, 147–171.

It is on land surfaces that evaporation has its most direct
effect on human well-being and, here, man's activities are likely
to have the most direct effect on the rate of evaporation. Extensive study has generated a range of physical and empirical
descriptions of the evaporation process which are 'causal' in the
sense that they attempt to predict the evaporation that will occur
in response to changing meteorological input. In § 3, these
various causal models are reviewed in order to elucidate their
inter-relationship.

In recent years, attempts have been made to describe evaporation from land surfaces using meteorological records as a measure
of actual evaporation rates on a regional scale. Such consequential models or developments of them could well prove useful as *a
posteriori* measures of actual evaporation, and are outlined in § 5.

The later sections draw on the review of the causal models to
discuss the interactions between evaporation and climatic change.
The likely effect of changes in meteorological variables on evaporation rate is considered in § 6 , while § 7 explores how such
variables might themselves respond to changes imposed by the
influence of Man on land-surface characteristics.

2.   STANDARD OR 'POTENTIAL' EVAPORATION RATES

The rate of natural evaporation from land surfaces is a complex function of atmospheric, soil and vegetation factors. In
trying to bring some order into this complexity, it is useful to
define certain idealized standard rates of evaporation, which are
designed to give a measure of the meteorological or climatological
control over the evaporation process at a particular location.
Penman (1948) created the concept of *potential evaporation,* λE_o,
which might be defined as 'the quantity of water evaporated, per
unit area and unit time from an idealized, extensive free water
surface under prevailing atmospheric conditions'.

On the basis of the experimental evidence available, it was
believed for many years that the type and form of vegetation cover
on the Earth's surface (and even its presence or absence) had
little effect on the natural rate of evaporation, *providing* this
was limited by the heat supplied to the surface and not by the
availability of surface water. This belief made it reasonable to
propose a concept, *potential evapotranspiration,* λE_T, solely
determined by meteorological conditions, and defined (Gangopadhyaya
et al. 1966) as the "maximum quantity of water capable of being
lost as water vapour, in a given climate, by a continuous, extensive stretch of vegetation covering the whole ground when the soil
is kept saturated". The concept was, and is, used as a standard
'scale' which can be modified to include the influence of surface
characteristics, usually by applying a multiplication factor. The
term remains in common use today, though it is recognized that
vegetation type and even its local position affect λE_T and that

empirical formulae used to evaluate the latter require local
calibration (Tanner 1967).

It is clear now that these initial assumptions, namely that
potential evapotranspiration is independent of vegetation cover
and that it represents a maximum rate, may be ascribed to the
fact that many of the original studies took place over short crops.
With such crops, plant factors are of less importance whereas the
control exerted by the atmosphere itself is maximized, since it
dictates not only the driving potential in the diffusion process,
but also generates the dominant and, possibly, controlling resistance to vapour transfer. This realization, coupled with the
continuing desire to develop a conservative concept independent of
surface characteristics led to the emergence of a more precisely
defined standard evaporation rate, *reference crop evapotranspiration* E_{RC} (Doorenbos & Pruitt 1977). This is defined as 'the rate
of evapotranspiration from an extensive surface of 8 to 15 cm tall,
green grass cover of uniform height, actively growing, completely
shading the ground and not short of water'.

3.   A CLASSIFICATION OF CAUSAL MODELS OF EVAPORATION

An attempt has been made recently to classify a broad group
of inter-related 'causal descriptions' in terms of the meteorological input they require (Shuttleworth 1979). This hierarchial
classification is summarized here as a basis for understanding why
only some techniques can realistically provide direct estimates of
one of the standard evaporation rates rather than of actual evaporation. This understanding is essential to the choice of models
for quantifying the effect of climate change on the evaporation
term in the global water budget. It is a common procedure to
estimate actual evaporation by first calculating one of these conceptual standard rates, subsequently adjusting it to give actual
evaporation using an empirical factor. This procedure is discussed
further in § 4. More recent techniques that are capable of estimating actual evaporation directly require a closer description
of reality by explicitly acknowledging the effect of the surface
characteristics.

The classification scheme outlined in Table I lists the
various evaporation models together with their data requirements,
the type of estimate they might legitimately provide, and an
assessment of their current usage. The scheme is meant to be
fairly general in terms of the broad classes described, but only
a single (usually recent) example is given to typify the numerous
different forms within each class.

In amplifying the various techniques listed in Table I, the
additional assumptions involved in progressing from one class to
the next are described.

Table I. Types of models used for evaporation estimates and their data requirements.

Model class (see § 3)	Evaporation estimate	Data requirement	Current usage
(a) Simulation (1) Numerical (SPAM) (2) Analytic (Shuttleworth)	Actual	(i) Detailed models of physiological response (ii) Detailed information on canopy exchange processes (iii) Detailed information on canopy structure (iv) Short-term measurements of meteorological parameters	Small: increasing interest
(b) Single source (1) Transpiration (Penman & Monteith) (2) Interception (Rutter) (3) Unified (Shuttleworth)	Actual	(i) Submodels of surface resistance (ii) Coarse measurements of canopy structure (iii) Short-term measurements of meteorological parameters	Small: increasing usage
(c) Intermediate	λE_T	(i) Daily meteorological data (ii) Coarse measurements of canopy structure (iii) Information on rainfall pattern	Minimal: increasing interest
(d) Energy balance (e.g., Penman)	λE_{RC} λE_T (short crop)	Daily meteorological data (T, R_N, u, D)	Large: stable
(e) Radiation (e.g., Priestley & Taylor)	λE_{RC} λE_T (short crop)	Daily meteorological data (T, R_N)	Medium: increasing
(f) Humidity (e.g., Dalton)	λE_{RC} λE_T (short crop)	Daily meteorological data (T, u, e)	Small: decreasing
(g) Temperature (e.g., Blaney & Criddle)	λE_{RC} λE_T (short crop)	Daily meteorological data (T)	Medium: stable

3.1 Simulation Models

Clearly, the best way to describe actual evaporation successfully is to build a model which simulates, as nearly as possible,
the physical and physiological processes occurring in reality:
(a) *Numerical Simulation Models*. There are several one-dimensional models which attempt a numerical simulation of the
evaporation process by dividing the vegetation into a finite
number of horizontal layers. The Soil-Plant-Atmosphere Model
(SPAM) (Sinclair *et al*. 1976) is an excellent example.
(b) *Analytical Simulation Models*. Shuttleworth (1976*b*) demonstrated that numerical multilayer models (see above) can in fact
be rewritten in analytical form by taking the limit corresponding
to an infinite number of levels. He showed that the use of such
models is equivalent to using a formula similar to that relevant
to single-source models (§ 3.2), with the form:

$$\lambda E = [\Delta A' + (\rho c_p D + \delta)/r_H]/[\Delta + \gamma_E(r_H + r_c)/r_V] \tag{1}$$

As the definition of the terms r_c and δ is too complex to include
here, see Shuttleworth (1976*b*); for the other terms, see § 3.2(a).

3.2 Single-Source Models

Single-layer or 'big-leaf' models of plant canopies (e.g.,
Monteith 1965, Thom 1972) assume that the overall effect of the
whole canopy on the energy fluxes above the canopy is reasonably
approximated by assuming that all the elements making up the
vegetation are exposed to the same microclimate. The sensible
heat and latent heat (evaporation) fluxes flow through resistances
r_H and r_V, respectively, in moving from the surface of the leaves
at a given level to some higher level above the canopy, namely
'the screen height' at which measurements of temperature and
vapour pressure are made. In dry conditions, the latent heat flux
(which arises inside the stomatal cavities when transpiration is
taking place) is subject to the additional diffusive resistance
r_{ST} encountered in negotiating the stomatal opening. This resistance is usually called the bulk stomatal resistance. If the
canopy is completely wet, both fluxes arise at the surface of the
vegetation, and the evaporation process may be modelled as an
interception process. When the canopy is only partially wet, the
situation is more complex and the simple extension of previous
ideas has proved inadequate (Shuttleworth 1976*a*). Recent work
(Shuttleworth 1978) has suggested that an effective 'surface'
resistance can be used to describe the situation, but this formulation has been tested only for tall (forest) vegetation.
   The totally dry canopy, totally wet canopy and universal
cases are discussed separately below.
(a) *Transpiration Models*. The single-source representation of
transpiring canopies is a well established and tested model. The
predicted λE is given by

$$\lambda E = [\Delta A' + (\rho c_p D)/r_H]/[\Delta + \gamma(r_V + r_{ST})/r_H] \qquad (2)$$

where A' is the energy available for evaporation (often approximated by the net radiation input, R_N over long time periods), D is the vapour-pressure deficit measured at screen height, Δ is the rate of change of the saturated vapour pressure with temperature, ρ is air density, c_p is the specific heat of air at constant pressure, and γ is the 'psychometric' constant $\simeq 0.66$ mb K^{-1}.

In applying Equation (2), it is commonly assumed that r_H and r_V are equal and are related to r_M the equivalent resistance for momentum. The assumption

$$r_H = r_V = r_M = u(z)/u_*^2 \qquad (3)$$

(where $u(z)$ is the wind speed at a height z above the ground) has been used with considerable success and will usually suffice. The friction velocity u_* can be estimated in practice by

$$u_* = k \, u(z)/\ln[(z - d)/z_o] \qquad (4)$$

where k is Von Karman's constant (~ 0.41), d is the zero-plane displacement and z_o the roughness length; d and z_o are commonly found to have an approximate relationship to h, the height of the vegetation, namely, d = 0.75 h and z_o = 0.1 h.
(b) *Interception Models*. When the source of water vapour is a completely wet canopy, the term r_{ST} in Equation (2) is zero. The effect of a change in this   parameter on the observed evaporation is most marked for vegetation types in which $r_{V,H}$ is small in comparison with r_{ST}, that is for tall (forest) vegetation, although the effect is still present to some degree for other vegetation types.

The most successful model of rainfall interception in forests (Rutter *et al.* 1971, 1975) has been tested successfully by Gash and Morton (1978) and Calder (1977); the conceptual framework of this model is shown in Figure 1. In essence, it calculates a running water balance for the canopy and trunks of a forest stand using, as inputs, hourly rainfall and the meteorological variables necessary to estimate evaporation. It computes the rate of evaporation of intercepted water as well as the amount of water reaching the ground in the form of drips from the canopy ('throughfall') abd down the tree trunks ('stemflow'). The value of the potential evaporation, E_p, is calculated from Equation (2), using the assumptions $r_H = r_V = u(z)/u_*^2$ and $r_{ST} = 0$. The model uses C and C_t, the effective depth of water currently standing on the canopy and trunks, respectively, as 'state variables' and requires both a knowledge of canopy structure and a parameterization of canopy drainage.
(c) *Unified Single Source Models*. The overlap between the evaporation of intercepted water and transpiration is poorly defined in the model just described. To unify the two processes

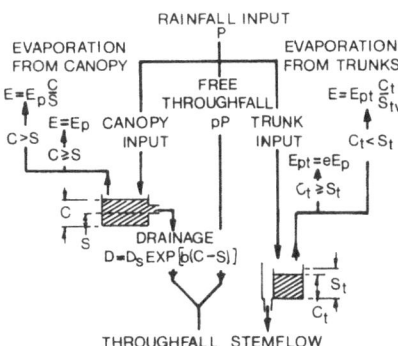

Fig. 1. Conceptual framework of the 'Rutter' model of rainfall
interception by forest vegetation.

correctly, a physically-continuous treatment of the transition
between wet and dry canopies is required. Equation (2) forms the
basis of this unified model, in which r_{ST} is replaced by a
redifined surface resistance r_S, equal to r_{ST} in dry conditions
and to zero in wet conditions, with a smooth transition between
the two depending on the fractional surface wetness.

A recent theoretical analysis (Shuttleworth 1978), so far
tested only over tall vegetation, suggested an expression of the
form:

$$r_S = \{W/r_b (\Delta/\gamma+1)+(1-W)/[r_{ST}+r_b (\Delta/\gamma+1)]\}^{-1} -r_b (\Delta/\gamma+1) \qquad (5)$$

where $W = 1$ when $C \geqslant S$ $\qquad\qquad\qquad\qquad\qquad\qquad (6)$
and $\quad W = (R-1)/(R-S/C)$ when $C < S$

In Equation (6), C is the amount of water stored in the canopy,
S the storage capacity of the canopy, and

$$R = (r_{ST}/r_A)(r_A-r_b)[r_{ST}+(\Delta/\gamma+1)r_b]^{-1} \qquad (7)$$

where $r_A = r_H = r_V$ is estimated as in the Rutter model and r_b is
an estimate of the mean *boundary layer* resistance of the
vegetation elements in the canopy given by $r_b \sim 1.1\, u_*^{-1} + 5.6\, u_*^{\frac{1}{3}}$
(with u_* in m s^{-1}).

Although simulation and single-source methods are superior to
all other techniques both conceptually and in that they provide a
direct estimate of actual evaporation, their use is limited to
locations with short-term meteorological data, and by the adequacy
of submodels describing stomatal resistance. The remaining
techniques are used primarily to estimate one of the two standard
rates (see § 2) which is then adjusted empirically to give actual
evaporation (§ 4).

3.3 Intermediate Models

    'Intermediate' models represent an attempt to extend the
usefulness of the more rigorous models described above by adding
terms representing physical characteristics of vegetation and
climate to the Penman equation (see § 3.4). Rijtema (1965) has
demonstrated that, providing the measured or estimated net radiation used in the Penman equation is that appropriate to the
vegetation itself, the equation gives an estimate of *potential
evapotranspiration* λE_T for *short, green vegetation*. Hence, the
Penman formula provides an estimate of *reference crop evapotranspiration* λE_{RC} (see § 2). Thom and Oliver (1977) attempted to modify
the original Penman equation to include additional terms, with a
physical basis, that can be adjusted empirically to yield an
estimate of λE_T for different crops. Their equation takes the
form:

$$\lambda E = (\Delta Q_N + \gamma E_a')/[\Delta + \gamma(1 + n)] \tag{8}$$

where E_a' is a modified version of the equivalent term in the Penman
equation, namely

$$E_a' = 13.8(e_s - e)(1 + U/100)/\ln^2(z/z_o) \tag{9}$$

where e and e_s are the actual and saturation values (mb) of vapour
pressure measured at a height z above d and U is the corresponding
wind run (miles per day). In Equation (8), the term Q_N is an
estimate of the net energy available for evaporation at the evapo-
rating surface, n is a measure of the stomatal control exerted by
the crop, and is defined as $n = r_s/r_a$ in which r_a is the effective
aerodynamic resistance, implicit in E_a', and given by the
expression:

$$r_a = 4.72[\ln^2(z/z_o)](1 + 0.54 u) \tag{10}$$

and r_s is the effective surface resistance estimated by

$$r_s = (1 - I/E)r_{sd} \tag{11}$$

In Equation (11), r_{sd} is an estimate of the average stomatal resistance of the crop, which is assumed to be constant, and I is that
part of the total evaporation, E, occurring directly from the
intercepted precipitation.
    Gash (1978) pointed out that Equation (8) could be rewritten
in a form in which it can be used in conjunction with separate
estimates or measurements of interception loss. Currently, such
interception estimates are only available for tall vegetation.
The equation he proposed has the form:

$$\lambda E = (\Delta Q_N + \gamma E_a')/[\Delta + \gamma(1 + r_{sd}/r_a)] + I(1 - c) \tag{12}$$

where I is the interception loss, and the correction term c, which compensates for including transpiration under wet conditions, is given by

$$c = (\Delta + \gamma)/[\Delta + \gamma(1 + r_{sd}/r_a)] \tag{13}$$

3.4 Energy Balance Models

The Penman equation is the original and typical example of an energy balance model and can take the form:

$$\lambda E_{RC} = [\Delta Q_N + \gamma f(u)(e_s - e)]/(\Delta + \gamma) \tag{14}$$

The above analysis of Thom and Oliver (Equation 8) provides the link between single-source and energy-balance models. For example, Equation (14) can provide estimates of potential evaporation λE_o if Q_N is a measurement of net energy input over a free water surface, or of reference crop evaporation λE_{RC}, if Q_N has the appropriate value for short, green vegetation (Rijtema 1965). Thom and Oliver (1977) also demonstrated that the wind function

$$f(u) = 0.26(1 + U/100) \tag{15}$$

used with Equation (14) implicitly serves two purposes, namely: (i) It contains a reasonable, average description of the effect of thermal stratification on a rigorous formula for aerodynamic resistance with an assumed value of sensible heat flux of $H = +50$ W m^{-2}; (ii) When compared to physically rigorous formulae for aerodynamic resistance in neutral conditions, the small value of z_o it implies is such as to provide some compensation for the absence of an effective surface resistance in the denominator. Equation (14) is the simplest physically-based equation which can be used to estimate λE_{RC}. The other models for estimating evaporation that are based on meteorological data (Table I, e, f, g) implicitly require additional assumptions as to the relationship between the meteorological parameters appearing in the Penman equation. Such correlations do have some intuitive physical basis, but there must always remain doubt as to the universality of such empiricism in moving from one climatic region to the next.

3.5 Radiation Models

Of the remaining classes, those that are most likely to preserve some of the observed universality of the Penman equation are the empirical equations which relate reference crop evaporation to radiation (or, more correctly, to a combination of temperature and radiation). A representative equation is that due to Priestley and Taylor (1972) which has the form:

$$\lambda E_{RC} = \alpha [\Delta / (\Delta + \gamma)] Q_N \tag{16}$$

where α is an empirical constant ~ 1.26.

The conceptual basis for equations of this type is the existence of a locally-conservative relationship between the two terms in the numerator of Equation (14).

3.6 Humidity Methods

It is possible, of course, to express the Penman equation differently by establishing empirical relationships between reference crop evaporation and vapour pressure deficit. Such expressions, which are modifications of the Dalton equation, have the general form:

$$\lambda E_{RC} = f'(u)(e_s - e) \tag{17}$$

in which f' is an empirical factor for the location of interest. Clearly, such an equation is likely to be less universally applicable than those involving radiation since, in this case, the smaller of the two terms in Equation (14) is being used to estimate the greater.

3.7 Temperature Methods

Many empirical formulae relate λE_{RC} to temperature. The widely-used Blaney and Criddle (1950) equation is one such example, designed to provide daily estimates of evaporation averaged over a month, and taking the form:

$$\lambda E_{RC} = C_u d_\ell (0.46T + 8) \tag{18}$$

where C_u is the so-called 'consumptive use' factor, reported in engineering literature, but best determined locally, d_ℓ is the fraction of daylight hours occurring in a month, and T is the temperature in $^\circ$C.

The physical basis for this equation is simply that both the 'radiation' and 'aerodynamic' terms in the Penman equation are likely to have some relationship, albeit ill-defined, with temperature. Since the radiation term is generally the larger of the two, it is the correlation between radiation and temperature which is more important. There is a lag in the yearly temperature cycle with respect to the yearly radiation cycle, and the monthly 'consumptive use' factor often includes, unwittingly, some allowance for this thermal lag.

4. RELATION OF ACTUAL TO POTENTIAL EVAPORATION RATES IN CAUSAL MODELS

In practical applications, actual evaporation is the rate most often required. However, many of the techniques used in estimating evaporation (including those in most common use) provide one of the 'standard' rates, generally λE_{RC}. The conventional practice is to multiply this standard rate by a factor as in the equation

$$\lambda E = K_C \; \lambda E_{RC} \tag{19}$$

The factor K_C in Equation (19) is called the *crop coefficient*. Studies have been made of the variation in K_C in response to decreasing soil water content. While the results differ considerably in detail, as is to be expected for this type of empirical parameter, the overall behaviour of the crop coefficient during a drying cycle generally follows the pattern shown in Figure 2. This behaviour seems to be generally similar for both crops and soil: during the first, fairly constant, stage of the drying cycle, K_C remains quite close to its initial value until a 'wilting point' is reached, when it begins to decrease in response to declining soil water content. In conditions of prolonged drought, the crop (if present) begins to die and the evaporation rate is then controlled by soil characteristics, especially hydraulic conductivity.

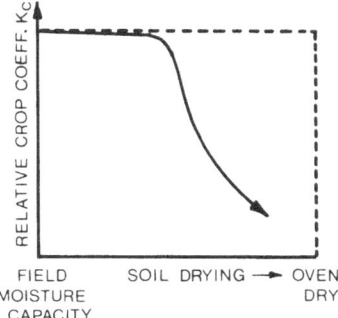

Fig. 2. Variation in relative crop coefficient K_C in response to changing soil moisture content.

5. CONSEQUENTIAL DESCRIPTIONS OF EVAPORATION

The control by meteorological parameters of the evaporation rate at a particular location was discussed in § 3. However, atmospheric temperature and humidity are, in some measure, a reflection of the fluxes of sensible and latent heat integrated over an extensive area. Recently, attempts have been made to

develop measurement procedures with a conceptual basis which incorporate the effect on meteorological parameters of the actual
surface energy exchange. In this respect, they are distinct from
the causal models discussed above and are categorized here as
consequential descriptions.

At least in its present form, the consequential approach is
fundamentally a conceptual theory of evaporation *measurement*, in
that it attempts to estimate the evaporation which has occurred
in the recent past from its consequences on actual meteorological
measurements. The approach makes no attempt to describe the
mechanism giving rise to this 'measured' evaporation rate, and is
not suitable for predictive applications in medium- or long-term
meteorological models. Such models are required to compute the
evolution of surface fluxes interactively as a response to (model-
generated) meteorological inputs, and must therefore be causal.

In previous sections, it was noted that it is possible to
provide reasonable estimates of actual evaporation under potential
conditions for free water surfaces (potential evaporation) and
well-watered short crops (reference crop evaporation) using
energy-balance equations of the Penman type, i.e.,

$$\lambda E_p = \Delta Q_N/(\Delta + \gamma) + \gamma E_a/(\Delta + \gamma) \tag{20}$$

Equation (20) can also be evaluated as a defined combination of
measured meteorological parameters under non-potential conditions.
In the advection-aridity approach (Bouchet 1963, Morton 1965,
Brutsaert & Stricker 1979), a distinction is made between the
potential evaporation which would occur over a region with freely-
available water λE_{po}, and the value of potential evaporation λE_p
calculated from Equation (20) using actual meteorological data
with water in limited supply.

Under non-potential conditions, less energy is converted into
latent heat. If the actual regional evaporation is λE, the additional energy q is given by

$$q = \lambda E_{po} - \lambda E \tag{21}$$

This additional energy will increase the apparent potential evaporation. Bouchet (1963) assumed that, if there was no energy
advection into or out of a region at a scale larger than regional,
the increase in the apparent potential evaporation λE_p, above that
which would occur in potential conditions λE_{po}, corresponded to
this extra energy, i.e.,

$$\lambda E_p = \lambda E_{po} + q \tag{22}$$

Combining Equations (21) and (22) gives an expression for the
actual regional evaporation:

$$\lambda E = 2\lambda E_{po} - \lambda E_p \tag{23}$$

Morton (1976) proposed that E_{po} be calculated from the equation
proposed by Priestley and Taylor (1972), namely

$$\lambda E_{po} \simeq \alpha \, \Delta Q_N / (\Delta + \gamma) \tag{24}$$

where $\alpha \simeq 1.26$. Brutsaert and Stricker (1979) used a comparison
with measured evaporation in non-potential conditions to suggest
the use of the Penman (1948) version of Equation (20) for λE_p.

6.   THE RESPONSE OF EVAPORATION TO CHANGES IN CLIMATE VARIABLES

In earlier sections, the meteorological variables primarily
responsible for controlling evaporation were identified as: *Net
radiation* (strictly, R_N–G); *Temperature* (through Δ); *Vapour
Pressure Deficit, Windspeed* (through E_a); *Rainfall* (directly
through interception loss and indirectly via stomatal control).
The evaporative response to climate change is complicated by
the fact that meteorological variables influencing evaporation may
realistically be expected to change in a related way. Moreover,
present-day vegetation is strongly linked to prevailing climate
(e.g., Figure 3) and, therefore, will alter in response to major,
long-term climatic change, particularly changes in temperature and
precipitation. It is naive to suggest simple rules for predicting
changes in evaporation rate in response to large-scale and long-
lived changes in climatic factors. There are, already, many global
circulation models and, increasingly, attempts are being made to
describe the effect of land-surface interactions. However, if
further progress is to be made, it will be necessary to establish
a quantitative basis for describing the transitions between biomes
which are either adjacent or overlapping in any climatic classification. The general observations made below apply only to small-
scale or short-term climatic changes in which the type of
vegetation prevalent at the location of interest *does not alter*
significantly.
It is helpful to begin by considering the effects of climatic
changes on the standard evaporation rates described earlier.
Potential evaporation is best described by the version of the
Penman equation (Rijtema 1965) which has the form:

$$\lambda E_o = (\Delta Q_n + \gamma E_a) / (\Delta + \gamma) \tag{25}$$

in which $E_a = (e_s - e)(1 + U/100)$ $\tag{26}$

Reference crop evaporation is arguably described best by the
equation suggested by Thom and Oliver (1977) namely,

$$\lambda E_{RC} = (\Delta Q_n + 2.5\gamma E_a) / (\Delta + 2.4\gamma) \tag{27}$$

while potential evapotranspiration can be defined as

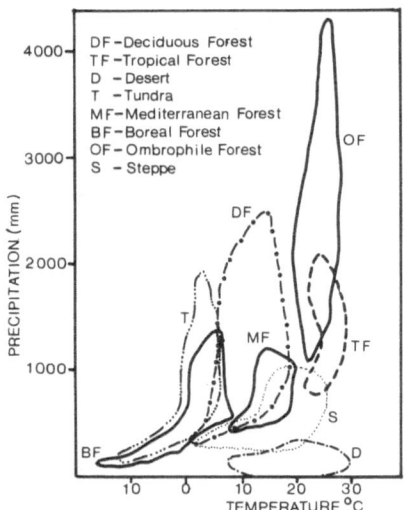

Fig. 3.   Distribution of major biomes in relation to precipitation
          and temperature (from Perrier 1981).

$$\lambda E_T = (\Delta Q_N + m\gamma E_a)/[\Delta + (1+n)\gamma] + I(1-c) \quad \text{for } T > 0°C$$

$$\qquad = (\Delta Q_N + \gamma E_a)/(\Delta + \gamma) \qquad\qquad\qquad \text{for } T \geqslant 0°C$$

(28)

in which I is the interception loss, related to rainfall amount
and pattern and the storage capacity of the vegetation, and

$$m = 53\ln^2(20/h + 2.5) \tag{29}$$

$$n = r_s[m(1 + U/100)]/250 \tag{30}$$

$$c = (\Delta + \gamma)/[\Delta + (1 + n)\gamma] \tag{31}$$

In Equations (29, 30), h is the vegetation height, and r_s is an
average value of bulk stomatal resistance for the   vegetation type
when soil moisture is not a strong limiting factor. Table II
gives typical values of r_s for some important vegetation types
(Perrier 1981).

     The introduction of Equation (28) as a description of poten-
tial evapotranspiration is speculative, but it is a reflection of
our present understanding of how different vegetation types affect
potential evaporation. It is a development of previous work (Gash
1978, Thom & Oliver 1977, Shuttleworth & Calder 1979) assuming a
reference height of 2 m above the vegetation, a zero-plane

Table II.   Typical values of r_s (m s^{-1}) for some important vegeta-
            tion types (from Perrier 1981).
            (Note: Values measured in field experiments exhibit
            variability of around × 2.)

	Crop		Biome
Alfalfa	40	Coniferous forest	200–300
Barley	70	Deciduous forest	100–150
Citrus	250	Tropical forest	100–300
Cotton	130	Grassland (temperate)	100
Maize	80	Grassland (subtropical)	200
Potato	70	Tundra	400
Rice	80		
Sugarbeet	50		
Sunflower	40		
Wheat	60		

displacement of 0.75 h and an aerodynamic roughness of 0.1 h.
The parameters m, n and I are crop-dependent: typical values for
contrasting vegetation types are:

Short grass n ∿ 2.5, m ∿ 3.5, I ∿ 0.2P
Forest   n ∿ 30, m ∿ 25, I ∿ 0.3P (temperate conditions)
                              ∿ 0.15P (tropical conditions)   (32)

where P is the precipitation rate.
        These standard evaporation rates all show a dependence on
both vapour pressure deficit and wind speed through the term E_a
(Equation 26), but there is some doubt as to the true role of
this term in the description of evaporation. As a measured entity
at a specific location, its role is evidently causal. At the same
time, it is a combination of meteorological parameters which is
very sensitive to modification by surface energy exchange and, as
such, can be considered, with some justification, a reflection of
the evaporation input to the atmosphere (Brutsaert & Stricker
1979). In light of this uncertainty, it may be convenient that,
when the atmosphere is passing over a *moist* surface, there is a
positive correlation between the terms γE_a and ΔQ_n. Equations
(25), (27) and (28) can be rewritten, therefore in the *approximate* form:

$$\lambda E_o = [1.26/(1 + \gamma/\Delta)]Q_N \tag{33}$$

$$\lambda E_{RC} = [1.65/(1 + 2.4\gamma/\Delta)]Q_N \tag{34}$$

and $\lambda E_T = \{(1 + 0.26 \text{ m})/[1 + (1+n)\gamma/N]\}Q_N + I(1-c)$ for T $>$ 0°C
$$\tag{35}$$
$\quad\quad = [1.26/(1 + \gamma/\Delta)]Q_N$ for T $\leqslant$ 0°C

6.1 Evaporative Response to Changes in Radiation

On the basis of Equations (33 to 35), it is clear that all
the standard evaporation rates show a marked dependence on radia-
tion input and that, for λE_o and λE_{RC}, the dependence is approximately linear. The exact nature of the relationship for λE_T is
less obvious because of the interception term in Equation (35).
When applying Equation (35) to short vegetation and moist soil,
however, λE_T equals λE_{RC} (Thom & Oliver 1977) so that an approximately linear relationship is regained. Shuttleworth and Calder
(1979) showed that potential evapotranspiration for forest vegetation also bears an approximately linear relationship to radiation,
although the correlation coefficient has a marked dependence on the
local precipitation rate.
The standard (potential) evaporation rates will change,
therefore, in a direct and approximately proportional way in
response to a change in radiation, and this will be reflected in
the actual evaporation rate.

6.2 Evaporative Response to Changes in Temperature

Equations (33, 34 and 35) incorporate an implicit dependence
on temperature through the parameter Δ. As temperature (and Δ)
increase, the fraction of energy received at the surface which is
converted into latent heat, $\alpha = \lambda E/Q_N$, increases. Figure 4 shows
the variation in this fraction as a function of temperature for
λE_o, λE_{RC} and for the leading (transpiration) term in Equation (35),
for the two extreme limits of short and tall vegetation (see
Equation 32). These results are only approximate, and the
assumption $\gamma E_a = 0.26 \Delta Q_N$ cannot remain valid as α approaches
unity since the atmosphere would rapidly saturate. Nor is this
assumption likely to hold at low temperatures which usually are
associated with low radiation, enhanced advection and an increase
in E_a. Even so, the fraction of radiation converted into latent
heat under potential conditions increases with temperature by
about 2% K^{-1} at normal temperatures, and this will be reflected in
the actual evaporation rate.

6.3 Evaporative Response to Changes in Precipitation

Potential evaporation and reference crop evaporation involve
no explicit recognition of the role of precipitation under potential conditions. In the more general situation, our present
understanding of potential evapotranspiration suggests some
dependence on precipitation input which varies with crop type and
is a direct consequence of rainfall interception and its evaporation from the surface of the vegetation. The actual evaporation
rate is also related to cumulative precipitation by way of the
surface (stomatal) response to soil moisture deficit. This control tends to moderate the rate of depletion of soil moisture.

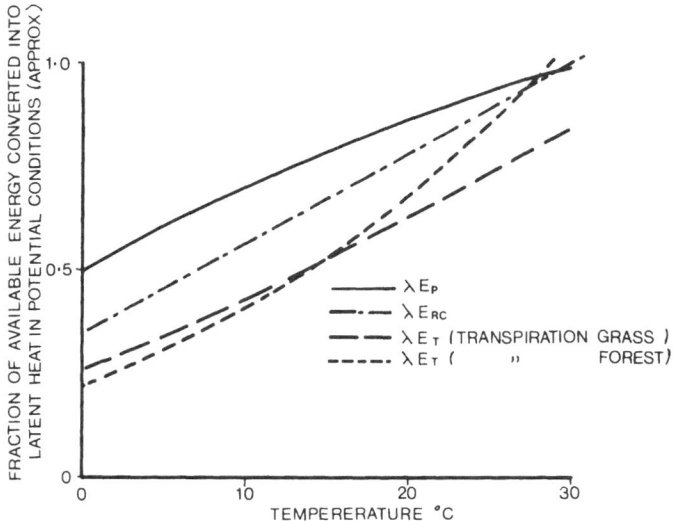

Fig. 4. Fraction of energy available at the Earth's surface used in evaporation in potential conditions for different 'standard' evaporation rates.

In climatic regions where water is adequate or abundant, the dominant vegetation is commonly tall. In such conditions, changes in precipitation will generate a related change in interception loss, while the influence through soil moisture on stomatal control is small. It is commonly observed that, at a particular location, the interception loss I over a given time period displays an approximately-linear relationship with the equivalent precipitation input P of the form:

$$I = A + BP \qquad\qquad (36)$$

the coefficients A and B varying with vegetation type and climate. Equatorial and ombrophilous forest biomes, usually associated with climatic regions where rainstorms are short and intense, lose only 10-20% of incoming precipitation by interception. Forests in temperate or Mediterranean climates, where rainstorms are less extreme, have an interception loss of about 30% (occurring over just part of the year in deciduous forests). Coniferous forests in northern latitudes can have interception losses as high as 50% (Halldin *et al.* 1979, Leyton *et al.* 1967).

In climatic regions where water is readily available throughout the year and forest vegetation is commonly found, potential evapotranspiration is a realistic estimate of actual evaporation and will change in response to small changes in precipitation as a direct consequence of changes in interception loss. To a first approximation, the size of the change is given by the expression:

$$\Delta(\lambda E) = (1-c) \, B \, \Delta \, P \tag{37}$$

where c and B are defined by Equations (31 and 36).

Where water is always in short supply throughout the year and annual evaporation is strongly (and, approximately, linearly) related to the water available in the form of soil moisture, the actual relationship between precipitation and the components of the hydrological cycle (evaporation, infiltration and surface run-off) varies considerably and can only be described by a detailed, hydrological model calibrated for local conditions. However, the present discussion is only concerned with changes in evaporation in response to changes in precipitation input, and it is possible to exploit the approximate linearity of the model parameters under water-limiting conditions. A small change in precipitation will generate, under the postulated conditions of regular water shortage, an approximately proportional increase in evaporation; thus

$$\Delta(\lambda E)/\lambda E = \Delta P/P \tag{38}$$

The extreme case of this last situation is the desert biome. Hot deserts are situated in climatic zones with annual rainfall totals of less than, say, 100 mm. Vegetation is sparse. When it does rain, runoff is important: the surface dries out almost immediately and evaporation quickly becomes negligible (Logan 1960). No doubt there will be some proportional evaporative response to increased precipitation, but the absolute values involved are small. Cold deserts are anomalous. They are characterized by long periods of permanent snow or ice with low sublimation of water vapour and generally only little precipitation. Potential evaporation is a good description of the actual evaporation rate for most, if not all of the year, so that a change in precipitation, which usually falls as snow, will have little effect.

In situations intermediate between the extremes of water abundance and shortage, the evaporative response to a climatic change in precipitation is more complex and more difficult to predict. This is particularly true for climatic zones in which there is marked seasonal variation of vegetation cover. It is fairly certain that an increase in annual precipitation will, in general, cause an increase in annual evaporation, but the actual relationship between the two will usually require empirical determination or the use of a physically-based and locally-valid hydrological model (e.g., Lockwood & Sellers: this volume). The situation is further complicated by the fact that such climatic regions are commonly those which are most extensively influenced by human intervention in the form of agriculture.

Figure 5 expresses the above comments in the form of a schematic diagram relating the evaporative response of the relevant biome to precipitation input.

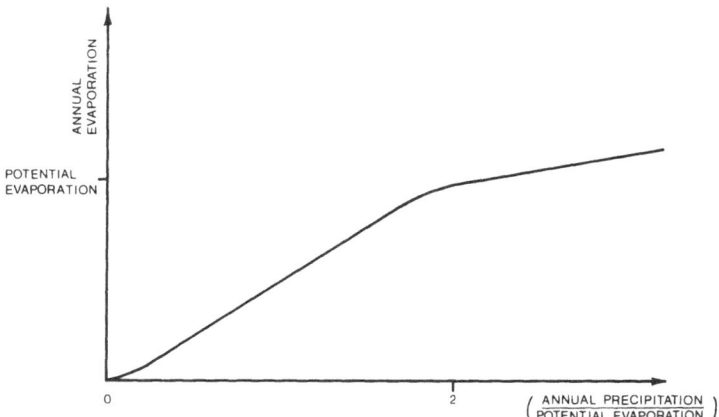

Fig. 5.   Actual annual evaporation as a function of the ratio
           between annual precipitation and potential evaporation
           (with some recognition of the correlation between
           precipitation and biome).

6.4   Comments on the Evaporative Response to Correlated Changes in
       Climatological Variables

       Previous discussion describes changes in evaporation in
response to changes in individual meteorological variables,
although changes in these controlling parameters are almost
certainly highly correlated. A climatological increase in atmospheric temperature is very probably related to an increase in
radiation, and such a change will have the dual effect of increasing the energy available for the evaporation process, as well as
the fraction of this energy converted into latent heat. If such
changes occur on a global scale, the global rate of evaporation
will increase (as two-thirds of the Earth's surface is water-
covered), with a commensurate increase in the global precipitation
rate (cf. Flohn: this volume). In this way, a correlated change
in precipitation will also occur, with indirect consequences for
the evaporation process over land surfaces.
       The effects of a decrease in radiation and the related
decrease in temperature are likely to be less stable because there
is a discontinuity in potential evapotranspiration at $0°C$. The
energy available for evaporation (sublimation) is altered significantly at this temperature since the surface albedo changes from
around 0.25, typical of vegetation, to around 0.8, typical of
snow. Thus, the evaporation rate at a specific location may be
significantly reduced once freezing occurs. About 50% of the
Earth's land surface is covered by snow or ice in the northern
hemisphere winter, falling to 12% in summer (Kuhn 1981). A fall
in mean global temperature which dramatically altered this ratio
could cause noticeable changes in the global hydrological cycle,
particularly in the northern hemisphere.

7.   EFFECT OF LAND SURFACE CHANGES ON CLIMATIC VARIABLES

The measured value of such meteorological variables as
temperature, humidity and wind speed is determined by the dynamic
mechanisms involved in the energy, mass and momentum conservation
of the atmosphere. Surface changes influence the measured temperature and humidity of the atmosphere by changing either total energy
exchange, or the way this energy is partitioned. Near-surface
measurements can be influenced by local surface conditions; more
extensive surface changes may have an integrated effect at higher
levels in the atmosphere.

7.1 Influence of Local Surface Conditions on Near-Surface
     Meteorological Measurements

It is well known that surface energy and momentum fluxes are
transferred into the atmosphere by the process of turbulent
diffusion and that gradients occur near the surface. Extensive
sets of synoptic meteorological data rely on measurements taken
within a few metres of the surface. Such data will be influenced
by surface exchange processes. Pearce *et al.* (1980) demonstrated
such a difference between grassland and forest data, with the
most extreme differences occurring during or immediately after
rainfall; they quote the following regression equations from air
temperature T and wet-bulb depression ΔT_w:

$$T \quad (forest) = 0.01 \quad T \quad (grassland) - 0.29 \quad (dry\ conditions)$$
$$\qquad\qquad\ = 0.99 \quad T \quad (grassland) - 0.63 \quad (wet\ conditions)$$

$$\Delta T_w \ (forest) = 0.85\Delta \ T_w \ (grassland) + 0.109 \quad (dry\ conditions)$$
$$\qquad\qquad\ = 0.36\Delta \ T_w \ (grassland) + 0.114 \quad (wet\ conditions)$$

These differences reflect the relative partitioning of energy
between these two vegetation types under potential conditions.

7.2 Influence of Surface Conditions on Total Energy Exchange

On reaching the ground, part of the short-wave radiation from
the Sun is reflected by the surface. The fraction reflected, the
reflection coefficient or albedo, depends on the angle of incidence of the solar beam and the type of vegetation. Differences
arise because tall vegetation is better able to absorb the solar
beam by multiple reflections within the canopy. Table III gives
some typical values of albedo expressed as a daily mean for a
broad range of vegetation types, while Figure 6 shows the relationship between measured albedo and crop height for several types of
crop. Changes in the solar radiation input of around 5-10% are
quite possible if surface vegetation changes significantly. In
general, a decrease in vegetation height increases the fraction of
solar energy reflected and, hence, reduces the total energy transferred into the atmosphere as latent or sensible heat.

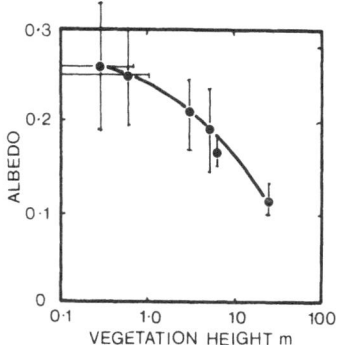

Fig. 6. Relation between height of vegetation and albedo (from
         Stanhill 1970).

Table III. Typical values of the mean albedo for several
            different crop types (from Monteith 1973).

Vegetation Type (at maximum ground cover)	Daily Mean Albedo (%)	Vegetation Type (at maximum ground cover)	Daily Mean Albedo (%)
Farm crops		Natural vegetation and forest	
Grass	24		
Sugar Beet	26	Heather	14
Barley	23	Bracken	24
Wheat	26	Gorse	18
Beans	24	Maquis, evergreen scrub	21
Maize	22	Natural pasture	25
Tobacco	24	Derived savanna	15
Cucumber	26	Guinea savanna	19
Tomato	23		
Wheat	22	Forests and orchards	
Pasture	25		
Barley	26	Deciduous woodland	18
Pineapple	15	Coniferous woodland	16
Maize	18	Orange orchard	16
Tobacco	19	Aleppo pine	17
Sorghum	20	Eucalyptus	19
Sugar Cane	15	Tropical rain forest	13
Cotton	21	Swamp forest	12
Groundnuts	17		

     The albedo of bare soil depends on organic content, water
content, particle size and angle of incidence. Arable farming in
middle latitudes, in which bare soil is exposed for an extended
period of time, commonly involves soil with significant organic

content, and with quite a high surface water content during the
dormant period. The albedo of such soil is usually less than the
albedo of most farm crops and of natural vegetation.

7.3   Surface Effects on the Partition of Energy

Mention has been made, earlier, of the fact that natural
vegetation is strongly, but not perfectly, related to climatic
zone (Figure 3). Even in natural conditions, overlaps between
biomes with respect to climatic variables distort this relationship, but arguably the most noticeable perturbation at present is
due to inadvertent human modification of the atmosphere and biosphere. The consequences of such modifications are still largely
speculative even though the effect on agriculture and forestry
has received much attention over the last few decades.

Hydrologists and agriculturalists can readily develop
'engineering solutions', calibrated against local measurements, to
the problem of predicting the effect of changes in surface cover
on the energy and water budget. This approach is expensive and
time consuming and thus tends to be concentrated in developed
countries. Nevertheless, as its descriptive ability is assured,
its results find ready acceptance.

The need to predict changes in water balance in regions where
traditional hydrological data bases do not (or cannot) exist and
to develop widely-applicable models has generated increased
research into the detailed processes involved in the hydrological
cycle. The problem of providing a suitable physical model is
difficult because of the complexity and diversity of the problem;
because of this, physically-based models tend to evolve towards an
engineering solution. In this way, descriptive equations known to
be realistic and relevant in particular locations are applied,
even though the parameters involved are optimized with respect to
data which are spatially and temporally extensive. Equation (28)
is an excellent example. It conveys the essence of a physical
understanding from local studies though, in hydrological application, r_s and the parameters involved in an interception submodel
are best determined from a relevant hydrological data base.
Once calibrated, equations of this type can generate worthwhile
descriptions of the evaporation process. Shuttleworth and Calder
(1979) used an earlier version of this approach to model the
ratio of coniferous forest-to-grassland potential evapotranspiration, the magnitude of which changes from near unity in the
eastern UK to about 2.5 in the extreme west due to the effect of
precipitation. The model has been extended to predict the likely
effect of afforestation in other areas of Great Britain (Calder &
Newson 1979).

Less extreme changes in surface vegetation will generate
similar, but less marked, changes in the water balance and give
rise to consequential changes in the atmosphere. For example,
the enhanced runoff fraction, due to artificially-improved

drainage, reduces the size of the soil moisture store, limits evaporation in periods of water shortage, and reduces downwind precipitation. The opposite practice, irrigation, is likely to have a pronounced effect on local meteorological parameters, but various limitations (engineering, energy, political and financial) are likely to restrict its application to a scale where it is unlikely to have a noticeable effect on global meteorology.

In general, the influence of Man has its most obvious effects in circumstances where land-surface characteristics are intimately related to climate, transitions between overlapping or adjacent biomes (see Figure 3) being particularly sensitive. In this respect, the evolution between steppe and hot desert, in response to a change in water balance, and between boreal forest and polar desert, in response to a change in temperature (see Miller: this volume), is likely to be particularly important.

REFERENCES

Blaney, H.F. & Criddle, W.O. 1950 Determining water requirements in irrigated areas from climatological and irrigation data. *Soil. Cons. Tech. Paper* 96, 48 pp. U.S. Dept of Agriculture.

Bouchet, R.J. 1963 Evapotranspiration, réele et potentielle, signification climatique. *Int. Ass. Sci. Hydrol. Publ.* 62, pp. 34-142.

Brutsaert, W. & Stricker, H. 1979 An advection-aridity approach to estimate actual evapotranspiration. *Water Resour. Res.*, 15, 443-450.

Calder, I.R. 1977 A model of transpiration and interception loss from a spruce forest in Plynlimon, central Wales. *J. Hydrol.*, 33, 247-265.

Calder, I.R. & Newson, M.D. 1979 Land-use and upland water resources in Britain. *Water Res. Bull.*, 15, 1628-1639.

Doorenbos, J. & Pruitt, W.O. 1977 Crop water requirements. *FAO Irrigation and Draining Paper* No. 24. Rome: FAO.

Gangopadhyaya, M., Uryvaev, V.A., Omar, M.H., Nordenson, T.J. & Harbeck, G.E. 1966 Measurement and estimation of evaporation and evapotranspiration. *WMO Tech. Note* No. 83. Geneva: WMO.

Gash, J.H.C. 1978 Comment on the paper by A.S. Thom & H.R. Oliver: On Penman's equation for estimating regional evaporation. *Q. Jl R. met. Soc.*, 104, 532-533.

Gash, J.H.C. & Morton, A.J. 1978 An application of the Rutter model to the estimation of interception loss from Thetford Forest. *J. Hydrol.*, **38**, 49-58.

Halldin, S., Grip, H. & Perttu, K. 1979 Model for energy exchange of a pine forest canopy. In: *Comparison of Forest Water and Energy Exchange Models* (ed. S. Halldin), pp. 59-75. Copenhagen: ISEM.

Kuhn, M. 1981 Vertical flux of heat and moisture in snow and ice. In: *Proc. Study Conf. on Land Surface Processes in Atmospheric General Circulation Models*. Geneva: WMO/ICSU.

Leyton, L., Reynolds, E.R.C. & Thompson, F.B. 1967 Rainfall interception in forest and moorland. In: *Forest Hydrology* (ed. W.E. Sopper & H.W. Lull), pp. 163-178. Oxford: Pergamon Press.

Logan, R.F. 1960 The central Namib desert. *South West Africa Publ.* No. 758, 162 pp.

Monteith, J.L. 1965 Evaporation and the evnironment. *Symp. Soc. exp. Biol.*, **19**, 205-234.

Monteith, J.L. 1973 *Principles of Environmental Physics*, 241 pp. London: Edward Arnold.

Morton, F.I. 1965 Potential evaporation and river basin evaporation. *J. Hydraul. Div. Am. Soc. civ. Engrs*, **91**, HY6, 67-97.

Morton, F.I. 1976 Climatological estimates of evapotranspiration. *J. Hydraul. Div. Am. Soc. civ. Engrs*, **102**, HY3, 275-291.

Pearce, A.J., Gash, J.H.C. & Stewart, J.B. 1980 Rainfall interception in a forest stand estimated from grassland meteorological data. *J. Hydrol.*, **46**, 147-168.

Penman, H.L. 1948 Natural evaporation from open water, bare soil and grass. *Proc. R. Soc. Lond.*, **A193**, 120-145.

Perrier, A. 1981 Vegetation. In: *Proc. Study Conf. on Land Surface Processes in Atmospheric General Circulation Models*. Geneva: WMO/ICSU.

Priestley, C.H.B. & Taylor, R.J. 1972 On the assessment of surface heat flux and evaporation using large scale parameters. *Mon. Weath. Rev.*, **100**, 81-92.

Ritjema, P.E. 1965 An analysis of actual evaporation. *Agric. Res. Repts* No. 659, 107 pp. Wageningen: PUDOC.

Rutter, A.J., Kershaw, K.A., Robins, P.C. & Morton, A.J. 1971
    A predictive model of rainfall interception in forests. I:
    Derivation of the model from observations in a plantation of
    Corsican Pine. *Agric. Met.*, 9, 367-384.

Rutter, A.J., Morton, A.J. & Robins, P.C. 1975 A predictive model
    of rainfall interception in forests. II: Generalization of
    the model and comparison with observations in some coniferous
    and hardwood stands. *J. appl. Ecol.*, 12, 367-380.

Shuttleworth, W.J. 1976a Experimental evidence for the failure of
    the Penman-Monteith equation in partially wet conditions.
    Bound.-Layer Met., 10, 91-94.

Shuttleworth, W.J. 1976b A one-dimensional theoretical description
    of the vegetation-atmosphere interaction. *Bound.-Layer Met.*,
    10, 273-302.

Shuttleworth, W.J. 1978 A simplified one-dimensional theoretical
    description of the vegetation-atmosphere interaction. *Bound.-
    Layer Met.*, 14, 3-27.

Shuttleworth, W.J. 1979 Evaporation. In: *Inst. Hydrol. Rept* No.
    56. Wallingford, UK: Institute of Hydrology.

Shuttleworth, W.J. & Calder, I.R. 1979 Has the Priestley-Taylor
    equation any relevance to forest evaporation. *J. appl. Met.*,
    18, 639-646.

Sinclair, T.R., Murphey, C.E. & Knoerr, K.R. 1976 Development and
    evaluation of simplified models simulating canopy photosyn-
    thesis and transpiration. *J. appl. Ecol.*, 13, 813-829.

Stanhill, G. 1970 Some results of helicopter measurements of
    albedo. *Sol. Energy*, 13, 59-64.

Tanner, C.B. 1967 Measurement of evapotranspiration. In: *Irriga-
    tion of Agricultural Lands: Agronomy II*, pp. 534-575.

Thom, A.S. 1972 Momentum, mass and heat exchange of vegetation.
    Q. Jl R. met. Soc., 98, 124-134.

Thom, A.S. & Oliver, H.R. 1977 On Penman's equation for estimating
    regional evaporation. *Q. Jl R. met. Soc.*, 103, 345-357.

THE USE OF LONG-TERM RIVER LEVEL AND DISCHARGE RECORDS IN THE
STUDY OF CLIMATIC VARIATIONS IN THE FEDERAL REPUBLIC OF GERMANY

H.J. Liebscher

Federal Institute of Hydrology
Koblenz
Federal Republic of Germany

ABSTRACT. In the Federal Republic of Germany, there are several
sources of long-term water level data both at tide gauges in the
North and Baltic Seas and on inland waters. This paper discusses
the problems of using the level and discharge records from the
River Rhine, many of which date back to the early 19th century, to
infer climate variations over this period. The results of the
investigation show how climate variations affect different months
in different ways and affect low flows differently to high flows.
Regional differences in the effect of climatic variations on the
streamflow regime of different portions of the basin can also be
recognized.

1. INTRODUCTION

     The hydrologist, the water resource designer, engineer and
manager all share a major interest in the consequences of climatic
variation on available water supplies, in particular on changes in
volume and seasonal distribution of river levels and river flows.
     Long-term records of these elements, if they are homogeneous,
can be studied for evidence of climate variability, though, in most
cases, river level records cannot be used because of changes in
river channel morphology. For the same reason, maximum flood
levels are also unsuitable. It is necessary, therefore, to turn
to more complex measures of river flow that are less affected by
anthropogenic changes. The examination of long-term precipitation
records may assist with such studies, but there are significant
constraints on their use:
(a) Few long-term rain gauge records are available;
(b) The validity of such data as are available is often affected
     by changes in location, observational practices and exposure;

173

A. Street-Perrott et al. (eds.), Variations in the Global Water Budget, 173-184.

(c) Rain gauges only provide point information.
     Runoff, being the residual between precipitation and evapo-
transpiration, may also be regarded as a climatic variable. In
contrast to rainfall, runoff is an integrated variable representing
the average over the catchment area, and is thus suitable for
investigating climatic variations. The main problem in using runoff data is the likelihood of anthropogenic changes within the
catchment. Examples are:
(i)   Water resource projects: reservoir construction; harnessing
       of lakes; inter-basin transfers; water withdrawal for
       industrial, domestic and agricultural purposes;
(ii)   River works: deepening for improvements for navigation and
       drainage, bank raising and flood plain impoundment; dredging
       for river gravels; bridge construction; river training;
(iii) Land use changes: urbanization; afforestation; irrigation
       and other agricultural practices; drainage of swamps.
Each of these changes is capable of affecting the flow regime and
many of them have a primary influence on the low flows and on the
relationship between the flow in the river and the river level.

2.   AVAILABLE DATA

     In the Federal Republic of Germany, regular water level
recording at gauges located on the larger rivers (as well as on the
coast) extend back to the early years of the last century - the
first discharge measurements on the Rhine were conducted in the
late 18th century. Despite the errors affecting these early
measurements, they have been used with success together with more
recent discharge measurements in stable-bed river sections to
construct discharge series for some gauges which extend back to
the mid-19th century and, in a few cases, to the beginning of that
century.
     This paper concentrates on the river flow records of the Rhine.
Table I lists the stations for which daily mean discharge data are
available and Figure 1 gives their locations.

Table I. Details of available river flow data.

Station	Lat.	Long.	Period of record	Catchment area (km^2)
Basle	47°33'	7°36'	1801-1978	36 035
Maxau	49°02'	8°18'	1821-1978	50 343
Worms	49°32'	8°22'	1821-1978	68 936
Kaub	50°05'	7°46'	1821-1978	103 729
Andernach	50°27'	7°35'	1821-1978	139 795
Cologne	50°56'	6°58'	1821-1978	144 612

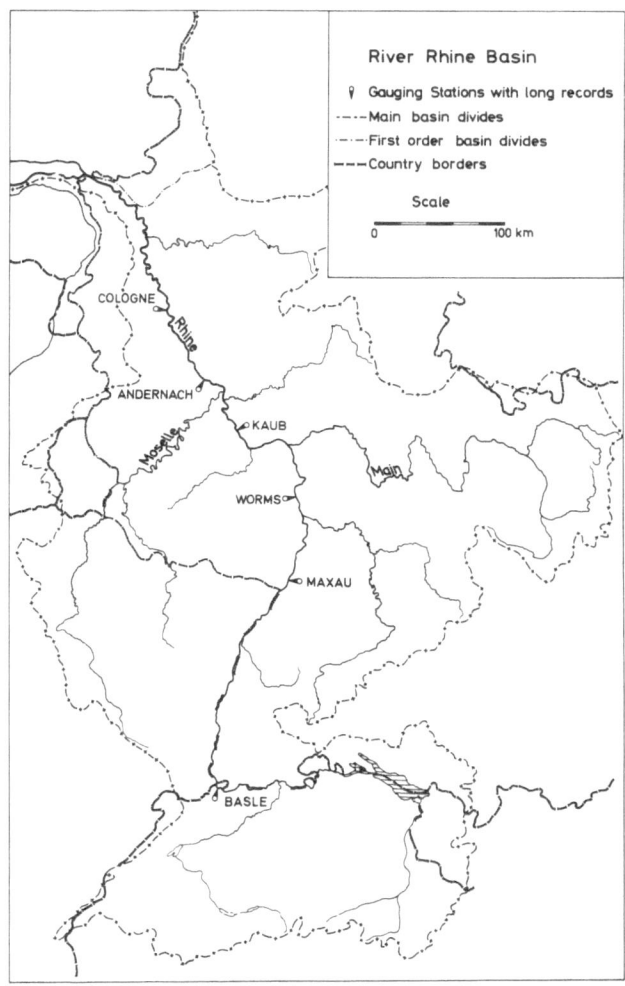

Fig. 1.   Location of stream flow gauges in the Rhine basin.

3.   PROBLEMS OF USING FLOOD DATA

Apart from the anthropogenic effects on the flow record, additional uncertainties result from the conversion of water level to discharge. Figure 2 illustrates the changes in level that have taken place over time at the Worms gauge for a given discharge. In particular, it shows that the impact of river works has been most marked for high discharges. The same effect is demonstrated in a different way, in Figure 3, which simulates the effect at Maxau of the 1882/1883 extreme flood under various stages of river improvement (including the removal of 90% of the natural flood detention

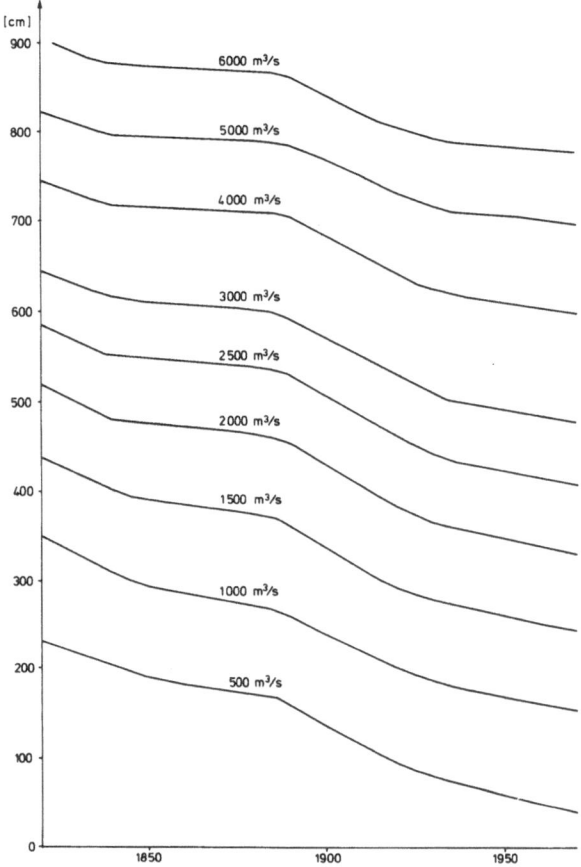

Fig. 2. Variations in water level (cm) at constant discharges
         (m^3 s^{-1}) at the Worms stream flow gauge, from data for
         1821 to 1970.

areas). The trend, to increase the size of the peak and to reduce
the catchment's response time, is quite clear. Developments in the
Upper Rhine since 1955 have resulted in the following increases in
the estimated 100-year return period flood: 300 m^3 s^{-1} at Strasbourg (8%); 700 m^3 s^{-1} at Karlsruhe Maxau (15%) and 900 m^3 s^{-1} at
Worms (16%). These examples show that data on river levels and
short-term high water periods cannot be used to ascertain climatic
variations. The following discussion examines the changes in total
runoff on a monthly, seasonal and annual basis, and the extent to
which these changes may be ascribed to climatic variations.

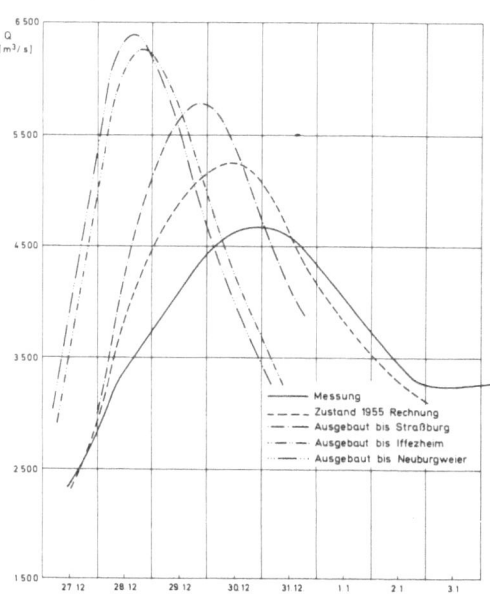

Fig. 3. Effect of river improvement works on the flood hydrograph
at the Maxau gauging station (Engel 1977): ——, flood
hydrograph for period 27 December 1882 to 3 January 1883;
---, computed hydrograph for the same flood given river
improvements up to 1955; -·-, computed hydrograph following further river improvements as far as Strasbourg;
-···-, same but for improvements as far as Iffezheim;
-·····-, same but for improvements as far as Neuburgweier.

4. VARIATIONS IN DISCHARGES

The results of a trend analysis of runoff for each station
(between 1821 and 1978) for individual months as well as for the
winter and summer halves of the years and for whole years is given
in Figure 4. The trend is given as an annual average rate of change
over the period of the record. Figure 4 also shows the 95% and 99%
confidence intervals which, in most cases, include the zero change.
However, there is a large measure of consistency in the behaviour
between stations. In particular, the annual runoff shows an increase
for every gauge except that at Cologne. This increase may be
ascribed to an increase in winter discharges especially during
February, which month's change comes closest to individual statistical significance. Cologne's aberrant trend (Figure 4) is probably
due to a change in the level-discharge relationship in the last
century. Another point of consistency is the trend along the course

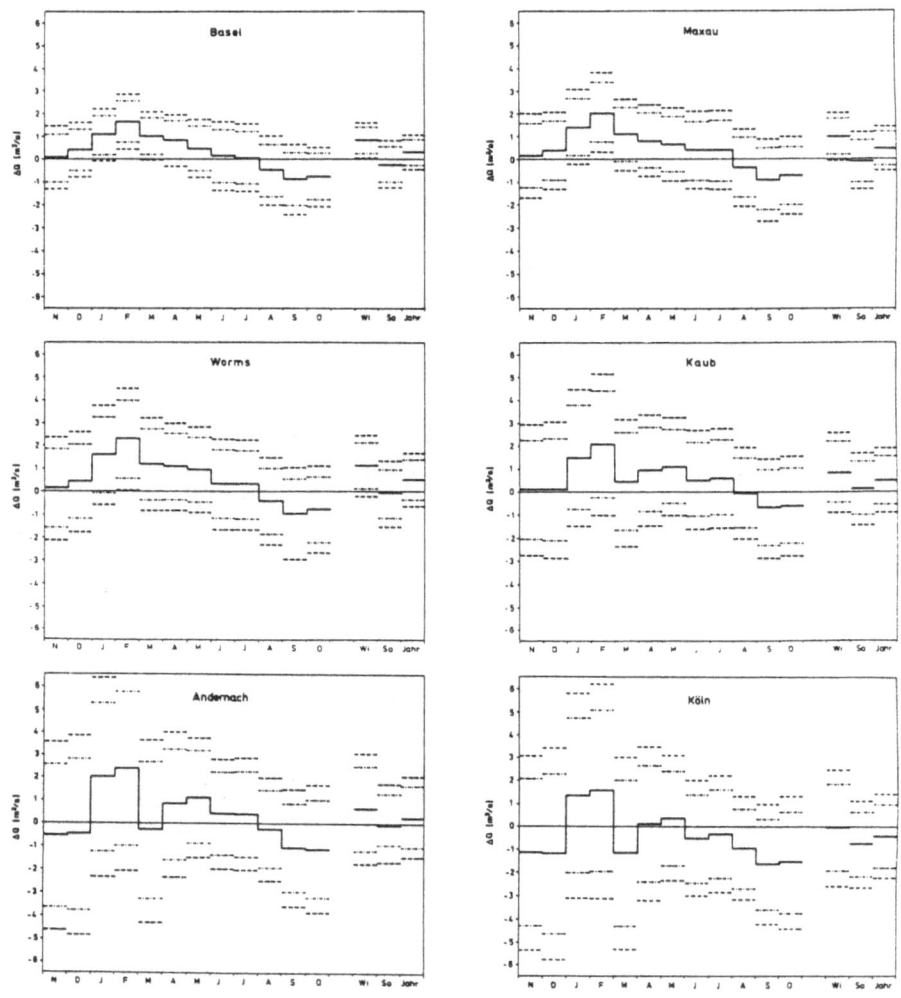

Fig. 4.   Trends in mean monthly, seasonal and annual discharges at
          six stream gauge stations from 1821 to 1978 (after
          Herlitze 1981) at 95% and 99% levels of significance.

of the Rhine which is also visible in Figure 4.   The regularity of
the monthly pattern with a maximum positive trend in February and
maximum negative trend in September is most clearly seen at Basle,
the most upstream station.   As one moves down the course, this
smooth monthly pattern is increasingly perturbed although the
January and February maxima are augmented.   In contrast, the
increases in the other winter months, especially March, diminish.

Thus, Basle shows most clearly a trend towards an enhanced seasonal variation in the flow.

An alternative way of demonstrating temporal trends of flow characteristics is through the study of running means; Figure 5 shows a sample. As well as showing the same time and space trends as Figure 4, the running means permit epochs of particular change to be identified. The most noteworthy results are:

(a) The periods of very high November runoff before 1840, from 1870 to 1895, and from 1915 to 1945 on the lower Rhine stations;

(b) Between 1900 and 1910, January flows increased dramatically, the augmented flows mostly having been maintained subsequently;

(c) The increase in the February runoff mainly took place after 1920.

While these results could be interpreted as evidence of anthropogenic change, might they not be indicative of a small change in the climate?

During the past 70 years, several reservoirs have been built in the Rhine basin to store water in time of surplus for subsequent release (to supply) in times of shortage. Times of water surplus correspond, in the Alpine area, to the period of snow and glacier melt between April and July. In non-Alpine catchments, the winter (November to April) is the period of water surplus. The period of water shortage is generally from August to October. Accordingly, the influence of reservoirs should be to occasion a decrease during winter and spring discharges and a corresponding increase in the late summer. The trend in Figure 4 is the precise opposite.

Water resource exploitation of the Rhine does not lead simply to a redistribution of flows within the years - there are losses due to agricultural and industrial use. This has been estimated to be a little over 1% of the Rhine's runoff at the German-Dutch border and has been increasing at about $0.15 \text{ m}^3 \text{ s}^{-1}$ per annum over recent years. As much of this use occurs below Cologne, the influence of this source of disturbance lies below the limit of evidence at all the stations used in this study. There are no significant water transfers to or from the Rhine basin while urbanization affects only a small proportion of the area. Therefore, the differences observed in Figure 4 must be due to climatic variation.

It is possible that the recession of the glaciers between 1850 and 1950 may be responsible. However, glacier melt contributes only 3.4% of the discharge at Basle; any increase due to the recession of the glaciers could only be a small fraction of this and, thus, would also lie below the limits of the evidence.

The responsibility of a climate change for the observed trends can be tested more directly by making use of the rather limited raingauge network. Ambs (1979) constructed a time series of rainfall records over the northern part of the Upper Rhine plain. As this area corresponds to the catchment between Basle and Kaub, it is possible to compare the rainfall series with the areal runoff obtained as the difference between the observations at the two gauges. Both series have been expressed in terms of 31-year running

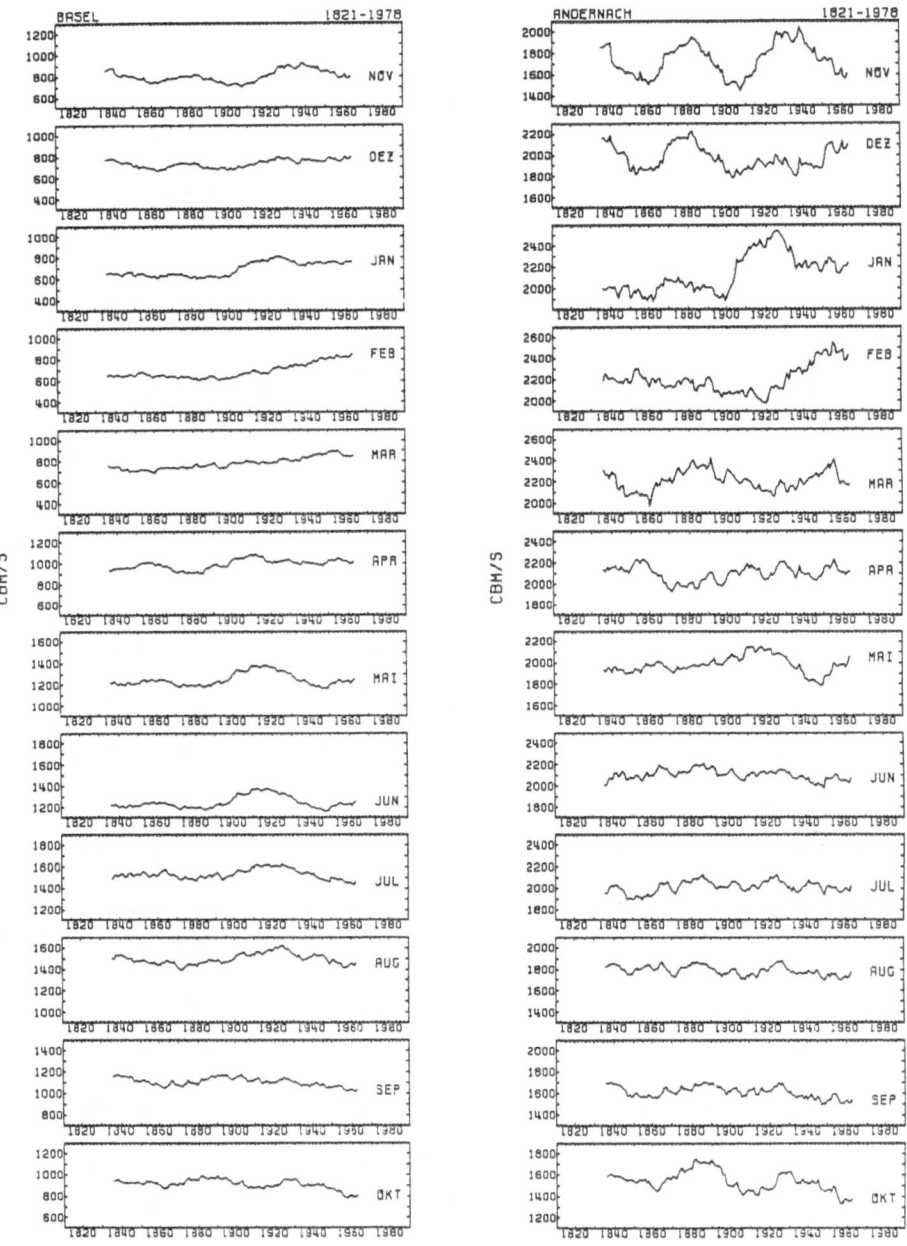

Fig. 5. Running monthly mean discharges over 31 years at the
Basle and Andernach stream gauge stations, from data for
1821-1978 (after Herlitze 1981).

means of runoff for the winter and summer halves of the year. The
results, shown in Figure 6, are rather inconclusive. There is a
weak upward trend in the winter rainfall mostly concentrated in the
present century and this, possibly, may accord with the runoff
increase referred to above. However, there is no evidence for an
increase in the runoff between the Basle and Kaub gauges because
the runoff from both increases by a similar amount. Discharges,
of course, are not influenced by precipitation alone. Evaporation
enters the equation and there are important lag effects so that a
season's rainfall does not necessarily make its effect in the river
during that season. A long-term temperature series ought to be
included in the investigation and, perhaps, a more sophisticated
transfer mechanism.

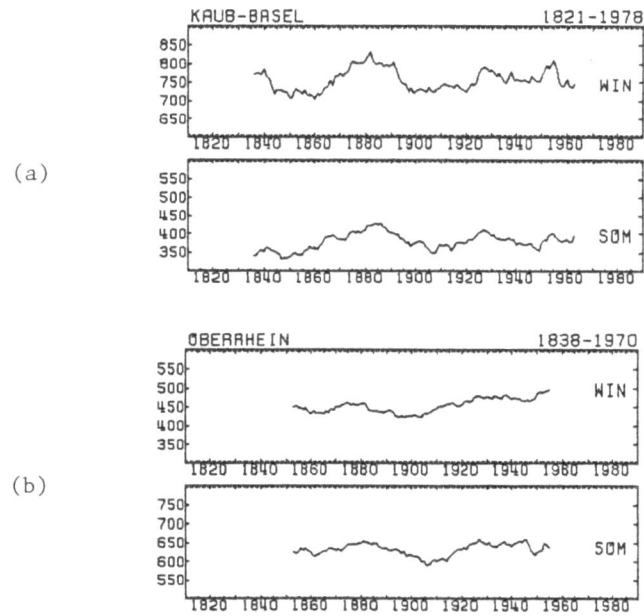

Fig. 6.   Running semi-annual mean differences over 31 years between
           (a) the discharges ($m^3 s^{-1}$) at the Kaub and Basle stream
           gauge stations, and (b) precipitation (mm) in the upper
           Rhine plain; from data for 1821 to 1978 (after Herlitze
           1981).

    The tests so far have been based on the supposition that the
variations would be most evident in particular seasons of the year.
However, it is conceivable that climatic changes could influence
the duration and volume of the runoff during periods of high and

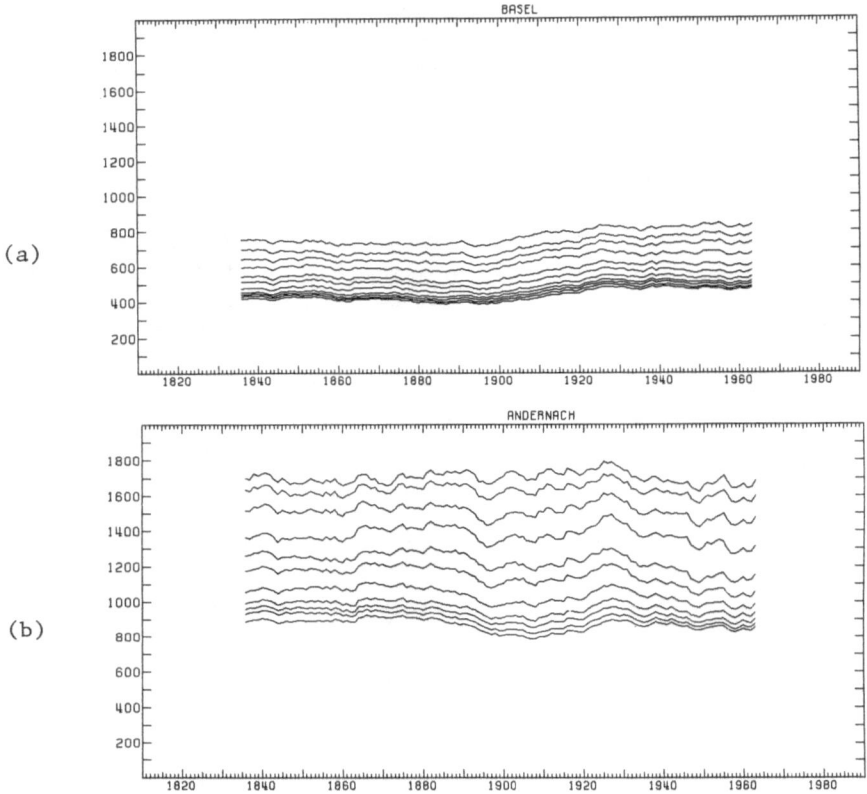

Fig. 7. Daily mean minimum discharges (m³ s⁻¹) over 31 years,
         from data for 1821 to 1978 (after Herlitze 1981) at:
         (a) Basle; (b) Andernach. Averaging periods: 5 days
         (lowest lines), 10, 15, 20, 30, 45, 60, 90, 120, 180 days
         (uppermost lines).

low flows. To test this possibility, the flow data were analysed
to extract the maximum 1, 2, 3, 5, 10, 15, 20, 25, 30, 45 and 60
consecutive day flows in each year; each of these series was in
turn subjected to the 31-year running mean process referred to
above. Corresponding annual minimum flows over 5, 10, 15, 20, 30,
45, 60, 90, 120, 150 and 180 consecutive days were also extracted.
Figures 7 and 8 show examples of this analysis for Basle and
Andernach. The most noticeable feature is the smooth course of the
curves of the annual minimum flows for Basle. An overall increase
is observable in the low flow, but is mainly concentrated between
1900 and 1920. This is in accord with the previous discussion, the
low flow period at Basle being around January and February. For

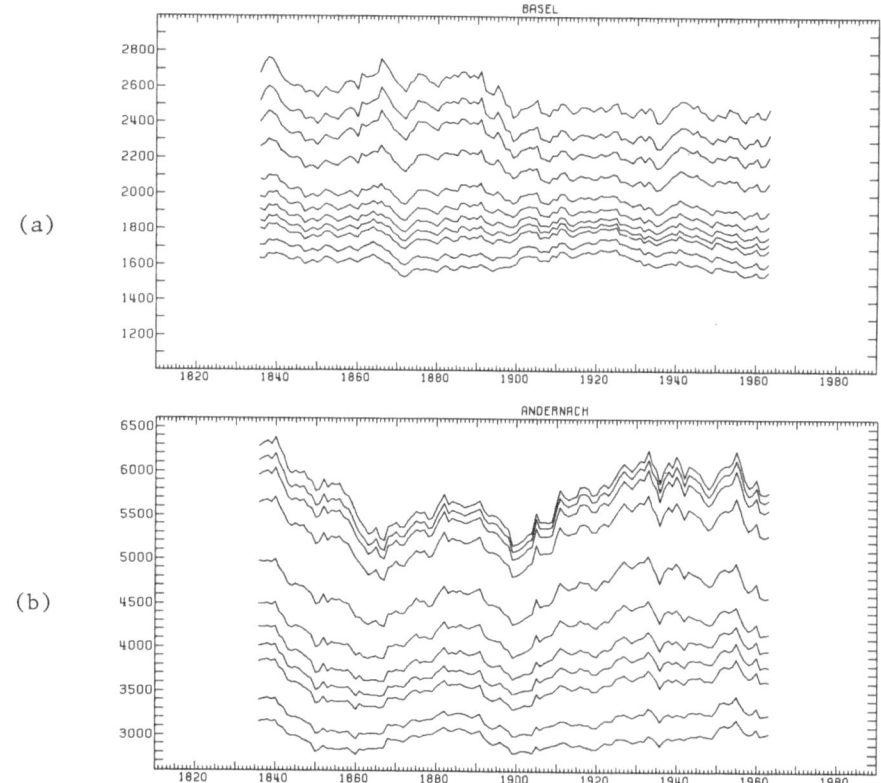

Fig. 8. Daily mean maximum discharges (m^3 s^{-1}) over 31 years,
from data for 1821 to 1978 (after Herlitze 1981) at:
(a) Basle; (b) Andernach. Averaging periods: 1 day
(uppermost lines), 2, 3, 5, 10, 15, 20, 25, 30, 45, 60
days (lowest lines).

both the minima and maxima, the variation over time at the downstream gauge is much less consistent than that of the upstream
gauge, evidence possibly of the simpler hydrological regime upstream and the larger size of catchment area downstream. Another
contrast between the trends at upper and lower gauge sites occurs
in the shape of the low flow hydrograph, as is apparent in the
spacing of the 5, 10, 15, 20 and 30 day curves. At Basle, the
short duration curves appear to have opened up over the recording
period indicating 'peakier' hydrograph shapes or an increased
tendency for low flow periods to be interrupted. The opposite
tendency is apparent at Andernach.
     Difficulties in identifying trends from short duration high
flows have already been referred to, the traces being decidedly more

irregular than those for low flows. Over the longer durations
(e.g., 45 and 60 days), the trends are not particularly marked
although, at Andernach, the pre-1840 period stands out as one of
high runoff, but a rising trend is observed throughout this century.
Basle displays neither of these trends.

Diagrams similar to those of Figure 4 were prepared to test
the significance of the apparent trends shown in Figures 7 and 8.
These tend to show that individual curves could arise by chance
except for the minimum flow series for Upper Rhine stations. However, in contrast to the trends in Figure 4, there is less
coherence between stations to support the visual impression of
uniformity so it is possible that variations are better studied
using seasonally-defined data than by using extreme values.

5.   CONCLUSIONS AND RECOMMENDATIONS

Investigations of climate variation through the study of long-
term series of water levels and discharges are only just beginning,
but do appear to be capable of yielding useful results. This paper
has presented a number of suggestions for analysing flow records
based upon trend analysis and running means. The availability of
a series of gauge stations down the course of a single river was of
considerable value for identifying individual trends. Some aspects
of the flow regime appeared more useful than others: analyses
of monthly and seasonal running mean values appeared very worthwhile, but flood peaks were too much influenced by river works.
Future prospects include a more sophisticated form of hydrological
modelling incorporating concurrent rainfall and temperature series
and the extension of the study to the Weser and Danube regions.
Data from these rivers still require verification - a vital precursor to a study of this type though a time-consuming process.

REFERENCES

Ambs, A. 1979 Statistische Analyse von Zeitreihen monatlicher
    Niederschlagssummen im Oberrheingebiet und Bildung von
    Gebietsmitteln. Diploma Thesis: Univ. Bonn, 89 pp.
    [Unpublished.]

Engel, H. 1977 Anthropogene Einflüsse auf die Abflussverhältnisse
    des Oberrheins zwischen Basel und Worms. Kurzdokumentation
    der Arbeiten der Hochwasserstudienkommission für den Rhein,
    Koblenz.

Herlitze, R. 1981 Untersuchungen langer Abflussreihen am Ober- und
    Mittelrhein. Diploma Thesis: Univ. Bonn, 146 pp.
    [Unpublished.]

PLANT AND SOIL WATER STORAGE IN ARCTIC AND BOREAL FOREST ECOSYSTEMS

P.C. Miller

Systems Ecology Research Group
San Diego State University
San Diego, California 921821900, USA

ABSTRACT. The amount of water stored in plants and soil in Arctic
and boreal forest was calculated from recent estimates of the
amounts of carbon stored in organic matter above and below ground.
The total water content of the biomass and seasonally-thawed layer
of the Arctic and boreal forest was estimated to be 5700 Gt, most of
which is stored in trees, peatlands, and that uppermost layer of
the soil subject to seasonal thawing. The water held in the
organic layer of the soil in northern ecosystems should increase
by about 1.6 Gt yr^{-1} from the expected increase in organic matter
due to CO_2-induced climatic change. Measured evapotranspiration
rates range from less than 0.5 mm per day on the driest sites to
4.5 mm per day on sites with standing or flowing water, indicating
an evaporation rate of 3400 Gt yr^{-1} from the total Arctic and
boreal forest region. The overall effect of climatic warming on
the water balance of northern ecosystems is complex and difficult
to predict because of the interactions of physical and biological
processes which affect and are affected by local water balance.

1.   INTRODUCTION

     The evaluation of the storage amounts and fluxes of water in
the global hydrological cycle has proceeded slowly despite the
potential importance of this information. Much of the emphasis has
been on the oceans and polar ice caps where most of the water is
stored. As understanding of the global water budget becomes more
detailed in terms of the amounts of water and of mechanisms, both
the quantities and fluxes of water in the terrestrial biomass,
soils and subsoils also need to be estimated. Northern latitudes
are important because of their sensitivity to climatic change and

A. Street-Perrott et al. (eds.), Variations in the Global Water Budget, 185–196.

of the stabilizing effect of freezing and thawing in these regions.
In this paper, the amount of water in the biomass and in the
seasonally-thawed soil layer of the ice-free circumpolar terrestrial
Arctic and northern boreal forest are estimated from the results of
a workshop on the carbon balance of northern ecosystems held in
San Diego in March 1980 (Miller 1981). The objective of the workshop was to refine extant estimates of the areal extent of northern
ecosystems and to estimate their carbon content, current carbon
flux, and future carbon flux assuming a world-wide CO_2-induced
climatic warming. The assumption is made that these estimates of
carbon in biomass and in dead organic matter can be converted into
estimates of water content by suitable conversion factors. The
water content of living plant material varies within narrow limits
(usually 800-1000% of dry weight), while that of the organic soil
layer is usually 300-1000% of dry weight (Larcher 1980, Miller 1981).

2.   METHODS

      At the workshop, two methods were used to estimate the areal
extent and carbon content of the Arctic: one based on geographic
zones and the other on ecosystem types; both methods gave similar
results. Since available data on water relations are based on
ecosystem type, only this method is considered here.
      The terrestrial Arctic was divided into ice-covered land and
six ecosystem types: (1) polar desert, (2) semi-desert, (3) wet
sedge-moss tundra, (4) cotton-grass tussock tundra, (5) low-shrub
tundra and (6) tall-shrub tundra. The area of the terrestrial
Arctic and of each ecosystem type was estimated from information in
Tikhomirov *et al*. (1981), Andreev and Aleksandrova (1981) and
Bliss (1977). Estimates of plant standing crop and net production
in these ecosystem types were based on Wielgolaski *et al*. (1981),
Tikhomirov *et al*. (1981), Bliss and Svoboda (unpublished data),
and Komárková and Webber (1980). The boreal forest was divided
into three zones (northern, middle and southern), after an initial
attempt to divide it into five ecosystem types had been abandoned
owing to insufficient data. For each zone, estimates were made of
the areal extent of the carbon content of the biomass, forest floor
and soil, and of carbon fluxes.
      Although the polar deserts and ice caps cover 42% of the total
Arctic land surface, their biological production is insignificant.
Plant cover in the polar deserts averages 1-2% of the surface area,
with vascular plants predominating. The semi-desert ecosystem is
dominated by cushion plants, lichens and mosses. The abundance of
lichens and mosses results in a total plant cover of 70-90%; the
vascular plant cover is generally 15-20%. The wet sedge-moss
tundra is dominated by up to four species of sedge and one or two
species of grass in low Arctic areas, but usually only contains one
sedge species, of limited distribution, in the high Arctic. This
type of vegetation dominates the coastal plain of Alaska and large

areas of the northern sub-Arctic and southern Arctic zones of the
USSR; it is only found in limited areas of the Canadian Arctic.
Cotton-grass tussock tundra is characterized by cotton grass,
three to five species of dwarf heath shrubs, scattered low shrubs
of birch and willow, and abundant lichens and mosses. This ecosystem is common in Alaska and the USSR, but occurs only infre-
quently in Canada. In low-shrub tundra, shrub heights average
0.5-1.0 m. The dominant species are willow and birch, with dwarf
heath species, cotton grass, sedges, lichens and mosses. Low-shrub
tundra dominates the southern sub-Arctic in the USSR and is common
in the low Arctic of eastern Canada. Tall-shrub tundra (where the
average shrub height is 2-4 m), mainly comprising willow and alder,
occurs in riparian habitats and on steep slopes that are well
drained.

For the circumpolar boreal forest, estimates of carbon content
were based on data from the major vegetation types near Fairbanks,
Alaska. Stands of quaking aspen, paper birch, white spruce, black
spruce and balsam poplar were analysed for the standing crop of
the forest floor (Van Cleve & Noonan 1975, Van Cleve & Viereck:
unpublished data) and for tree biomass, including tops and roots,
following the methods of Barney *et al.* (1978). The carbon content
of the trees, forest floor and the mineral soil of black spruce
stands were used to estimate the carbon content of northern boreal
forest. The carbon content of the middle boreal forest was estimated from data for birch, aspen and poplar stands, and that of
the southern boreal forest from data for white spruce stands. In
other words, the forest types with the lowest biomass and standing
crop of carbon were used to estimate the carbon in northern boreal
forest, the types with the highest biomass and standing crops were
used for southern boreal forest, and intermediate values were used
for middle boreal forest.

Carbon content was converted to water content by assuming a
carbon:biomass ratio of 0.45 and a carbon:dead organic matter ratio
of 0.58 (Ajtay *et al.* 1979). The water contents (g H_2O per g dry
weight) of the biomass, organic soil layer, and of mineral soil was
assumed to be, respectively, 9, 5, and 0.2-0.35. The depth of the
annual maximum thaw, thickness of the organic layer and bulk
densities were estimated for each type of ecosystem based on the
methods of Everett (1981) (Table I). The difference between the
maximum depth of thaw and the organic matter thickness gave the
depth of the mineral soil in the seasonally-thawed layer. The
product of this depth and the bulk density gave the weight of
mineral soil per unit area. The water content of the permafrost
(perenially frozen ground) was not estimated.

Evapotranspiration measurements are rare for Arctic ecosystems.
Evapotranspiration estimates are given in this paper for tussock
tundra (based on the work of Stuart *et al.* 1982), semi-desert
(Addison 1977), boreal forest (Pavlov 1976), and wet meadow (Weller
& Holmgren 1974, Addison 1977, Koranda *et al.* 1978, Miller *et al.*
1976, and Ng & Miller 1975, 1977). Future water fluxes and

Table I. Summary of parameters and fluxes related to water content in the seasonally-thawed layer (or the uppermost 1 m in the absence of permafrost) in the Arctic and boreal forest ecosystems.

PD, Polar desert; SD, Semi-desert; WM, Wet meadow; TT, Tussock tundra, LS, Low shrub; TS, Tall shrub; NT, Northern boreal forest; MT, Middle boreal forest; ST, Southern boreal forest; P, Peatland.

	PD	SD	WM	TT	LS	TS	NT	MT	ST	P
Parameters										
Depth of thaw (cm)	58	58	30	50	50	80	50	all	all	all
Depth of organic layer (cm)	0	2	20	30	5	5	30	10	5	–
Depth of thawed mineral soil (cm)	58	56	10	20	45	75	20	100	all	–
Bulk density of mineral soil (g cm^{-3})	1.3	1.3	1.4	1.3	1.3	1.3	1.3	1.3	1.3	–
Weight of thawed mineral soil (kg/m^2) (kg m^{-2})	754	728	140	260	585	975	260	1300	1300	–
Water content of mineral soil (kg kg^{-1} dw)	0.2	0.2	0.35	0.2	0.2	0.2	0.35	0.2	0.2	0.35
Water content of organic layer (kg kg^{-1} dw)	5	5	5	5	5	5	5	5	5	5
Fluxes										
Precipitation (mm yr^{-1})	175	175	200	250	300	325	350	400	450	450
Evapotranspiration (mm yr^{-1})	75	75	185	125	175	200	240	240	240	240
Runoff, including blowing snow (mm yr^{-1})*	100	100	15	125	125	125	110	160	210	210
Runoff (mm yr^{-1})†	100	100	125	125	125	125	150	200	250	–
Runoff: precipitation ratio	0.57	0.57	0.08	0.50	0.42	0.39	0.31	0.40	0.47	0.47

*summer season only
†estimated from Hare and Hay 1974

quantities were estimated by postulating a $4°C$ warming in the
Arctic with increased levels of atmospheric CO_2 and were based on
the conclusions of the workshop on changes in carbon content from
global climatic changes (Wigley *et al.* 1980).

3.   RESULTS

3.1   Areal Exent

The Arctic and boreal forest ecosystems cover about
17×16^6 km^2, 11% of the total world terrestrial area (Table II)
and 3.4% of the total world area. Of this area, 60% is forested,
34% is in the Arctic, and 6% comprises peatlands in the boreal
forest region. This estimate of the extent of the northern ecosystems differs only from earlier estimates (Whittaker & Likens
1973, 1975, Woodwell *et al.* 1978, Ajtay *et al.* 1979) in recognizing
that a large area of the Arctic (2.76×10^6 km^2), that had been
included in previous estimates, is ice-covered and therefore should
be excluded from consideration of the carbon balance.

3.2   Carbon Content

The amount of carbon stored in the biomass is small in peatlands and Arctic ecosystems, but relatively large in the boreal
forest. Overall, the biomass in northern ecosystems is about 14%
of the total world biomass. Most of the carbon is contained in
dead organic matter in the permafrost and seasonally-thawed layer
of wet sedge-moss tundra, tussock tundra and boreal forest ecosystems. The carbon in the dead organic matter is almost 400 Gt
(1 Gt = 10^{12} kg or $\simeq$ 1 km^3) or 23% of the dead organic carbon in the
terrestrial biosphere. About 120 Gt are held in the permafrost
and about 273 Gt in the seasonally-thawed layer. The overall
carbon content of Arctic and boreal forest ecosystems is about 21%
of the total world terrestrial carbon.

3.3   Water Content

The total water content of the biomass and of seasonally-
thawed layer of the Arctic and boreal forest is estimated to be
about 5700 Gt (Table II). About 75% of this amount is stored in
the forested areas and peatlands of the boreal forest, and 31% is
in the thawed organic layer of the wet meadow tundra, tussock
tundra, and boreal forest peatlands. Only 21% of the water stored
in northern ecosystems is contained within the biomass. The total
water content of the Arctic and boreal forest is equivalent to only
about 0.07% of the global groundwater reservoir or about 38% of the
water in the atmosphere. According to ^{14}C dates of the surface
organic mat (Everett 1981), the water stored in the organic layer
has accumulated during the last 6000-10 000 years.

Table II. Summary of areal extent, carbon content and water content of the seasonally-thawed layer (or uppermost 1 m in the absence of permafrost) and water fluxes in Arctic and boreal forest ecosystem types.

For explanation of the types of ecosystems, see Table I.

	PD	SD	WM	TT	LS	TS	NT	MT	ST	P	Total
Area (10^6 km^2)	0.80	1.50	1.00	0.90	1.28	0.23	3.34	3.57	3.18	1.1	16.9
Carbon content (Gt)											
Biomass	0.00	0.43	0.95	3.00	0.98	0.60	11.9	24.3	34.7	3	80
Dead organic layer	0.02	10.8	13.4	26.1	4.86	0.09	11.5	6.5	10.6	122	206
Water content (Gt)											
Biomass	0.00	9	19	60	20	12	159	326	465	60	1160
Dead organic layer	0.2	93	115	225	42	1	99	56	91	1049	1770
Mineral soil	120	219	49	82	150	45	304	928	827	–	2724
Total	120	321	183	367	212	58	562	1310	1383	1109	5655
Fluxes (Gt yr^{-1})											
Precipitation	140	263	200	225	384	75	1169	1428	1431	495	5810
Evapotranspiration	60	113	185	113	224	46	802	857	763	264	3427
Runoff	80	150	15	112	160	29	367	571	668	231	2383
Runoff: precipitation ratio	0.57	0.57	0.08	0.50	0.42	0.39	0.31	0.40	0.47	0.47	0.41

Assuming the carbon content is increasing by about 0.19 Gt yr^{-1} (Miller 1981), the water content in the organic layer should be increasing by about 1.6 Gt yr^{-1} (equivalent to about 0.1 mm yr^{-1}). With a global warming due to increased atmospheric CO_2, the rate of carbon accumulation is expected to increase additionally by about 0.011 Gt yr^{-1}. Thus the rate of increase of water storage should accelerate by about 0.09 Gt yr^{-2}. This predicted rate of accumulation of water should be compared with the estimated recent loss of 85 Gt yr^{-1} for the Greenland ice cap (Orvig 1970).

3.4 Precipitation

The quantity of water stored on the Earth's surface is less important than the fluxes of water between the atmosphere and the surface. Unfortunately, the fluxes are poorly known in northern ecosystems because of the paucity of measurements, the sparse distribution and, generally, low elevation of weather stations, and the technical problems associated with the measurement of pre-cipitation including blowing snow. Hare and Hay (1974) stated that water budgets for northern ecosystems could not be formulated on the basis of existing information. A few direct measurements of evapotranspiration have been made since 1974; however, data are still scarce.

Precipitation is low in Arctic ecosystems, ranging from negligible amounts in the polar desert, around 140 mm yr^{-1} in the high Arctic to 200-350 mm yr^{-1} in the low Arctic. In the boreal forest, precipitation is higher at about 350-450 mm yr^{-1}. Large variations occur about these averages especially in the boreal forest. The total precipitation over northern ecosystems is about 5800 Gt yr^{-1} or about 1% of the global total.

3.5 Evapotranspiration

Although few measurements have been made and only at widely-scattered sites, evapotranspiration appears to range from < 0.5 mm per day on the driest sites, < 3 mm per day on moist sites, to 4-5 mm per day on wet sites with free standing water or sites receiving water from snowmelt early in the thawing season (Table III). In the boreal forest, evapotranspiration is about 2 mm per day or about 240 mm yr^{-1} (Pavlov 1976). Hence, the total evapo-transpiration from northern ecosystems is estimated to be about 3400 Gt yr^{-1} which is equivalent to about 0.7% of the global evapotranspiration or about 5% of the atmospheric water content. Over 75% of this flux emanates from the boreal forest; about 25% occurs in May and June during snowmelt.

3.6 Runoff

The difference between precipitation and evapotranspiration implies that water equivalent to about 2400 Gt yr^{-1} flows directly

Table III.   Evapotranspiration rates measured in different Arctic
             tundra ecosystems.   Sites: a, Eagle Sumit, Alaska;
             b, Devon Island, Canada; c, Hardangervidda, Norway;
             c, Eagle Creek, Alaska; e, Barrow, Alaska.

Ecosystem	Site	Rate		Author
		mm per day	mm yr^{-1}	
Fellfield & lichen heath	a	0.3-0.5	30-50	Miller 1981
Raised beach ridge	b	0.6-1.2	74	Addison 1977, Ryden 1977
Lichen heath	c	1.7		Skartveit *et al.* 1975
Cotton-grass tussock tundra	d			Stuart *et al.* 1981
Tussock area		0.4-1.2		
Moss-covered intertussock area		0.6-1.9		
Dry meadow	c	2.0		Skartveit *et al.* 1975
Wet meadow (mid-summer)	e	2.7		Weller & Holmgren 1979, Dingman *et al.* 1980
Wet meadow	b	3.2-3.4	185	Addison 1977, Ryden 1977
Wet meadow	e	4.6-5.6		Koranda *et al.* 1978

into rivers and sea from the northern ecosystems, i.e., an average
runoff of about 140 mm yr^{-1}. The resulting runoff:precipitation
ratio is about 0.41.

3.7 Prediction of water contents in Arctic and boreal forest eco-
     systems

     If global warming occurs due to the effects of rising CO_2
levels, it is predicted that the capacity of the Arctic as a carbon
sink will increase and the boreal forest will remain nearly constant. However, within the boreal forest, there will be a reduction of carbon stored in the forest floor and soil and a gain of
carbon stored in the biomass. This predicted change in carbon
budgets implies that water will increasingly be retained in the
Arctic ecosystems, but there will be a net loss from boreal forest
ecosystems. In addition, with a more favourable water balance in
the Arctic, moss growth would increase in those ecosystems which
currently have a low vegetation cover, subsequently increasing the
areal extent of organic soils. This change in cover should
decrease surface temperatures by about 2°C and increase evapotranspiration by about 30%, according to Stuart *et al.* (1981).

At any given site, the local water balance controls the growth
of mosses which, in turn, controls the water storage. This relationship introduces considerable biological complexity into the
hydrological system. Paradoxically, because of the biological
response of moss growth, prolonged global warming can be expected
to increase the total amount of ice stored in perennially-frozen
ground. Although rising temperatures will undoubtedly tend to
increase the depth of the seasonally-thawed layer, the predicted
increase in the growth of mosses could raise the surface of the
organic layer by more than the increased depth of thaw. As organic
soils have a relatively high ice content this effect would be
apparent especially where a mineral soil becomes moss covered. The
difference in ice content results in shallower depths of thaw in
organic soils than in mineral soil. Thus the height of the permafrost could rise and water could be sequestered from the global
water balance into a permanently frozen state. The short-term
response of the thawed layer to global warming is complex and hard
to predict as changes in the physical and biological systems,
which have different response times, would be involved. Changes
in thaw depth, which depend on the climate and the physical properties of the soil, should occur rapidly, i.e., within a year. The
change in moss growth and surface height depends on physical,
chemical and biological parameters and should change more slowly,
depending on several environmental factors.

4.   DISCUSSION

The quantity and dynamics of water held in ground ice, permafrost, lakes and ponds need to be estimated more precisely, since
reassessment of these factors could change the conclusions of this
analysis. The errors associated with the estimates of water
storage in northern ecosystems are large. Original data are few
and have to be applied to large areas. Errors of ± 50% or more
can therefore be expected in the estimates made. The lack of data
creates large uncertainties in the prediction of the possible
future water balance of northern ecosystems. Given the predicted
increase in global temperature, current evidence indicates that
northern ecosystems, especially in the Arctic, will store more
water. Water in the boreal forest may decrease slightly. It is
difficult to make direct measurements of the annual changes in
water storage as the changes are small in relation to the amount
stored or exchanged. Indirect techniques, such as the use of
carbon accumulation rates or isotope ratios, seem to be the most
accurate methods available. The inter-dependence of the physical
and biotic components of the environment, including the water
balance, heat exchange, thaw depth, vegetation cover and organic
content, requires that both components be considered together in
northern ecosystems.

ACKNOWLEDGEMENTS. This paper is based in part on the results of
a workshop on carbon balance in northern ecosystems supported by
the U.S. Department of Energy and attended by P.C. Miller,
J. Andrews, D. Billings, L. Bliss, J. Brown, C. Cooper, K. Everett,
O.W. Heal, A. Johnson, D.R. Klein, J. Kummerow, A.E. Linkins,
G. Marion, H. Nichols, W.C. Oechel, D. Parkinson, W. Schlesinger,
G. Shaver, B. Sveinbjornsson, L. Tieszen, K. Van Cleve, L. Viereck,
P. Webber and R. White. I thank Patsy Miller and Rhonda Watson for
their help in preparing this paper.

REFERENCES

Addison, P.A. 1977 Studies on evapotranspiration and energy budgets
    on Truelove Lowland. In: *Truelove Lowland, Devon Island,
    Canada: A High Arctic Ecosystem* (ed. L.C. Bliss), pp. 281-300.
    Edmonton, Canada: University of Alberta Press.

Ajtay, G.L., Ketner, P. & Duvigneaud, P. 1979 Terrestrial primary
    production and phytomass. In: *The Global Carbon Cycle* (ed.
    B. Bolin, E.T. Degens, S. Kempe & P. Ketner), pp. 129-181.
    (*SCOPE Rept* 13). New York: John Wiley.

Andreev, V.N. & Aleksandrova, V.D. 1981 Geobotanical division of the
    Soviet arctic. In: *Tundra Ecosystems: A Comparative Analysis*
    (ed. L.C. Bliss, O.W. Heal & J.J. Moore), pp. 25-34.
    Cambridge, England: Cambridge University Press.

Barney, R.J., Van Cleve, K. & Schlenter, R. 1978 Biomass distri-
    bution and crown characteristics in two Alaskan *Picea mariana*
    ecosystems. *Can. J. For. Res., 8*, 36-41.

Bliss, L.C. (ed.) 1977 *Truelove Lowland, Devon Island, Canada:
    A High Arctic Ecosystem,* 714 pp. Edmonton: University of
    Alberta Press.

Dingman, S.L., Barry, R.G., Weller, G., Benson, C., LeDrew, E.F.
    & Goodwin, C.W. 1980 Climate, snow cover, microclimate, and
    hydrology. In: *An Arctic Ecosystem: The Coastal Tundra at
    Barrow, Alaska* (ed. J. Brown, P.C. Miller, L.L. Tieszen &
    F.L. Bunnell), pp. 30-65. Stroudsburg, Penn.: Dowden,
    Hutchinson & Ross.

Everett, K. 1981 Soil landscapes at selected sites along the Yukon
    River-Prudhoe Bay haul road and within NPR-A. Progress Rept
    to U.S. Army (Department of Defence).

Hare, F.K. & Hay, J.E. 1974 The climate of Canada and Alaska. In:
    Climates of North America (ed. R.A. Bryson & F.K. Hare), pp.
    49-192. New York: Elsevier.

Komárková, V. & Webber, P.J. 1980 Two low arctic vegetation maps along the Meade River at Atkasook, Alaska. *Arct. Alp. Res.*, 12, 447-472.

Koranda, J.J., Clegg, B. & Stuart, M. 1978 Radio-tracer measurement of transpiration in tundra vegetation, Barrow, Alaska. In: *Vegetation and Production Ecology of an Alaska Arctic Tundra* (ed. L.L. Tieszen), pp. 359-370. New York, Heidelberg, Berlin: Springer-Verlag.

Larcher, W.L. 1980 *Physiological Plant Ecology*, 303 pp. New York, Heidelberg, Berlin: Springer-Verlag.

Miller, P.C. 1981 Carbon dioxide effects research and assessment program: carbon balance in northern ecosystems and the potential effect of carbon dioxide induced climatic change. Rept of Workshop, San Diego, California, 7-9 March 1980: CONF-8003118, 109 pp. Washington, D.C.: U.S. Dept of Energy.

Miller, P.C., Stoner, W.A. & Tieszen, L.L. 1976 A model of stand photosynthesis for the wet meadow tundra at Barrow, Alaska. *Ecology*, 57, 411-430.

Ng, E. & Miller, P.C. 1975 A model of the effect of tundra vegetation on soil temperatures. In: *Climate of the Arctic* (ed. G. Weller & S.A. Bowling), pp. 222-226. Fairbanks: University of Alaska Geophysical Institute.

Ng, E. & Miller, P.C. 1977 Validation of a model of the effect of tundra vegetation on soil temperatures. *Arct. Alp. Res.*, 9, 89-104.

Orvig, S. 1970 The hydrological cycle of Greenland and Antarctica. In: *Proc. Symp. World Water Balance, Reading, 1970*, Vol. 1. Gentbrugge: Int. Ass. Sci. Hydrol. (Publ. 92.)

Pavlov, A.V. 1976 Heat transfer of the soil and atmosphere at northern and temperature altitudes. CRREL Draft Translation No. 511, 296 pp.

Rydén, B.E. 1977 Hydrology of Truelove Lowland. In: *Truelove Lowland, Devon Island, Canada: A High Arctic Ecosystem* (ed. L.C. Bliss), pp. 107-136. Edmonton, Canada: University of Alberta Press.

Skartveit, A., Rydén, B.E. & Karenlampi, L. 1975 Climate and hydrology of some Fennoscandian tundra ecosystems. In: *Fennoscandian Tundra Ecosystems, Part 1: Plants and Microorganisms* (ed. F.E. Wielgolaski), pp. 41-56. New York, Heidelberg, Berlin: Springer-Verlag.

Stuart, L., Oberbauer, S. & Miller, P.C. 1982 Evapotranspiration
    measurements in *Eriophorum vaginatum* tussocks tundra in
    Alaska. *Holarc. Ecol.*, 5, 145-149.

Tikhomirov, B.A., Shamurin, V.F. & Aleksandrova, F.D. 1981 Phyto-
    mass and primary production of tundra communities, USSR. In:
    Tundra: Comparative Analysis of Ecosystems (ed. L.C. Bliss,
    O.W. Heal & J.J. Moore). Cambridge, England: Cambridge
    University Press.

Van Cleve, K. & Noonan, L.L. 1975 Litterfall and nutrient cycling
    in the forest floor of birch and aspen stands in interior
    Alaska. *Can. J. For. Res.*, 5, 626-639.

Weller, G. & Holmgren, B. 1974 The microclimate of the Arctic
    tundra. *J. appl. Met.*, 13, 854-862.

Whittaker, R.H. & Likens, G.E. 1973 Carbon in the biota. *Brook-
    haven Symp. Biol.*, 24, 281-302.

Whittaker, R.H. & Likens, G.E. 1975 The biosphere and man. In:
    Primary Productivity of the Biosphere (ed. H. Lieth & R.H.
    Whittaker), pp. 305-328. New York, Heidelberg, Berlin:
    Springer-Verlag.

Wielgolaski, F.E., Bliss, L.C., Svoboda, J. & Doyle, G. 1981
    Primary production of tundra. In: *Tundra Ecosystems: A
    Comparative Analysis* (ed. L.C. Bliss, O.W. Heal & J.J. Moore),
    pp. 187-225. Cambridge, England: Cambridge University Press.

Wigley, T.M.L., Jones, P.D. & Kelley, P.M. 1980 Senario for a warm
    high CO_2 world. *Nature, Lond.*, 283, 17.

Woodwell, G.M., Whittaker, R.H., Reiners, W.A., Likens, G.E.,
    Delwiche, C.C. & Botkin, D.B. 1978 The biota and the world
    carbon budget. *Science, N.Y.*, 199, 141-146.

RECENT FLUCTUATIONS OF ALPINE GLACIERS AND THEIR METEOROLOGICAL
CAUSES: 1880-1980

Louis Reynaud

Laboratoire de Glaciologie et Geophysique de
l'Environnement du CNRS
Grenoble, France

ABSTRACT. Rather than looking directly at the response of the
glacier termini to meteorological factors such as temperature and
precipitation, variations in glacier mass balance are used here to
give a climatic explanation for glacier fluctuations. In this way,
it is possible to clarify the relationship between net mass
balance and climate, to investigate the reaction of glaciers to
changes in the rate of accumulation and to provide information on
the climatic significance of variations in glacier length. An
analysis of data derived from regular surveys carried out on seven
glaciers, covering a distance of 500 km along the Alps, indicates
that mass-balance fluctuations are similar over a wide area and
therefore represent a sensitive high-altitude indicator of regional
climate.

1. INTRODUCTION

The water budget of glacierized areas like the Alps is
strongly related to glacier fluctuations. The amount of water
stored or released is dependent on subtle climatic variations that
glaciers, because of their unusual sensitivity, record and amplify.
Investigators who have attempted to relate glacier variations
directly to meteorological data have produced results which are
conclusive only in qualitative terms (Hoinkes 1970, Grard 1971,
Posamentier 1977). Such an approach does not account for two
important factors: the influence of meteorological conditions on
the mass balance of a glacier and the mechanisms of the glacier's
reaction to variations in that balance. For this reason, the
Laboratoire de Glaciologie in Grenoble has concentrated on

A. Street-Perrott et al. (eds.), Variations in the Global Water Budget, 197–205.

improving our understanding of both the relationship between net
mass balance and meteorological factors (Martin 1974, 1977), and
the spatial and temporal variability of the glacier balance, using
the linear statistical model developed by Lliboutry (1974). This
model, later extended to allow a comparison of the balance of
several glaciers (Reynaud 1977), has provided detailed information
on the reaction of the Mer de Glace at different altitudes to mass-
balance variations (Martin 1977) as well as on the mechanisms
involved in glacier fluctuations. In particular, it has helped to
determine the significance of variations in glacier length as an
indicator of climatic fluctuations.

The linear model has been used successfully to analyse mass-
balance measurements from ten glaciers in the French, Swiss,
Austrian and Italian Alps (Reynaud 1980), in order to study how
the relationship between climatic conditions and glaciers varies
over the entire Alpine range. This paper describes Lliboutry's
linear model and its applications, as well as the dynamic response
of a glacier to fluctuations in its mass balance. It concludes
with an analysis of seven extended series of mass-balance measurements, some extending back to 1954.

2.   GLACIER MASS BALANCES

Using the annual balances measured over 16 consecutive years
at 30 different sites on the ablation zone of the St-Sorlin Glacier,
Lliboutry (1974) verified the following linear statistical model of
the annual balance at site j for year t on an individual glacier:

$$b_{jt} = a_j + b_t + \varepsilon_{jt} \tag{1}$$

where a_j is a geographic parameter independent of the year and
characteristic of the site, b_t is a parameter depending only on
the year and not on the site, with a mean value of zero, ε_{jt} is a
random deviation from the mean with a standard deviation equal
to 0.20 m of ice on the St-Sorlin Glacier.

This model has been tested using a group of sites on the
St-Sorlin Glacier (Lliboutry & Echevin 1974, Vallon & Leiva 1982)
and data from the ablation zone of the Mer de Glace (Reynaud 1977).
It has also recently been tested successfully on measurements over
31 years made, since 1949, on the Sarennes Glacier (Bienvenu 1981).

3.   CORRELATION BETWEEN b_t AND METEOROLOGICAL VARIABLES

If b_t is assumed to be randomly distributed, its standard
deviation on the St-Sorlin Glacier is equivalent to 0.88 m of ice.

Martin (1977) looked first for a multiple correlation between
b_t and the variables measured at neighbouring meteorological
stations, and then with those measured at the Lyon-Bron airfield

130 km away. A homogeneous series of temperature and precipitation
data back to 1881 was established by making use of a very high
correlation between the Lyon-Bron measurements and those of a
neighbouring station. The most significant variables were found
to be:

(a) The mean maximum daily temperature for July and August (T_{7-8})
which accounts for 58% of the variance in b_t. According to a
previous study on the ablation zone of the St-Sorlin Glacier
(Martin 1974), 43% of summer ablation is caused by convection and
conduction and 57% by the radiation balance. However, these two
variables are highly correlated since high air temperatures correspond to sunny days. Although solar radiation becomes the most
important factor at high altitudes (De la Casinière 1974), it can
always be estimated from the maximum daily air temperature.
Furthermore, when snowfall temporarily whitens the glacier, thereby
reducing the absorbed radiation, the maximum air temperature is
also lower.

(b) June precipitation (P_6), which accounts for 16% of the
variance. In reality, both June precipitation (still in the form
of snow) and temperature are important, but there is a strong
negative correlation between the two.

(c) Total precipitation from October to May (P_{10-5}), which
accounts for 5% of the variance. The correlation between winter
accumulation on the St-Sorlin Glacier and Lyon precipitation data
is much weaker than for temperatures.

In all, these three variables accounted for 77% of the
variance in b_t, according to the following multiple regression
equation:

$$b_t = 3.0 \, \Delta P_{10-5} + 6.5 \, \Delta P_6 - 0.343 \, \Delta T_{7-8} \qquad (2)$$

where Δ signifies the deviation with respect to the mean, b_t, P_{10-5}
and P_6 are in metres water equivalent and T_{7-8} is in °C. A
regression coefficient of 3.0 does not mean that precipitation
totals at St-Sorlin are three times higher than those at Lyon. In
reality, when there is a great deal of snow in winter, it remains
on the glaciers longer in summer thereby increasing the albedo.
The result is that much less solar radiation is absorbed by the
snow than would have been by the bare ice, resulting in reduced
ablation.

The above correlation, based on data from a distant meteorological station, shows that mass-balance fluctuations are similar
over a wide region. This is supported by the balances measured on
the Mer de Glace and on the Argentière Glacier, both of which
reflect the same b_t fluctuations as the St-Sorlin and Sarennes
Glaciers. Although weather can vary greatly from one location to
another in the Alps on a given day, the mass balance for an entire
year shows much less local diversity.

4.   GLACIER REACTIONS TO BALANCE FLUCTUATIONS

On the Mer de Glace, waves comprising 4 to 5 m high bulges on
the glacier's surface extending over several kilometres occurred
during the periods 1891-96, 1921-27, 1941-45 and 1970-78. The
waves moved much faster than the glacier itself (450-500 m yr^{-1}
$cf.$ 70-130 m yr^{-1} at the centre of the glacier's surface), though -
contrary to the usual assumption - no simultaneous acceleration
occurred. The velocities of the glacier varied in a synchronous
manner along its 8 km length. Although waves related to its surface move down the glacier, the glacier itself moves as a block
in accord with its bulk velocity (Lliboutry & Reynaud 1981).
Martin (1977) looked for correlations between the levels
measured along various transverse profiles during year t and the
fluctuations b_t of the mass balance in years t, t-1, t-2, ... t-p.
There was not only a high correlation for p = 0, as would be
expected, but also for various other values of p, as shown below.
The distance x was measured downstream from a point in the zone of
wave formation:
        Trélaporte profile (x = 4000 m) p = 3 yr
        Echelets profile (x = 6030 m)   p = 8-9 yr
        Montenvers profile (x = 7030 m) p = 10 yr
        Mottets profile (x = 7600 m)   p = 11 yr
This confirms that the waves are due to years with above-average
balances and that the wave velocity below Trélaporte is about
500 m yr^{-1}. It can be seen that in order to disentangle the
reaction of a glacier to balance fluctuations, the surface elevation at various cross profiles must be monitored (around 1 October
each year), along with annual measurements of velocity, in order
to eliminate seasonal fluctuations. The data are much more difficult to analyse if only terminal variations are measured. The
snout position is, in fact, the resultant of glacier advance (the
front advancing if the velocity increases) and summer ablation of
the ice (the front advancing if the glacier snout is swollen by
the arrival of a wave).
Figure 1 shows the variations in length of four glaciers on
the north face of the Mont Blanc range: Bossons, Argentière,
Trient and the Mer de Glace. Note that the response of the termini
is marked by a variable time lag. For the maxima clearly observed
at the snouts of three glaciers (1890, 1920 and 1970, with 1940
only clearly visible for the Bossons glacier), the Bossons Glacier
reacted first, with very marked fluctuations, while the Argentière
and Trient Glaciers only reacted 4 to 7 years later, and the Mer de
Glace 15 years later with a much less marked variation.
The time constants were similar for the first two maxima (1890
and 1920), but longer for the 1970 maximum. Probably, this is
related to different glacier conditions, i.e., the fronts of these
four glaciers were in very similar positions in 1890 and 1920,
whereas the strong retreat during the 1940s and 1950s left them
with a very different shape at the time of the 1970 advance, due

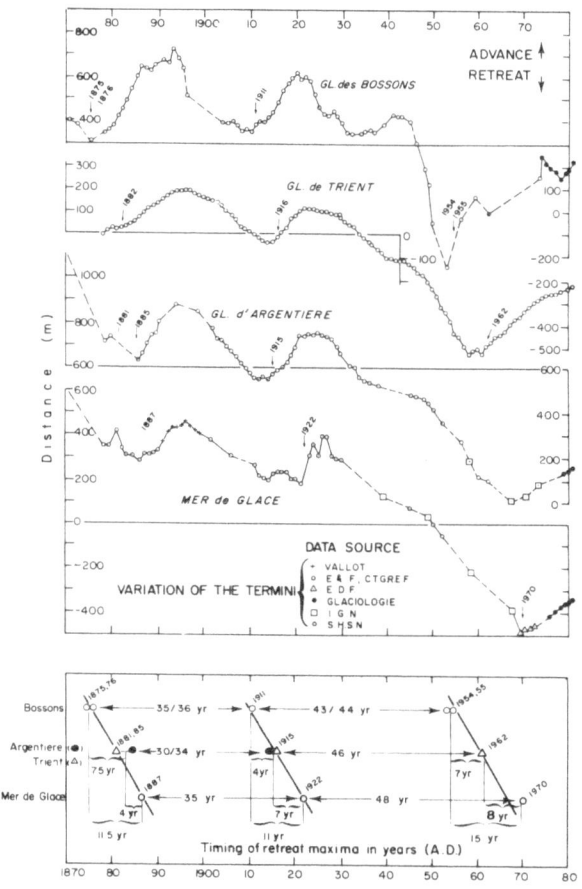

Fig. 1.   Comparison of the terminal variations of the Bossons,
Trient, Argentière and Mer de Glace glaciers.

to a net recession of 500 to 700 m since 1920. These differences
in reaction time reduce the value of any climatic analysis of
glacier fluctuations in which all types of glaciers are lumped
together, as was emphasized by Kuhn (1978) with respect to the
study carried out by Posamentier (1977). The amalgamation of snout
records, in order to calaculate the proportion of glaciers advancing or receding, results in even greater smoothing of the glacier
response to climatic variations.

5.   THE DISTRIBUTION OF MASS BALANCE OVER THE ALPS AS A WHOLE

The excellent results obtained by Martin (1977) for the reconstruction of the Sarennes balances from the Lyon-Bron meteorological

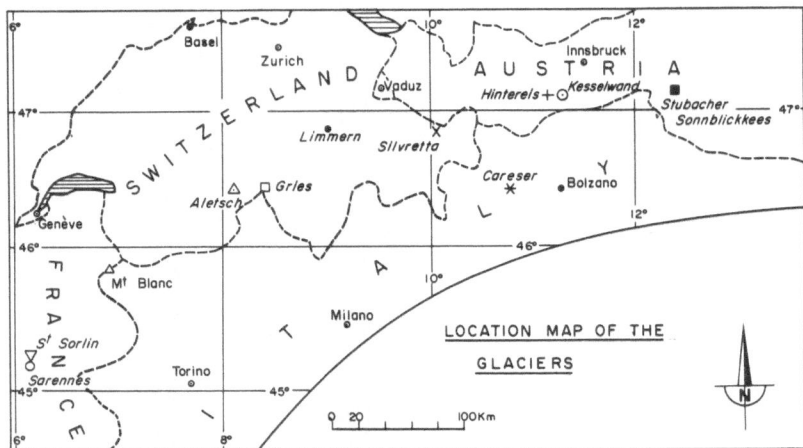

Fig. 2. Location map of the Alpine glaciers studied.

data and for the correlation between mass-balance variations and
surface levels, led us to extend the investigation to 10 glaciers
in the Alps (Figure 2): two in France, four in Switzerland, three
in Austria and one in Italy (Reynaud 1980). The results of this
comparison showed that the linear model of mass-balance variations
(Lliboutry 1974) can be extended to the entire Alps with a mean
residual of 30 cm of water for the 23-yr period with comparable
data, except for 1974 when the two French glaciers behaved
differently from those in the rest of the Alps. This analysis
utilized mass-balance data up to 1975. As data up to 1978 are
available now for seven glaciers, the two series can be analysed:
(i)   Four glaciers for 25 years (1954-1978): Aletsch, Sarennes,
Limmern and Hintereis.
(ii) Seven glaciers for 17 years (1962-1978): the above four
plus St-Sorlin, Silvretta and Gries.
     Table I gives the means and standard deviations for these two
series. It should be noted that the glaciers reflect different
regimes, some having a positive net mass balance and others a
negative balance.
     On the other hand, the standard deviations of the 2 distribu-
tions are much less varied: from 63 to 88 cm H_2O equivalent for
all seven glaciers and from 55 to 77 cm for the four glaciers with
the longest records. In order to provide a general picture of
mass-balance variations over the Alps, factor analysis was applied
to each series in turn, in an attempt to reduce the number of
initial significant variables (4 or 7) in such a way as to account
for the maximum proportion of the variance. These new variables
are the factor axes. The results are shown in Figure 3 for both
series (with or without 1974) in order to assess the influence of
this particular year on the results of the analysis.

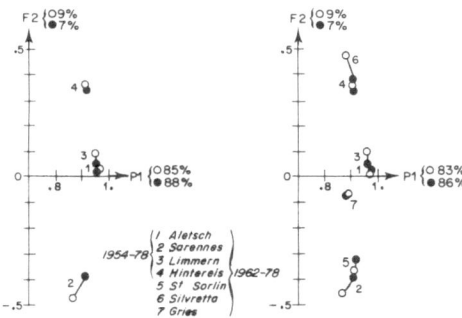

Fig. 3.   Factor analysis of the two series of mass balances for
          1954-78 (left) and 1962-78 (right) showing the proportion
          (%) of the variance explained by each axis: O, including
          1974; ●, excluding 1974.

Table I.   Mean values and standard deviations (SD) of mass balances
           for Alpine glaciers. Values in Roman refer to the seven-
           glacier series over 17 years (1962-78); values in italics
           to the four-glacier series over 25 years (1954-78).

	Aletsch	Sarennes	Limmern	Hinte-reis	St-Sorlin	Silv-retta	Gries
Mean (m H$_2$O equivalent)	0.17	-0.42	-0.16	-0.17	0.07	0.00	-0.08
	0.10	*-0.42*	*-0.17*	*-0.22*			
SD (m H$_2$O equivalent)	0.83	0.86	0.79	0.63	0.81	0.88	0.77
	0.75	*0.77*	*0.73*	*0.55*			

     There is practically no distinction between the glaciers along
the first axis, Fl, which accounts for 85% of the variance for the
four glaciers and 83% for the seven glaciers. The second axis, F2,
accounts for 9% of the variance, leaving 6 to 7% for the other axes.
It is interesting to note that axis Fl distinguishes between the
glaciers according to their geographic distribution along the Alps.
     These results are interpreted to mean that axis Fl is related
to the sample size effect. In other words, the extended version of
Lliboutry's linear model accounts for 83 to 85% of the total
variance whereas geographical differences explain only 9%, the
remaining 7%, with no particular structure, being regarded as noise.
     The analysis of the same series, omitting the anomalous year
1974, increases the variance explained by axis F2 (+ 3%) and
reduces that on axis Fl by an insignificant amount (-2.5%). Thus,
this particular year is not responsible for the observed geographical differences between the glaciers.

7. CONCLUSION

The value of the linear model developed by Lliboutry is two-fold: first, it can be used to analyse the distribution of balance over a single glacier without requiring measurements of area as do conventional methods; secondly, when extended to a regional scale, it makes a comparison of glaciers possible and provides information on the relative importance of time and geographical location for an understanding of the observed pattern of mass-balance variations. In the future, the possibility of relating b_t to meteorological factors offers an index of climate on the annual time-scale. Although this measure is not perfect, because summer temperature variations are the predominant influence on the mass-balance fluctuations of Alpine glaciers, the continuation of such measurements and the development of longer series will lead to better estimates of glacier balance and also of the climatic variations that cause them.

REFERENCES

Bienvenu, T. 1981 Analyse de la série de bilans de masse relevée sur le Glacier de Sarennes de 1949 à 1980 et la corrélation de ces bilans avec les facteurs météorologiques. Diplôme d'Etudes Approfondies: Université Scientifique et Médicale de Grenoble.

De la Casinière, A. 1974 Heat exchange over a melting snow surface. *J. Glaciol*, 13, 55-72.

Grard, R. 1971 Essai d'un modèle explicatif des variations des glaciers par le climat. S.H.F. Section Glaciologie, 18 pp. [Unpublished.]

Hoinkes, H. 1970 Methoden und Möglichkeiten von Massenhaushats-studien auf Gletschern. *Z. Gletsch. Glazialgeol.*, 6, 38-85.

Kuhn, M. 1978 On the non-linearity of glacier length response to climatic changes: comments on a paper by H.W. Posamentier. *J. Glaciol.*, 20, 443-445.

Lliboutry, L. 1974 Multivariate statistical analysis of glacier annual balances. *J. Glaciol.*, 13, 317-392.

Lliboutry, L. & Echevin, M. 1974 Mesure des bilans annuels en zone d'accumulation. *Z. Gletsch. Glazialgeol.*, 10, 71-88.

Lliboutry, L. & Reynaud, L. 1981 Global dynamics of a temperate valley glacier, Mer de Glace, and past velocities deduced from Forbes bands. *J. Glaciol.*, 27, 207-226.

Martin, S. 1974 Corrélations bilan de masse - facteurs météorologiques dans les Grandes Rousses. *Z. Gletsch. Glazialgeol.*, 10, 89-100.

Martin, S. 1977 Analyse et reconstitution de la série des bilans annuels du Glacier de Sarennes, sa relation avec les fluctuations du niveau de trois glaciers du Massif du Mont-Blanc (Bossons, Argentière, Mer de Glace). *Z. Gletsch. Glazialgeol.*, 13, 127-153.

Posamentier, H.W. 1977 A new climatic model for glacier behaviour of the Austrian Alps. *J. Glaciol.*, 18, 57-85.

Reynaud, L. 1977 Glaciers fluctuations in the Mont Blanc area (French Alps). *Z. Gletsch. Glazialgeol.*, 13, 155-166.

Reynaud, L. 1980 Can the linear balance model be extended to the whole Alps? *Int. Ass. Sci. Hydrol. Publ.* No. 126, pp. 273-284.

Reynaud, L. 1981 Reconstitution of past velocities of Mer de Glace using Forbes bands. *Z. Gletsch. Glazialgeol.*, 15, 149-163.

Vallon, M. & Leiva, J.C. 1982 Bilan de masse et fluctuations récentes du Glacier de Saint-Sorlin, Alpes françaises. *Z. Gletsch. Glazialgeol.*, 17. [In the press.]

RADIOMETRIC CHRONOLOGY OF SOME HIMALAYAN GLACIERS

N. Bhandari and V.N. Nijampukar

Physical Research Laboratory
Ahmedabad
India

C.P. Vohra

Geological Survey of India
Lucknow
India

ABSTRACT. Surface ice samples collected from three Himalayan glaciers, Nehnar (Kashmir Valley), Gara (Himachal Pradesh) and Changme Khangpu (Sikkim) were dated using the radionuclides ^{32}Si and ^{210}Pb. The ages of the snout ice of these three glaciers are about 500, 200 and 100 years, respectively. The bulk flow rates over the past few centuries on Nehnar, Gara and Changme Khangpu (CK) Glaciers are calculated as 6, 20 and 40 m yr^{-1}, respectively. As expected, these rates are smaller than the present-day surface flow rates of Nehnar and Gara Glaciers. However, in the case of CK Glacier, there is a significant disagreement indicating that this glacier is not in a steady state.

Oxygen isotopes and ^{210}Pb in cores from the ablation zones of the Nehnar and CK Glaciers, respectively of 102 and 12 m depth, were measured and related to dust content, total β activity and chemical tracers as a function of depth. The oxygen isotope ratio is characteristically different in Nehnar (0–46 m, $\delta^{18}O = -9.5\pm1\text{‰}$) and CK (0–12 m, $\delta^{18}O = -18\pm1\text{‰}$) glacier ice. This is probably due to different sources of precipitation in the two regions, the western Himalaya mainly receiving rain in winter from depressions moving from west to east and the eastern Himalaya mainly receiving monsoon rain in summer. In the top 46 m of the Nehnar core, five cycles of $\delta^{18}O$ varying between -8.5 and -10.5‰ were observed. An attempt is made to interpret these data in relation to precipitation and temperature cycles.

A. Street-Perrott et al. (eds.), Variations in the Global Water Budget, 207–216.
Copyright © 1983 by D. Reidel Publishing Company.

1. INTRODUCTION

The Himalayas form one of the most important glacier systems
on Earth. The northern rivers of the Indian subcontinent are fed
by rain and snow falling on this mountain range and the mass
balance of the glaciers plays a critical role in the annual water
budget of northern India. The water reserves contained in Himalayan glaciers are estimated as about 10^{12} m^3, comparable to the
groundwater reserves of India. In spite of their importance, not
much is known of the past behaviour of Himalayan glaciers. Historical records (Mayewski & Jeschke 1979, Vohra 1980) are qualitative
and scanty, but show that the Himalayan glaciers, by and large,
have been receding since AD 1870. Over the past decade, a systematic study of the topography and mass balance of a few glaciers
from selected basins has been carried out by the Glaciology
Division of the Geological Survey of India. The observations give
the accumulation and ablation contours and flow rates of these
glaciers, but were made for limited periods, thereby only providing
seasonal and annual variability. Thus, although these data are
useful, they do not give an average long-term picture. Information
about the secular behaviour of the glaciers can best be obtained
by radiometric dating (Nijampurkar et al. 1982).

Three glaciers in different basins, the Nehnar, Gara and
Changme Khangpu (CK) were selected for radiometric dating; their
locations are shown in Figure 1. The dating is based on cosmogenic
^{32}Si with a half-life of 105 yr (Elmore et al. 1980, Kutschera et
al. 1980) and the radiogenic isotope ^{210}Pb (half-life 22.3 yr).
The purpose of dating the ice samples is to obtain glacier flow
rates averaged over periods ranging from decades to centuries.
A comparison with modern observations then allows any significant
secular changes to be estimated. Simultaneous study of the stable
isotopes of oxygen permits this information to be related to
palaeoclimatic conditions. The results have important implications
for the understanding of glacier history, past variations in mass
balance and the nature of the mass transfer between the accumulation area and the terminus. The estimation of the long-term mass
balance of glaciers is also of prime importance for the development of water resources.

Surface ice samples collected during the summers of 1977 and
1978 were analysed for ^{32}Si and ^{210}Pb. In addition, core samples
and fresh precipitation were analysed for ^{210}Pb, β and γ activity
and oxygen isotope ratios. The results are discussed in terms of
glacier flow, snout ice ages and palaeoclimatic conditions.

2. DESCRIPTION OF THE GLACIERS

The location and characteristics of the three glaciers are
given in Table I. The Nehnar Glacier, in the Kashmir valley of the
western Himalaya, is a small valley glacier (Figure 2) which has

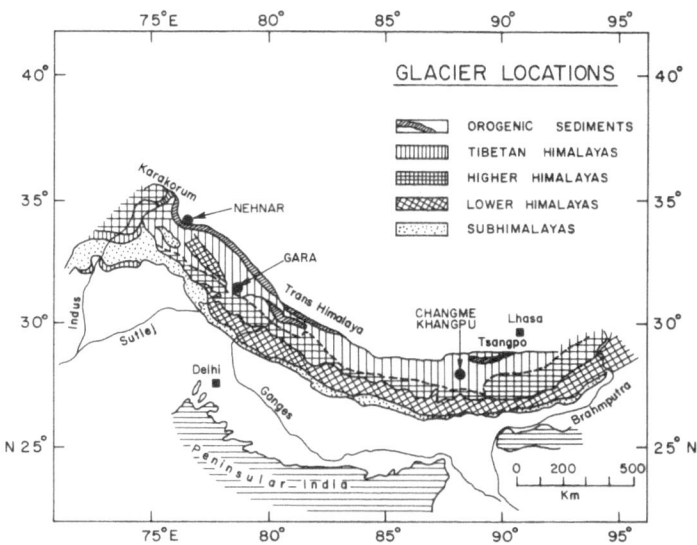

Fig. 1. Location of glaciers studied in the Himalayas.

Table I. Physical characteristics of the glaciers with calculated snout ages and average flow rates.

	Nehnar	Gara	Changme-Khangpu
Latitude	34° 09'N	31° 30'N	27° 58'N
Longitude	75° 31'E	78° 26'E	88° 42'E
Altitude range	3920-	4710-	4850-
(m a.s.l.)	4925	5600	5800
Dimensions			
length (km)	3.4	6	5.8
width (km)	0.35	1.5	0.8
area (km^2)	1.68	2.5	4.6
^{32}Si model age (yr)	500	200	100
Past flow rate (m yr^{-1})	6	20	40
Modern flow rate (m yr^{-1})	⩾12 (1977)	≃60 (1975)	⩽13 (1981)

been described in detail by Vohra *et al.* (1977) who studied its surface ice dynamics. The terminus of this glacier has been thinning and receding for the last few years. In 1975-76, it showed an average retreat of 9 m (Vohra *et al.* 1977). Evidence for six retreat stages are present in the moraine sequence.

The Gara Glacier is a small two-component valley glacier situated in Himachal Pradesh (Figure 3). It flows in the NE direction, being steep (1:3) in the upper 2 km and sloping gently (1:12)

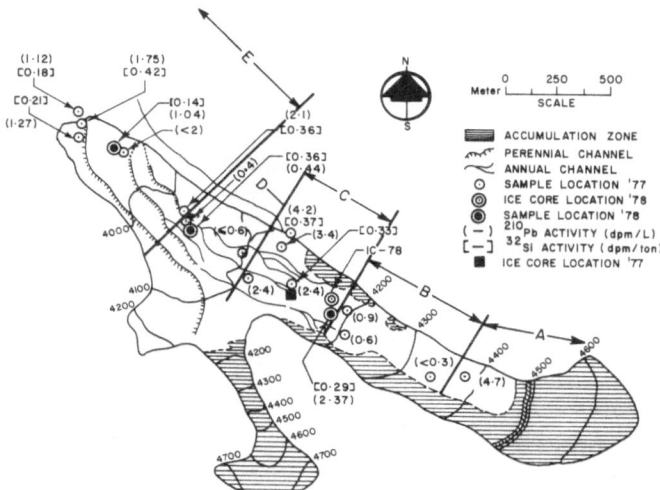

Fig. 2.   Distribution of ^{32}Si activity (dpm t^{-1}) and ^{210}Pb (dpm l^{-1})
          in surface samples from Nehnar Glacier. Five zones A, B,
          C, D and E with alternately high and low ^{210}Pb activity
          are shown (from Nijampurkar *et al*. 1982).

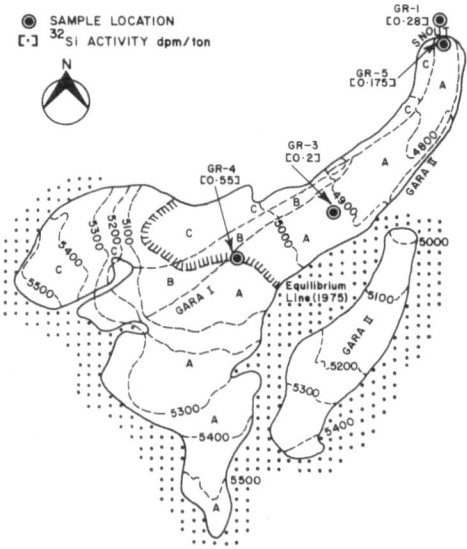

Fig. 3.   Distribution of ^{32}Si activity (dpm t^{-1}) in surface samples
          from Gara Glacier (from Bhandari *et al*. 1981).

lower down, and is fed by three tributary ice bodies (A, B, C in
Figure 3) with independent accumulation areas. The main active ice

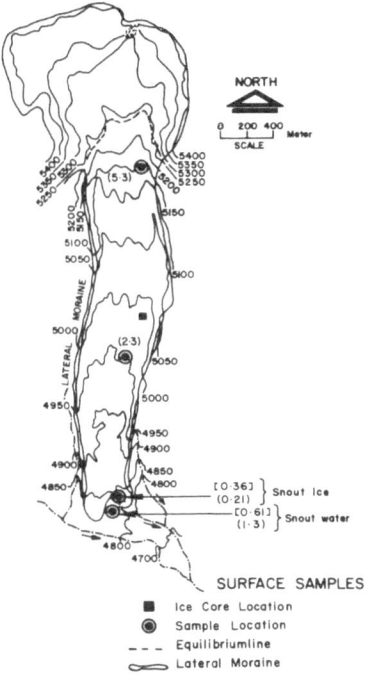

Fig. 4.   Distribution of isotope activity in surface samples of
          Changme Khangpu Glacier (from Nijampurkar *et al.* 1981):
          [ ], ^{32}Si; ( ), ^{210}Pb.

stream is shown as Gara I in Figure 3. The Gara Glacier retreated
by about 24 m between 1973 and 1975, but its terminus was practically static during 1976 (Vohra 1980).
       Changme Khangpu (CK) Glacier is a young, small valley glacier
in the Sikkim valley of the eastern Himalaya (Figure 4). There is
strong geological evidence that this glacier has gone through at
least four cycles of advance and recession in the past.

3.   SAMPLING PROCEDURES AND ANALYTICAL RESULTS

       Several surface ice samples (typically 2 tons) from the Nehnar
and Gara Glaciers, and snout ice samples from CK Glacier, were
collected and processed for ^{32}Si, which was estimated *via* its
daughter product ^{32}P on a low-level β detector. Surface ice
samples from Nehnar and CK Glaciers and fresh snow samples from the
latter were also analysed for ^{210}Pb *via* its daughter ^{210}Bi. The
procedures have been outlined by Nijampurkar *et al.* (1982). Two
ice cores, 12 m and 102 m long, were raised from the ablation area

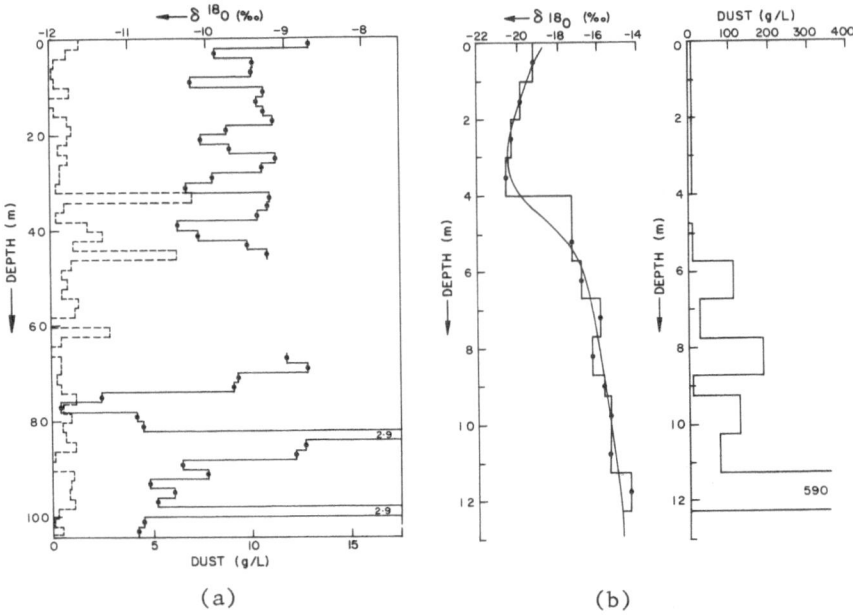

Fig. 5. Variation of the oxygen isotope ratio and abundance of
         suspended dust as function of depth (m) in the cores from
         (a) the ablation zone of Nehnar Glacier (102 m core; for
         location, see Figure 2) excluding section from 44–66 m
         due to oil contamination during coring; (b) Changme
         Khangpu Glacier (12 m core; for location, see Figure 4):
         ——, $\delta^{18}O$ (‰); – – –, dust (g l^{-1}). Snout ice value for
         Nehnar Glacier = -8 ‰.

of Nehnar Glacier in 1977 and 1978 and were divided into 1 m and
2 m sections, respectively. A core, about 12 m long, was raised
from CK Glacier in 1978 and divided into 1 m sections. The
sections were analysed for ^{210}Pb, β and γ activity, $\delta^{18}O$, dust and
pollen content, and chemical composition; the detailed results have
been reported elsewhere (Nijampurkar et al. 1981, 1982, Bhandari et
al. 1981). Based on these data (see Figures 2 to 5), the following
observations can be made on each of the glaciers.
(a) Nehnar Glacier
     (i) The ^{32}Si activity decreased from the equilibrium line
to the snout, first gradually and then abruptly, from 0.37 to 0.18
dpm t^{-1} (Figure 2), indicating an increase in the age of the ice
towards the terminus.
     (ii) The ^{210}Pb in surface ice showed complex behaviour with
five zones having alternately high and low values of ^{210}Pb (see
Figure 2).
     (iii) There was a well-defined peak of ^{210}Pb, total β acti-

vity and dust content at 2 to 3 m depth in the 1977 core taken at 4150 m a.s.l. A similar horizon also occurred at 10 to 12 m in the 1978 core, located 180 m further upstream at 4180 m a.s.l. The ^{210}Pb activity of 30 dpm l^{-1} in this horizon was four times higher than the expected fresh fallout value (Nijampurkar et al. 1982, Bhandari et al. 1981).

(iv) In the upper 46 m of the 1978 core, the δ^{18}O values ranged from -8 to -11‰ and showed several cycles from high to low values. The amplitude as well as the period increased below 66 m (see Figure 5a).

(b) Gara Glacier: The ^{32}Si activity decreased from the equilibrium line to the snout (Figure 3) from 0.55 to 0.175 dpm t^{-1}.

(c) Changme Khangpu Glacier

(i) The ^{32}Si value in the snout ice sample was 0.36 dpm t^{-1}.

(ii) The ^{210}Pb activity in samples collected from different altitudes during a single precipitation event on 28 August 1978 showed a gradual increase with altitude from 1.8 dpm l^{-1} at 4650 m to 8 dpm l^{-1} at 5450 m (Nijampurkar et al. 1981).

(iii) The δ^{18}O values in the core correlated well with the dust content. The values of δ^{18}O ranged from -14 to -21‰ and was distinctly different from the values in the Nehnar core. The δ^{18}O values were lower above 4 m than between 4 and 12 m in the CK core (Figure 5b). Surface ice samples collected between 4900 and 5450 m a.s.l. showed δ^{18}O values ranging from -19 to -13‰, whereas samples of recent snow from the same altitude range gave values between -21 and -17‰ (Nijampurkar et al. 1981).

(iv) The ^{210}Pb activity showed a peak value of 7.5 dpm l^{-1} at 2 to 4 m, whereas its concentration was negligibly small at other levels of the core, apart from the 10.25-11.25 m section, which showed an activity level of 0.07 dpm l^{-1}.

4.   DISCUSSION

The average fallout values of ^{32}Si and ^{210}Pb in fresh snow are not accurately known. Based on available data, Nijampurkar et al. (1982) adopted values of 0.7 dpm t^{-1} for ^{32}Si and 8 dpm l^{-1} for ^{210}Pb. Using these values, the 'ages' of ice at various locations can be calculated, provided that the 'closed box' assumption is valid. In the CK Glacier, the ^{32}Si and ^{210}Pb ages are concordant, but this is not true for the Nehnar Glacier. Discordant ages might arise if young ice was being mixed continuously with older ice in the ablation zone of the Nehnar Glacier so that the 'closed box' assumption would not be valid. A simplified two-component model, in which the young component was estimated from the ^{210}Pb activity, was proposed to estimate the young and old components (Nijampurkar et al. 1982). The fraction of the young component present in various ice samples was calculated using this model assuming that all the ^{210}Pb is young (< 10 yr). A correction for this was made in the observed ^{32}Si activity by

subtracting the activity attributed to the young component. The
residual activity represents the 'old' component from which the
age of the ice was calculated. Following this procedure, it is
possible to estimate the 'real' age of the ice. These model ages
are given in Table I. The ages of ice collected from various
locations and their separation can be used to determine past rates
of ice flow. The *average* flow velocities from the accumulation
zone to the terminus for the three glaciers are given in Table I.
On the Nehnar Glacier, the extensive sampling network makes it
possible, therefore, to obtain the flow rates between different
positions along the glacier. The calculated rates are 4 m yr^{-1}
from the accumulation zone down to an altitude of 4175 m a.s.l.
and 2.1 m yr^{-1} in the lower part of the glacier between 4050 and
3900 m a.s.l. The situation is more complicated in the intervening
region, because of an irregular pattern of accumulation (Figure 2).
In fact, an inversion in ages is observed near 4050 m a.s.l., where
young ice is fed in from the secondary lateral accumulation zones.
An upper limit of 17 m yr^{-1} from the accumulation zone to an altitude of 4050 m a.s.l. can be calculated from these data.

The average surface flow rates of recent ice on all three
glaciers have been measured (by the Geological Survey of India)
using stakes. The rates found were approximately 12 m yr^{-1} (1977-
78) for Nehnar (Vohra *et al.* 1977), 60 m yr^{-1} (1974-75) for Gara
(Geological Survey of India 1977) and 13 m yr^{-1} (1980-81 for CK
Glacier (R. Nahak, personal communication). The surface flow
rates were averaged over the uppermost few metres of the glacier
whereas the radiometric ages were averaged over the entire glacier
body and would be expected, therefore, to be lower. A comparison
indicates that the modern flow rates are indeed larger than past
flow rates by a factor of 2 or 3 for Nehnar and Gara. However,
for the CK Glacier, the modern flow rates are smaller, being only
one third of past flow rates. Moreover, there is evidence for
several cycles of advance and retreat (Nijampurkar *et al.* 1981)
possibly due to tectonic activity. The discrepancy between the
modern and past flow rates indicates that this glacier is not in
a steady state and may continue to retreat in the near future.

The cyclic fluctuations of the $\delta^{18}O$ values in the ice and
their correlation with the dust content seem to reflect past variations in climate. The Himalayan glaciers are very dusty. Dust
concentrations of 20 mg l^{-1} to 5 g l^{-1} are quite common in the
Nehnar Glacier, while CK Glacier occasionally shows even higher
values, up to 600 g l^{-1}. The annual fallout of dust is significant and is particularly high during the summer. The dust is
deposited usually between the snow layers because the precipitation
is strongly seasonal.

The accumulation of debris in Himalayan glaciers can be
visualized in the following way. The climate in the Himalayas is
quite erratic with high snowfall leading to a positive mass balance
in some years and a negative mass balance in others. This results
in a fluctuating equilibrium line from year to year. The mass

balance, mapped for the Nehnar (Vohra *et al*. 1977) and CK Glaciers, also varies at different locations on the glacier.

During phases of negative mass balance, some or all of the snow and some of the ice in the accumulation zone will melt away leaving behind a layer of concentrated debris on the surface, whereas during phases of positive mass balance, the dust will be buried interspersed in the ice. The concentration of dust is related directly therefore to the degree of ablation and hence to the climatic conditions. Of the various factors, in addition to the net mass balance, which determine $\delta^{18}O$ the most important are the mean annual air temperature and mean annual precipitation (Dansgaard 1964). As air temperature determines the $\delta^{18}O$ value in the fresh snow, so it will influence, through ablation, the dust content of the ice. The core data thus reflect the ambient temperature and will reveal any periodicity in the climatic conditions prevailing over the time span represented by the ice. Although no definite time scale is available for the core sequences, rough estimates based on the snout age (500 yr) indicate that the Nehnar core represents accumulation over 100 to 350 years. For the CK core, the accumulation rate of snow in the accumulation area over the past decade is known to be 0.7 m H_2O yr^{-1} with a snout age of 100 yr (Bhandari *et al*. 1982). This yields an estimated age of about 50-65 yr for the basal ice.

The difference in $\delta^{18}O$ between the Nehnar and CK Glaciers can be understood in terms of their different precipitation sources. Nehnar Glacier receives most of its precipitation in winter mainly from depressions moving eastwards, whereas CK Glacier receives summer monsoon precipitation originating in the Arabian Sea and the Bay of Bengal. It is not surprising that the precipitation at the two sites differs in isotopic composition since this depends on the degree of fractionation (Sonntag *et al*.: this volume).

In summary, the $^{32}Si-^{210}Pb$ dates indicate that the ice in the termini of Nehnar, Gara and Changme Khangpu Glaciers is 500, 200 and 100 years old, respectively. The *average* flow velocity was calculated to be 6 m yr^{-1} for Nehnar, 20 m yr^{-1} for Gara and 40 m yr^{-1} for the CK Glacier. These flow rates are slightly lower than the modern rates for Nehnar and Gara Glaciers, as expected, whereas the modern rates are lower than the past flow rates for CK Glacier, suggesting that the CK Glacier is not in a steady state at present. Variations in climatic conditions, operating through surface air temperature or annual precipitation, seem to have been responsible for several cycles of high and low $\delta^{18}O$ values and would also have resulted in variations in ablation reflected in the dust content of the ice cores.

ACKNOWLEDGEMENTS. We are grateful to several colleagues of the Geological Survey of India for help with the field work, to Professor D. Lal, Shri V.S. Krishnaswamy, Dr W. Ambach, Dr H. Moser and Dr P.R. Pisharoty for useful comments and to Dr W. Dansgaard for laboratory facilities for the oxygen isotope analysis.

REFERENCES

Bhandari, N., Bhatt, D.I., Nijampurkar, V.N., Singh, R.K.,
    Srivatsava, D. & Vohra, C.P. 1981 Radiometric age of the
    snout ice of Nehnar Glacier. *Proc Indian Acad. Sci. (Earth
    Planet. Sci.)*, 90, 227-235.

Bhandari, N., Nijampurkar, V.N. & Shukla, P.N. 1982 Deposition of
    Chinese nuclear debris in Changme-Khangpu Glacier. *Current
    Sci.* [In the press.]

Dansgaard, W. 1964 Stable isotopes in precipitation. *Tellus*, 16,
    436-468.

Elmore, D., Anantaraman, N., Fullbright, H.W., Gove, H.E., Hans,
    H.S., Nishiizumi, K., Murrel, M.T. & Honda, M. 1980 Half-life
    of ^{32}Si using tandem accelerator mass spectrometry. *Phys.
    Rev. Lett.*, 45, 589.

Geological Survey of India 1977 Report of the inter-departmental
    expedition to Gara Glacier during 1974-75. Lucknow: Geolo-
    gical Survey of India (Glaciology Division).

Kutschera, W., Henning, W., Paul, M., Smither, R.K., Stephensen,
    E.J., Yntema, J.L., Alburger, D.E., Cumming, J.B. & Harbottle,
    G. 1980 Measurements of the ^{32}Si half-life *via* accelerator
    mass-spectrometry. *Phys. Rev. Lett.*, 45, 592.

Mayewski, P.A. & Jeschke, P.A. 1979 Himalayan and Trans-Himalayan
    glacier fluctuations since AD 1812. *Arct. Alp. Res.*, 11,
    267-287.

Nijampurkar, V.N., Bhandari, N., Borole, D.V. & Bhattacharya, U.
    1981 Radiometric chronology of the Changme-Khangpu Glacier,
    Sikkim. Physical Research Laboratory Ahmedabad Rept GLAC-
    81-03.

Nijampurkar, V.N., Bhandari, N., Krishnan, V. & Vohra, C.P. 1982
    Radiometric chronology of the Nehnar Glacier, Kashmir. *J.
    Glaciol.*, 28, 91.

Vohra, C.P. 1980 Glacier resources of the Himalaya and their
    importance to environmental studies. Lucknow: Geological
    Survey of India (Glaciology Division). [Unpublished.]

Vohra, S.K., Krishnan, V., Pathak, C.S., Tiwari, R.A. & Bedi, A.K.
    1977 *Report on the Panjtarni Glacier Expedition 1975.*
    Lucknow: Geological Survey of India (Glaciology Division).

Secular Variability:
Interactions and Teleconnections

INTRODUCTION TO SECULAR VARIABILITY: INTERACTIONS AND
TELECONNECTIONS

R.A.S. Ratcliffe

Royal Meteorological Society
Bracknell
Berkshire, UK

In this section, variability over time-scales from about a
month up to several 100 years is considered and the connection
between unusual meteorological events separated by vast distances
(teleconnections) is demonstrated.

Many years ago, Walker discovered that when surface pressure
was low over Indonesia it was high in the Pacific near Easter
Island and *vice versa*; this so-called 'Southern Oscillation' has
since been shown to account for much of the variation of some
meteorological parameters (particularly rainfall) in many parts of
the tropics.

It has gradually been realized that the ocean plays an
important part in the mechanism of teleconnections. Bjerknes
showed that, during episodes of the El Niño off the west coast of
South America, ocean temperatures became higher than usual in the
tropical East Pacific. He went on to suggest that not only were
these higher sea temperatures the primary cause of heavier rainfall in the East Pacific, but that there was evidence that the
Hadley cell was strengthened leading to a stronger subtropical
anticyclonic belt and increased westerly circulation on the poleward side of this belt. Such ideas received considerable support
from work by Rowntree. He used numerical model techniques to
demonstrate that changes in tropical ocean temperatures, such as
were common in the east Pacific, appeared to have effects on the
atmospheric circulation similar to those expected by Bjerknes.
Rowntree was also able to show by similar means that ocean
temperatures off tropical West Africa appeared to effect the mid-
latitude westerly regime in the North Atlantic. In particular,
he indicated that high sea temperatures in that region may have
been a factor in the very cold European winter of 1963 when the
Atlantic westerlies were much further south than usual.

218

A. Street-Perrott et al. (eds.), Variations in the Global Water Budget, 218–220.
Copyright © 1983 by D. Reidel Publishing Company.

There has always been a good deal of controversy in the literature as to what is the exact mechanism of these teleconnections and, in particular, whether in general the atmosphere controls the sea temperature or *vice versa*. It is certain that these processes are complex: the atmosphere may well initiate sea temperature anomalies but, once established, these are more persistent than atmospheric anomalies and therefore can react back on the atmosphere. Reiter's contribution to this section throws new light on the mechanism of the Southern Oscillation and El Niño. He gives evidence, for instance, that negative 500 mb anomalies in January in the central North Pacific are related to tropical rainfall anomalies, the latter maximizing at 1-month lag. The 500 mb anomalies in the North Pacific are themselves correlated with SST anomalies in that region but, in this case, with the atmospheric anomalies showing a tendency to lead the oceanic ones by about a month. Deep troughs in the central North Pacific have a tendency to push the ITCZ towards the equator and, once in this position, decreases in the wind stress may initiate warming of the tropical ocean, hence inducing equatorial surges of precipitation and other phenomena associated with El Niño events. The possible importance of the semi-permanently very warm ocean in the equatorial West Pacific in controlling the Southern Oscillation is also discussed by Reiter.

Weare's contribution compliments that of Reiter by considering the changes in moisture flux related to El Niño. He, too, points to the importance of circulation changes in initiating El Niño events and draws attention to the fact that the weakening of the South Pacific high is strongly related to the higher ocean temperatures of the El Niño. However, it is important to realize that the flux of latent heat to the atmosphere depends on both ocean temperature and wind speed and, therefore, an increase of ocean temperature accompanied by a decreased wind speed may have complicated effects on the flux of latent heat from the ocean to the atmosphere. Indeed, this may be the cause of the lag correlations observed between ocean temperature and latent heat flux.

The role of ocean temperatures in temperate latitudes in influencing the weather is more controversial, but the contribution of Zhang Yan lends considerable support to the ideas put forward a decade ago by Ratcliffe and Murray. She shows that a certain pattern of ocean temperature anomalies in the Pacific preceded the anomalous atmospheric situation which resulted in the abnormally heavy rains in China in 1980. There is a suggestion that similar floods in earlier years were accompanied by similar anomalies of ocean temperature and that very different, or even opposite, SST anomaly patterns occurred in very dry years.

Nicholson and Chervil have considered rainfall variations over virtually the whole of the African continent: the data set used, comprising about 1100 stations with records in general starting about 1920, is unique. They looked particularly at the drought-prone sub-Saharan region and demonstrated that, on some

occasions at least, such droughts appear to be related to an
intensified Hadley cell. In these cases, there is greater rainfall near the equator and on the poleward side of the subtropical
zone with increased subsidence in the drought area. There is
evidence that similar anomalies occur contemporaneously in
southern Africa. More controversial is their suggestion that in
some past epochs (notably 1820-1840 and 1895-1920 approximately)
negative anomalies of rainfall occurred over practically the whole
of the African continent suggesting some quite different mechanism.
In this context, it is probable that rainfall anomalies covering
large areas have a positive feedback effect as demonstrated by
Rowntree in this volume. Therefore, once a negative rainfall
anomaly had become established over Africa, it would tend to
persist.

Kininmonth discusses two anomalous rainfall events in
Australia, one an unusually dry period in the north and the other
unusually heavy rain in the dry interior. There is evidence that
the first case is related to increased subsidence in the subtropical anticyclonic region (strengthened Hadley circulation).
The second case illustrates a phenomena only recently recognized,
namely the connection which can occasionally occur between the
equatorial moisture source and a dynamic system in higher
latitudes.

Mooley and Parthasarathy report an excellent study of droughts
and floods over India over the last 100 years. This study is based
on over 300 rainfall station reports and throws a good deal of
light on the variability of the Indian monsoon. Their conclusion
that the occurrence of droughts and floods in India is random in
time is in marked contradiction to some authors who aver that the
frequency of failures in the Indian monsoon is increasing.

The last paper in this section concerns a study by Chiu,
using satellite and other data, of variations in the extent of
Antarctic sea ice from 1973-1980. Over this period, an apparent
decrease of 2.5×10^6 km^2 of total sea ice has occurred. Since
the area covered by sea ice has been shown by earlier research to
be related to the location of the maximum cyclonic activity in the
southern ocean, such a decrease may be expected to have a small
effect on southern ocean climate. However, the temperature change
associated with the observed decrease of Antarctic sea ice over
the 7-yr period is less than 1°C and it is believed that this
apparent decrease may be only part of an oscillation of longer
period. There is, therefore, no evidence that these changes are
as yet indicative of a global warming.

RECENT RAINFALL FLUCTUATIONS IN AFRICA - INTERHEMISPHERIC
TELECONNECTIONS

S.E. Nicholson

Graduate School of Geography
Clark University
Worcester, Mass. 01610, USA

R.M. Chervin

National Center for Atmospheric Research
Boulder
Colorado 80307, USA

ABSTRACT. This paper deals with the spatial and temporal
characteristics of rainfall fluctuations, particularly droughts,
in the sub-Saharan region and in the semi-arid zones of southern
Africa. The investigation, based on records from 1100 stations,
commenced with 84 regionally-averaged rainfall series from which
anomalous years were selected for studying both the changes of
rainfall seasonality and the spatial patterns associated with major
drought episodes and wetter years. The most important results are:
(a) the markedly synchronous occurrence of droughts in the subtropics of both hemispheres; (b) the extreme spatial coherence of
rainfall anomalies; (c) synchronous fluctuations along both
tropical and temperate margins of the semi-arid/arid zones;
(d) persistence of anomalies in the sub-Saharan zones; (e) the
apparently 'normal' location of the Intertropical Convergence Zone
(ITCZ) during most drought years. These results suggest that
major rainfall fluctuations in Africa are associated primarily with
factors modifying the intensity or frequency of disturbances,
e.g., changes in the intensity of the Hadley circulation or factors
modifying energetics and stability.

A. Street-Perrott et al. (eds.), Variations in the Global Water Budget, 221–238.

1. INTRODUCTION

Arid and semi-arid lands, where interannual rainfall variability is inherently extreme, dominate the African continent;
severe droughts are a common occurrence in nearly every region.
The meteorological causes of such droughts are still unclear, but
numerous studies of the recent one in the Sahel suggest that major
changes in large-scale tropical circulation patterns were involved
(Krueger & Winston 1975, Kanamitsu & Krishnamurti 1978, Kidson
1977). Some studies (Hastenrath 1978, Fleer 1981, Doberitz 1969,
Berlage 1966, Bjerknes 1969) have shown that such large-scale
changes in tropical circulation result in synchronous rainfall
anomalies throughout large areas of the tropics, but few studies
have considered the more specific question of inter-hemispheric
teleconnections in the tropics. The semi-arid regions of Africa
provide an excellent opportunity for such a study. Moreover, the
northern and southern extremes of the tropics are essentially part
of the same circulation system, a complete understanding of which
and of its relationship to rainfall fluctuations (especially
Sahelian droughts) requires an examination of the system's
northern and southern components.
    The relationship between rainfall fluctuations in the semi-
arid subtropics of Africa was investigated using data from 1100
African stations (Figure 1) for which records generally commence
in the 1920s or earlier. These include the northern hemispheric
station network described previously (Nicholson 1979) and southern
African data compiled from weather service reports and agricultural
bulletins found in various libraries in England, Germany, Portugal
and the USA (a complete catalogue of this data set is in preparation). This account summarizes the results of previous research
focused on the Sahel and nearby semi-arid zones, and compares the
conclusions reached with the results of preliminary analyses of
the southern African data.

2. RECENT RAINFALL FLUCTUATIONS IN NORTHERN AFRICA

To compare stations with diverse means and variabilities, data
were standardized by dividing departures from annual means at each
station by the standard deviation σ of annual totals at that
station. Regional departures (Figure 2) represent the arithmetic
average of standardized departures for all stations available in
that region in a given year (Nicholson 1979, 1980). The standardized annual departure patterns for the entire area north of $10°S$
were 'typed', using a map classification scheme by Lund (1963),
which Blasing and Fritts (1976) modified to examine winter pressure
anomaly patterns for North America. The results (Nicholson 1980)
include four dominant patterns, depicted in Figure 3 in simplified form. Finally, rainfall seasonality corresponding to each of
the four types was examined using cross-sections of monthly rain-

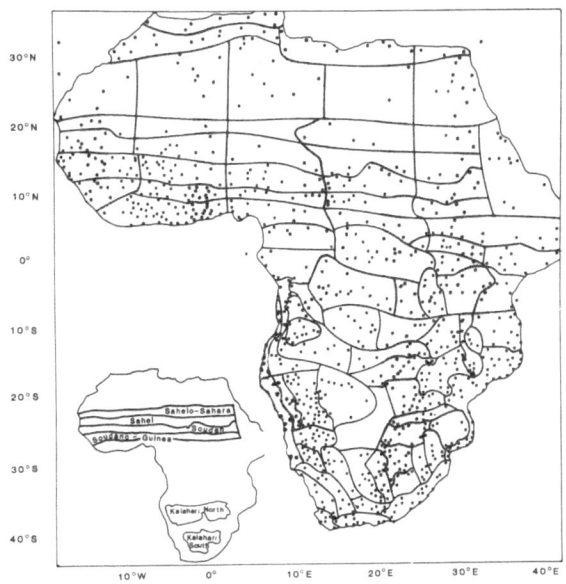

Fig. 1. Network of stations, location of meridional cross-sections
         (shaded areas indicate stations used in cross-sections)
         and regional divisions for rainfall analyses. Inset:
         location of regions averaged to form rainfall departure
         series for Sahelo-Saharan, Sahel, Soudan and Soudano-
         Guinean zones and for northern and southern Kalahari.

fall as a function of latitude for the five longitudinal sectors
shown in Figure 1; those for 0°W are summarized in Figure 4. The
results of these analyses suggest several characteristics of
recent rainfall fluctuations in northern Africa.

     In the four semi-arid sub-Saharan zones (Sahelo-Saharan,
Sahelian, Soudanian and Soudano-Guinean) droughts occurred in the
1910s, 1940s and from about 1967 to 1973, while it was abnormally
wet in the 1950s (Figure 2). Although the drought only became
extreme in the late 1960s, a dry period had already begun in 1960;
thus, if the drought resulted from anomalous tropical circulation
patterns, the major circulation change probably occurred several
years before the onset of drought (Nicholson 1979). The rainfall
departure series for the four sub-Saharan zones exhibit remarkable
persistence of rainfall anomalies, with several 8- to 10-yr runs
of departures below or above the median (Figure 2). For the
Sahelian series, autocorrelations with 1-, 2- and 3-yr lags, calculated as 0.33, 0.28 and 0.26, respectively, give further evidence
of this persistence (Nicholson 1979). Walker and Rowntree (1977)

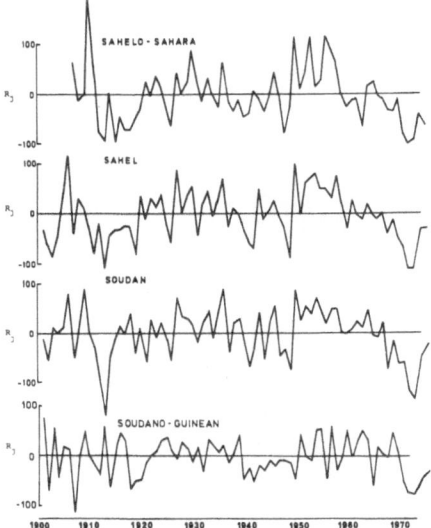

Fig. 2.   Standardized annual rainfall departures R_j ($\cong$ % standard
           departure) for the Sahelo-Saharan, Sahel, Soudan and
           Soudano-Guinean zones (from Nicholson 1980).

similarly noted this characteristic of Sahelian rainfall, especially
in areas nearest to the desert margin.
     Clearly, the similarity of the four curves in Figure 2
indicates that rainfall variations are generally coherent throughout the four semi-arid sub-Saharan zones (Nicholson 1980) although,
within each zone, fluctuations are not consistently uniform.
Additional analysis suggests that this coherence, especially in
the case of extreme fluctuations, extends over much larger areas of
the continent. Figure 5 represents composites of three extreme
years corresponding to four 'rainfall anomaly types' frequently
occurring in West Africa. The complete set of anomaly types
includes several in which longitudinal variation exists across the
east-west expanse of the sub-Sahara (e.g., wet in the west, dry in
the east), but the dominant types represent minimal variation within
these zones, as shown in Figure 3. With the 'a' types, departures
in the equatorial region strongly contrast with those in the subtropics; 'b' types represent departures of the same sign in nearly
all regions, i.e., uniformly dry or wet in both equatorial and subtropical regions. The areal extent of anomalies is highlighted by
the regional departures for the abnormally wet 1950s and the
drought period from 1968 to 1973 (Nicholson 1981b). Few regions
differed from the pattern prevailing over the continent.
     Meridional cross-sections (Figure 4) of monthly rainfall as
a function of latitude illustrate rainfall seasonality at 0°W for

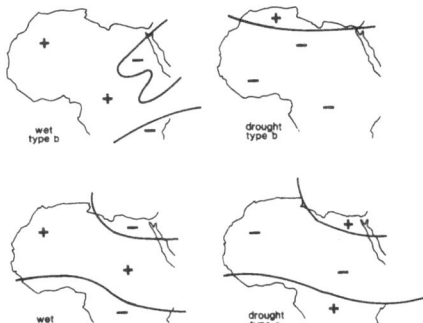

Fig. 3. Schematic representation of the major types of rainfall
anomaly patterns observed in tropical and subtropical
Africa: 'a' types, departures in the equatorial region
contrast markedly with those in adjacent subtropical zones;
'b' types, departures principally of the same sign in both
the equatorial and subtropical zones; +, above average
rainfall; -, below average rainfall.

3-yr composites. The positions of the 25 mm isohyet and the
'center of gravity' of the rain belt (a term used by Tanaka *et al.*
(1975) to mean the latitudinal zone of maximum rainfall) are considered relative measures of changes in the position of the ITCZ;
the number of months with rainfall $\geqslant$ 100 mm is a measure of the
intensity of the rainy season, which is defined here as those months
with $\geqslant$ 25 mm precipitation (Nicholson 1981*a*); see Figure 5 legend
for classification of the rainfall anomaly types.

The position of the 25 mm isohyet suggests that the ITCZ
advances and retreats normally during drought years, except
possibly in the months of July and August. Thus, an anomalous
position of the ITCZ is at most a critical factor in the drought
years in regions where rainfall is limited to the months of July
and August, i.e., areas north of *ca*. 18°N. This is only a small
part of the usual drought-affected area, which extends equatorward
to at least 10°N. If the 'center of gravity' of the rain belt
(regarded as areas with $\geqslant$ 200 mm per month) is taken as a measure
of the ITCZ position, it appears that the convergence zone was
displaced from normal only in the set of years (1950, 1953, 1956)
which were wet in the sub-Saharan zone and unusually dry in the
equatorial regions below 10°N. Furthermore, between 18°N and 10°N
the rainy season appeared to be of normal length during drought
years (e.g., in the 0°W sector, 5 months at 15°N), but in most
cases there were fewer months in which rainfall exceeded 100 mm
(e.g., in the 0°W sector, 1 or 2 months at 15°N, compared with a
normal of 3 months). In summary, from the evidence of the wet and
dry composites, it seems that an abnormal displacement of the ITCZ
is probably not a critical factor in drought except possibly in
areas between 18°N and the Saharan margin. The main difference
between wet and dry years appears to be the intensity of the rainy

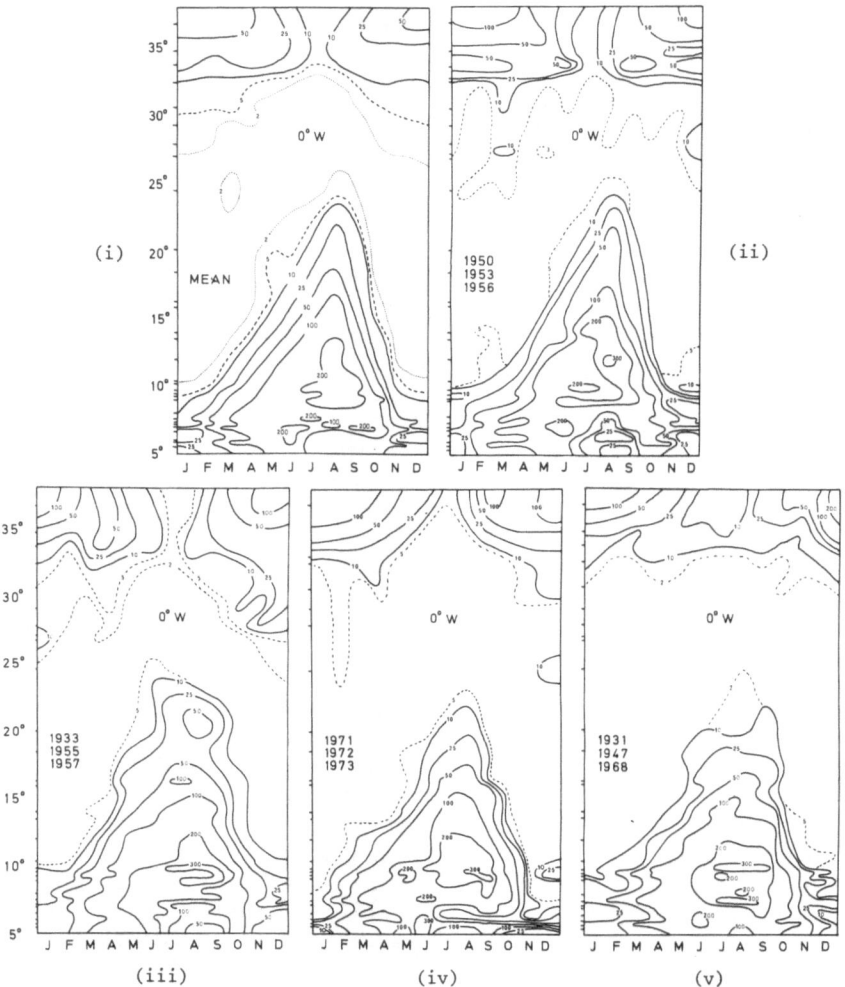

Fig. 4.   Rainfall (mm per month) as a function of months and
          latitudes at 0°W for 3-year composites (meridional cross-
          sections) for the four rainfall anomaly types (see Figure
          3): (i) overall mean; (ii) wet type 'a'; (iii) wet type
          'b'; (iv) drought type 'b'; (v) drought type 'a'. Marks
          on the ordinates denote latitudes of stations from which
          data were used.

season; drought years are characterized by a rainy season of usual
length, but with fewer months having > 100 mm rainfall (see also
Nicholson 1981a).
     A number of analyses (Figures 3 to 5) shows that rainfall

fluctuations generally are manifested as expansions and contractions of the arid area in the Saharan region (Nicholson 1980). In
most cases, fluctuations of the same sign occur both north and
south of the desert. This characteristic is seen also on historical and palaeo-time scales. Rainfall along the northern
margin, unlike that to the south, is of extratropical origin, falls
mostly during the winter season, and is unrelated to the ITCZ:
synchronous fluctuations along both margins therefore suggest that
tropical circulation changes on a larger scale, rather than merely
a displacement of the ITCZ and/or the subtropical high, are
responsible for major rainfall anomalies. The common factor in
aridity along both tropical and temperate margins of the desert is
the descending branch of the Hadley circulation; hence major
changes in that system are likely to be involved.

3.   ANALYSIS OF RAINFALL FLUCTUATIONS IN SOUTHERN AFRICA

     The semi-arid Kalahari and the Namib desert represent the
southern-hemisphere analogue of the Sahara: the subtropical aridity
relates to the South Atlantic High and the southern, descending,
branch of the Hadley circulation; the tropical margins of these
regions mark the limit of penetration of the ITCZ into the southern
hemisphere. Thus, the arid and semi-arid zones of the two hemispheres are manifestations of the same circulation system. In
order to understand fully the changes in the northern subtropics,
the entire system must be investigated.
     Rainfall fluctuations in southern Africa over the last few
centuries are being reconstructed using a variety of 'proxy' data
(e.g., drought and lake-level records) and landscape descriptions,
combined with historical references to weather and climate and
actual rainfall measurements (Nicholson 1981b). The resulting
chronology suggests that, on a scale of decades or longer, fluctuations tend to be synchronous over most of the continent. Summary
maps for three historical periods (Figure 6) clearly indicate this
characteristic: extreme and persistent drought affected the
Saharan margins and the Kalahari-Namib region in the 1830s and
between ca. 1895 and 1920, whereas rainfall was probably well
above the present-day normals in both regions in the last decades
of the nineteenth centry.
     Modern instrumental records are being used to test the
hypothesis of synchronous fluctuations in the African subtropics
of both hemispheres. A rainfall data set is being compiled for
680 stations in southern Africa (see Figure 1) and includes
monthly totals for each year from the beginning of the record (usually in the 1910s or 1920s) to 1973. The results, though preliminary, do indicate some interesting relationships between southern-
and northern-hemisphere Africa. The analyses are similar to those
described (above) for northern Africa. However, the regional and

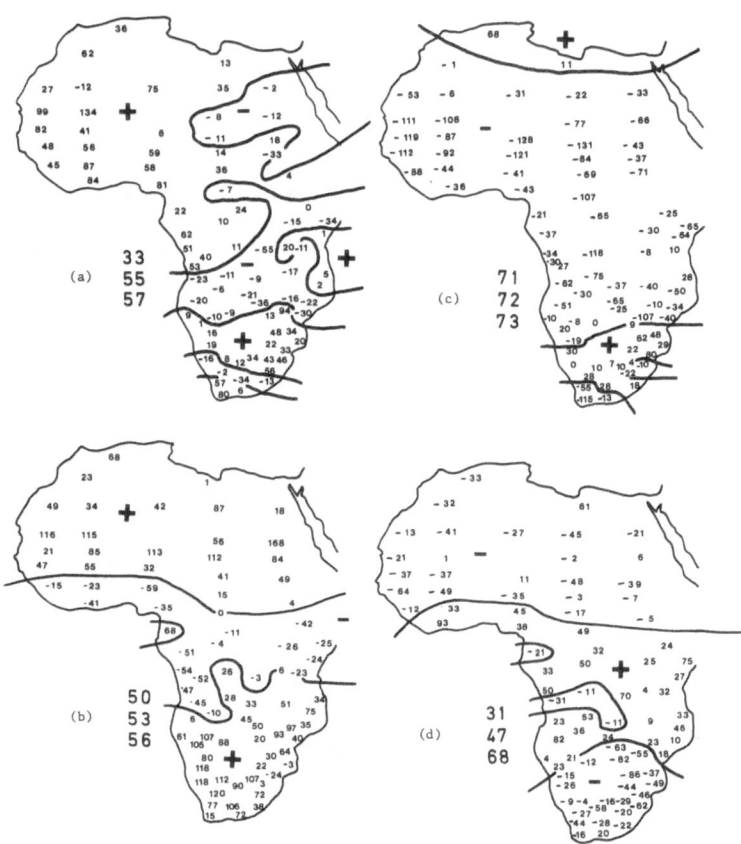

Fig. 5.   Regional rainfall departures for 3-year composites for the
          four rainfall anomaly types (described in Figure 3):   wet
          type (a), 1950, 1953, 1956; wet type (b), 1933, 1955, 1957;
          drought type (a), 1931, 1947, 1968; drought type (b), 1971,
          1972, 1973.   Location of regions is shown in Figure 1.

zonal divisions, although based on climatic characteristics, are
still arbitrary.   There may also be inhomogeneities in the data
series.   The number of stations comprising a regional average
varies greatly in time, the variability decreasing as station
number increased.   The series are homogeneous for northern Africa
after about 1920, but not until much later for southern Africa.
This must be considered in examining the two Kalahari series
(Figure 7) which are derived from several arbitrarily-grouped
regions north and south of the Kalahari (see Figure 1).   Furthermore, the southern Africa data have not been subjected to independent 'typing' analysis.   Instead, data for these stations were
merely merged for the same three-year sets used in the northern
African analysis (see Figure 5).   These were not the most extreme

c. 1820-1840

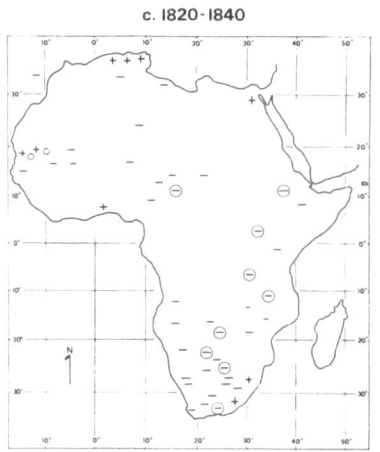

c. 1870-1895   c. 1895-1920

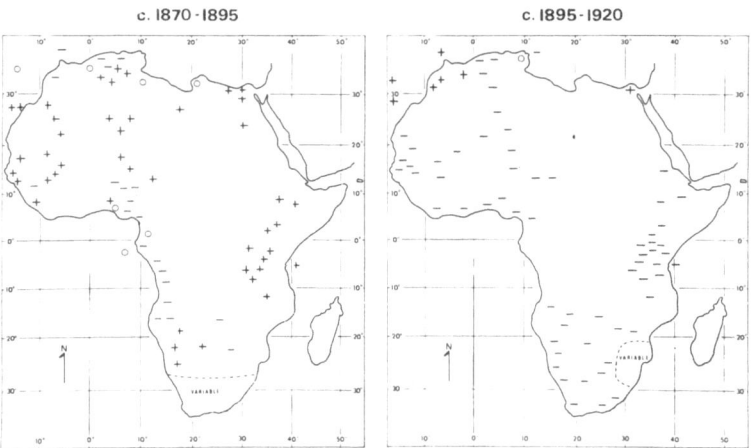

Fig. 6.   Schematic representation of rainfall anomalies (Nicholson
          1981*b*) for (a) *ca.* 1820-1840; (b) *ca.* 1870-1895; (c) *ca.*
          1895-1920; +, > normal; 0, normal; -, < normal; o, regional
          indicators such as lakes and rivers.

years nor necessarily the years of anomalous rainfall in southern
Africa; thus, the exercise can only show whether any coherent
patterns prevailed over that area during years that were particularly anomalous in sub-Saharan areas. Meridional cross-sections
for more southerly latitudes are similarly derived for the same
three-year composites (Figure 8). These preliminary analyses

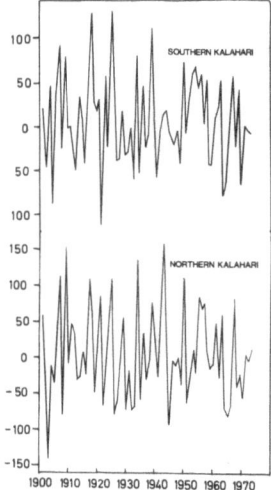

Fig. 7. Standardized annual rainfall departures (≅ % standard
departure) for northern and southern Kalahari.

only serve for a general comparison of rainfall fluctuations in
northern and southern Africa.

The series of rainfall departures for the southern and
northern margins of the Kalahari show some features of the sub-
Saharan series, but with notable differences. As in the sub-
Sahara, rainfall tended to be above average in the 1950s and was
generally below normal in the 1940s; these features were most
pronounced south of the Kalahari, but not along the northern margin, the region very similar to the Sahel climatically. During
the 1910s, a period of extreme and persistent drought south of the
Sahara, rainfall departures were only slightly negative along the
Kalahari margins. However, large areas of Namibia experienced
extreme drought during most of that decade (Nicholson 1981*b*).
The Kalahari was affected by drought in the early 1960s when a
drier period set in in the Sahel, but, during the period of most
extreme drought in northern Africa (*ca.* 1968-1973), rainfall was
only moderately below normal in the Kalahari.

Although the rainfall departure series do not appear to contain many long runs of wet and dry years (Figure 7), such a conclusion needs to be supported by further analysis. It seems
reasonable to conclude that the degree of persistence noticed in
the four series for northern Africa is not present in the Kalahari
series. A possible explanation for the persistence in the sub-
Saharan region is a feedback mechanism, such as that proposed by
Charney (1975) and since supported by a number of studies, in
which rainfall fluctuations impose changes in surface parameters
(e.g., albedo, soil moisture) which, in turn, reinforce the rainfall anomaly. Such a mechanism requires an extensive area subject

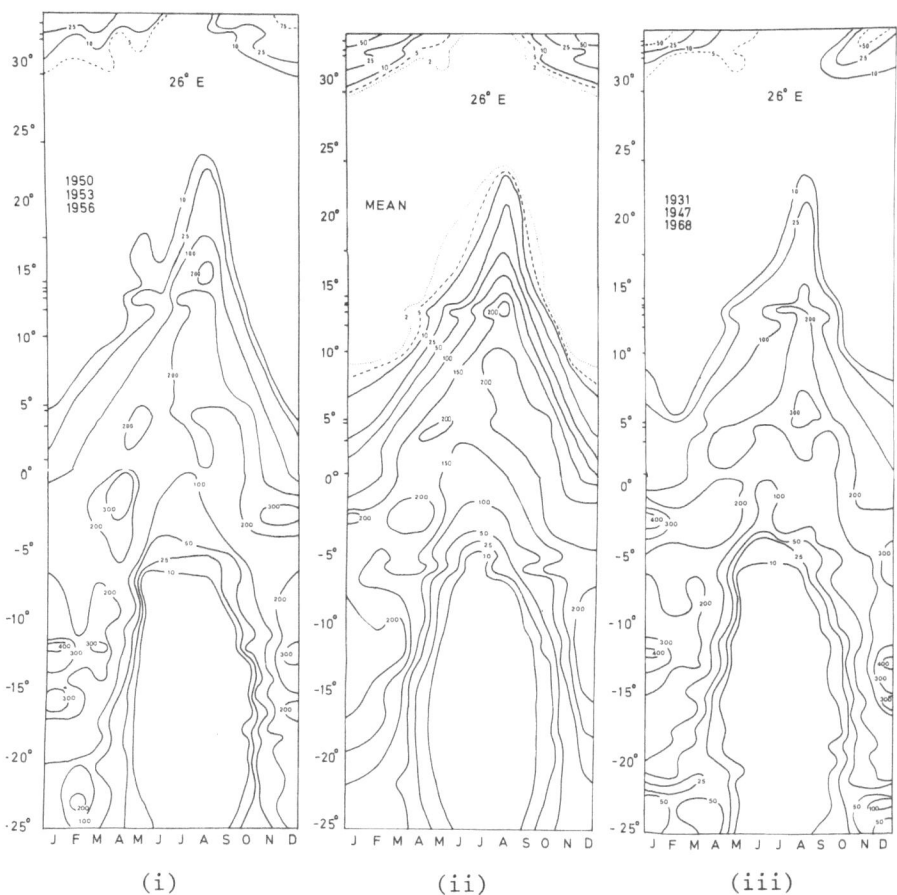

Fig. 8.   Rainfall (mm per month) as a function of months and lati-
          tudes at 26°E for 3-year composites (meridional cross-
          sections) for (i) rainfall anomaly wet type 'a'; (ii) over-
          all mean; (iii) drought type 'a'. Marks on ordinates
          denote latitudes of stations from which data were used.

to large, uniform rainfall departures, a sharp gradation to
extreme desert conditions and, as originally formulated, a negative net radiation balance in summer. The Kalahari, which is a
relatively small area and nowhere a true desert, and which has a
positive radiation balance, does not meet these criteria.
     The three-year composites of regional anomalies (Figure 5)
and the regional anomalies for 1950 and 1970 (Figure 9) reveal
more spatially-coherent rainfall variation over the African continent than is suggested by comparison of the departure series for

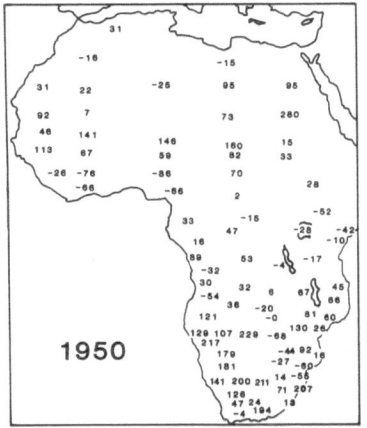
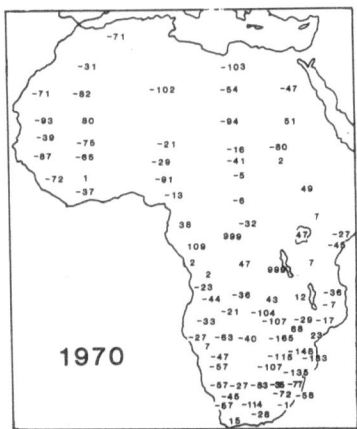

Fig. 9. Regional rainfall departures for 1950 and 1970 (regions
          shown in Figure 1).

the sub-Saharan region and the Kalahari. The coherence (in a non-
statistical sense) of rainfall anomalies apparent in the northern
African analyses is almost continental in character in many
extreme years. The two sub-Saharan drought composites (Figure 5)
indicate strong negative regional departures over large regions of
southern Africa. In the case of drought type 'b' (1971-73), negative departures extended throughout the equatorial region southward to the northern Kalahari; above average rainfall was mostly
confined to a small area south and east of the Kalahari. The
pattern of negative regional departures dominating the entire continent was even stronger in 1970 (Figure 9). In the case of
drought type 'a' years (1931, 1947, 1968), the pattern over
southern Africa closely resembles that over the northern part of
the continent: positive regional departures prevailed from $10°N$
to $20°S$ (i.e., throughout the equatorial regions) separating two
large areas of negative anomalies including both tropical and
extra-tropical margins of the Kalahari and Sahara. The reverse of
this pattern prevailed in the sub-Saharan wetter years 1950, 1953
and 1956 (wet type 'a'), and positive anomalies were as large and
spatially coherent in southern as in northern Africa, but with
below average rainfall in the equatorial region. Again, departures of the same sign prevailed both north and south of the arid/
semi-arid zones. Wet type 'b', which appeared to represent wetter
years in both sub-Saharan and equatorial regions, is seen with the
addition of southern African data to represent a variant of wet
type 'a'. However, there were individual years, such as 1950
(Figure 9), in which positive departures prevailed over almost
the entire continent. In the wet type 'b' pattern, there were
small areas with departures differing from the large-scale pattern.
However, the main differences between the two types appear to be

that, in type 'b' years, the negative anomaly in the equatorial region was further south, and the net departures in the composite were small and negative over the eastern Sahara.

The relatively frequent occurrence of years, such as those represented by drought type 'b' (Figure 5) or 1950 and 1970 (Figure 9), in which departures of one sign were nearly continental in character, suggest that there is large interannual variability in the absolute amount of rainfall over the continent. This, in turn, implies large fluctuations in the contribution of latent heat flux to the tropical energy balance in the African sector at least. No conclusion can be drawn for the global tropics because there may be equally large sectors elsewhere where anomalies opposite to those over the African continent prevail, but the question merits further investigation.

Analyses discussed in § 2 suggest that rainfall fluctuations over northern Africa, especially in the 20th century, take the form of an expansion or contraction of the arid zone, as during the 1950s or ca. 1968-73. Such a tendency is less often seen in southern Africa; only in composites corresponding to wet type 'a' and drought type 'a' do anomalies of the same sign occur both north and south of the Kalahari, positive in one case, negative in the other. Wet type 'b' and drought type 'b' seem instead to represent a southward or northward displacement, respectively, of the semi-arid region of southern Africa.

There is no 'rainless' area in the core of the Kalahari by which a measure might be derived of the poleward limit of penetration of the ITCZ. However, if the 'center of gravity' of the rain belt is a valid measure of relative changes in the position of the ITCZ, the cross-sections in Figure 9 suggest that, in the austral summer during type 'b' drought years, the ITCZ was located in its usual position, but penetrated further poleward into the southern hemisphere during wet years of type 'a'; this conclusion as to the position of the ITCZ also applies to northern Africa in the boreal summer. Positive annual departures in both northern and southern Africa seem instead to be related to a more 'intense' rainy season (e.g., the large areas with up to 400 mm per month rainfall, with more months at a given latitude having rainfall in excess of 200 mm).

4.   DISCUSSION

Because African upper air data are scarce before 1960 and pressure patterns are difficult to interpret in the tropics, rainfall patterns provide the only clue to circulation differences between wetter periods prior to the 1960s and the subsequent drier episode (Nicholson, 1981). The important points are:
(a)   The tendency for subtropical droughts to be associated with less intense rainy seasons of usual length, a normal position of the ITCZ over .the sectors studied, and abnormally intense rainfall in the equatorial regions;

(b) The continental coherence of rainfall anomaly patterns,
     including those in areas where rainfall is not associated
     with ITCZ, and the occurrence of many years with anomalies
     of the same sign in both equatorial and subtropical regions;
(c) The tendency, particularly in northern Africa, for rainfall
     departures of the same sign to prevail on both poleward and
     equatorward margins of the arid/semi-arid subtropical zone,
     i.e., synchronous fluctuations in regions where 'summer'
     tropical rains associated with the ITCZ prevail, and in
     extra-tropical regions where rains fall primarily in the
     winter and transition seasons (see Figures 3 to 5).

A diminished northward penetration of the ITCZ is suggested
by numerous authors (Bryson 1973, Winstanley 1973, Kraus 1977a,b,
Lamb 1978) as a factor in sub-Saharan droughts. Such a condition
would mean increased rainfall in the equatorial regions during
subtropical droughts, because a diminished annual excursion implies
that the convergence zone lies in equatorial latitudes for a longer
period during the year. In view of that, type 'a' drought patterns
could conceivably be associated with an anomalous position of the
ITCZ. However, evidence presented above suggests that, during
sub-Saharan drought years, the ITCZ is situated in its normal
position, although it appears to penetrate abnormally far northward in some wetter years. Furthermore, variation of the ITCZ
position, as described above, could not explain droughts in which
rainfall is reduced in both subtropical and equatorial regions
(e.g., drought type 'b'). Finally, a mere displacement of the
ITCZ could not explain the changes of rainfall intensity, the
continental scale of the anomalies, and the synchronous fluctuations of the same sign north and south of the arid/semi-arid zones.
Rainfall deficits north or the Sahara during years or Sahel drougnt
are in conflict with one aspect of this hypothesis, namely, the
expansion of the circumpolar vortex and the corresponding equatorward displacement of the subtropical highs. An anomalous excursion
of the ITCZ is unlikely, therefore, to be a factor in the major
droughts discussed here. This generalization cannot yet be made
for longer-term climatic fluctuations and historical droughts nor
for southern Africa, since the major droughts which occurred there
(e.g., in the early 1960s and the 1910s) have yet to be analysed.

What circulation changes are compatible with the rainfall
fluctuations described here? The casual mechanism, which is not
likely to be the same for each drought, must involve a combination
of factors. Several different anomaly patterns characterize the
rainfall fluctuations over Africa in the present century: each
may relate to a different combination of casual events. These
probably include major changes in the Hadley circulation and/or
the upper level wave patterns over the continent, but additional
factors, such as sea-surface temperature changes, may also be
important.

Situations in which opposing departures prevail in equatorial
and subtropical regions (e.g., 'a' types) may be related to

changes in the intensity of the Hadley circulation. A more
intense circulation cell in the transition seasons (greater subsidence poleward and greater upward flow in the equatorial regions)
could account for the decreased 'intensity' of the rainy season
during sub-Saharan droughts and simultaneous intensification and
increase of equatorial rains. A slight contraction of the ascending branch in June to August could account for the extreme
dryness between *ca.* 18°N and 25°N in these months and the contraction of the equatorial rain belt at 12°E at that time, as noted
elsewhere (Nicholson 1981*a*). Such a contraction might also
account for situations in which annual rainfall is simultaneously
reduced in both subtropical and equatorial regions (e.g., type 'b'
droughts). In such cases, rainfall appears to be intensified in
equatorial regions (i.e., abnormally high rainfall for several
months), although annual totals there are below normal (Nicholson
1981*a*); a longer dry season (i.e., a narrower belt of maximum
rainfall) could account for the reduced annual totals. Analogous,
synchronous changes in the southern Hadley circulation might produce the coherent fluctuations in northern and southern Africa.

Changes in the long waves of the upper troposphere westerlies
might evoke rainfall fluctuations over Africa, especially those
which occur synchronously in northern and southern Africa or which
result in similar anomalies in both subtropical and equatorial
regions. Occasionally, deep troughs form which approach or even
cross the equator (Namias 1963); over Africa, these interact with
disturbances in the low-level easterly flow (Flohn 1975, Kutzbach:
this volume) to create eastward-propagating depressions over the
Sahara, which appear to influence rainfall over Africa in areas
from the Mediterranean coast equatorward to 10°N. An increased
frequency of such depressions could partially explain the tendency
for rainfall to increase along both the tropical and temperate
borders of the Sahara; in fact, the type 'b' wet composite
pattern over West Africa appears to relate to such a system
(Nicholson 1981*a*). One further observation, seen in the 0°W cross-
section (Figure 4) and a similar one for 12°W (Nicholson 1981*a*),
supports the role of such disturbances in producing rainfall
fluctuations over northern Africa. In these sectors, rainfall
fluctuations north of the Sahara, where winter rains dominate,
appear to be determined by events in the transition seasons and in
early or late summer - the season when the disturbances described
above are most active. However, relatively weak Hadley circulation
at these times, as suggested could also explain both the increase
of rainfall north and south of the Sahara and the apparent
influence of rainfall in the transition seasons.

Direct evidence for some of the proposed circulation changes
is available. An analysis of the velocity potential field over
the tropics during the 1972 drought (Kanamitsu & Krishnamurti 1978)
reveals increased vertical circulation in the meridional plane
(i.e., Hadley-type overturning) at the expense of east-west overturning in the zonal plane, strong divergence over the Cameroons
in West Africa (implying rising motion there and compensating

descent in surrounding regions including the sub-Sahara) and an
equatorward shift of the Hadley cell. This is compatible with the
rainfall pattern of 1972 (when the drought extended from 10°N to
35°N) and more 'intense' rainfall in several months, despite below
normal annual totals for equatorial rainfall at 12°E. The study
concludes that a weaker easterly jet over Africa, weaker vertical
wind shear and, hence, weaker barotropic-baroclinic instability
were also factors in the sub-Saharan drought. Kidson (1977) provides further evidence of a weaker easterly jet during drought
years as well as evidence for the complete disappearance of the
850 mb trough near 8°N and reduced moisture convergence into the
sub-Saharan region. These features are consistent with the
changes in the Hadley circulation shown by Kanamitsu and Krishnamurti (1978) and the circulation changes suggested here on the
basis of rainfall patterns. While these two studies were not concerned with southern Africa, they nevertheless present evidence of
analogous changes over the southern sub-continent. In particular,
the Hadley-type overturnings (Kanamitsu & Krishnamurti 1978:
Figure 4) markedly intensified in 1972. This might, in part,
explain the dry conditions over southern Africa which occurred
synchronously with the recent sub-Saharan drought.

More definite conclusions concerning the relationship between
rainfall fluctuations in northern and southern Africa, and their
relationship to atmospheric parameters, must await the completion
of additional analyses based more directly on the data from the
southern hemisphere. It appears that rainfall changes in the sub-
Saharan region are part of large-scale changes in the tropical
atmosphere, changes which act on at least a continental scale and
affect both hemispheres. The continental-scale anomalies of the
years 1950 and 1970, for example, emphasize this point. Both the
large magnitude and spatial extent of the anomalies lead to the
possibility of a feedback between the rainfall fluctuations and
the general circulation.

ACKNOWLEDGEMENTS. The authors would like to acknowledge the
generous help of numerous librarians who made the data collection
process considerably easier. Invaluable assistance was provided
by Dennis Joseph (NCAR) who created the tape from the collected
data. Funding for the project was provided in part by the Climate
Dynamics Program of the National Science Foundation (Grant ATM
8019009). The National Center for Atmospheric Research, also
sponsored by the NSF, provided computing support.

REFERENCES

Berlage, H.P. 1966 The southern oscillation and world weather.
    Med. Verhandel., 88, 152 pp.

Bjerknes, J. 1969 Atmospheric teleconnections from the equatorial Pacific. *Mon. Weath. Rev.*, 97, 163-172.

Blasing, T.J. & Fritts, H.C. 1976 Reconstruction of past climatic anomalies in the North Pacific and western North America from tree-ring data. *Quat. Res.*, 6, 563-580.

Bryson, R.A. 1973 Drought in Sahelia: Who or what is to blame? *Ecologist*, 3, 366-371.

Charney, J.G. 1975 Dynamics of deserts and drought in the Sahel. *Q. Jl R. met. Soc.*, 101, 193-202.

Doberitz, R. 1969 Cross spectrum and filter analysis of monthly rainfall and wind data in the tropical Atlantic region. *Bonner Met. Abhandl.*, 11, 53 pp.

Fleer, H. 1981 Large-scale tropical rainfall anomalies. *Bonner Met. Abhandl.*, 26, 114 pp.

Flohn, H. 1975 Tropische Zirkulationsformen im Lichte der Satellitenaufnahmen. *Bonner Met. Abhandl.*, 21, 82 pp.

Hastenrath, S. 1978 On modes of tropical circulation and climate anomalies. *J. atmos. Sci.*, 35, 2222-2231.

Kanamitsu, M. & Krishnamurti, T.N. 1978 Northern summer tropical circulations during drought and normal rainfall months. *Mon. Weath. Rev.*, 106, 331-347.

Kidson, J.W. 1977 African rainfall and its relation to the upper air circulation. *Q. Jl R. met. Soc.*, 103, 441-456.

Kraus, E.B. 1977a Subtropical droughts and cross-equatorial energy transports. *Mon. Weath. Rev.*, 105, 1009-1018.

Kraus, E.B. 1977b The seasonal excursion of the intertropical convergence zone. *Mon. Weath. Rev.*, 105, 1052-1055.

Krueger, A. & Winston, J. 1975 Large-scale circulation anomalies over the tropics during 1971-72. *Mon. Weath. Rev.*, 103, 465-473.

Lamb, P.J. 1978 Case studies of tropical Atlantic surface circulation patterns associated with sub-Saharan weather anomalies. *Tellus*, 30, 240-251.

Lund, I.A. 1963 Map-pattern classification by statistical methods. *J. appl. Met.*, 2, 56-65.

Namias, J. 1963 Interactions of circulation and weather between hemispheres. *Mon. Weath. Rev.*, 93, 482–486.

Nicholson, S.E. 1979 Revised rainfall series for the West African subtropics. *Mon. Weath. Rev.*, 107, 620–623.

Nicholson, S.E. 1980 Saharan climates in historic times. In: *The Sahara and the Nile* (ed. M.A.J. Williams & H. Faure), pp. 173–200. Rotterdam: A.A. Balkema.

Nicholson, S.E. 1981*a* Rainfall and atmospheric circulation during drought periods and wetter years in West Africa. *Mon. Weath. Rev.*, 109, 2191–2208.

Nicholson, S.E. 1981*b* The historical climatology of Africa. In: *Climate and History: Studies in Past Climates and their Impact on Man* (ed. T.M. Wigley, M.J. Ingram & G. Farmer), pp. 249–270. Cambridge, England: Cambridge University Press.

Tanaka, M., Weare, B.C., Navato, A.R. & Newell, R.E. 1975 Recent African rainfall patterns. *Nature, Lond.*, 255, 201–203.

Walker, J. & Rowntree, P.R. 1977 The effect of soil moisture on circulation and rainfall in a tropical model. *Q. Jl R. met. Soc.*, 103, 29–46.

Winstanley, D. 1973 Rainfall patterns and general atmospheric circulation. *Nature, Lond.*, 245, 190–194.

DROUGHTS AND FLOODS OVER INDIA IN SUMMER MONSOON SEASONS 1871-1980

D.A. Mooley and B. Parthasarathy

Indian Institute of Tropical Meteorology
Pune 411005
India

ABSTRACT. Utilizing the monthly rainfall of 306 stations with fairly even distribution over India and rational criteria, the incidence of droughts and floods over the country and regions thereof in the summer monsoon season has been examined for the period 1871-1980.
    Droughts over India were severe in four years: 1877, 1899, 1918 and 1972. Floods over India were severe in three years: 1878, 1892 and 1961. The occurrence of droughts and floods has been found to be random. The only years when none of the regions of India experienced either drought or flood were 1930, 1931 and 1957. Having identified individual years of monsoon rainfall excess and deficit, differences in depression track behaviour and in the location of the monsoon trough were examined.

1. INTRODUCTION

In the tropics, rainfall is the climatic parameter that most concerns the farmer. Interannual (year-to-year) variability is an inherent characteristic of rainfall which affects a variety of economic activities.
    With agriculture as the predominant occupation of the population and the monsoon contributing 80-95% of annual rainfall over a large part of the country, the activity of the summer monsoon (June to September) is the most important factor in the Indian economy. The seasonal rainfall in different parts of India varies markedly from year-to-year. Almost every year sees some region experiencing drought or flood, though the area so affected is generally a small part of the total land mass. However, occasionally, the variations are large and a large part of the country suffers from drought or flood, with the inevitable dislocation of the economy

239

A. Street-Perrott et al. (eds.), Variations in the Global Water Budget, 239–252.
Copyright © 1983 by D. Reidel Publishing Company.

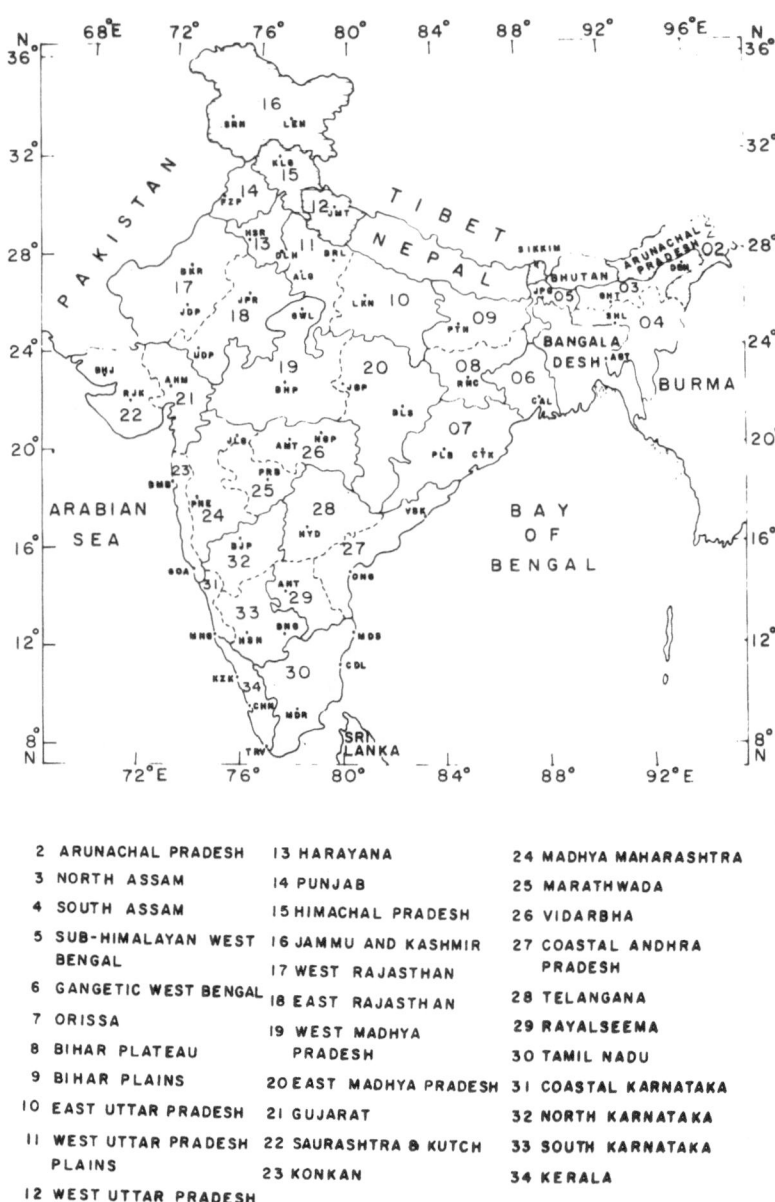

2 ARUNACHAL PRADESH   13 HARAYANA   24 MADHYA MAHARASHTRA

3 NORTH ASSAM   14 PUNJAB   25 MARATHWADA

4 SOUTH ASSAM   15 HIMACHAL PRADESH   26 VIDARBHA

5 SUB-HIMALAYAN WEST   16 JAMMU AND KASHMIR   27 COASTAL ANDHRA
   BENGAL   PRADESH
                        17 WEST RAJASTHAN
6 GANGETIC WEST BENGAL 18 EAST RAJASTHAN   28 TELANGANA

7 ORISSA   19 WEST MADHYA   29 RAYALSEEMA
                           PRADESH
8 BIHAR PLATEAU   30 TAMIL NADU

9 BIHAR PLAINS   20 EAST MADHYA PRADESH   31 COASTAL KARNATAKA

10 EAST UTTAR PRADESH   21 GUJARAT   32 NORTH KARNATAKA

11 WEST UTTAR PRADESH   22 SAURASHTRA & KUTCH   33 SOUTH KARNATAKA
   PLAINS   34 KERALA
                        23 KONKAN
12 WEST UTTAR PRADESH
   HILLS

Fig. 1.   Meteorological regions of India.   Area considered:
          Contiguous India excluding hilly areas (hatched).

and acute suffering of the people (Indian Famine Commission 1880, 1898, 1901, Bhatia 1967, Srivastava 1968, Indian Meteorological Dept. 1962, Nedungadi 1966, Ramdas 1960, 1976, Mooley 1975, 1976).

The droughts and floods occurring during the summer monsoon season over India as a whole and over different regions of the country during the long period 1871-1980 are identified. Utilizing the same well-distributed network of rain-gauges for the whole period and suitable objective criteria, the seasonal rainfall distribution over India during some of the worst droughts or floods is examined and possible causes for the extremes are discussed.

2.   RAINFALL DATA

The meteorological regions into which the Indian subcontinent has been divided are shown in Figure 1. The hatched areas, being mountainous and without a sufficient network of rain-gauges to give reliable areal averages have not been considered. Thus, the total area of India is 3.29×10^6 km^2, but the area considered is only 2.88×10^6 km^2.

The number of rain-gauges in the country increased from about 50 in 1850 to 5000 in 1970. To avoid inhomogeneity, a reduced subset network of rain-gauges has been selected in such a way as to give at least one rain-gauge in each district of each region while retaining, as far as possible, long-term stations. This 'master' network comprises 306 stations for the whole period from 1871 to 1978. Data for all these stations were not available for 1979 and 1980 for which data from about 250 rain-gauges were used.

Depths of the rainfall during the summer monsoon for the regions were computed for each year from 1871 to 1980, from monthly station data, after weighting each station value by the area of the district represented by the station.

3.   STATISTICAL TESTS FOR HOMOGENEITY AND NORMALITY

The individual regional rainfall series and an 'all-India' series were subjected to a number of tests to detect departures from simple random behaviour (e.g., trends and oscillations). Other tests were applied to check the normality of the various series.

The trend and oscillation tests revealed isolated cases of patterns, but, (a) as there was no consistency between the identified regions, and (b) the frequency did not exceed that expected from 29 regions, it was concluded that the regional series were random.

The tests for normality again gave marginally contradictory results. The sample skewness values were those which most often gave significant departures. Both positive and negative signifi-

(a)

(b)

(c)

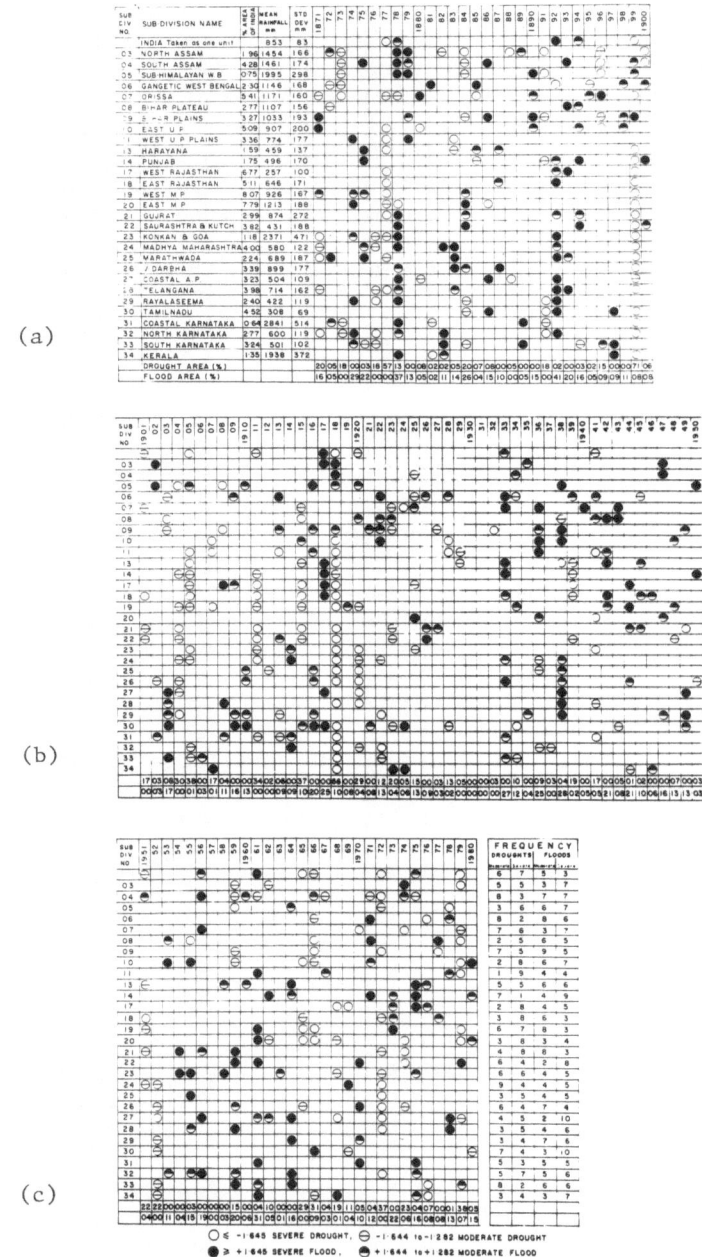

Fig. 2. Droughts and flood over different meteorological regions
of India between: (a) 1871 and 1900; (b) 1901 and 1950;
(c) 1951 and 1980.

cant values were found; four series exceeded the 1% level, one of
them being the all-India data set. However, in three of these
series, the cause was traced to a single record which was regarded
as anomalous and therefore removed from the analysis (Mooley 1973a).
However, the Chi-square and Kolmogrov tests did not support this
apparent departure from normality for any of the four. Overall,
the monsoon rainfall series is inferred to be normally distributed.
Use was made of this conclusion in selecting a criterion for
defining a flood or a drought year.

4. CRITERION FOR IDENTIFICATION OF DROUGHTS AND FLOODS

It can be assumed that water-dependent economic activities
adjust themselves to the water normally available and the variability of that water. Hence, it is natural to express the rainfall
of any season in a way which takes into account normal seasonal
rainfall and its variability. Such a criterion is the standard
deviate, t_i, given by

$$t_i = (X_i - \bar{X})/s \tag{1}$$

where X_i is the seasonal rainfall in year i
      $\bar{X}$ is the long-term average rainfall
      s is the standard deviation of the rainfall series.
The criteria adopted for identifying a season of drought or
of flood are a standard deviate ≤ -1.282 or ≥ 1.282, respectively.
These correspond to the 10% points under the normal distribution,
i.e., in 10% of the years floods (or droughts) will be experienced.
Further classification of statistical expectations is given below:
(i)   Moderate drought: $-1.282 > t_i \geq -1.645$
(ii)   Severe drought: $t_i < -1.645$
(iii) Moderate flood: $1.282 < t_i \leq 1.645$
(iv)   Severe flood: $t_i > 1.645$

5. INCIDENCE OF DROUGHTS AND FLOODS OVER INDIA

The seasonal rainfall series for India and its regions were
converted into series of standard deviates. Figure 2 shows, for
each year and region, the different levels of floods and droughts
which were experienced. The figure also gives the mean and
standard deviation of the seasonal rainfall and the frequency of
moderate or severe droughts and floods for each region during the
period 1871 to 1980, and the proportion (%) of areas suffering
floods and droughts.
In 1877, 1899, 1918 and 1972, more than 40% of the total area
experienced drought, while more than 40% of the area was flooded
in 1982.

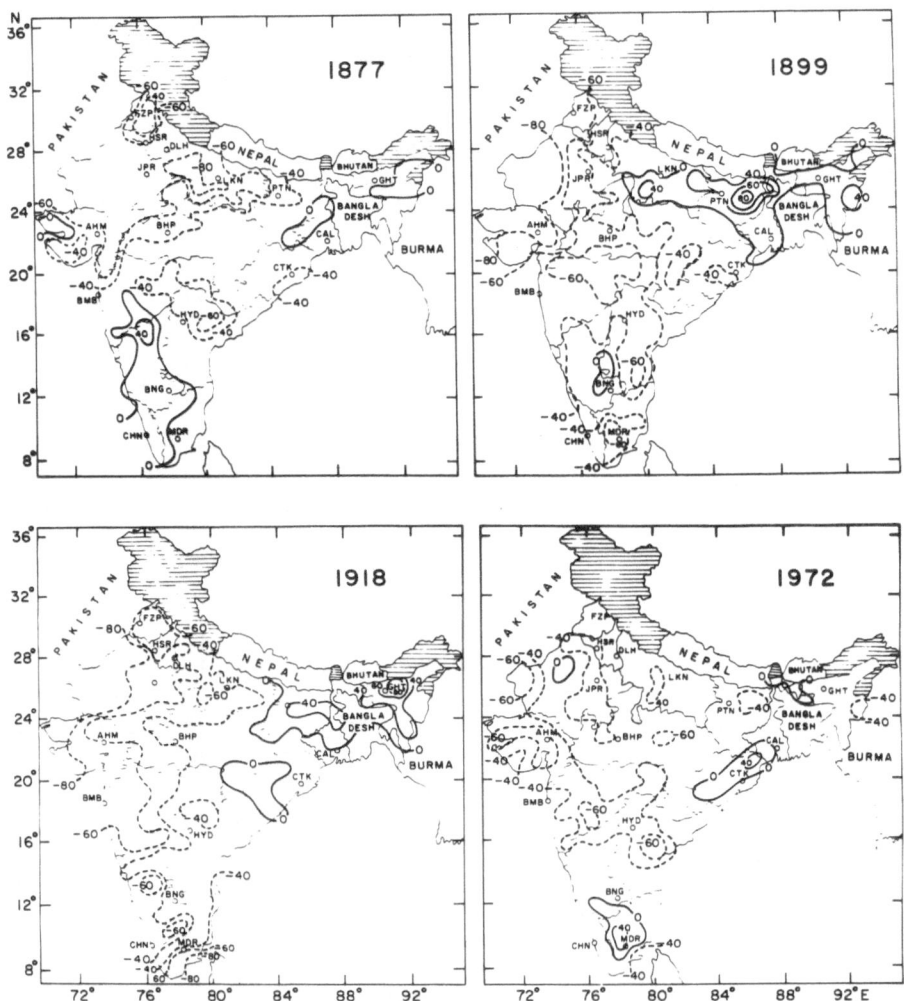

Fig. 3.   Percentage monsoon rainfall departure from normal over
            India in the monsoon season during the years of the most
            intense droughts.

5.1   Droughts

        Figure 3 shows the percentage seasonal rainfall departure from
the normal at different rain-gauge stations during the four years
of severe drought. Quite a large part of the country experienced a
departure of < -60% during 1877, 1899 and 1918. However, in 1972,

the area with a departure of $< -60\%$ was very small. In both 1899
and 1918, east Uttar Pradesh (see Figure 1) recorded excessive
monsoon rainfall. During 1899, there were some small pockets in
west Rajasthan and Gujarat where the seasonal rainfall was $< 5\%$ of
normal.

The two 20-year periods, 1901-20 and 1961-80, are characterized
by the high incidence of drought in the all-India series, the former
having five droughts and latter, four. The study by Mooley &
Parthasarathy (1979) showed six weak monsoon years between 1841
and 1860. From the overall behaviour of the monsoon over this
total time span, it appears that 20-yr periods of high drought
incidence have been separated by 40-yr periods with a low incidence
of drought. Within the 40-yr period, 1861-1900, the last 30-yr
period had two droughts while from 1921 to 1960 there were only
two droughts. Two 20-yr periods were free from drought, i.e.,
1878-97 and 1921-40. The largest proportion of the total area
experiencing drought was 17.5% - during 1911-20, and the smallest
was 4.25% - during the two decades 1931-50.

Care must be taken in inferences as to apparent cyclicity
because application of the Swed and Eisenhart test to the time span
between successive droughts suggests that the occurrence of droughts
is random.

5.2 Floods

The years identified by the all-India series as experiencing
severe floods, as indicated by the monsoon rainfall excess, were
1878, 1892 and 1961; the flood affecting the largest area (41%)
took place in 1892. Figure 4 shows the percentage seasonal rainfall departure from normal during the years of most severe flood
(1878, 1892, 1938, 1961). During 1878, the Saurashtra-Kutch,
peninsula and Gujarat experienced particularly excessive rainfall.
In 1892, the area west of 80°E generally experienced rainfall
excess. Although, in 1961, a large part of India was subject to
excess rainfall (with the greatest excess over western parts of
west Rajasthan and Saurashtra, north Tamil Nadu and coastal (south)
Andhra), there were a few pockets deficient in rainfall. During
1938, excess rainfall occurred in the northern peninsula.

Application of Swed and Eisenhart's test to the time interval
between successive flood years suggests that their occurrence is
random.

The country was free of floods during the decades ending 1890,
1910, 1930 and 1950, but floods occurred in two years during the
decades 1871-80 and 1891-1900. In the remaining decades, between
1871 and 1980, there was only one flood year. The mean area
affected ranged from 6% from 1921-30 to 13% between 1871-80. 1930,
1931 and 1957 were remarkable in being years during which no sub-
division of India experienced drought or flood.

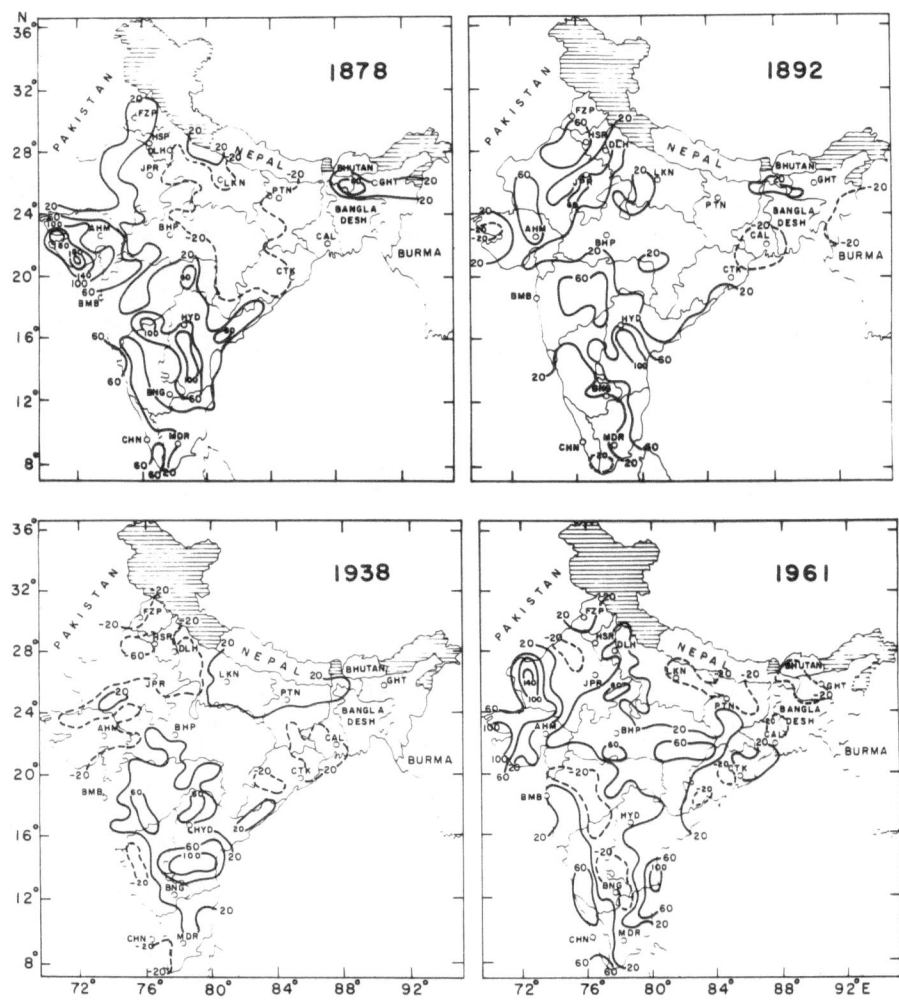

Fig. 4. Percentage monsoon rainfall departure from normal over
         India in the monsoon season during the years of the most
         intense floods.

6.   DEPRESSIONS IN YEARS OF DROUGHT OR FLOOD

     Much the larger majority of monsoon depressions form over the
Head Bay with just a small number forming over the land area of
north India. While a majority of the depressions move in a W-NNW
direction across central and northern India, a few move N-NE. At

times, some of the depressions, after moving W-NW up to Uttar
Pradesh and Rajasthan, veer N-NE under the influence of westerly
troughs before meeting the western Himalayas and producing heavy
or very heavy rain over the area. Rainfall associated with monsoon
depressions during July and August has been investigated by Mooley
(1973*b*). These systems generally produce plentiful rainfall over
northern and central India, and the west coast, and play a promi-
nant role in the country's water resources.

The behaviour of depressions during drought and flood years
can be contrasted in terms of the frequency of depression formation over the area north of 15°N and east of 80°E, the westernmost
longitude reached, the area of dispersion and proportion of depressions reaching 80°E or further west. Typical of the type of
analysis that has been carried out is Figure 5 which shows the end
positions of depressions for drought and flood years. A number of
important points emerge from the analysis depicted in this figure.
In June
(i)   The scatter around the normal track is much larger during
        drought years than in flood years;
(ii)   The proportion of depressions retracing their course or
        moving N-NE was greater in drought years than flood years;
(iii)   The number of depressions dying out east of 85°E was larger
        during drought years than during flood years.
In July
(i)   While depressions rarely reached west of 73°E during drought
        years, several travelled west of this longitude during flood
        years.
(ii)   The number of depressions dispersing east of 80°E was
        larger during drought years than in flood years.
In August, the main point of difference is that a much larger
number of depressions dispersed east of 85°E during drought years
than during flood years.
In September, those depressions which had moved west of 80°E in
drought years mostly ended up in positions north of 27°N while
depressions during flood years mainly died out in positions south
of this latitude.

Table I indicates the westward penetration of the depressions.
It can be seen that the proportion of depressions reaching 80°E in
each of the months is higher in flood years than in drought years,
particularly so in August, and that the westward penetration in
each month is further in flood years than drought years, particularly in July.

This is supported by the observation that the mean number of
days with depressions west of 80°E in each month is more in a flood
year than in a drought year, while for the season it is more than
double.

During drought years, a large number of depressions either
move N-NE from North Bay (20°N) and the adjoining land area or
disperse before reaching longitude 80°E whereas, during flood
years, such events are few. Mulky and Banerji (1960) found that

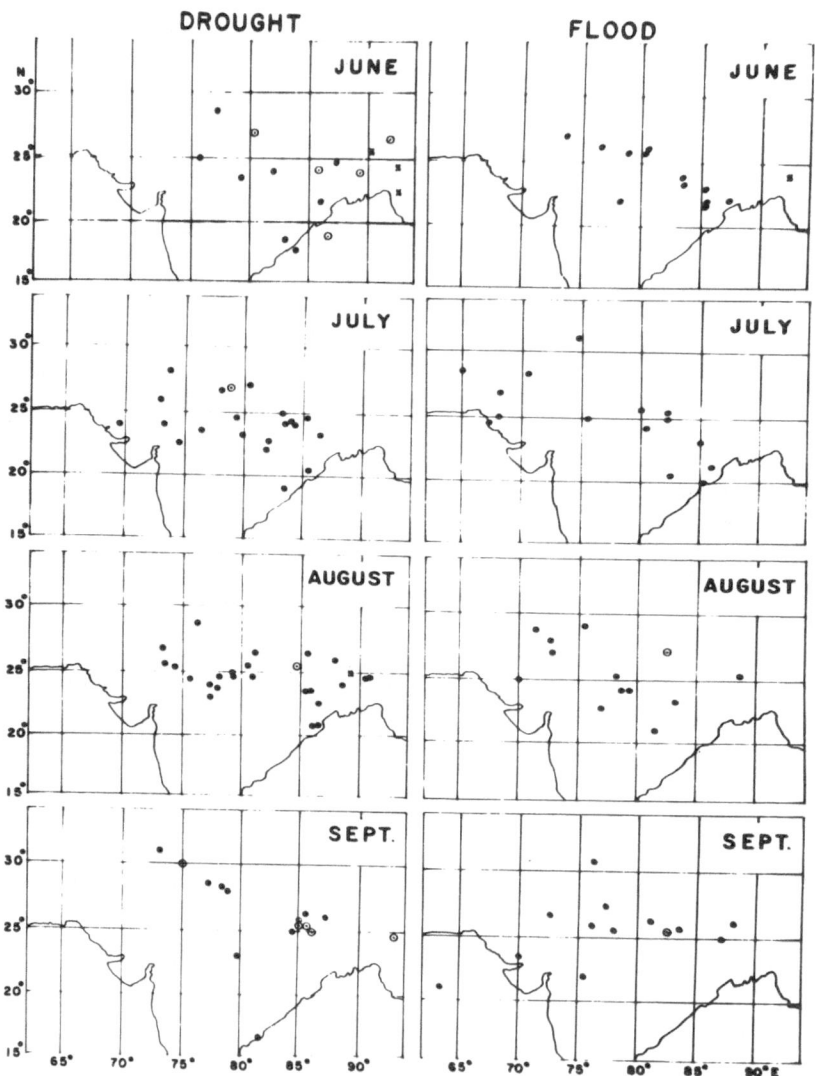

Fig. 5.   End positions of depression tracks over India during
          flood and drought years: ×, depression with N-NE motion;
          ⊙, depression veering N-NE; ●, depression with W-NW motion.

the monsoon depressions moved in a direction parallel to the wind
direction at the 9 km level in the area to the right of the track.
This would suggest a SSE-S wind over NE India at about 9 km and
may be due to displacement of the Tibetan anticyclone under the

Table I. Total number of depressions (N), proportion of depressions moving west of 80°E (P) and their mean westernmost longitude (WL) during drought and flood years between 1877 and 1930.

	Drought Year			Flood Year		
	N	P(%)	WL(°)	N	P(%)	WL(°)
June	16	19	77.4	13	31	76.7
July	21	48	75.5	16	56	70.9
August	26	42	76.3	13	69	74.9
September	18	55	76.9	15	73	74.8
Season	81	42	76.4	57	58	74.0

influence of the eastward moving troughs in the westerlies during the drought years. The westerly troughs bring in relatively colder air in their train which leads to the weakening of the monsoon depression. Murakami (1978) has shown that the Tibetan anticyclone at 200 mb is shifted southeastwards during intense summer monsoons.

7. MONSOON TROUGH LOCATION

The location of the monsoon trough exercises a profound influence on the rainfall distribution. In its normal position, the trough runs from south Bengal to northwest Rajasthan, but can move to the north and south of its normal location. When it lies over the foothills of the Himalayas, there is a marked decrease in the rainfall over most parts of India and a marked increase in the rainfall over the sub-Himalayan region. This situation is referred to as a 'break' in the monsoon. Trough locations over the foothills of the Himalayas have been examined for drought and flood years using the record of the number of days of 'break' and the length of the longest spell of 'break', provided by Ramamurthy (1969) for the period 1888-1967, and similar information for the period 1968-1980 obtained from the Indian Daily Weather Reports and Weather Charts. Table II clearly brings out the much higher frequency of the days of 'break' and the extended longest spells of 'break' during drought years. The mean frequency during drought years is about three times that during flood years and the mean longest spell during droughts is two and a half times that during floods.

8. SUMMARY

A large network of long-term rainfall records were used to construct a series of monsoon season rainfall for 29 regions of India. From this series, years of 'flood' and 'drought' were

Table II.   Number of days of 'break' (N_B) and the length of the
            longest spell of 'break' in July and August (L) during
            drought and flood years.

Drought year	N_B	L	Flood year	N_B	L
1899	23	8	1892	0	0
1901	4	4	1894	9	3
1905	13	9	1917	10	7
1911	11	11	1933	8	8
1918	23	17	1956	4	4
1920	12	6	1961	0	0
1941	9	9	1975	4	4
1951	15	6	Mean	5.0	≈4
1965	15	12			
1966	15	10			
1972	14	14			
1979	24	17			
Mean	14.8	≈10			

identified by a criterion which takes into account the normal
monsoon rainfall and its variability. These series were used to
identify the years of greatest areal extent of flood and drought
and years of maximum departure. Tests indicated that such events
occurred randomly in time although certain decades displayed
excess droughts and others few or none. The years 1930, 1931 and
1957 were unusual in that none of the Indian regions experienced
flood or drought.

     Having identified years of monsoon excess and deficit, it was
possible to set up objective tests to contrast depression track
behaviour and monsoon trough location. In comparison to the
depressions during the flood years, depressions during the drought
years are characterized by lesser westward activity. Monsoon
troughs lay over the Himalayan foothills more often during drought
years than during flood years.

ACKNOWLEDGEMENTS. The authors are grateful to the Director for
facilities to pursue this work and to the Deputy Director-General of
Meteorology (Climatology & Geophysics) for making the rainfall data
available. They would also like to thank Mrs N.A. Sontakke and
A.A. Munot for assistance in computations and Mrs S.P. Lakade for
typing the manuscript.

REFERENCES

Bhatia, B.M. 1967 *Famine in India - A Study in Some Aspects of the
     Economic History of India (1860-1965)* (2nd edn), 389 pp. Asia
     Publishing House.

Indian Famine Commission 1880 Famine Relief Part I - Measures of
    protection and prevention. Blue Book C-2591 & C-2735, London.

Indian Famine Commission 1898 Report of Government of India, 371 pp.
    Calcutta: Central Printing Office.

Indian Famine Commission 1901 Report of Government of India, pp.
    205-207. Calcutta: Central Printing Office.

Indian Meteorological Dept 1962 Rainfall and floods during 1961
    southwest monsoon period. *Ind. J. Met. & Geophys.*, 13,
    147-156.

Mooley, D.A. 1973*a* Gamma distribution probability model for Asian
    summer monsoon monthly rainfall. *Mon. Weath. Rev.*, 101,
    160-176.

Mooley, D.A. 1973*b* Some aspects of Indian monsoon depressions and
    associated rainfall. *Mon. Weath. Rev.*, 101, 271-280.

Mooley, D.A. 1975 Vagaries of the Indian summer monsoon during the
    last ten years. *Vavu Mandal*, 5, 65-66.

Mooley, D.A. 1976 Worst summer monsoon failures over the Asiatic
    Monsson Area. In: *Proc. Drought in Asiatic Monsoon Area
    Symp., Pune, December 1972*, pp. 34-43. Indian National Science
    Academy (Bulletin No. 54).

Mooley, D.A. & Parthasarathy, B. 1979 Poison distribution and years
    of bad monsoon over India. *Arch. Met. Geoph. Biokl. B*, 27,
    381-388.

Mulky, G.R. & Banerji, A.K. 1960 The mean upper wind circulation
    around monsoon depressions in India. *J. Met.*, 17, 8-14.

Murakami, T. 1978 Regional energetics of the 200 mb summer circu-
    lations. *Mon. Weath. Rev.*, 106, 614-628.

Nedungadi, T.M.K. 1966 A diagramatic representation of monsoon
    rainfall and floods with particular reference to southwest
    monsoon of 1961. *Ind. J. Met. & Geophys.*, 17, 373-378.

Ramamurthy, K. 1969 Some aspects of the 'break' in the Indian
    southwest monsoon during July and August. *Forecasting Manual,
    Part IV*, 13 pp. Pune: Indian Meteorological Dept.

Ramdas, L.A. 1960 The establishment, fluctuations and retreat of
    southwest monsoon over India. In: *Proc. Monsoons of the
    World Symp., New Delhi*, February 1958, pp. 251-256. New
    Delhi: Manager of Publications.

Ramdas, L.A. 1976 Droughts and floods in India and some other
    countries near and far from India. In: *Proc. Droughts in the
    Asiatic Monsoon Area Symp., Pune, December 1972*, pp. 91-101.
    Indian National Science Academy (Bulletin No. 54).

THE HEAVY RAINFALL IN CHINA IN 1980 AND A COMPARISON WITH EARLIER
EXTREMES

Zhang Yan

Research Institute of Weather and Climate
Academy of Meteorological Science
Beijing, China

ABSTRACT. A long period of heavy rainfall occurred in the middle
and lower Yangtze Valley in the summer of 1980. The main features
of the precipitation, the atmospheric circulation and the sea surface temperature related to the precipitation are discussed. For
comparison, some important features of five cases of extremely heavy
flood and drought in the last fifty years are analysed. The
extremely heavy rainfall of 1980 was related to atmospheric and
oceanic factors over wide areas. Among these, the effects of the
abnormal variation of the Subtropical High over the NW Pacific and
the 'cooling in the north and warming in the south' pattern of sea
surface temperature in neighbouring seas east of China were important factors. The characteristic features related to heavy rainfall
appeared first in the ocean, then in the weather systems over the
ocean and finally in the weather systems over the continent.

1. INTRODUCTION

A relatively continuous and extremely long period of heavy
rainfall occurred in a wide area of the Yangtze River and the Huai
River Basin, particularly in the middle and lower Yangtze Valley,
between June and August 1980. The atmospheric circulation and sea
surface temperature (SST) relevant to the onset of this heavy rainfall are discussed, and a comparative analyses of earlier cases of
extremes of precipitation are given.

2. PRECIPITATION RECORD

Heavy rain occurred frequently in the 1980 months of June to

253

August; total precipitation in these months exceeded 1000 mm over
a wide area. The total amount of precipitation at Sangzhi and the
Yellow Mountain were 1473 and 1160 mm, respectively. The concentrated heavy rain belt extended in a W-E direction along the
middle and lower Yangtze Valley over a front 300-400 km in width
(Figure 1).

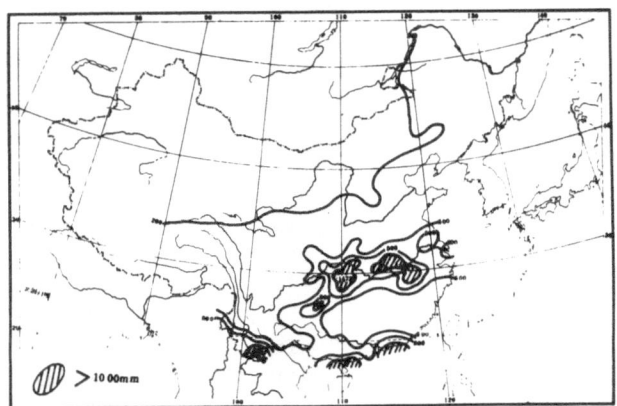

Fig. 1   Total precipitation between 100° and 120°E longitude over
         China, June to August 1980.

     The precipitation over such a wide area in the 1980 summer
was exceeded only in the heavy flood year of 1954, which was the
wettest summer in recent centuries. However, the duration of the
rainy season in 1980 was longer than that of 1954 and the precipitation in some localities in the relevant months was even greater
than in 1954.
     The heavy precipitation in 1980 occurred in the well-known
'Plum Rain' season[1] in the middle and lower Yangtze River Valley.
In 1980, the 'Plum Rain' began 7 days earlier than usual and ended
53 days later than the average. This cessation was about a month
later than in 1954, which was itself the latest recorded ending of
'Plum Rain'. The duration of the 'Plum Rain' season of 1980 was
longer than the two longest previously recorded, those of 1896 and
1954, by about 20 days.
     In August, there is normally a drought throughout the middle
and lower Yangtze Valley, but in 1980 prolonged heavy rainfall
continued in this area - an event rarely seen in the historical
record. The number of 'rainy days' (with daily precipitation ≥
0.1 mm per day) between June and August in many places in the
Yangtze Valley and Huai River Basin was *ca.* 15-24 days more than
normal.
     To gain an insight into the causes of this extremely heavy
rainfall in the middle and lower Yangtze Valley in 1980, which are

undoubtedly complicated, both the atmospheric circulation and oceanographic features have been examined.

3.   ATMOSPHERIC CIRCULATION PATTERNS

Many abnormal weather conditions occurred in the northern hemisphere in the summer of 1980, including heavy rainfall in India and Pakistan, abnormally low temperatures in Japan and Korea, a persistent drought in North China, a very hot summer in the central and southern parts of North America. Many features which occurred in the atmospheric circulation patterns were closely related to the extremely heavy rainfall in the middle and lower Yangtze Valley, particularly:
(i)   A strong and active cold northerly airstream;
(ii)   Persistent warm moist airstreams over southwest China;
(iii) An intense Northwest Pacific Subtropical High.

3.1 The Cold Northerly Airstream

A relatively strong cold northerly airstream located north of the Yangtze River was associated with an abnormal atmospheric circulation from June to August 1980, namely, a persistent, large and intensive trough stationary over western Siberia near the Urals, and accompanied by a strong intrusion of cold polar air (Figure 2). It is very rare for such a strong stationary trough to persist for three months.
Strong intrusions of cold air frequently came up against the warm moist air along the middle and lower Yangtze Valley resulting in heavy precipitation.
To the east of the Ural trough, a blocking high remained over Eastern Siberia and the Okhotsk Sea. The duration of the blocking high was also abnormally long, as it had been in 1954. It is noteworthy that, in 1954, the blocking high disappeared at the end of July and the heavy rain that year ended at the same time. In 1980, the blocking high disappeared at the end of August at which time the heavy rainfall also ceased.

3.2 Warm Moist Airstreams over SW China

From May 1980 onwards, strong and large troughs occurred over India and the Bay of Bengal. Between June and August 1980, the India-Bengal trough was the largest and strongest in the middle and lower latitudes (10°-35°N) of the northern hemisphere (see Figure 2). In 1980, the centres of the trough were around 3° and 10° north of the average position for July and August thereby effectively redistributing the heat and energy balance between lower and higher latitudes. In front of this trough, intense warm and moist air currents extended north to the Yangtze Valley, with frequent SW jet streams carrying abundant moisture and energy to

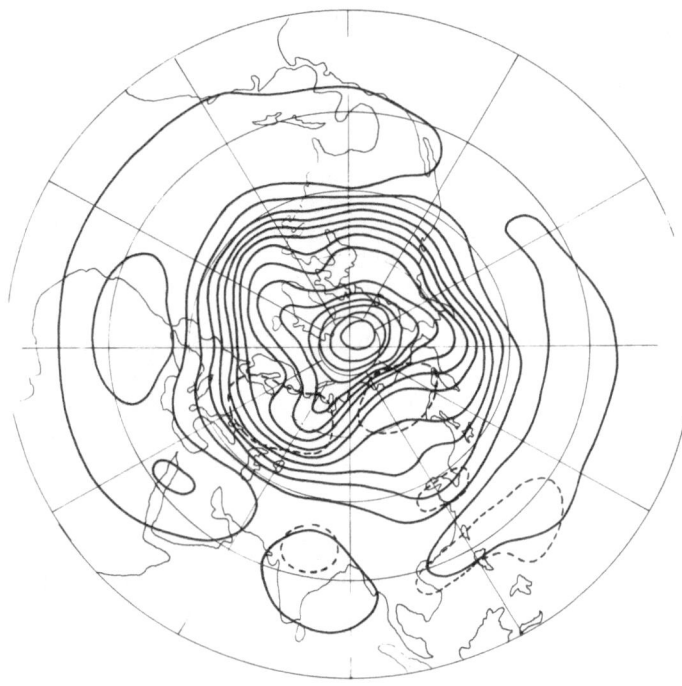

Fig. 2   Mean height of the 500 mb level in the northern hemisphere,
         August 1980.

the Yangtze Valley (Zhang Yan *et al.* 1978) and providing a
necessary precondition for extremely heavy rainfall.
       Meanwhile, a large area of cloud clusters formed over the
tropical ocean in advance of the India-Bengal trough, spreading
continuously eastwards to the Yangtze Valley.   These cloud clusters
were related to the trans-equatorial flow and the cloud systems of
cold outbreaks into the low latitudes of the southern hemisphere
(Figure 3).   It is evident that the heavy rainfall in China in the
summer of 1980 was related to the low-latitude oceans and even to
the airflow pattern in the southern hemisphere.

3.3 The NW Pacific Subtropical High in Summer 1980

       The NW Pacific Subtropical High (110°-180°E, north of 10°N)
exercises a major control on the 'Plum Rain' of China.   During the
abnormal 'Plum Rain' of 1980, the High showed abnormalities in its
areal coverage, intensity and position.
       3.3.1 <u>Coverage and intensity of the High</u>.   The NW Pacific
Subtropical High covered a very wide area, was unusually strong
and very persistent in the summer of 1980.   Comparison of the areal

Fig. 3   Satellite cloud map, centred on 140°E, 1200Z 31 July 1980.

index (A.I.) and height index (H.I.) for 1980 and the previous 30
years gives some indication of the exceptional circumstances that
occurred in 1980 (A.I. being the area enclosed within the 5880 geopotential metre contour of the NW Pacific Subtropical High computed
from grids of 5° latitude and 10° longitude; H.I. being the sum
of the heights above the 5880 geopotential metre contour, at the
same grid points). Both indexes remained very high in July and
August, and showed large positive anomalies in many months in 1980.
The summation of A.I. and H.I. over the three summer months of 1980
(J, June; A, August) was, respectively:

$$\sum_{J-A} A.I. = 85$$

$$\sum_{J-A} H.I. = 190$$

these being the highest values since 1951.

   3.3.2 <u>Position of the High</u>. In general, the ridge lines of
the NW Pacific Subtropical High in June, July and August lie along
20°N, 25°N and 28°N respectively. After the northward shift of the
High in July and August, the monsoon rain belt normally shifts
northward too, while the North China rainy season sets in. However,

in 1980, the NW Pacific Subtropical High and the corresponding
rain belt did not shift northward as usual, but maintained the June
position. Indeed, the monthly mean position of the ridge line of
the High in August 1980 was along 21°N, its most southerly recorded
position.

Along the W-NW flank of such a strong Subtropical High, a
large amount of moisture, kinetic and heat energy would be transported continuously into the rain belt (Zhang Yan et al.1978). The
advection of cold air from the north and NW of China to the same
area formed a strong convergent zone, heavy rainfall resulting.

In July and August 1980, the ridge line of the Subtropical
High remained to the south of the middle and lower Yangtze Valley.
The abnormal position of the High in August 1980 was the main
factor in producing the heavy continuous rainfall which occurred
in August in these areas. Correspondingly, it caused the abnormal,
severe drought in North China and affected the behaviour of
typhoons in the SE China Sea during 1980.

4.   OCEANOGRAPHIC FEATURES

The oceans are one of the most important components of the
terrestrial climate system (Newell 1979, Bjerkness 1969) with sea
surface temperature (SST) an important variable intimately related
to atmospheric variations (Namias 1979, Ratcliffe & Murray 1970,
Ratcliffe 1973). The prolonged duration of the extreme monsoon
rainfall of 1980 prompts the study of the characteristic features
of the SST field.

There was a very strong meridional gradient of SST to the east
of China from June to August 1980 (Figure 4). The ten-day average
difference of SST between 30°N and 34°N across the oceanic front
was as large as 9°C (Zhang Yan et al. 1980). The width of the
oceanic front was about 300 to 400 km, and extended east-west almost
parallel to the usual position of the atmospheric front along the
middle and lower Yangtze Valley. Owing to the longer time constant
of the ocean, the oceanic front was more stable than the atmospheric
fronts, the latitudinal shifting of which was governed partly by
the oceanic front. Ordinarily, when the cold fronts move across
the warmer lands from NW Asia to the middle and lower Yangtze River,
they would be weak or disappear altogether in July and August. But,
in 1980, when the cold fronts moved to this area, rather than disappear they intensified and became quasi-stationary, resulting in
enhanced precipitation. From Figure 3, it can be seen that a
frontal cloud system extended from the middle and lower Yangtze
River eastwards to the ocean; the existence of the oceanic front
clearly played an important role.

After January 1980, SST had been lower than average in the
Japan Sea, Okhotsk Sea and Yellow Sea north of the oceanic front,
while sea ice in the Okhotsk Sea was relatively prevalent in early
1980. In February, the Okhotsk Sea was roughly 90% ice covered,

which is 12% above normal. The sea ice area extended in March to
130% of the normal, the largest coverage for that month since 1970
(Ocean-Meteorological Group 1980).

Meanwhile, on the south side of the oceanic front between
Taiwan province and the south of Japan, SST was higher than normal.
Near the Kuroshio, SST showed positive anomalies of 1-2°C with a
central maximum departure from the mean of about 3°C. Such a
'cooling in the north, warming in the south' pattern was favourable
to the formation of the oceanic front.

When considering air/sea interaction it is usually not clear
which is the driving factor. In some cases, the atmosphere is the
active factor while, in others, the ocean is dominant (Zhang Yan &
Bi Muying 1977). As, in 1980, the oceanic front set in before the
atmospheric front, the ocean may have been the major factor,
although the atmospheric circulation played an important role in
forming the oceanic front. The 'cooling in the north, warming in
the south' SST pattern appeared about 6 to 10 months early in 1980,
and promoted the formation of the oceanic front. In the June-August
period, the position and intensity of the oceanic front was relatively stable in contrast to the atmospheric fronts. Furthermore,
the cooling at the northern side of the oceanic front was greatly
influenced by the prevalence of sea ice. In this case, the
oceanic front was the factor which helped to maintain the atmospheric front and, thus, the persistence of the heavy rain. The
presence of the oceanic front may serve as a useful tool in long-
range weather forecasting since it is based on an understanding of
physical relationships and not on a purely statistical relationship.

The NW Pacific Subtropical High, the India-Bengal Trough and
the trans-equatorial airflows are generally active over the ocean
and their abnormal variations are also related to the ocean. These
abnormalities might be the result of an anomaly in the exchange of
heat between the ocean and atmosphere over a wide area. Moreover,
in 1980, the characteristic features of the NW Pacific Subtropical
High and the India-Bengal Trough (the weather systems over the
ocean) appeared early (in February and May, respectively), one to
six months earlier than the development of the characteristic
features of the Ural trough and the East Siberian blocking high
(the weather systems over the continent). Thus, the factors
governing the heavy rainfall in 1980, appeared first in the ocean,
then in the weather systems over the ocean, and finally in the
weather systems over the continent.

4.   SOME HISTORICAL PRECEDENTS

Over the past 50 years, there have been three heavy flood years
in the middle and lower Yangtze Valley: 1931, 1954 and 1980. In
these years, the rain belts were very similar, all being in a zonal
W-E direction. The heavy flood in 1931 chiefly occurred in July,
while the flood in 1954 spanned June and July; both were monsoon

rain in the 'Plum Rain' period. On the other hand, in 1958 and
1965, the 'Plum Rains' failed. The relationships between flood and
drought years in east central China and the NW Pacific Subtropical
High have been studied previously (Zhang Yan *et al.* 1975, Zhang Yan
1979). For these earlier cases, comparative analyses of the SST
field and the water vapour transport, in front of the India-Bengal
trough, have been made.

4.1 Characteristic Features of SST Field

In the heavy flood years of 1931, 1954 and 1980, there were
distinct oceanic fronts in the seas to the east of the Yangtze
Valley (Figure 4). In 1954, the oceanic front disappeared in late
July when the heavy rainfall also ended. In 1980, in contrast, the
oceanic front prevailed throughout August, disappearing in September
when the heavy rainfall (during August) diminished. Although
monthly SST data are not available for 1931, the charts of 3-monthly
SST averages and of monthly surface air temperature anomalies
indicate that the SST pattern in 1931 was similar to those in 1954
and 1980 (Institute of Geography & Shanghai Observatory 1980,
Meteorological Group 1931).
      In these three flood years, there was a distinct 'cooling in
the north, warming in the south' of the SST field in the seas east
of China, the pattern having a relatively early onset (Figure 4)
even to the extent of appearing in the autumn of the previous year.
In 1980, the oceanic warming to the south of the Kuroshio appeared
in October 1979; similarly, in 1954, the 'northern cooling and
southern warming' phenomenon initially appeared in early December
1953.
      In summary, this 'cooling' and 'warming' effect was relatively
stable and persistent. In these three flood years, the characteristic features of the SST field were very similar, and the SST
pattern appeared about 6 months earlier than the excessive rains.
      In contrast, in 1958 and 1965 when the 'Plum Rains' were
deficient, the SST fields were quite different from those in flood
years (Figure 5). There were no distinct oceanic fronts east of the
Yangtze Valley, the SST gradients were rather weak between early
spring and late summer, and the 'northern cooling, southern warming'
pattern was either absent or even reversed.

4.2 Water Vapour Flux and the India-Bengal Trough

The India-Bengal trough was very deep and persistent in 1980.
The water vapour flux (F) has been calculated along 110°E longitude
in front of this trough for the June to August period in 1954, 1958,
1965 and 1980, using the formula

$$F = (1/g)\ \bar{v}q$$

where $\bar{v}$ is the monthly mean wind velocity and q is the specific
humidity.

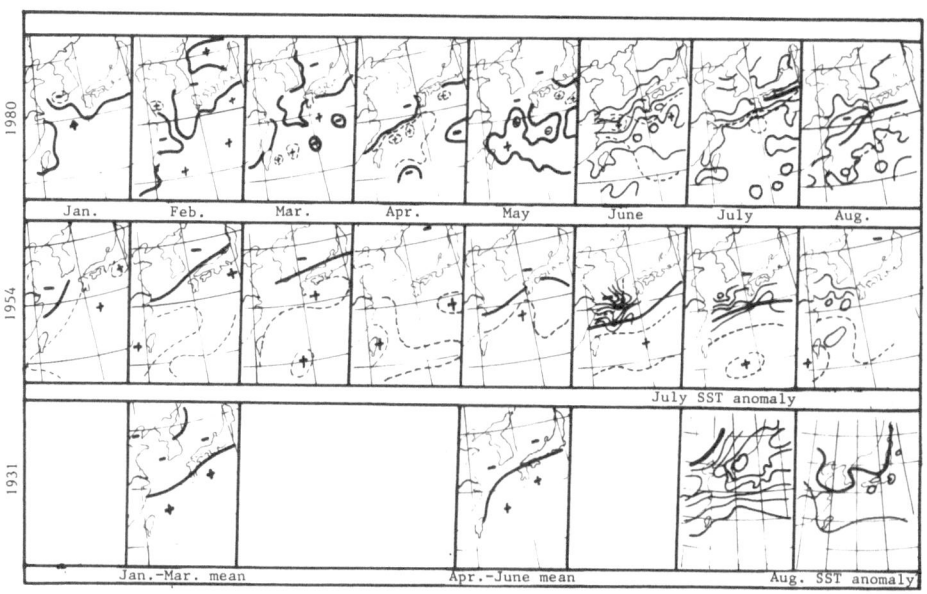

Fig. 4   Sea surface temperatures in 1931, 1954 and 1980 (flood years): ---, monthly mean anomaly; ——, SST isotherm; ——, zero departure.

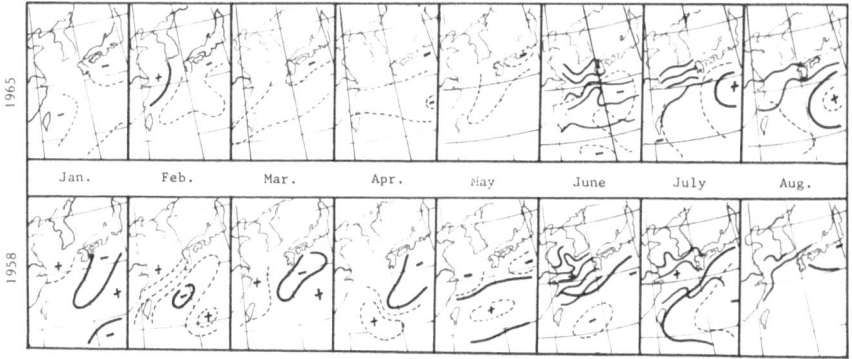

Fig. 5   Sea surface temperature in 1958 and 1965 (drought years): key as Figure 4.

From Table I, it can be seen that, south of the Yangtze River (south of 30°N), the monthly mean vapour flux (F) from the south was relatively high along 110°E in these three months in 1980. It can also be seen that the flux of water vapour at 500 mb in the

flood years (1980, 1954) was much larger than in 1958 and 1965
when the 'Plum Rains' were light. This suggests that the vapour
transfer in 1980 was associated with a deep monsoon airflow. This
result is in agreement with previous studies (Zhang Yan & Xiong
Shuichun 1978), i.e., the moist layer in heavy rainfall years is
much deeper than in the drier years.

Table I.   Monthly mean vapour flux (F) for June to August (g cm^{-1}
           mb^{-1} s^{-1}).

		Changsha 28°12'N 113°04'E			Zhijiang 27°27'N 109°28'E			Guangzhou 23°08'N 113°09'E			Nanning 22°49'N 108°21'E		
level (mb)		850	700	500	850	700	500	850	700	500	850	700	500
year													
flood	1954	–	–	–	–	–	–	–	–	–	22.0	12.5	3.2
	1980	20.0	11.9	4.7	23.9	12.7	4.6	22.3	11.7	1.8	23.8	15.6	3.3
dry	1958	15.5	9.6	3.6	20.2	10.7	2.7	14.4	6.2	0.7	18.0	7.4	1.0
	1965	11,5	4.7	1.1	18.1	7.8	0.8	18.1	10.0	1.5	17.6	8.2	2.1

5.   SUMMARY AND CONCLUSION

The extremely heavy rainfall in China from June to August 1980
was part of a global abnormality. The causes were complicated,
but included continental cold northwesterlies stronger than usual,
deep and persistent warm moist flow in the SW trough, abnormal
variations in the NW Pacific Subtropical High, and a marked
'northern cooling, southern warming' effect in the ocean to the
east. Among these, the last two factors are related to the
extremely heavy rainfall more directly and some of their distinctive
features can be seen several months beforehand. It may be useful
in long-range weather forecasting to note that the characteristic
features related to the heavy rainfall in 1980 appeared first in
the sea surface temperature, then in the weather systems over the
ocean and, finally, in the weather systems over the continent.

NOTE

(1), Plum rain: continuous heavy rainfall in the period when plums
    are ripe and yellow, a kind of monsoon rain in China.

REFERENCES

Bjerknes, J. 1969 Atmospheric teleconnections from the Equatorial
     Pacific. *Mon. Weath. Rev.*, 97, 163-172.

Institute of Geography & Shanghai Observatory 1980 *Monthly Mean Sea Surface Temperature Departures over North Pacific Ocean,* 324 pp. Beijing: Oceanography Publisher.

Meteorological Group 1931 Monthly mean surface temperature departures. In: *Geophysical Review,* pp. 1868-1873. Tokyo: Japan Meteorological Agency.

Namias, J. 1969 Use of sea-surface temperature in long-range prediction. In: *Sea-surface Temperature (WMO Technical Note* No. 103) pp. 1-18.

Newell, R.E. 1979 Climate and the ocean. *American Scientist,* 67, 405-416.

Ocean-Meteorological Group 1980 Oceanographic summary. In: *Geophysical Review,* No. 956, pp. 26-33; No. 966, pp. 20-27; No. 967, pp. 24-32. Tokyo: Japan Meteorological Agency.

Ratcliffe, R.A.S. 1973 Recent work on sea surface temperature anomalies related to long-range forecasting. *Weather,* 28, 106-117.

Ratcliffe, R.A.S. & Murray, R. 1970 New lag associations between North Atlantic sea temperature and European pressure applied to long range weather forecasting. *Q. Jl R. met. Soc.,* 96, 226-246.

Zhang Yan 1979 On the finale of the rainy season in Hunan and Zeijiang Provinces and its long range forecasting. *Scientia Atmospherica Sinica,* 3, 363-369. (In Chinese.)

Zhang Yan & Bi Muying 1977 Air-sea inter-relationship between 700 mb atmospheric circulation and sea surface temperature over the North Pacific. *Scientia Atmospherica Sinica,* 4, 273-281. (In Chinese.)

Zhang Yan & Xiong Shuichun 1978 Charts of 'Wind-column' and its application on the analyses and forecast of meso-storms. In: *Memoir of Research Institute of Weather and Climate,* Vol. 1, pp. 1-10. Beijing: Academy of Meteorological Science.

Zhang Yan, Chen Qilu, Hu Nanfeng & Tien Senlie 1978 Low level jets and heavy rainfall at the End of Rain seasons. In: *Symposium of Heavy Rain, Dalien 1978,* pp. 196-203. Changchun: Jilin Publisher.

Zhang Yan, Li Yuehong & Bi Muying 1975 Preliminary studies on the
      interactions of sea surface temperature between the 'Plum Rain'
      and the Subtropical High of the NW Pacific in early summer.
      In: *Symposium on Long-Range Hydro-Meteorological Forecasting
      of the Yangtze Valley, 1975,* Vol. 1, pp. 164-177. Wuhan:
      The Office of Yangtze Valley Project.

Zhang Yan *et al.* (Heavy Rain Group) 1980 A preliminary study on the
      extremely heavy rain over the Yangtze Valley in summer 1980.
      Meteorological Monthly, (12), pp. 1-5. (In Chinese.)

VARIABILITY OF RAINFALL OVER NORTHERN AUSTRALIA

William R. Kininmonth

Bureau of Meteorology
Melbourne, Victoria 3001
Australia

ABSTRACT. This paper looks at two distinct regions of the tropics, the high rainfall north coast and the dry interior of northern Australia. In particular the incidence of an abnormally dry period on the coastal margin and an abnormally wet period over the interior are examined. The anomalies are discussed in terms of the seasonal regimes.

1. INTRODUCTION

Tropical rainfall is associated with various meteorological systems: monsoons, tropical cyclones, the equatorial trough and active covergence zones to name the most important. Each major system has been studied in detail and extensive bibliographies are available. It is generally possible, particularly in equatorial latitudes, to relate the cause of individual rainfall events to one of the conventional rainfall models.

The tropical cyclone is an important rainfall-producing mechanism in summer, but the area of impact is generally limited and its occurrence, at a particular location, infrequent. An active monsoon, with associated disturbances, is the most general rain-producing regime, though it is neither steady nor is its penetration to higher latitudes consistent.

Extreme winter rainfall over tropical Australia is rare in the north and west, although showers from the southeast trade winds are not uncommon on the east coast. Sporadic rain is experienced regularly over inland areas as polar fronts penetrate into low latitudes and these form the basis of the winter rainfall regime. However, showers are isolated and usually only generate a small total rainfall.

A. Street-Perrott et al. (eds.), Variations in the Global Water Budget, 265–272.
Copyright © 1983 by D. Reidel Publishing Company.

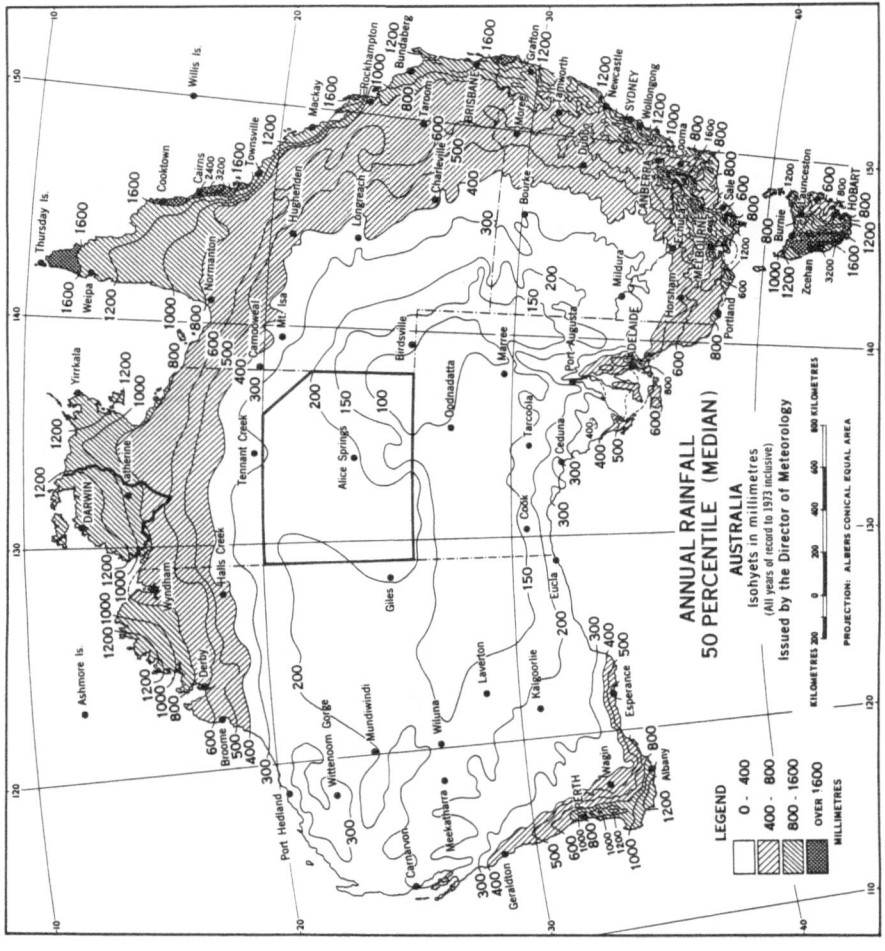

Fig. 1. Median annual rainfall for Australia. The Darwin/Daly and Alice Springs rainfall districts are enclosed by solid lines.

   Two aspects of the variability over northern Australia are
discussed here. Firstly, the changes to the broad-scale flow
associated with persistent dry spells over the high rainfall
northern area; secondly, the meteorology of spasmodic high rainfall events of the arid interior.

2.   CLIMATOLOGY

   The significant features of the geographic distribution of the
median annual rainfall for the Australian continent, shown in
Figure 1, are the arid interior and the wetter north coastal
margins. A detailed description of the seasonal march of rainfall in tropical and sub-tropical latitudes (Riehl 1954) is
largely based on the movement of the equatorial trough and of the
associated rain systems. The seasonal distribution over northern
Australia is consistent with Riehl's description, particularly
with respect to the brief wet season in the far north that attends
the proximity of the active equatorial trough.
   Table I illustrates the seasonal rainfall distribution over
northern Australia and gives a measure of the intra-seasonal
variability. Monthly values of the first decile, the median and
the ninth decile are presented for the high-rainfall Darwin/Daly
district and for the arid Alice Springs district (see Figure 1).
The characteristic summer rainfall is pronounced in the former
district and discernible in the latter. Winter rainfall is
suppressed to the extent that median rainfall is near zero.

Table I. First, median and ninth decile values of monthly rain-
          fall (mm) for the Darwin/Daly and Alice Springs rainfall
          districts (enclosed by solid boundaries in Figure 1).

	J	F	M	A	M	J	J	A	S	O	N	D
(a) Darwin/Daly District												
1st Decile	169	156	109	12	0	0	0	0	0	7	45	128
Median	327	295	256	47	6	1	0	0	5	33	103	201
9th Decile	438	413	402	141	35	10	7	3	18	75	155	368
(b) Alice Springs District												
1st Decile	3	1	1	0	0	0	0	0	0	0	1	3
Median	20	20	13	4	5	4	1	1	3	14	17	20
9th Decile	90	114	85	42	54	36	32	28	22	48	49	88

   The magnitude of periodic rainfall events typically shows a
skewed distribution with averages biased by infrequent events.
Relatively rare events, particularly in the continental interior,

can produce rainfall totals that are many times the median. The
occasional heavier falls in the Alice Springs district are evident
in the ninth decile data.

3.   SUMMER EXTREMES

     Charney (1975) has described how continental desert regions
can locally reinforce the existing large-scale subsidence. The
sub-tropical deserts are regions of net radiative heat loss, even
in summer, due to the high surface albedo. The radiative cooling
in the troposphere is balanced by advection of warm air and subsidence heating, except in the boundary layer where dry convection
from the hot surface prevails during the day-time. The large-
scale subsidence over the desert regions dries the troposphere
and inhibits rain bearing systems.
     In summer, a positive feedback mechanism operates as the arid
and semi-arid grasslands dry out under the influence of high
temperatures. As the surface albedo (which is a function of the
vegetation cover, dry sparse vegetation having a high albedo)
increases, the radiative cooling and tropospheric subsidence is
enhanced and rain-producing systems are increasingly inhibited.
     Against the background of the subsidence over the continental
interior, the coastal margins provide a contrast. Northern
Australia experiences a typical, albeit erratic, monsoon in summer.
Mean winds below 1.5 km are westerly at Darwin between December
and February, becoming southeasterly in March (Maher & Lee 1977).
During these summer months, moisture in the lower troposphere is
advected over the coastal margins and convective rain accompanies
synoptic disturbances.
     The monsoon season is interspersed with periods of exceptional
meteorological activity, including tropical cyclones or contrasting
periods of low rainfall, each for varying durations. It is only
when an abnormality is exceptionally strong or persists for an
extended period that it is evident in the monthly rainfall data.
Wet or dry periods overlapping a calendar month are also masked in
the data.

Table II. Monthly values of median and actual rainfall (mm) and
           decile rank for the Darwin/Daly rainfall district for
           the 1964/65 summer.

	1964						1965					
	J	A	S	O	N	D	J	F	M	A	M	J
Median	0	0	5	33	103	201	327	295	256	47	6	1
Actual	0	3	18	75	252	183	200	161	497	11	14	5
Decile	1	6	9	9	10	4	2	2	10	1	8	5

A summer when marked contrasts in the decile rank can be identified was that for the rainy season of 1964-1965. An extended dry period during January and February was followed by an abnormally wet March. Table II shows the Darwin/Daly district rainfall and decile rank for each month of that summer, together with the median for those months. A marked change in activity occurred at the end of February and allows a comparison between the mean conditions of January/February and of March. The extent of the departure from the long-term mean of the circulation pattern for February 1965 was most marked in the wind data for Darwin at the 700 mb level. The dry period was associated with strong advection of dry continental air as the usual (3 m s^{-1}) southeasterly winds strengthened to 12 m s^{-1}. The airflow below 700 mb was weak in magnitude, but also with a mean southeasterly origin. The break in the monsoon was extensive in influence and is consistent with an equatorward movement of the equatorial trough and strengthening of the meridional Hadley circulation. The return of the monsoon in March was a dramatic reversal of the circulation. Again the contrast was most marked at the 700 mb level where the March mean flow over Darwin was 1 m s^{-1} from the northeast compared with the long-term norm of 4 m s^{-1} from the southeast. The expected dry southeast flow was absent at all levels below 700 mb. Moisture-laden air advected over the coastal margin acted as a source for the extensive convective rainfall.

The extent of the circulation reversal and the duration of the preceding and following anomalies was indicative of possible bi-modal stability. The monsoon normally fluctuates in activity, but these data provide the suggestion that it can, on occasions, become locked into a particular state for a lengthy period.

4.   A WINTER ANOMALY

The suppression of winter rainfall over northern Australia by the subsidence of the meridional circulation has been discussed. The expected number of days with rain over the Alice Springs district is 2 to 3 per month; further to the north, the expected incidence of days with rain and the median totals decrease. Despite this broad-scale control, periods with heavy rain are not entirely absent.

The synoptic systems that bring rain can be likened to a description of the extended linkage between the tropics and higher latitudes discussed by Riehl (1950). Despite the dominance of the mean meridional circulation, the tropical atmosphere is not steady in the middle and upper troposphere. When equatorial disturbances combine with troughs in the polar westerlies, the resultant extended flow will reach from the tropics into high latitudes.

Interaction between a mid-latitude trough and a tropical system over Australia can bring very heavy and extensive rain to the inland areas. One such system occurred in May 1981 and is

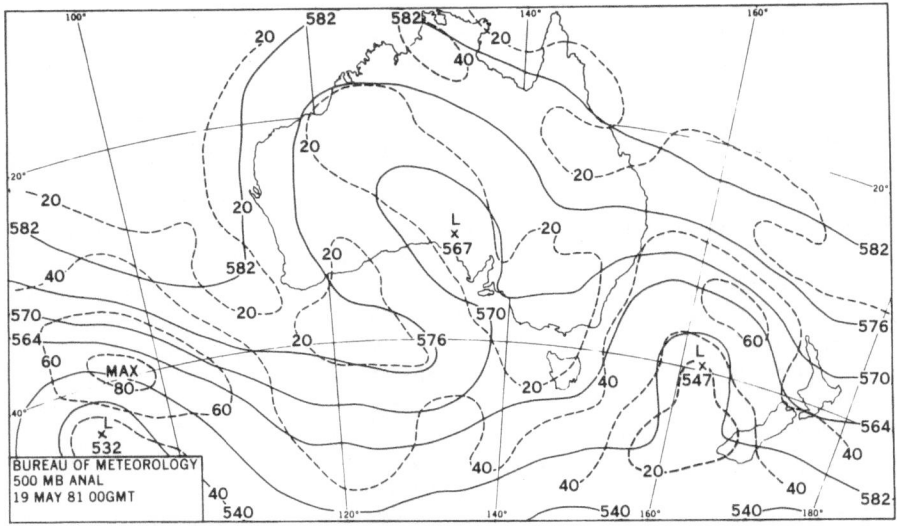

Fig. 2   500 mb geopotential and wind fields for 00 GMT 19 May 1981.

described below. Although significant rain was recorded in the
eastern Alice Springs district, the heaviest rain was further to
the southeast with many storm totals between 50 and 100 mm over
the region between Birdsville and Bourke, adjoining the driest
part of the continent.

The precursor to the major rainfall was paradoxical. The MSL
chart for 00 GMT 19 May 1981 showed a very intense anticyclone
directing a strong flow of dry southeasterly air over the Australian sub-tropics in the lower troposphere. However, in the middle
troposphere, the flow was quite different. At 500 mb, the airflow
was northwesterly with winds of 40 m s^{-1} near Darwin on the eastern
flank of a mid-latitude trough. A cold vortex was located over
southern Australia near longitude 135°E at the time. These middle
tropospheric features are shown in Figure 2.

A GMS infrared photograph at 00 GMT 19 May 1981 (Figure 3)
clearly shows the interaction between the low-latitude moisture
source and the higher-latitude dynamic system. An extensive cloud
sheet marks out the region of dynamic lifting in the moist northwesterly airflow. During the next 24 hours, the system developed
further and moved eastwards.

The heaviest rainfall from the system was to the southeast
of Alice Springs during the period 19-23 May. Scattered light
falls were recorded over most districts across northern Australia
on 20 May. The extended meridional moisture flow and the slow
eastward propagation of the trough system are indicative of the
broad-scale impact of the rain system. A more detailed description

Fig. 3 GMS infrared photograph at 00 GMT 19 May 1981.

of a similar system occurring in May 1979, but with the heaviest rain to the northwest of Alice Springs, has been given by Downey *et al.* (1981).

5. DISCUSSION

The seasonal march of rainfall of northern Australia largely conforms to the well-known tropical progression. However, around the median patterns, there is significant variability. Although definitive conclusions cannot be drawn from these two case studies, they are indicative of the processes in play. The extremes of rainfall associated with large-scale anomalies of the circulation and the magnitude of the forcing give grounds for optimism that future studies will lead to useful and reliable prediction methodologies for these extremes.

ACKNOWLEDGEMENT. This paper is published with the permission of the Director of Meteorology.

REFERENCES

Charney, J.G. 1975 Dynamics of deserts and droughts in the Sahel. *Q. Jl R. met. Soc.*, 101, 193-202.

Downey, W.K., Tsuchiya, T. & Schreiner, A.J. 1981 Some aspects of a northwestern Australian cloudband. *Aust. Met. Mag.*, 29, 99-113.

Maher, J. & Lee, D.M. 1975 *Meteorological Summary: Upper Air Statistics Australia*, 202 pp. Canberra: Australian Govt Publishing Service.

Riehl, H. 1950 On the role of the tropics in the general circulation of the atmosphere. *Tellus*, 2, 1-23.

Riehl, H. 1954 *Tropical Meteorology*, 392 pp. New York: McGraw-Hill.

MOISTURE VARIATIONS ASSOCIATED WITH EL NIÑO EVENTS

Bryan C. Weare

Department of Land, Air and Water Resources
University of California
Davis, CA 95616, USA

ABSTRACT. El Niño, a period of unusually warm surface water in the eastern tropical Pacific Ocean, is shown to be related to variations in several components of the atmospheric moisture budget over much of the basin. Changes in both the ocean temperatures and the moisture terms are diagnosed using an analysis of about 5×10^6 individual marine weather reports for the period 1957-76. From these data, year-month means were calculated for each 5° latitude by 5° longitude grid of sea surface temperature, surface latent heat flux, and surface horizontal vapour flux divergence.

The analyses indicate that El Niño events are associated with increased surface latent heat fluxes of up to about 40 W m^{-2} over much of the eastern Pacific basin. The increases tend to follow those for sea surface temperature by up to several months, the shortest lags being for regions near the equator. El Niño is also found to be associated with an equatorward movement of the eastern Pacific convergence zone which has its mean near 8°N. There are some suggestions of systematic variations in the position of the Southern Convergence Zone near 25°S over several parts of the basin. The relationship of these changes to observed variations in the strength of the South Pacific High is briefly discussed.

1. INTRODUCTION

The eastern tropical Pacific Ocean occasionally becomes much warmer than average for periods of a year or more. This phenomenon, commonly referred to by its local name, El Niño, has recently been the subject of intense scientific interest as the clearest example of large-scale air-sea interaction. To assess the changes in several moisture variables associated with interannual variations in ocean surface temperatures, a recent analysis of the marine

273

A. Street-Perrott et al. (eds.), Variations in the Global Water Budget, 273–284.
Copyright © 1983 by D. Reidel Publishing Company.

weather reports (see Weare 1982, Weare *et al.* 1981) covering all of
the tropical Pacific for the period 1957-76 was used. Illustration
of both the spatial and temporal patterns of surface latent heat
flux and moisture divergence variations associated with El Niño is
complicated by the relative paucity of data in much of the tropical
Pacific region. To compensate for this, a combination of two
general diagnostic approaches was employed. The first involved
a comparison of maps of El Niño events for 'like' months of 1957-58,
1965, 1969, 1976. This method provided a general specification of
the spatial patterns of moisture changes associated with El Niño.
Correlations and coherence spectra were calculated using data
averaged over Pacific areas 10° in latitude and about 40° in longitude to provide satisfactory measures of the temporal relationships.

2.   MORPHOLOGY OF MOISTURE CHANGES ASSOCIATED WITH EL NIÑO

From the available data, it is possible to identify the
principal features concomitant with the development and decay of
the major El Niño events. As suggested by Horel and Wallace (1981),
the occurrence of El Niño was determined from the time coefficients
of the principal component of sea temperature computed by Weare
et al. (1976). The four El Niño periods were 1957-58, 1965, 1972
and 1976. Individual monthly means of particular parameters for
corresponding periods during each of these El Niño events were
averaged starting with the June preceding the primary El Niño year
and ending with the subsequent May. Thus, means for each 5° grid
were formed, for example, for the June months of 1964, 1971 and
1975, and the July months of 1957, 1965, 1972 and 1976. These
three- or four-year means were subtracted from the 20-year mean
for each specific location and calendar month to define the average
departures from the means for the 24 months preceding, during and
following each El Niño.
As expected, the sea temperature departure maps (not shown)
illustrate that temperature throughout the eastern Pacific is
higher in the June of an El Niño than in the corresponding month preceding it. On the other hand, in the western Pacific, sea temperature generally declines from the June preceding to the June within
an El Niño. The changes within the intervening year appear to have
nearly reversed the temperature departure patterns throughout the
entire tropical Pacific.
Figure 1 illustrates the 'average' departure from the mean of
surface latent heat flux for the months of June preceding and within an El Niño period. This flux is calculated from the marine
reports by using the well-known bulk formula:

$$Q_L = \rho L \cdot C_E V (q_s - q_a) \qquad\qquad (1)$$

where Q_L is the flux of latent heat <u>out</u> of the ocean and ρ, L, C_E,
V, q_s and q_a, respectively, are surface air density, latent heat

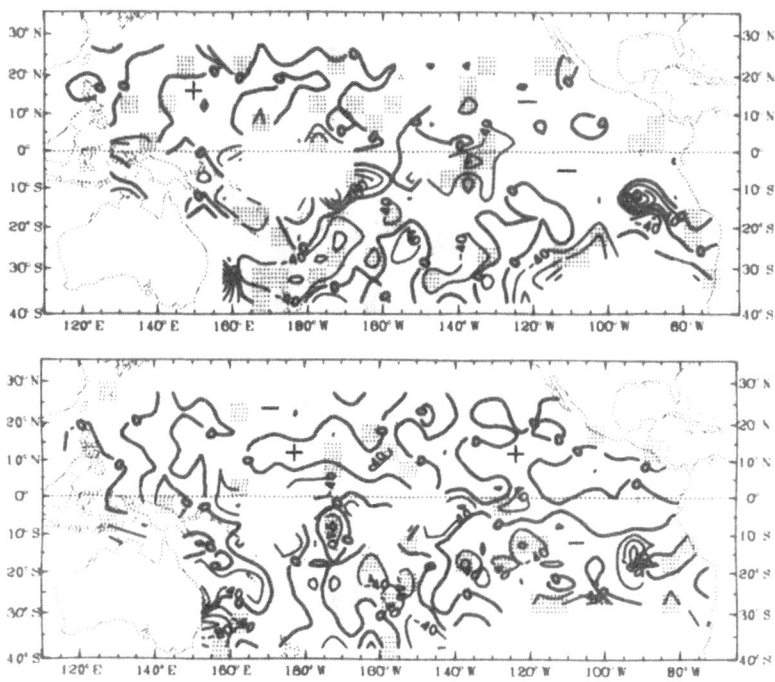

Fig. 1   'Average' El Niño surface latent heat flux departures
          (W m^{-2}) for June months: (a), 1964, 1971, 1975;
          (b), 1957, 1965, 1972, 1976.   The stippled regions
          identify the 5° grids for which departures from the
          long-term mean are more than two standard deviations.

of vaporization, exchange coefficient, wind speed, saturation
specific humidity at the sea surface and measured specific humidity.
The departure from the mean patterns of latent heat flux are
'noisier' than those of sea temperature, because Q_L is a function
of wind speed and far more observations of wind speed than of sea
temperature are necessary to derive a representative mean.   Figure
1 shows that:
(a)   Regions of warmer than average water masses are also regions
      of greater than average latent heat flux (and vice-versa);
(b)   Augmented latent heat fluxes are confined to a band adjacent
      to the equator narrower than that displaying enhanced sea
      temperatures;
(c)   The regions of the largest changes of latent heat flux are
      downstream (west) of areas showing the greatest departures
      from the mean of sea temperature.

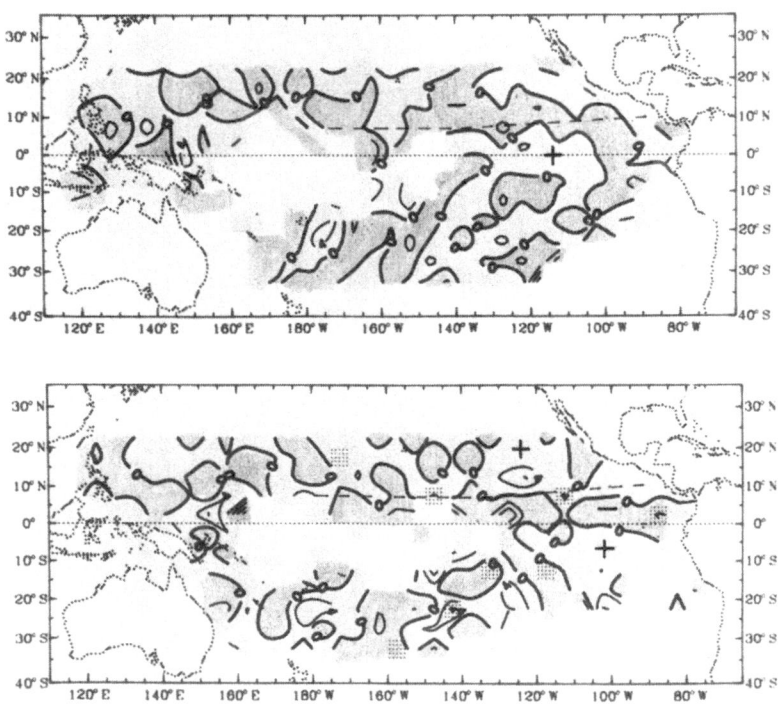

Fig. 2   'Average' El Niño moisture divergence departures
         (10^{-5} g kg^{-1} s^{-1}) for June months, as in Figure 1.

     The departures from the mean of surface moisture divergence
for the two June months are shown in Figure 2. The divergence is
calculated as the finite differences between the products of
surface spcefic humidity and wind components. Because this term
is a function of wind speed and direction (which both depend on
a relatively large number of observations to give representative
means) and requires good estimates at the four points over which
finite differencing takes place, the departure from the mean
patterns in Figure 2 are relatively 'noisy'. Nevertheless, systematic changes between the two June months (prior to and within an
El Niño) seem evident, especially near 0-10°N in the eastern
Pacific. The convergence zone in this region appears to have moved
toward the equator as El Niño developed. Changes in the strength
of the convergence zone were not resolved by this analysis.
     Figures 3 and 4 show maps similar to those in Figure 1 and 2,
but are for the December months preceding and towards the end of
the 'average' El Niño event. The sea temperature changes for the
December months (Table 1) were very similar to those between the
June months.

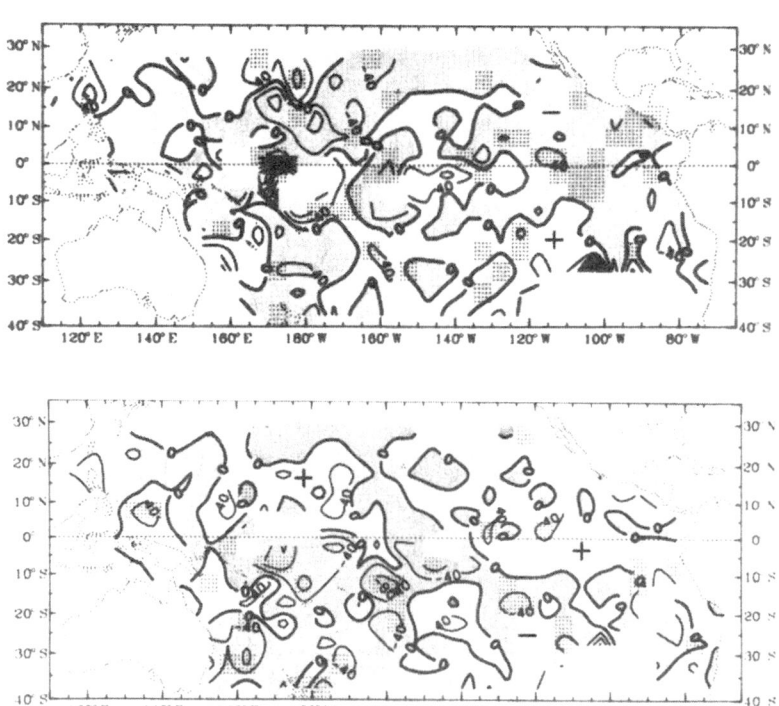

Fig. 3 'Average' El Niño surface latent heat flux departures
          (W m^{-2}) for December months, as in Figure 1.

Table 1. Comparative differences of sea surface temperature
          during the month of December in and before an El Niño
          event

location	temperature difference (°C)
Eastern equatorial Pacific	+ 2 to + 4 (warmer)
Eastern Pacific south of *ca*. 25°S	− 1 to − 2 (cooler)

Figure 3 shows the average departures from the mean of latent
heat flux for the two December months. As for June months, regions
of greater than average sea temperature correspond to regions of
larger than average Q_L. The largest changes are along the equator
and have a magnitude of about 80 W m^{-2}, i.e., about half the largest
changes calculated by Julian and Chervin (1978) in their general
circulation model simulation of January El Niño conditions.
    The differences in moisture divergence (Figure 4) between the
two December months are similar to those found for June months.
The convergence zone in the eastern Pacific is farther south in the

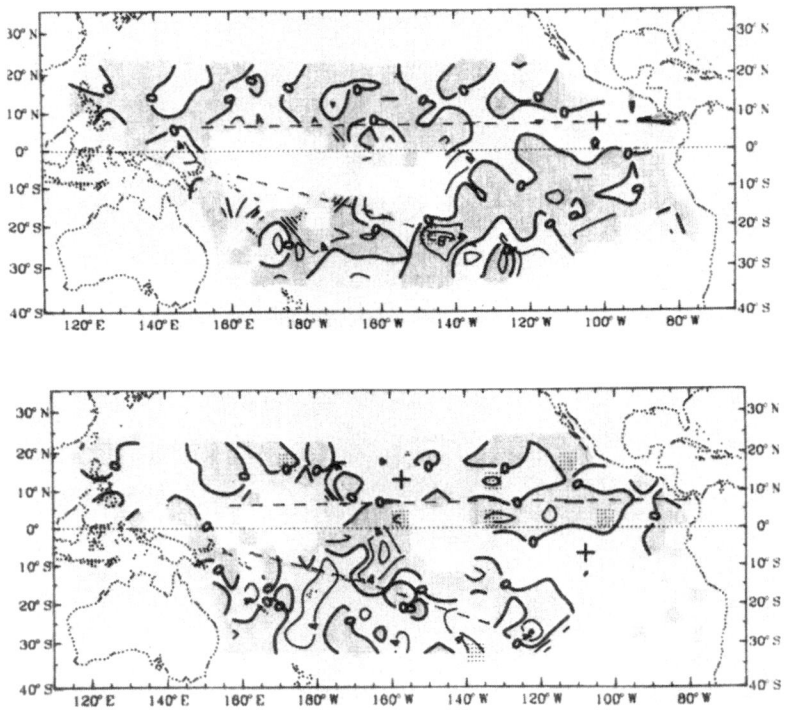

Fig. 4   'Average' El Niño moisture divergence departures
         (10^{-5} g kg^{-1} s^{-1}) for December months, as in Figure 1.

El Niño December. In addition, there appear to be coherent changes
in the area of the Southern Convergence Zone, about the Dateline
near $10°$-$40°$S. In the December before El Niño, the convergence
anomalies are to the south of the mean position (dashed line in
Figure 4) and north of that position during the El Niño December.
Both the eastern and southern Pacific changes in the moisture
divergence pattern suggest a weakening of the South Pacific High
and movement of the convergence zones toward that High during
El Niño. This is partially confirmed by pressure departure maps
(not shown).

     The departure patterns for sea temperature, latent heat flux,
surface moisture divergence, surface pressure and total cloudiness
have been examined for a 24-month sequence beginning in the June
before through to May following the 'average' El Niño. Figures 1
to 4 are quite representative. In general, the suggested sequences
of events, e.g., the surface warming or increased surface heat loss
in the eastern Pacific during the development of El Niño, can be
seen as fairly regular events in space and time during intervening
and subsequent months.

3.   TIME SERIES ANALYSES

    To understand the exact temporal relationships of moisture
variables within and between different oceanic regions, a number of
regional departure time series have been calculated. The series
comprise the average of the departures for a given 5° grid from its
long-term monthly mean for regions of 10° of latitude and about

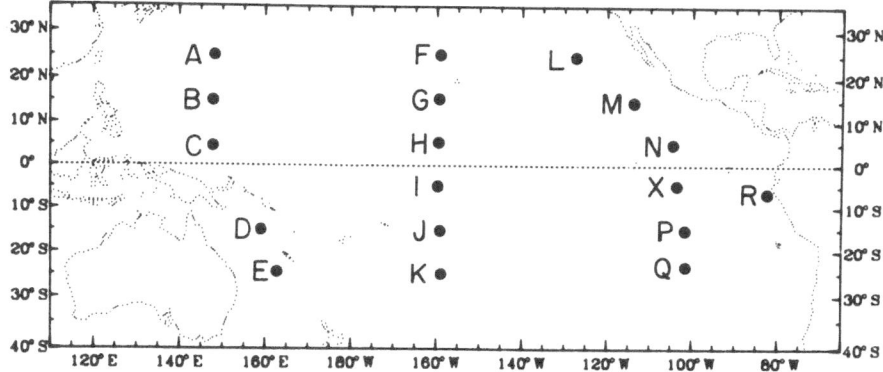

Fig.   5   The locations of the centres of the regions for which
           departures from the mean of various parameters were
           calculated.

40° of longtitude. For display purposes, all the time series have
been smoothed with a seven-point binomial filter. Figure 5
indicates the centres of these geographic regions (identified by
different letters). Generally, region X is considered to represent
the most significant aspects of El Niño as suggested by analyses of
ocean temperatures of Weare et al. (1976) and Weare (1982); the
departures from the mean of sea surface temperature of region X is
shown in each of the three Figures (6, 7, 8).
    The temporal evolution of surface latent heat fluxes for
regions B, N, X and P is illustrated in Figure 6. It is quite
evident that the variations in the latent heat fluxes for regions
N, X and P are quite similar to each other and to those of sea
temperature in reference region X. The amplitude of the variations
is up to about 30 W m^{-2}. In each case, sea temperature appears to
lead by several months, the lead being greater for regions N and P
than for X. These conclusions are corroborated in coherence spectra
between sea temperature in region X and latent heat flux for all
four regions (B, N, P and X). The spectra (not shown) indicate
that sea temperatures in region X are closely related to latent
fluxes in regions X, N and P at frequencies less than about 0.6
cycles/year. These frequencies are believed to correspond to those
of El Niño (Weare 1981). It is interesting that the coherence (not
illustrated) of region X sea temperature with region P latent heat

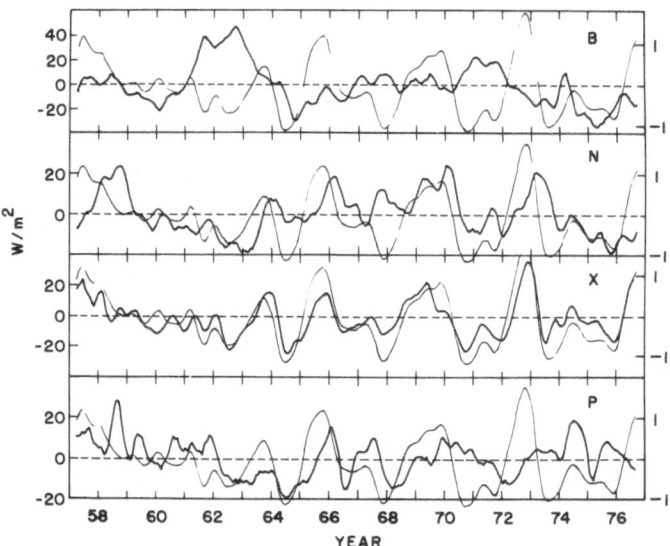

Fig. 6   Departures from the mean of (a), surface latent heat
flux (heavy lines, left hand scale) for regions B, N,
X and P (see Figure 5);   (b), sea surface temperature
(light lines, right hand scale) for region X.

flux is considerably greater than with region P sea temperature.
This is also true of the coherence between region X sea temperature
and the data for regions F, G, L, M and Q.

     The phase spectra show that region X latent heat flux changes
are nearly coincident with sea temperature changes. On the other
hand, region X sea temperatures lead the latent heat fluxes of N
and P by as much as 90°. The magnitudes of the leads are far
greater than those associated with the coherences of region X sea
temperatures with regions N and P sea temperatures (not shown).
These relations imply that latent heat changes in regions N and P
tend to follow local temperature changes by about one to four
months. Similar conclusions may be made for regions F to Q. These
apparent lags suggest that changes in latent heat flux are not
simply a direct thermodynamic response to higher sea temperatures,
since such a response should be nearly instantaneous.

     Latent heat flux variations over most of the eastern and
central Pacific are similar to those discussed above, though the
correlation with region X sea temperature generally diminishes with
distance. Further to the west, the changes in latent flux are
generally quite different: for instance region B has an enhanced
latent heat flux of up to 40 W m^{-2} for a *ca.* 3-year period centred
on 1962 (Figure 6). There is some suggestion that the major
variations in region B are related to oceanic mixed layer tempera-

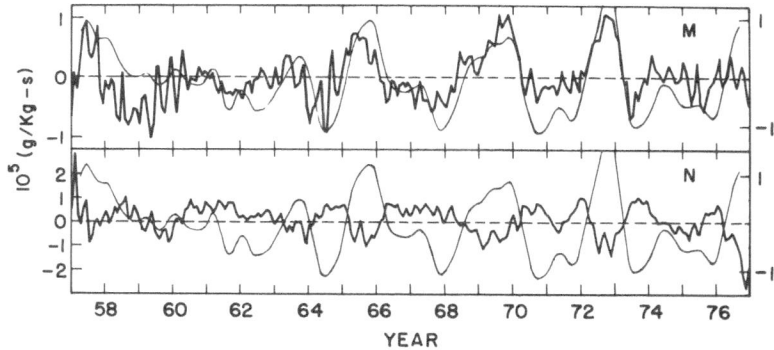

Fig. 7   Departures from the mean of (a), surface moisture
         divergence (heavy lines, left hand scale) for regions
         M and N (see Figure 5);   (b), sea surface temperature
         (light lines, right hand scale) for region X.

ture changes as indicated by dynamic height variations (White *et al*. 1979).

   Variations in moisture divergence in regions M and N are shown in Figure 7. Major features of the two series clearly appear to be negatively correlated, thus confirming the conclusion from the stratified map analysis that the position of the Inter-Tropical Convergence Zone (ITCZ) appears to be related to El Niño. Furthermore, the moisture divergence variations appear to lead those of sea temperature in region X by several months. Since the amplitudes of the opposing departures from the mean in regions M and N are quite similar ($ca.$ 10^{-5} g kg^{-1} s^{-1}), Figure 7 suggests that the magnitude of the ITCZ in this region is quite constant. The apparent shift of the Southern Convergence Zone (see § 2) is not evident in the time plots of divergence in regions J and K (not shown). The reason for this lies in the disposition of the convergence zone which runs diagonally through both regions: compensation for the changes, if they do occur, effectively obliterates the evidence of changes.

4.   DISCUSSION AND CONCLUSIONS

   Previous analyses have shown that spatially coherent and temporally persistent variations in the surface flux of latent heat over much of the tropical Pacific Ocean are associated with sea surface temperature variations in the eastern part of the region. The sea temperature variations are also shown to be related to regional changes in surface moisture divergence. In general, temperature changes lead those of the latent heat flux by several months. On the other hand, there is evidence that the moisture

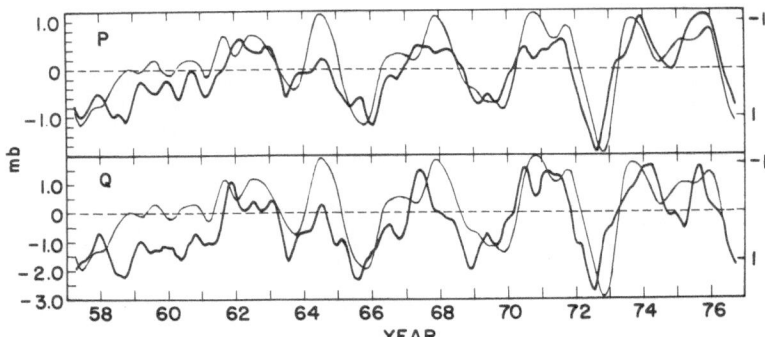

Fig. 8 Departures from the mean of (a), sea level pressure
        (heavy lines, left hand scale) for regions P and Q
        (see Figure 5); (b), sea surface temperature (light
        lines, right hand-inverted scale) for region X.

convergence changes lead El Niño events, as diagnosed by eastern
equatorial sea temperatures, by a few months.

    The illustrated relationships of the moisture variables could
be partially explained by the direct thermodynamic connection
between temperature and humidity. Higher sea temperatures are
associated with larger saturation humidities which, according to
Equation (1), should lead to higher latent heat fluxes if all else
were constant. This could explain the observation that higher
temperatures are generally followed by greater latent heat fluxes.
However, this simple argument does little to explain why variations
in latent heat fluxes lag one or more months behind sea temperature
changes. Furthermore, such an argument would seem to be totally
inadequate to explain why changes in moisture flux divergence near
10°N apparently lead changes in sea temperature in the equatorial
Pacific. It seems probable that the latter observations can only
be reconciled by assuming that circulation changes take place nearly
simultaneously with those of sea temperature. The analyses of the
moisture divergence suggest that the South Pacific High weakens
a short time before the sea temperatures rise in the eastern Pacific.
This is further illustrated in Figure 8, which shows the pressure
departures from the mean for regions P and Q along with the inverse
of the sea temperature for region X. As previously shown by Quinn
and Burt (1972) and Trenberth (1976), using Easter Island station
pressure measurements, the South Pacific High weakens a short time
before El Niño events.

    Many of the previously described observations are consistent
with the hypothesis that El Niño sea temperature changes are
strongly associated with the weakening of the South Pacific High.
For instance, if, as the eastern Pacific began to warm, the pressures in the High weakened but elsewhere remained nearly constant,

then weaker geostrophic winds about the High would result. These
weaker winds would tend to be accompanied by convergence zones to
the north and west which are geographically nearer the climatological centre of the High, as has been previously shown by the
moisture flux calculations. Furthermore, the weaker geostrophic
flow probably implies weaker overall wind speeds, which would tend
to diminish the surface latent heat loss (as given by Equation (1))
and counteract the tendency of the rising temperatures to increase
this flux. This conflict of influences might partly explain the
apparent lag between sea temperature changes and latent heat flux
variations. That sea temperature changes are negatively correlated
$(r \sim -0.3)$, to a moderate extent, with wind speed changes (except
in regions X and R) supports the concept of balancing influences.
As the El Niño continues, higher pressures near $30°S$, $100°W$ return,
strengthening the geostrophic flow. Thus, in the latter stages of
an El Niño event, higher sea temperatures and wind speeds tend to
increase the latent heat flux unequivocally. This may help to
explain why latent heat flux variations tend to follow significantly
the major El Niño features and why the suggested influences on
circulation at higher latitudes are mainly observable towards the
end of an El Niño event (Horel & Wallace 1981).
    While this general hypothesis may help to clarify the
relationship between various moisture parameters, it is not and was
not intended to be a complete theory of El Niño. In particular,
the hypothesis has little to offer as to the ultimate causes of
the eastern Pacific warming (Wyrtki 1975, 1979) or of the variations
in the South Pacific High. In addition, the hypothesis required
a number of simplifying assumptions, notably the constancy of
pressures in particular regions. Thus, verification or, better,
modification of the hypothesis will require much further study.

ACKNOWLEDGEMENT. Special thanks are due to P. Ted Strub who did
an excellent job programming much of the basic data management
and statistical analysis used in this study. The work was supported
by the Climate Dynamics Section of the US National Science
Foundation.

REFERENCES

Horel, J.D. & Wallace, J.M. 1981 Planetary scale atmospheric
    phenomena associated with the interannual variability of sea
    surface temperature in the equatorial Pacific. *Mon. Weath.
    Rev.*, 109, 813-829.

Julian, P.R. & Chervin, R.M. 1978 A study of the southern oscil-
    lation and Walker circulation phenomenon. *Mon. Weath. Rev.*,
    106, 1433-1451.

Quinn, W.H. & Burt, W.V. 1972 Use of the southern oscillation in weather prediction. *J. appl. Met.*, 11, 616-628.

Trenberth, K.E. 1976 Spatial and temporal variations of the Southern Oscillation. *Q. Jl R. met. Soc.*, 102, 639-653.

Weare, B.C. 1982 El Niño and Tropical Pacific Ocean surface temperatures. *J. phys. Ocean.* [In the press].

Weare, B.C., Navato, A.R. & Newell, R.E. 1976 Empirical orthogonal analysis of Pacific sea surface temperatures. *J. phys. Ocean.*, 6, 671-678.

Weare, B.C., Strub, P.T. & Samuel, M.D. 1981 Annual mean heat fluxes over the tropical Pacific Ocean. *J. phys. Ocean.*, 11, 705-717.

White, W.B., Hasunuma, K. & Meyers, G. 1979 Large-scale secular trend in steric sea level over the western North Pacific from 1954-1974. *J. Geodetic Soc. Japan*, 25, 49-55.

Wyrtki, K. 1975 El Niño - The dynamic response of the equatorial Pacific Ocean to atmospheric forcing. *J. phys. Ocean.*, 5, 572-584.

Wyrtki, K. 1979 The response of sea surface topography to the 1976 El Niño. *J. phys. Ocean.*, 9, 1223-1231.

SURGES OF TROPICAL PACIFIC RAINFALL AND TELECONNECTIONS WITH
EXTRATROPICAL CIRCULATION PATTERNS

Elmar R. Reiter

Department of Atmospheric Science
Colorado State University
Fort Collins
Colorado 80523, USA

ABSTRACT. Northern hemisphere 500-mb composite maps for episodes
of high and low precipitation in the so-called 'dry zone' of the
equatorial Pacific reveal significant changes in planetary wave
patterns, suggesting important extratropical connections of
tropical rainfall regimes. Anomalies in the Walker and Hadley
circulation systems are also revealed. Teleconnections between
these precipitation patterns and sea-surface temperature anomalies
in the northern and southern hemisphere (e.g., El Niño) and with
the Southern Oscillation can be demonstrated.
    Fluctuations in the equatorial Pacific rainfall regimes
appear to have undergone significant long-term trends which also
seem to be reflected in trends in the northern hemisphere extratropical circulation.

1. INTRODUCTION

    In a previous study, Reiter (1978) showed that there is an
excellent correlation between precipitation episodes in the so-
called 'dry region' of the equatorial Pacific as, for instance,
exemplified by the Line Islands rainfall records, and the convergence between the North and South Pacific trade-wind velocity
components, $\frac{1}{2}(-v_{NH} + v_{SH})$, averaged over the width of the Pacific,
v_{NH} and v_{SH} being the southerly components of the trade winds in
each hemisphere. It can also be shown that equatorial rainfall
peaks coincide with that phase of the Southern Oscillation (SO)
having positive surface pressure anomalies over Darwin (Australia)
and negative anomalies over Tahiti (Society Islands).

A. Street-Perrott et al. (eds.), Variations in the Global Water Budget, 285–299.
Copyright © 1983 by D. Reidel Publishing Company.

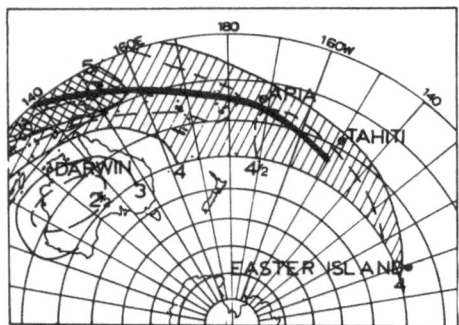

Fig. 1. Annual mean cloudiness (in oktas) from Atkinson and Sadler
         (1970), and the mean position of the South Pacific Conver-
         gence Zone (heavy line) from Gruber (1972). The shaded
         region indicates cloud cover greater than 4 oktas. Dia-
         gram derived from Trenberth (1976).

     Earlier work by Trenberth (1976) indicates that, during the
SO phase with low pressure anomalies at Darwin, the South Pacific
Convergence Zone (SPCZ) is displaced farther to the west and south
than normal, bringing abundant rains to Indonesia. (The mean
position of the SPCZ is shown in Figure 1.) During this phase of
the Southern Oscillation, more than normal rain falls in Indonesia,
eastern Australia and the New Hebrides-Fiji Islands area, in
accordance with a south-westward displacement of the SPCZ line
(shown in Figure 1). However, the equatorial area east of 160°E
tends to dry up and the Intertropical Convergence Zone (ITCZ) is
displaced away from the equator into the northern hemisphere.
Under such SO phase conditions, satellite cloud pictures give the
impression of two tropical convergence zones in the western and
central Pacific, one north and the other south of the equator.
     With positive pressure anomalies at Darwin, the SPCZ is
pushed from its mean position (Figure 1) towards the equator.
Cold upwelling water at the equator is replaced by positive sea
surface temperature (SST) anomalies in the central and eastern
equatorial Pacific, and the ITCZ moves into a near-equatorial
position (Ramage 1975). The equatorial Pacific dry zone, extending
approximately from Nauru through the Line Islands to the Peruvian
coast, experiences one of its prolonged surges in precipitation.
On satellite photographs, one extended convergence zone marked by
a cloud band spans almost the entire width of the Pacific Ocean
near the equator.

2. EQUATORIAL PRECIPITATION SURGES, SO AND EL NINO

     Several of the foregoing statements can be illustrated by the

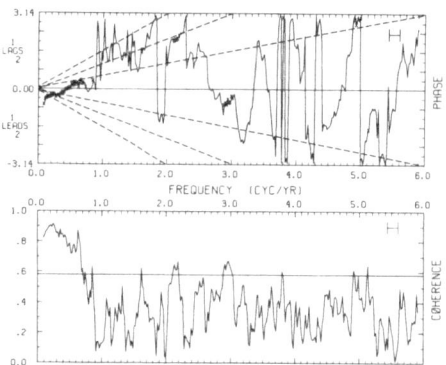

Fig. 2. Cospectra of Line Islands precipitation index and Darwin
         (12.5°S, 130.8°E) surface pressure anomalies, based on
         monthly data from 1935 to 1975 from which linear trends
         and annual cycles have been removed and to which cosine
         tapers have been applied. Cospectrum, quadrature spectrum
         and power estimates subjected to a nine-point smoothing;
         H, resolution bandwidth. Upper diagram: phase angles
         (radians); ×, points exceeding the 95% confidence interval;
         ---, lags of 3, 2, 1, -1, -2 and -3 months (reading clock-
         wise). Lower diagram: coherence with 16 degress of free-
         dom; ——, 95% significance level.

results of cospectral analyses. Figures 2 and 3 show that there
is a high coherence between the Line Islands precipitation
(expressed in the form of a monthly percentile index which
eliminates the seasonal cycle) and surface pressure anomalies at
Darwin and Tahiti. Spectrum computations were based on monthly
data from 1935 to 1975. In neither diagram is there evidence of
a significant cospectrum peak at the period of 26 months characteristic of the quasi-biennial oscillation (QBO). However, in a
different set of spectrum computations based on data between
1943 and 1975 (not shown here), such a peak is most prominent.
Indeed, inspection of the long precipitation records of the Line
Islands (Figure 4) shows that a quasi-biennial signal only
appeared to achieve significant amplitudes before 1930 and after
about 1963. Apparently, there are long-term climate variations
which affect the equatorial Pacific precipitation surges. Some
extratropical implications of these changes are discussed later.
     Figures 2 and 3 reveal a slight, but perhaps insignificant,
lead of Line Islands precipitation over Darwin surface pressures,
and an out-of-phase relationship with the Tahiti pressure
anomalies, at periods characteristic of the Southern Oscillation
(3-4 years). A significant lag appears in the covariance at low
frequencies between the Line Islands precipitation index and the

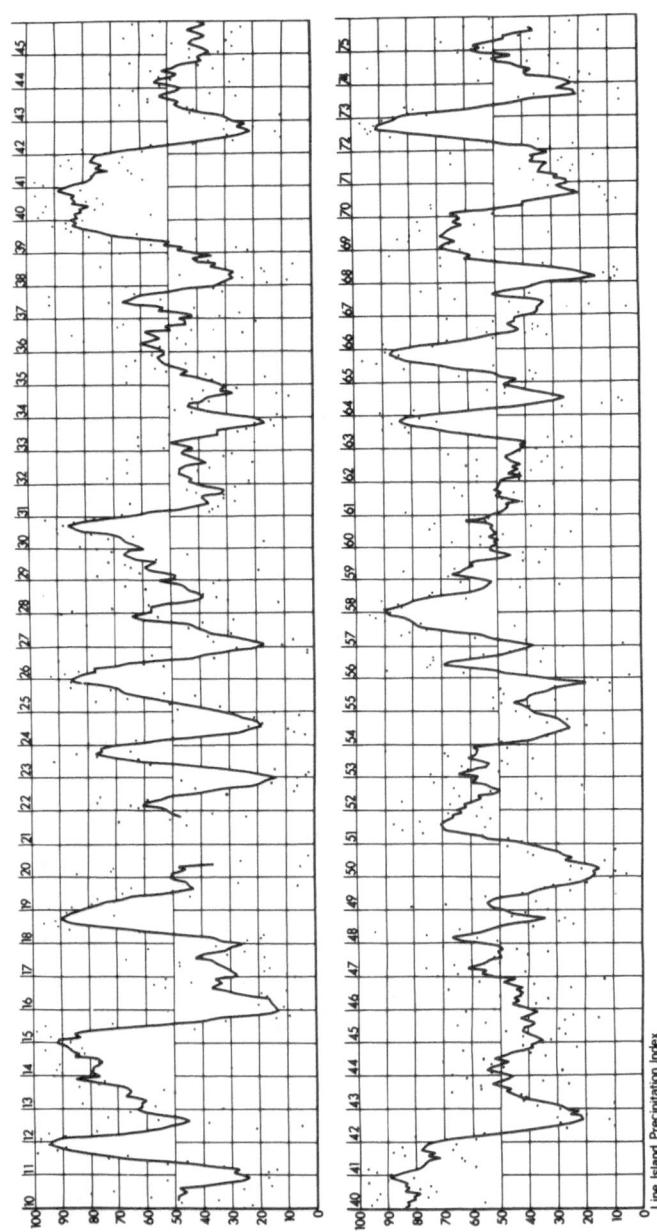

Fig. 4. Time series of Line Islands precipitation index: ·, monthly values; ——, smoothed seven-month running-mean.

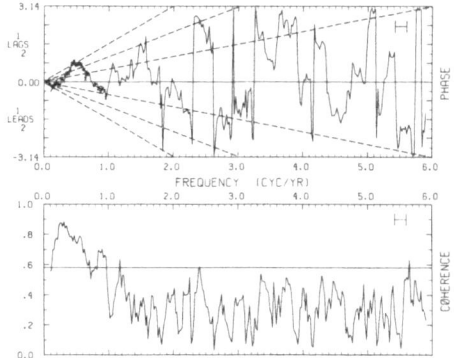

Fig. 3. Cospectra of Line Islands precipitation index and Tahiti
(17.6°S, 149.7°W) surface pressure anomalies, based on
monthly data from 1935 to 1975. For explanation, see
Figure 2.

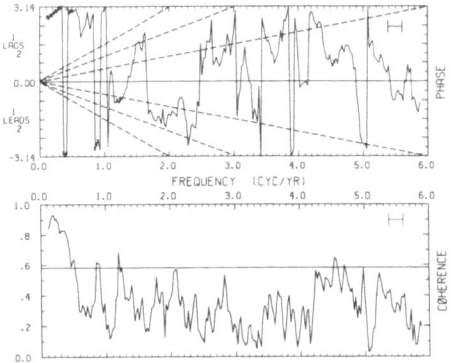

Fig. 5. Cospectra of Line Islands precipitation index and sea surface temperature anomalies from Puerto Chicama (7°S, 79°W),
based on monthly data from 1925 to 1975. For explanation,
see Figure 2.

Puerto Chimaca (Peru) SST anomalies (Figure 5). The coherence
between these two parameters is significantly high at periods
characterizing the Southern Oscillation. The time lag between
SST anomalies and Line Islands precipitation appears to become
smaller as one approaches the islands and virtually disappears
when SST anomalies and precipitation anomalies are compared in
their vicinity (Bjerknes 1966, Ramage 1977). The observed westward spreading of equatorial warm waters during El Niño periods
is consonant with such diminished time lags.

The migration of the ITCZ into a position close to the
equator during the SO phase with high pressure at Darwin can be
demonstrated by the high coherence between Line Islands precipitation and trade-wind convergence at long periods (~ 0.96, at
periods of 3 years) with virtually no phase lag (Figure 6). This
means that virtually all long-term variability of Line Islands
precipitation can be accounted for by the convergence between the
North and South Pacific trade-wind systems. Coherences are large
with the trade-wind convergences east as well as west of the Date
Line. (The North Pacific trade-wind box runs from 5°-19°N, the
South Pacific box from 1°-15°S.)

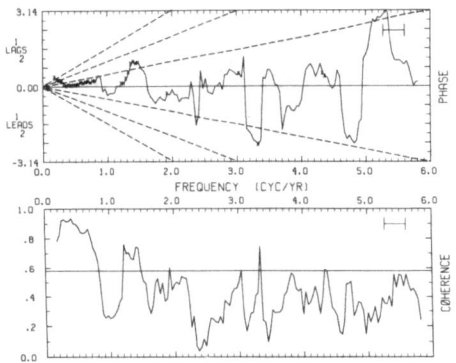

Fig. 6.   Cospectra of Line Islands precipitation index and trade-
          wind convergence, $\frac{1}{2}(-v_{NH} + v_{SH})$, between East Pacific
          regions north and south of the equator (at 5°-19°N, 180°-
          95°W and 1°-15°S, 180°-75°W), based on monthly data from
          1949 to 1975. For explanation, see Figure 2.

          Several authors (e.g., Cornejo-Garrido & Stone 1977) have
postulated that the equatorial Walker circulation, which changes
with the SO phase, is maintained by the release of latent heat in
the near-equatorial precipitation systems. As an indicator for
this Walker circulation, the differences between West and East
Pacific trade-wind convergence can be used. In Figure 7, these
differences are correlated with the surface pressure anomalies at
Darwin. Good coherence between these two parameters is found at
time scales characterizing the Southern Oscillation, in a sense
that a positive pressure anomaly at Darwin would lead excessive
convergence by a few months, hence upward motion would be
expected in the equatorial Walker cell over the West Pacific. This
conclusion runs counter to the schematic diagram by Julian and
Chervin (1978) which indicates downward motion in the Walker
circulation with high pressure anomalies at Darwin. However,

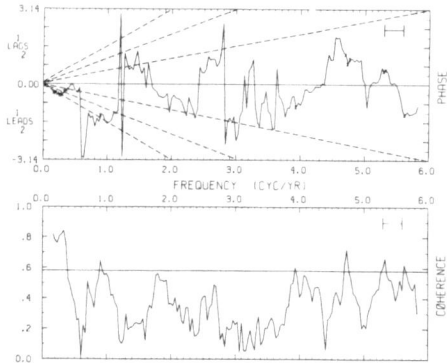

Fig. 7. Cospectra of Darwin surface pressure anomalies and trade-
wind convergence between the West and East Pacific, based
on monthly data from 1949 to 1978. With difference >0
between the West and East Pacific, vertical upward motions
in the equatorial West Pacific should dominate those in
the East Pacific. For explanation, see Figure 2.

Darwin is to the south of the normal position of the SPCZ (see
Figure 1) and of the trade-wind-convergence zone (0-5°N) specified
in our computations. Indeed, according to Figure 8 (which shows
the surface pressure correlation patterns with Djakarta), positive
pressure anomalies at Darwin would lead to an enhanced equatorward flow in the West Pacific in both hemispheres.
    Short-period surges of trade-wind convergence seem to have a
tendency to lead the Darwin surface pressure anomalies (Figure 7),
and therefore do not contradict possible maintenance of the Walker
circulation by latent heat releases. Only on long SO time scales
does the enhancement factor of surface confluence in the West
Pacific equatorial regions become important.

3. EXTRATROPICAL TELECONNECTIONS OF EQUATORIAL PRECIPITATION SURGES

    In §2, evidence was presented to support the idea that pre-
cipitation surges in the equatorial central Pacific are tied to the
SO phase with positive pressure anomalies at Darwin and to a near-
equatorial position of the ITCZ spanning the Pacific. Two processes seem to have some influence on the movement of the ITCZ into
such a position.
    One has to do with the establishment of warm SST anomalies
over the eastern equatorial Pacific, characteristic of El Niño
events. According to Hurlburt *et al.* (1976) and Busalacchi and
O'Brien (1981), such warmings are triggered by a relaxation of the
zonal easterlies mainly (but not exclusively) north of the equator
in the West Pacific. High-pressure anomalies at Darwin (Figure 8)

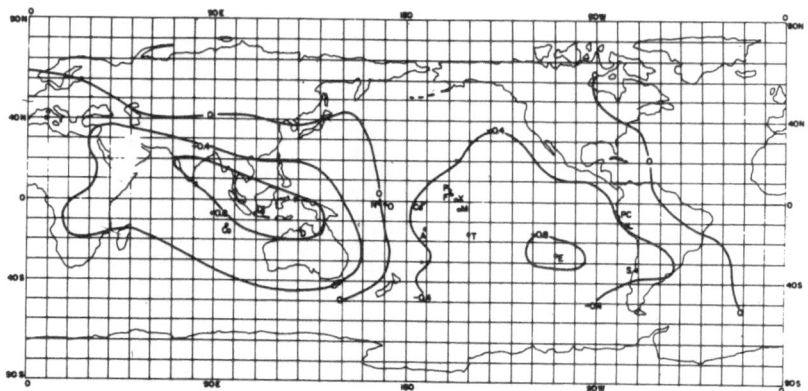

Fig. 8. Correlation of surface pressure anomalies with those at
         Djakarta, Indonesia (from Julian and Chervin (1978),
         after Berlage 1966).

would provide the mechanism for such a relaxation. Thus, the East
Pacific warming would be produced by an oceanic Kelvin wave that
travels along the equator from west to east.
     The most important process activating the Walker circulation-
Southern Oscillation-El Niño feedback mechanism may well be this
attraction of the ITCZ to the equatorial warm waters. However,
there are also extratropical implications. If the 500-mb
anomaly patterns for those, key, months with minimal precipitation
in the Line Islands are averaged and subtracted from their maximum
precipitation counterparts, pattern differences over the central
North Pacific are obtained that are significant at the 99%
confidence level (Figure 9). Low pressure anomalies dominate this
region during precipitation surges, maximizing one month before
the precipitation extremes. Lag correlations on a monthly basis
(Figure 10) show that 500-mb planetary wave anomalies in mid-
latitudes can serve as a trigger for tropical precipitation
patterns, especially during January. In that month, troughs over
the central North Pacific have a tendency to push the ITCZ equator-
wards. This process may help to activate the Southern Oscillation-
El Niño feedback mechanisms previously described.
     Atmospheric circulation and pressure anomalies over the North
Pacific reveal a high coherence with SST anomalies in that region,
with atmospheric anomalies showing a tendency to lead oceanic
anomalies by about one month over a wide region of the frequency
spectrum (Figure 11). Ocean-atmosphere feedback processes
(Hasselmann 1977) in that region may provide an important reflex
system which might also be responsible for extratropical-tropical
teleconnections. The possibility of such a low-frequency oceanic
involvement is indicated by high coherence and out-of-phase

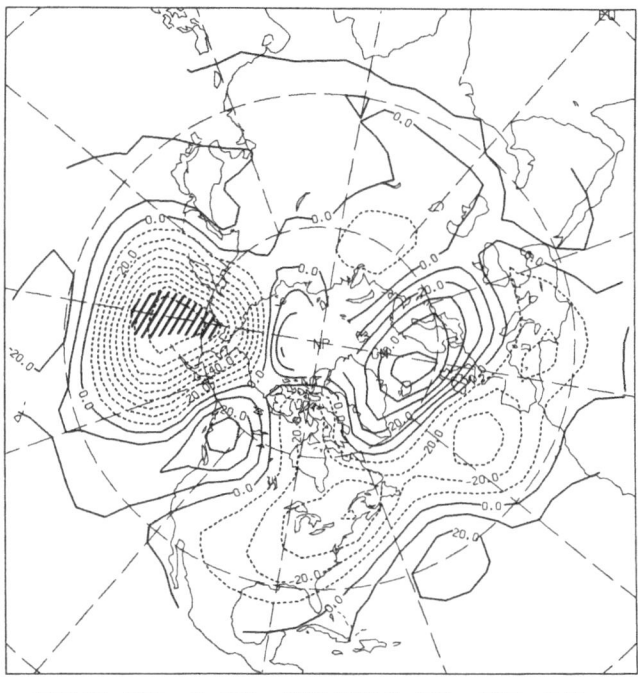

CONTOUR FROM -180.00   TO 180.00   CONTOUR INTERVAL OF   10.000   PT(3.3)=   20.105

Fig. 9.   Mean 500-mb anomalies averaged for months with precipita-
          tion minima in the Line Islands (Oct. 1952, May 1954, Feb.
          1956, Feb. 1957, Nov. 1958, Mar. 1961, June 1964, Apr.
          1967, May 1968, July 1970, July 1973, Aug. 1975) subtracted
          from like anomalies for months with precipitation maxima
          in the Line Islands (Aug. 1951, Nov. 1953, Apr. 1956, Dec.
          1957, Apr. 1959, Feb. 1961, Feb. 1964, Sept. 1965, Nov.
          1968, Nov. 1972), the 500-mb difference patterns being
          shown with 1 month lags with respect to the above key
          months. Contour (analysis) interval, 1 gpm; hatched
          region, area where difference between selected anomaly
          differences has level of confidence > 99%.

oscillations, at periods characteristic of the Southern Oscillation,
between the extension of the California current and the central
North Pacific (Figure 12). The SST anomalies in the California
current, on the other hand, show a behaviour similar to that of
the anomalies off the coast of Peru (Reiter 1979, Hurlburt *et al*.
(1976).

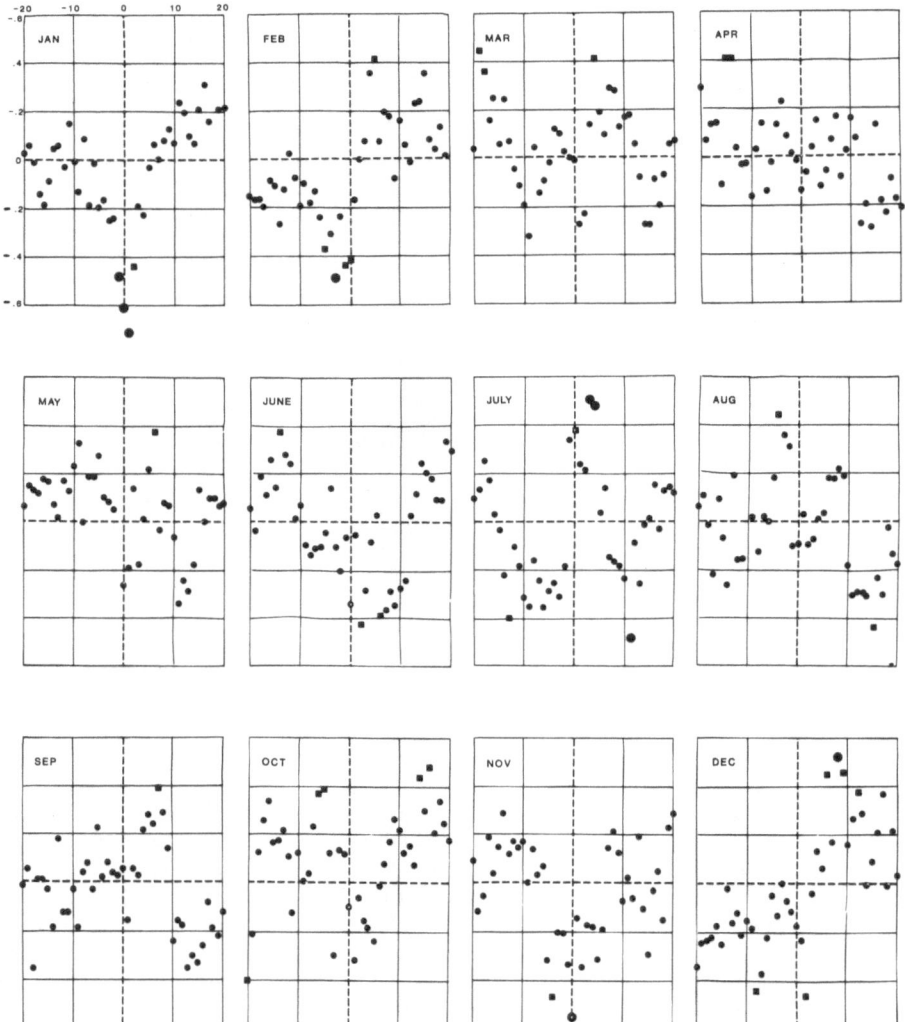

Fig. 10.   Monthly lag correlations between Line Islands precipita-
           tion index and 500-mb height anomalies at 50°-60°N, 180°-
           160°W, precipitation lag being for key months in respect
           of 500 mb data. Abscissae: time lags (in months);
           ordinates, correlation coefficients. Level of signifi-
           cance of correlation: ■, 95%; ●, 99%.

        The role of an equatorial position of the ITCZ, in activating
various tropical and extratropical feedback mechanisms, is
summarized in Figure 13.

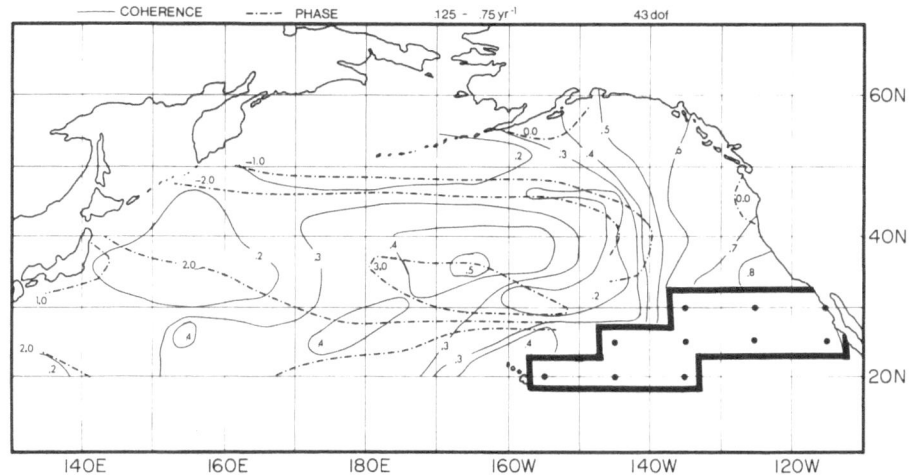

Fig. 11.   Cospectra of 500-mb anomalies in the region 50°-60°N,
           180°-160°W and sea surface temperature anomalies in the
           region 40°-50°N, 160°E-160°W, based on monthly data from
           1947 to 1978. For explanation, see Figure 2.

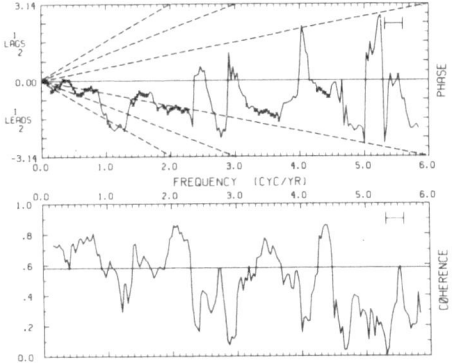

Fig. 12.   Coherence (——) and phase in radians (-·-) in the fre-
           quency band 0.125-0.75 cycles yr^{-1} between SST anomalies
           in the eastern Pacific (heavily bordered area) and at
           other grid points. Positive values of phase indicate
           that SST anomalies in the enclosed area lead those at
           the grid points (Middleton 1980).

4.   EFFECTS OF RECENT CLIMATE TRENDS ON EQUATORIAL PRECIPITATION
     SURGES

     Linear trend analysis of monthly isobaric height and thickness

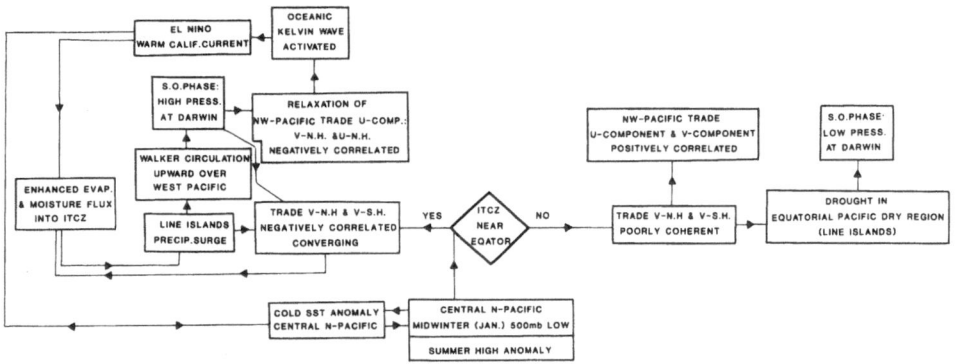

Fig. 13.   Schematic diagram showing feedback process with ITCZ
           near equator.

anomalies of the northern hemisphere (Reiter & Westhoff 1982)
reveals a significant pattern of cooling and warming, or of height
decreases and increases, suggesting long-term changes in planetary
wave configurations. During the winter season (December to March),
a cooling trend that is statistically significant (by more than
two standard deviations different from zero) is evident over the
central North Pacific, and a warming trend appears in the ridge
downstream over, or slightly to the west of, the west coast of
North America. Figure 14 shows, as an example, the linear 500-mb
trend slope pattern for January. The trends were computed from
the 1200 GMT upper air data between 1951 and 1978 supplied by the
National Meteorological Center (during the first four years, 1500
GMT data were used). The 0000 GMT data, albeit derived from a
shorter period (1955-1978), reveal the same behaviour with time.
      The trend patterns in Figure 14 closely resemble those in
Figure 9 which characterized the teleconnections of the Line
Islands precipitation surges. The northern hemisphere 500-mb
height trends during the past 30 years increasingly favoured conditions that should lead to frequent Southern Oscillation-El Niño
feedback processes (see Figure 13). Indeed, between 1930 and 1963,
there was a dearth of precipitation surges (see Figure 4), with
only two major events: in 1940/41 and 1957/58. Since 1963, such
precipitation surges have occurred frequently and with a superimposed quasi-biennial signal.

5. CONCLUSIONS

      An equatorial position of the ITCZ in the central Pacific
appears to be the critical element allowing certain ocean-
atmosphere feedback processes to take place that lead to
significant precipitation surges in the otherwise 'dry' equatorial

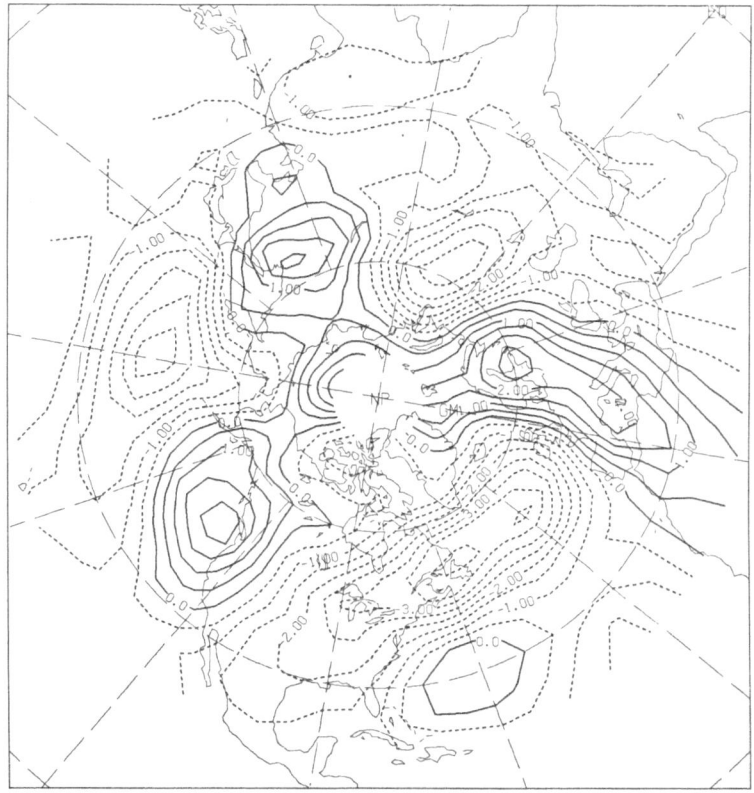

CONTOUR FROM -8.0000   TO 10.000   CONTOUR INTERVAL OF .50000   PT(3.3)= -.97500

Fig. 14. Slope (gpm yr^{-1}) of the linear trend of the 500-mb sur-
          face in January months from 1951 to 1978: ---, heights
          decreasing; ——, heights increasing. Note that the
          trends are greater than the zero slope by more than twice
          the standard deviation in the centre of the fall and rise
          areas over the central North Pacific, northern Canada,
          the eastern USA and northern Siberia.

Pacific. The Southern Oscillation and the El Niño phenomena are
involved in these processes in an important way. Even within the
time frame of recent climate changes, these processes and,
especially, the teleconnections with extratropical latitudes are of
considerable significance.
     One might speculate about the geographic constraints of the
Southern Oscillation. Why is it that the oscillation maximizes
between the South Pacific and Indonesia, and not in an area farther
to the west? In this context, it appears to be relevant that the
equatorial West Pacific shows the highest ocean surface temperatures (> 28°C) throughout the year, with only little annual and

interannual variability. This oceanic region provides, especially in January, convective energy to force the highest (coldest) tropopause levels in comparison with other regions on the globe (R.E. Newell, personal communication). Consequently, it seems that this oceanic region plays the role of an anchor position about which modulations in convective intensity, reflected for instance in the Southern Oscillation, are pivoting.

The existence of these inordinately warm SSTs is dictated by air-sea interaction in the tropical Pacific wind and ocean current systems. It appears, therefore, that only a detailed consideration of these interactions can lead to an improved understanding of the Southern Oscillation.

ACKNOWLEDGEMENT.   The research reported in this paper was supported by the National Science Foundation (NSF) under Grant ATM 80-16867 and by the Department of Energy under Contract DE-AS02-76EV01340. Some of the computations were performed at the Computing Facility of the National Center for Atmospheric Research, Boulder, Colorado: the provision of computing resources by NCAR (which is supported by the NSF Atmospheric Science Division) is gratefully acknowledged.

The author is indebted to Daniel R. Westhoff for his programing work and to Dr John W. Middleton for his help in developing computational procedures.

REFERENCES

Atkinson, G.D. & Sadler, J.C. 1970 Mean-cloudiness and gradient-level wind charts over the Tropics. Air Weather Service (MAC) Tech. Rept 215. US Air Force.

Berlage, H.P. 1966 The southern oscillation and world weather. *Mededeel. Verhandel. Kon. Ned. Meteor. Inst.*, No. 88, 152 pp.

Bjerknes, J. 1966 A possible response of the atmospheric Hadley circulation to equatorial anomalies of ocean temperature. *Tellus*, 18, 820-829.

Busalacchi, A.J. & O'Brien, J.J. 1981 Interannual variability of the equatorial Pacific in the 1960s. *J. geophys. Res.*, 86, 10901-10907.

Cornejo-Garrido, A.G. & Stone, P.H. 1977 On the heat balance of the Walker circulation. *J. atmos. Sci.*, 34, 1155-1162.

Gruber, A. 1972 Fluctuations in the position of the ITCZ in the Atlantic and Pacific Oceans. *J. atmos. Sci.*, 29, 193-197.

Hasselmann, K. 1977 The dynamic coupling between the atmosphere and the ocean. *WMO Marine Science Affairs Rept.* No. 472, pp. 31-44. Geneva: World Meteorological Organization.

Hurlburt, H.E., Kindle, J.C. & O'Brien, J.J. 1976 A numerical simulation of the onset of El Niño. *J. phys. Oceanogr.*, 6, 621-631.

Julian, P.R. & Chervin, R.M. 1978 A study of the southern oscillation and Walker circulation phenomenon. *Mon. Weath. Rev.*, 106, 1433-1451.

Middleton, J.W. 1980 A cross-spectral study of the spatial relationships in the North Pacific sea-surface temperature anomaly field. Environ. Res. Paper No. 23, 26 pp. Colorado State University

Ramage, C.S. 1975 Preliminary discussion of the meteorology of the 1972-73 El Niño. *Bull. Am. met. Soc.*, 56, 234-242.

Ramage, C.S. 1977 Sea surface temperature and local weather. *Mon. Weath. Rev.*, 105, 540-544.

Reiter, E.R. 1978 Long-term wind variability in the tropical Pacific, its possible causes and effects. *Mon. Weath. Rev.*, 106, 324-330.

Reiter, E.R. 1979 On the dynamic forcing of short-term climate fluctuations by feedback mechanisms. Environ. Res. Paper No. 21, 62 pp. Colorado State University.

Reiter, E.R. & Westhoff, D. 1982 Linear trends in the northern hemisphere tropospheric geopotential height and temperature patterns. *J. atmos Sci.*, 39. [In the press.]

Trenberth, K.E. 1976 Spatial and temporal variations of the southern oscillation. *Q. Jl R. met. Soc.*, 102, 639-653.

ANTARCTIC SEA ICE VARIATIONS 1973-1980

Long S. Chiu

Department of Meteorology and Physical Oceanography
Massachusetts Institute of Technology
Cambridge, Ma 02139, USA

ABSTRACT. Analysis of Antarctic marine ice[1] showed a decrease of 2.5 million km^2 in the total sea ice area between 1973 and 1980. While a decrease of 1.5 and 1.0 million km^3 is found in the Pacific and Indian Ocean sectors, respectively, over this period, the sea ice area in the Atlantic sector hardly showed any trend at all. Non-seasonal variations of sea ice area in the ocean sectors, with trends removed, showed no significant correlations. Significant correlations exist between monthly total ice area and an index of the Southern Oscillation.

1. INTRODUCTION

Sea ice is an important climatic variable. Because of its high albedo with respect to short wave radiation, it plays a major role in the global radiative energy budget. Being a good insulator, it inhibits air-sea turbulent energy exchanges and is probably, therefore, the most important factor controlling the interannual variations of the energy balance in polar regions (Weller 1980, Fletcher 1969). Because of the low salinity content of sea ice, salt tends to concentrate in the layer of water below the sea ice during the freezing process. The addition of salt to the upper layers of the water column reduces the column's stability and encourages convection. The process is thought to be important in the formation of Antarctic Bottom Water in the Weddell Sea (Gordon 1971).

The areal coverage of sea ice in both hemispheres varies in a complex way in space and time. Sea ice variations in the northern hemisphere have been examined by Walsh and Johnson (1979), Kukla (1978) and Zakharov and Strokina (1978). In the southern hemisphere,

301

A. Street-Perrott et al. (eds.), Variations in the Global Water Budget, 301–311.

sea ice observations were fragmentary until the late 1960s when
the advent of high resolution satellite imagery provided a
synoptic description of the ice field. The seasonal variations in
total Antarctic sea ice area from 1973 to 1975, and the variations
in the latitude of the northern pack ice boundary have been
investigated, respectively, by Zwally *et al.* (1979) and by Streten
and Pike (1980). As the latter pointed out the spatial limit of
sea ice is not an entirely satisfactory index of sea ice coverage
as the existence of polynas and large open water areas southward
of the boundary greatly affects the energy balance in this region.
        Variations of the sea ice area in the Antarctic between 1973
and 1980 are reported here, and the associated temperature and
surface pressure variations are examined.

2. DATA AND METHOD OF ANALYSIS

        Charts of sea ice information from 1972 onwards have been
issued by the Fleet Weather Facility of the US Navy. The data
were based primarily on visual and infrared scanning radiometer
imagery from the *NOAA* satellite and the microwave radiometer on
Nimbus 5. Satellite imagery from the *LANDSAT* and US Defence
Meteorological Satellite programmes as well as conventional
observations were used in the compilations. The analyses were
compared with climatological data where possible so that gross
errors were eliminated. The charts suffer from the one deficiency
that late reports were not normally included because of the time
constraints of operation (US Navy Fleet Weather Facility
1973-1980).
        The weekly charts of Antarctic ice display sea ice extent and
concentration (in tenths). The area where ice was present was
partitioned into 10° longitude by 5° latitude grids and the proportion of sea ice cover within each grid extracted. An area was
considered to be covered if the sea ice concentration exceeded
1 tenth. To examine its spatial variability, the sea ice cover
around Antarctica was divided into Atlantic (75°E-55°E), Pacific
(155°E-75°W) and Indian Ocean sectors (25°E-155°W). The areal ice
cover within each sector was obtained by summing the fractional
area cover, weighted by the appropriate area. For the 8-year
(416-week) period examined, data were missing for only six weeks;
these data were linearly interpolated.

3. VARIATIONS OF SEA ICE

        Figure 1 shows the seasonal variations of the total sea ice
area in the different ocean sectors as well as for Antarctica
overall. The total sea ice area reached a minimum in the seventh
to eighth week (i.e. late February). The sea ice area increased
from late March to September, at an average rate of about

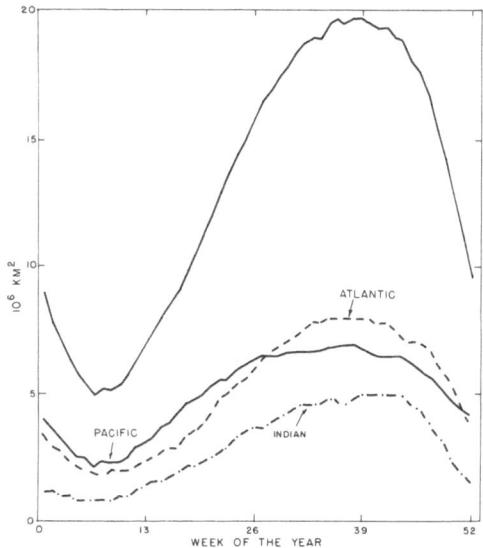

Fig. 1.   Seasonal variations of sea ice cover in Antarctica:
——— (upper), total for the Antarctic; ———, sea ice in the
Atlantic sector (75°W-25°E); ——— (lower), in the Pacific
sector (115°E-75°W) -·-, in the Indian Ocean sector (25°E-
155°E).

2.6 x 10^6 km^2 per month.   The maximum extent of 19.7 ± 1.26 x 10^6
km^2 occurred in September to October.   From November to January,
the sea ice area decreased rapidly - at an average rate of 5.4 x
10^6 km^2 per month.   These estimates are consistent with earlier
studies (Zakharov & Strokina 1978, Zwally *et al*. 1979).
   Although the Atlantic sector occupies only 100° of longitude,
the sea ice cover in this sector showed the largest seasonal
variation.   From a minimum of 1.8 x 10^6 km^2 in the eighth week, it
reached a maximum of 8.1 x 10^6 km^2 in the 38th week.   The seasonal
amplitudes (maximum to minimum) of sea ice areas in the Pacific
and Indian ocean sectors were 4.7 and 4.2 x 10^6 km^2, respectively.
The rate of ice increase in the Pacific sector showed a maximum
of about 0.3 x 10^6 km^2 per week between the 15th and 20th week,
whereas the rate of increase in the Atlantic sector did not reach
a maximum until the 20th week when the rate was about 0.4 x 10^6 km^2
per week.
   The autocorrelations of the total sea ice area with the sea
ice area in the three ocean sectors are shown in Figure 2.   Autocorrelation coefficients at lags of 4 and 8 weeks are, respectively,
0.56 and 0.30 for the total ice area.   These can be compared to
the autocorrelation coefficients at lags of 1 and 2 months,

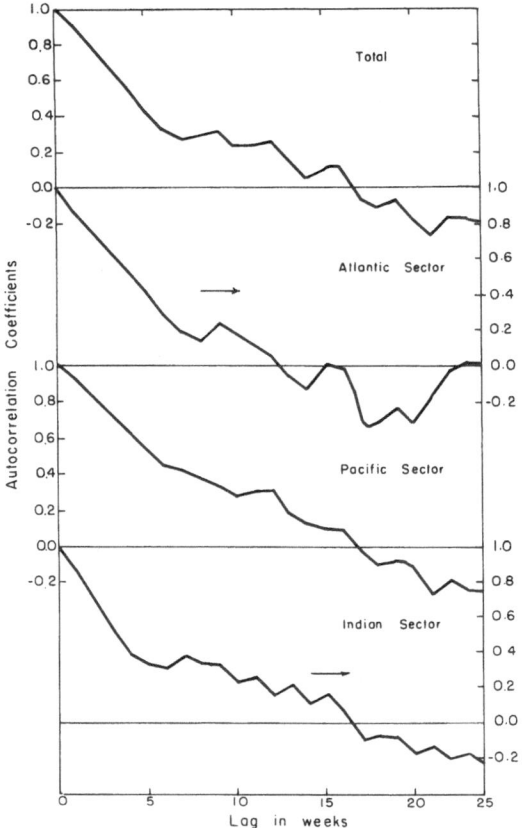

Fig. 2.   Autocorrelation coefficients of the total sea ice extent
          and sea ice area in the Atlantic, Pacific and Indian
          Ocean sectors.

respectively, of 0.64 and 0.38 for the sea ice area in the
northern hemisphere (Walsh & Johnson 1979).

The weekly departures of total sea ice area from the 1973-1980
means are shown in Figure 3.   In the first half of the record,
positive anomalies dominate whereas negative anomalies prevail in
the second half.   Linear regression analysis of the weekly departures showed a regression coefficient of -0.006×10^6 km^2 per week,
with a correlation coefficient of -0.67.   Thus, the total sea ice
area decreased by 2.5×10^6 km^2 between 1973 and 1980.

The significance of this trend can be estimated by examining
the s.d. of the slope.   However, because of the long autocorrelation in the time series, the number of independent samples is
less than 416.   The number of independent samples has been
estimated in the following way.   The variance of the estimated
autocorrelation coefficients of a stationary random normal process

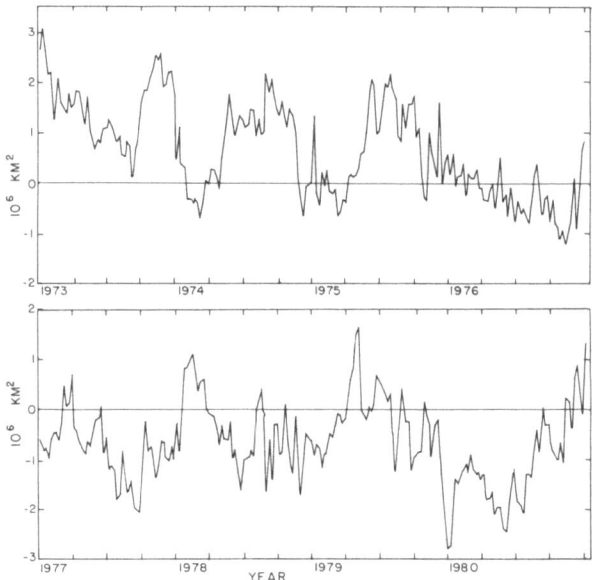

Fig. 3. Weekly departures of the total Antarctic sea ice extent
from the 1973-1980 means. Positions marked on the x-axis
denote the 13th, 26th, 39th and 52nd week of the year.

is 1/N, where N is the number of observations (Box & Jenkins 1976).
In this study, N equals 416. From Figure 2, it can be seen that
the autocorrelation coefficients are indistinguishable from zero
after about 15 weeks. The number of independent samples in this
series is therefore 416/15, namely about 27. A correlation
coefficient greater than 0.4 for the linear regression analysis is
significant at the 95% level of confidence (Bendat & Piersol 1971).
     The weekly departures of the sea ice area in the three ocean
sectors from the 1973-80 means are shown in Figure 4. For the
Indian, Atlantic and Pacific sectors, the standard deviations are
0.41, 0.56 and 0.64. x 10^6 km^2, respectively. Although the downward trend in sea ice area is evident both in the Pacific and
Indian ocean sectors, the Atlantic sector does not display a
straightforward trend. After a decrease in the first half of the
period, the sea ice area in the Atlantic sector shows an increase
in the second half. The weekly departures from the 1973-80 means
were converted into monthly departures and regression analyses
performed on the monthly series of anomalies in the different
sectors. The results are summarized in Table I. As can be seen,
the largest decrease occurred in the Pacific sector whereas there
was no trend at all in the Atlantic sector.

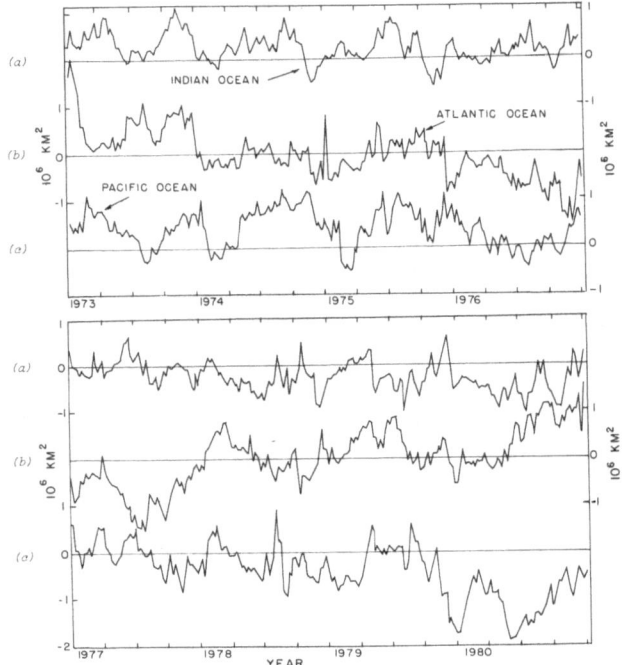

Fig. 4.   Weekly departures from the 1973-1980 means of sea ice area
in (a), the Indian Ocean sector $(25°E-155°E)$; (b), Atlantic
sector $(75°W-25°E)$; (c), Pacific sector $(155°E-75°W)$.
Positions marked on the x-axis denote the 13th, 26th, 39th
and 52nd week of the year.

Table I.   Trends of sea ice area variations in the ocean sectors

Ocean sector	Regression coefficient (10^6 km^2/month)	Total change (10^6 km^2)	Correlation coefficient
Atlantic	0.00015	0.01	0.008
Pacific	-0.016	-1.54	-0.76
Indian	-0.010	-0.96	-0.74

Variations in the Pacific sector contribute most to the
changes in total ice area, as these account for 34% of the
variance. Variations in the Atlantic and Indian ocean sector
account for 27% and 14% of the total variance, respectively.   The
remaining 25% is mainly attributed to a correlation between the
Pacific and Indian ocean sectors.

Calculation of the contemporaneous correlations between variations in the different sectors shows that there are no correlations between variations in the Atlantic sector with the other sectors (correlation coeffieient r < 0.1). Conversely, variations in the Pacific and Indian ocean sectors are highly correlated (r = 0.66). However, this high correlation is mainly due to the trends in the time series. If the trends are removed, r reduces to 0.23, which is not significant at the 95% level of confidence if the number of independent samples is 27.

The lag correlations between variations in the three sectors were examined for lags of up to 12 months. No significant lag correlations were found (r > 0.4). As the long-term trends in the time series may obscure correlations between variations on a shorter time scale, the calculation of the lag correlations were repeated using time series with the trends removed (detrended). For the detrended time series, variations in the Atlantic sector were found to have a correlation coefficient of -0.41 with variations in the Pacific 4 months later. However, this correlation is only marginally significant.

4. DISCUSSION

Trends in the Arctic ice area have been examined by Walsh and Johnson (1979) for the period 1953 to 1977. Over the past 20 years, the Arctic ice area reached a minimum in the early 1960s, increased to a maximum in the late 1960s and decreased again to its mid-1960s value in the mid-1970s. Analysis of a longer record (1936-76) showed slow variations in the summer Arctic ice extent with time scales of 5 to 10 years (Zakharov & Strokina 1978). A general decrease in Antarctic ice extent for the period 1973-78 was noted by Streten and Pike (1980) who suggested that this decrease is only part of the variations on a longer time scale. Because of the short duration of the record analysed in the current study, hasty conclusions as to the cause of the variations here demonstrated such as a global warming, are to be avoided, a point also stressed by Wigley *et al.* (1981).

From an analysis of the temperature records of Antarctic stations, Budd (1975) obtained a relation between Antarctic ice extent and annual mean temperature: namely, that a 1 K change in annual mean temperature corresponds to a 2° latitude change in the mean sea ice extent. In the latitude belt 60°-65°S, this can be expressed as 3 to 4 x 10^6 km^2 K^{-1}. Hence, a decrease of 2.5 x 10^6 km^2 in the ice extent can be associated with a temperature change of less than 1°.

A significant correlation exists between variations in the Pacific and Indian ocean sectors. This correlation is due mainly to the trends in the sectors. If the trends are removed from the time series, correlations cease to be significant. This is consistent with the findings of Streten and Pike (1980) on variations of sea ice extent.

308   L. S. CHIU

    The influence of Antarctic ice on global climate has been
discussed by Fletcher (1969). A significant relation exists
between the border of the pack ice and the band of maximum
cyclonic activity in the southern ocean (Schwerdtfeger &
Kachelheffer 1973). The seasonal variations of sea ice are
related to the seasonal wind field (Gordon & Taylor 1975, Streten
& Pike 1980). Streten and Pike found that the maximum ice extent
goes hand in hand with an index of the surface westerlies between
40° and 60°S on a seasonal basis, though this relationship does
not hold for non-seasonal variations.
    As the surface wind field is related to the surface pressure
field, the relationship between the pressure and ice fields has
been examined. From an eigenvector analysis of sea level pressure
over the globe, Kidson (1975) showed that the major pattern of sea
level pressure variations in the tropics resembles the so-called
'Southern Oscillation'. This term, coined some 50 years ago,
refers to "the tendency of pressure at stations in the Pacific,
and of rainfall in India and Java to increase, while pressure in
the region of the Indian ocean decreases" (Walker 1924). On
examining the lead and lag correlations of various meteorological
parameters between different parts of the globe, Walker (1923)
suggested that the Southern Oscillation may be of Antarctic origin.
    The correlations between total sea ice area and an index of
the Southern Oscillation, namely the departure from the mean
monthly pressure difference between Easter Island and Darwin
(Quinn & Burt 1972), are summarized in Table II.

Table II. Correlation coefficients between Southern Oscillation
           Index (SOI) and total ice extent for different months

Ice

SOI	Jan.	Feb.	Mar.	Apr.	May	June	July	Aug.	Sep.	Oct.	Nov.	Dec.
Jan.	-	-	-	-	-	-	-	-	-	-	-	-
Feb.	-	-	-	-	-	-	-	-	-	-	-	-
Mar.	-	-	-	-	-	0.70	0.71	0.74	0.75	-	-	-
Apr.	-	-	-	-	-	-	-	-	0.82	0.70	0.71	0.78
May	-	-	-	-	-	-	-	-	-	-	-	-
June	-	-	-	-	-	-	-	-	-	-	-	-
July	-	-	-	-	-	-	-	-	-	-	-	-
Aug.	-	-	-	-	-	-	-	-	-	-	-	-
Sept.	-	-	-	0.73	0.75	0.74	0.87	0.83	0.77	0.79·	-	-
Oct.	-	-	-	-	-	-	-	0.70	0.84	0.71	0.78	-
Nov.	0.72	-	-	-	-	-	-	0.71	0.72	0.88	0.82	0.70
Dec.	-	-	-	-	-	-	-	-	-	0.73	0.72	-

Note: Coefficients greater than 0.70 are significant at the 95%
       confidence level; - denotes coefficients less than 0.70.

The Southern Oscillation index chosen is not especially important as there is a relatively large 'sloshing' back and forth of pressure extremes, characterizing the oscillation between the continental loci (Julian & Chervin 1978). As the physics which controls the non-seasonal variations may be seasonally dependent, the correlations were calculated using seasonally stratified data. Pressure data are available back to the 19th century, but the total sea ice data are only available for eight years. As there are but eight independent observations, only correlations larger than 0.70 are significant at the 95% confidence level. It can be seen that there are significant correlations between the Southern Oscillation Index in March and April with the sea ice area in the following July to December. Large correlation coefficients ($r > 0.8$) are found between the ice area and the index for the months of July to November, with variations in the ice area preceding variations in the Index. This lead relation is consistent with Walker's original idea, the complete assessment of which must await, however, detailed analyses of the variations of the pressure, wind and temperature fields in the Antarctic region.

ACKNOWLEDGEMENTS. Thanks are due to Professor R.E. Newell, J. Anderson, M. Doherty and A. Navato for helpful discussions and comments, and to S. Ary and I. Kole for technical assistance. This research was supported by the Department of Energy through grant DE-AC02-76EV12195 and by the National Science Foundation through grant ATM-76-18016.

NOTE

[1] After this paper was presented in August 1981, a paper dealing with the same data, by G.J. Kukla and J. Gavin, was published in *Science, N.Y.*, 214, (4520).

REFERENCES

Bendat, J.S. & Piersol, A.G. 1971 *Random Data Analysis and Measurement Procedures*, 407 pp. New York: John Wiley.

Box, G.E.P. & Jenkins, G.M. 1976 *Time Series Analysis: Forecasting and Control*, 575 pp. San Francisco: Holden-Day.

Budd, W.F. 1975 Antarctic sea-ice variations from satellite sensing in relation to climate. *J. Glaciology*, 15, 417-427.

Fletcher, J.O. 1969 Ice extent in the Southern Oceans and its relation to world climate. Rand Corp Memorandum RM-5793-NSF, 108 pp. Santa Monica, California.

Gordon, A.L. 1971 Recent physical oceanographic studies of
    Antarctic waters. In: *Research in the Antarctic* (ed.
    L.O. Quam) – *Am. Ass. Adv. Sci. Publ.* no. 93.

Gordon, A.L. & Taylor, H.W. 1975 Seasonal change of Antarctic sea
    ice. *Science, N.Y.*, 147, 346–347.

Julian, P.R. & Chervin, R.M. 1978 A study of the Southern
    Oscillation and Walker circulation phenomenon. *Mon. Weath.
    Rev.*, 106, 1433–1451.

Kidson, J.W. 1975 Tropical eigenvector analysis and the Southern
    Oscillation. *Mon. Weath. Rev.*, 103, 187–196.

Kukla, G.J. 1978 Recent changes in snow and ice. In: *Climatic
    Change* (ed. J. Gribbin), pp. 114–129. Cambridge University
    Press.

Quinn, W.H. & Burt, W.V. 1972 Use of the Southern Oscillation in
    weather prediction. *J. appl. Met.*, 11, 616–628.

Schwerdtfeger, W. & Kachelheffer, St.J. 1973 The frequency of
    cyclone vortices over the southern ocean in relation to the
    extension of the pack ice belt. *Antarctic J.*, 8, 234.

Streten, N.A. & Pike, D.J. 1980 Characteristics of the broadscale
    Antarctic sea ice extent and the associated atmospheric
    circulation 1972–1977. *Arch. Met. Geoph. Biokl.*, A, 29,
    279–299.

U.S. Navy Fleet Weather Facility 1973–1980 *Antarctic Ice Charts.*
    Maryland: Suitland.

Walker, G.T. 1923 Correlation in seasonal variations of weather,
    VIII. *Mem. Ind. Meteor. Dept.*, 24, 75–131.

Walker, G.T. 1924 Correlation in seasonal variations of weather,
    IX. *Mem. Ind. Meteor. Dept.*, 24, 275–322.

Walsh, J.E. & Johnson, C.M. 1979 An analysis of sea ice
    fluctuation 1953–1977. *J. phys. Oceanogr.*, 9, 580–591.

Weller, G. 1980 Spatial and temporal variations in the south
    polar surface energy balance. *Mon. Weath. Rev.*, 108,
    2006–2014.

Wigley, T.M.L., Jones, P.D. & Kelly, P.M. 1981 Global warming?
    Nature, Lond., 291, 285.

Zakharov, V.F. & Strokina, L.A. 1978 Recent variations in extent of Arctic ocean ice cover. *Meteorologiya i Gidroloiya*, 7, 35-43.

Zwally, H.J., Parkinson, C.L., Carsey, F.D., Campbell, W.J. & Ramseier, R.O. 1979 Seasonal variation of total antarctic sea ice area, 1973-1975. *Antarctic J.*, 14, 102-103.

Long-Term Changes

INTRODUCTION TO LONG-TERM CHANGES

F. Alayne Street-Perrott

School of Geography
University of Oxford
England

The papers in this section deal with the nature and causes
of changes in the global water budget during the late Quaternary,
which is defined here as the last 127 000 years. Since this is
the most accessible part of the geological record, studies of
Quaternary climates form an important link between our understanding of the present-day global circulation and water budget,
and of the very different configurations existing in earlier
geological eras. The latter are well described in Frakes' book
Climates through Geological Time.
    The late Quaternary spans a full glacial - interglacial
cycle. Its climatic record, therefore, serves as a salutary
reminder of the range of possible climatic conditions in some of
the world's more densely populated areas. In examining the
distribution and alignment of fossil sand dunes in the United
States, Wells concludes that the upper tropospheric circulation
at mid-latitudes was characterized during the late-glacial by
three Rossby waves instead of the present four to five. The
increased wavelength of the large, anchored trough in the lee of
the Rockies resulted in pronounced aridity throughout the mid-
continent (a large sand desert developing in the High Plains),
but with consistently wetter conditions in the Southwest.
    At lower latitudes, a detailed picture of the changing
balance of precipitation and evaporation can be gained from the
fluctuations of closed lakes. Using a large data set, Street-
Perrott and Roberts endeavour to reconstruct the atmospheric
circulation patterns that gave rise to the variations in lake
level over Africa and western Eurasia during the last 18 000
years. They infer that the ITCZ and the mid-latitude westerlies
underwent significant changes in mean latitude and intensity
during this period.

A. Street-Perrott et al. (eds.), Variations in the Global Water Budget, 314–316.

Climatic reconstructions based on many different types of evidence all lead to the conclusion that the last glaciation was characterized by dry conditions over large parts of the tropics. Street-Perrott and Roberts place the maximum of aridity somewhat after the glacial maximum, at about 13 000 yr BP. Flohn argues that the general lowering of ocean temperatures reduced global evaporation and, hence, precipitation by ≈ 20%.

During the early post-glacial, Flohn estimates that higher sea temperatures led to an increase in global precipitation of ≈ 3-4%, though this increase was concentrated in the tropics. Kutzbach tabulates existing rainfall estimates for tropical Africa and India, based on the past extent of closed-basin lakes, and concludes that the change in precipitation ranged from a minimum of +14% to ⩾86% in the most arid areas. Two types of models, simple water-balance and combined water- and energy-balance, were used to derive these figures. Applying a combined model to the early Holocene Lake Megachad, Tetzlaff and Adams calculate that water losses through evaporation were also considerably reduced, partly due to increased cloud cover.

A new approach to the estimation of fluctuations in tropical precipitation in the Holocene is outlined by Gac *et al.* who use reconstructed changes in the salinity of the Senegal estuary as an indicator of variations in river discharge. The value of this method is crucially dependent on the assumption that the palaeo-salinity record has not been complicated seriously by fluctuations in sea level.

Several mechanisms of climatic change appear to have played an important role in influencing global patterns of precipitation, evaporation and water storage during the late Quaternary. Chief among them are the long-term cyclic variations in the amount and seasonal distribution of solar radiation received at different latitudes, resulting from changes in the geometry of the Earth's orbit. These astronomical variations were first discovered by Adhemar in 1842, but are often referred to as the Milankovitch variations, after the Serbian mathematician who investigated their climatic implications. Recent studies have revealed a striking coincidence in timing between the peaks and troughs in the Milankovitch radiation curves for the mid-latitudes of the northern hemisphere and the minima and maxima of such indicators of global ice volume as oxygen-isotope ratios in deep-sea cores.

Some uncertainty still surrounds the feedback mechanisms that induce the atmosphere, hydrosphere and cryosphere to respond globally to small variations in insolation over a restricted latitude band. Kutzbach reviews recent experiments with a low-resolution global circulation model that suggest an important linkage between the intensity of the summer monsoon over Africa and Eurasia (and, hence, the latent heat transport by the tropical circulation) and past variations in insolation in the northern hemisphere. Fastook proposes a quite different, and highly controversial, feedback mechanism to explain the near-synchronous response of glaciers in both hemispheres to changes in insolation between 45 and 60°N. He postulates an unstable response of the marine-based ice sheets, grounded on continental shelves, to the rise in sea level initiated by melting of terrestrial ice margins in mid-latitudes.

In contrast, Wells suggests that the northern hemisphere ice sheets may have been responsible for their own demise. By progressively modifying the atmospheric circulation, they may eventually have starved themselves of precipitation. Morner, taking a characteristically radical view, proposes that geophysical controls (changes in the shape of the geoid) may have been more important than climatic controls in causing fluctuations in sea level and in groundwater tables. His view, if correct, implies that future long-term changes in the volume of the large global water storages – the oceans, ice caps and groundwater – will be much more difficult to predict than elegant astronomical models would suggest.

The Milankovitch theory, because it deals with variations in insolation on a time scale of 20 000 to 100 000 years, is unable to account for any climatic event with a duration of 10^3 years or less. Flohn considers ways in which changes in the global energy budget, caused for example by variations in CO_2 levels or in the frequency of volcanic eruptions, may affect the global water budget by influencing the extent of equatorial and coastal upwelling. Upwelling is associated with cold water at the sea surface. This in turn results in high relative humidities, reduced oceanic evaporation and, hence, lower precipitation. Flohn's hypothesis accords well with oxygen-isotope data from ice cores, cave deposits and Saharan groundwater which indicate higher relative humidities than today over the oceans during the last glaciation (see also Sonntag's paper in this volume). The suppression of evaporation over areas of cold ocean may help to explain the correspondence in timing between the episodes of severe drought in tropical Africa described by Street-Perrott and Roberts, and the influxes of meltwater into the North Atlantic from the decaying ice sheets which have been revealed by deep-sea core studies.

Flohn's climatic feedback mechanism is a good example of a palaeoclimatic model with important implications for the prediction of future climates. The control exerted by air-sea interaction over the atmospheric content of the infrared-absorbing gases CO_2 and H_2O makes it imperative to include oceanic upwelling in any analysis of the climatic impact of increasing levels of atmospheric CO_2.

LATE-GLACIAL CIRCULATION OVER CENTRAL NORTH AMERICA REVEALED BY
AEOLIAN FEATURES

Gordon L. Wells

School of Geography
Mansfield Road
Oxford OX1 3TB
United Kingdom

ABSTRACT.   A continental survey of late-glacial aeolian features
(sand dunes, yardangs, deflation basins, etc.), based upon *LANDSAT*
imagery and high-altitude aerial photography, has made possible a
detailed reconstruction of the airflow over central North America
at the end of the Wisconsin Glaciation.   Over 450 sites with
late Pleistocene aeolian landforms have been identified.   The
trends of these features are compared with the trends of adjacent
modern sand dunes and with sand rose resultants calculated for 167
meteorological stations across the USA and southern Canada.
    In a number of regions, the direction of late-glacial surface
winds is 40° to 90° out of phase with the modern circulation.
Radiocarbon dates indicate that peak aeolian activity occurred
around 14 000 yr BP.   The continental distribution of relict
aeolian trends allows the late-glacial circulation pattern in the
upper troposphere to be reconstructed.   The dominant regime was
one of three planetary waves over the middle latitudes (40°-50°N)
whereas the modern circulation typically displays a four- or five-
wave configuration.   The climatic effects arising from the
persistence of three Rossby waves at the mid-latitudes include
increased aridity within the region of the vast orographically-
induced trough from the eastern Rocky Mountains to the Atlantic
coast and increased moisture supply to the Basin and Range
Province of the western USA.   The implications of this pattern for
the mass balance of the Laurentide Ice Sheet are discussed.

317

A. Street-Perrott et al. (eds.), Variations in the Global Water Budget, 317–330.

1. INTRODUCTION

     Over the past decade, several models have been put forward to
explain the general atmospheric circulation during different stages
of the last glaciation (Lamb & Woodroffe 1970, Williams *et al.*
1974, Gates 1976, Manabe & Hahn 1977). These models place special
emphasis on palaeoecological evidence (pollen and foraminiferal
assemblages) both for reconstructing possible values for glacial-
age summer and winter temperatures and global circulation patterns
and for verifying the results of computer simulations of circulation
during the last glaciation. While increasingly sophisticated
computer simulations of glacial climates have been developed little
attention has been paid to the direct evidence for changes in
circulation and water balance provided by wind-formed features.
     LANDSAT imagery and NASA U-2 high-altitude aerial photography
have made it possible to map aeolian features (sand dunes, yardangs,
deflation basins, etc.) on a continental scale. Late Pleistocene
aeolian activity over central North America left a distinctive
imprint on the landscapes of the High Plains, Eastern Seaboard and
Pacific Northwest. Across the USA and Canada, over 450 areas with
late-glacial aeolian landforms have been identified.
     After inspecting *LANDSAT* imagery of Mexico, the USA and Canada
(at a scale of 1:500 000) for evidence of past aeolian activity,
suspected areas were examined in detail using high-altitude aerial
photographs. Subsequently, extensive fieldwork was undertaken at
the most promising sites. These studies were complemented by an
exhaustive literature search to associate relict aeolian features
with radiocarbon dates, stratigraphic sections and palaeoecological
data in order to develop a chronology of events.

2. SAND DUNES, SAND ROSES AND MODERN CIRCULATION

     Before attempting to reconstruct the late-glacial circulation
pattern, the relationship between modern surface winds and the
resulting aeolian landforms must be understood. Sand grains with
an average diameter of 0.3 mm are entrained by winds moving in
excess of 6 m s^{-1} as measured by instruments mounted 10 m above
the surface in accordance with the World Meteorological Organization
standard (Fryberger 1979). Most sand movement in dune fields is
caused by winds of between 8 and 15 m s^{-1}. The trend of sand dune
movement is not therefore the simple resultant of daily prevailing
surface winds, but of the less frequent high-magnitude winds that
are often linked with the passage of fronts; Figure 1 illustrates
the types of sand dunes associated with different wind regimes.
Using wind frequency data, a sand rose can be constructed by summing
only the vector distribution of winds of greater than 6 m s^{-1},
according to the technique first described by Bagnold (1941) and
later refined by Fryberger (1979). Sand roses of this type have
been calculated for 167 meteorological stations across the United

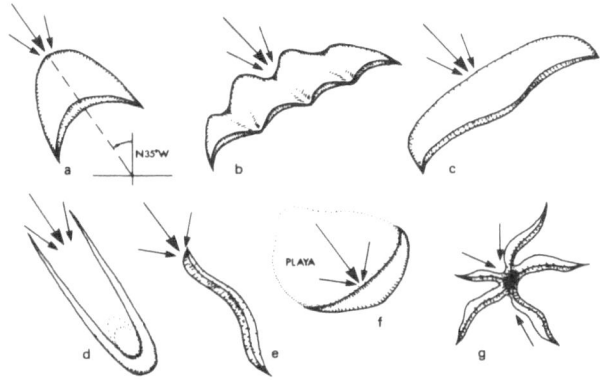

Fig. 1.   Types of central North American sand dunes showing asso-
          ciated wind regimes (small arrows) and direction of sand
          transport (large arrows). Formed by narrow-front unimodal
          wind regimes: (a), barchan dunes; (b), barchanoid ridges;
          (c), transverse ridges; (d), parabolic dunes. Formed by
          wide-front unimodal wind regimes: (e), longitudinal
          dunes; (f), lunette dunes. Formed by surface winds
          blowing in opposing directions: (g), star dunes, which
          do not migrate. Sizes of these landforms range from 10 m
          to > 2000 m (a, b, c, e), 10 m to 1000 m (d) and 100 m to
          > 2000 m (f, g).

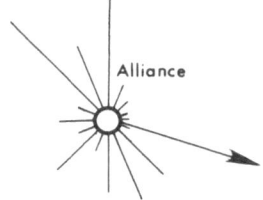

Fig. 2.   Sand rose calculated for Alliance, Nebraska, with arrow
          indicating the net direction of sand movement.

States and southern Canada; that for Alliance, Nebraska, is shown
in Figure 2. A comparison between the directions of sand movement
(as calculated from wind frequency data) and the trends of active
dune fields (as revealed by aerial photography) shows close agreement (Figure 3). Inland dune fields rarely deviate from the local
sand rose resultants by more than 5°.

     Several other characteristics of present-day large-scale
circulation are indicated by active dune trends and sand rose
resultants. The northern half of central North America is dominated by a westerly airflow, while the southeastern section
reflects the intrusion of tropical airflow around the western end
of the Bermuda High (Figure 4). The trends and resultants provide

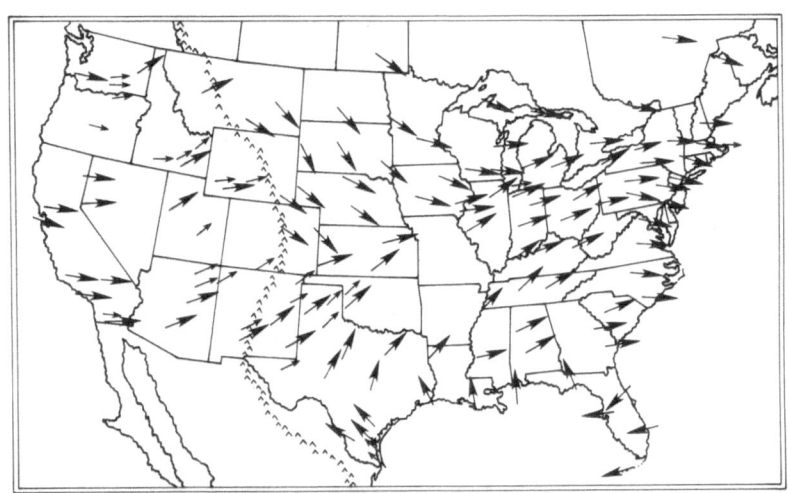

Fig. 3. Sand rose resultants (large arrows) and the trends of
         active sand dunes (small arrows) for central North America.

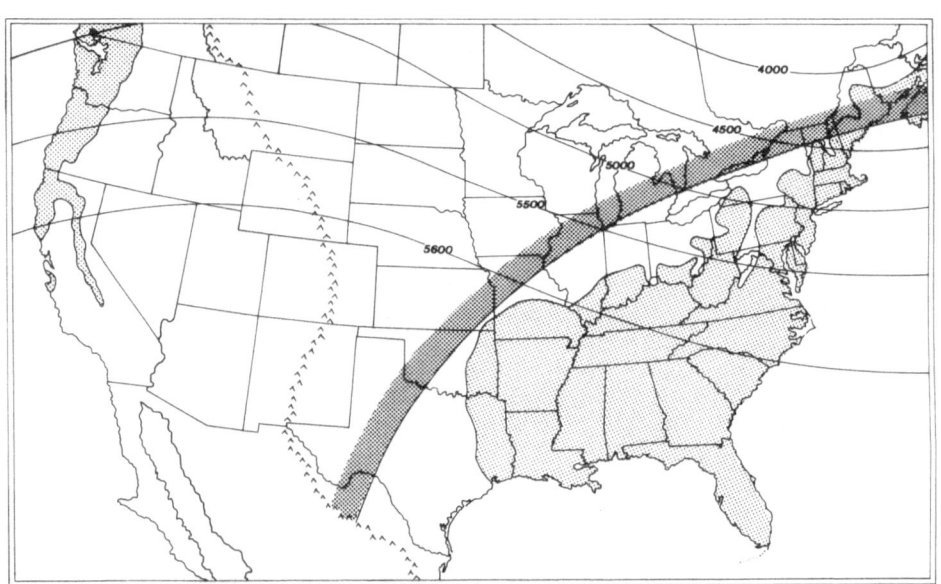

Fig. 4. Averaged planetary wave circulation (1000-500 mb thick-
         ness) for 1951-1966. Boundary zone of Arctic, Pacific
         and tropical airstreams indicated by the dark-shaded arc
         (after Bryson 1966); areas with > 1000 mm annual precipi-
         tation shown by lighter stipple.

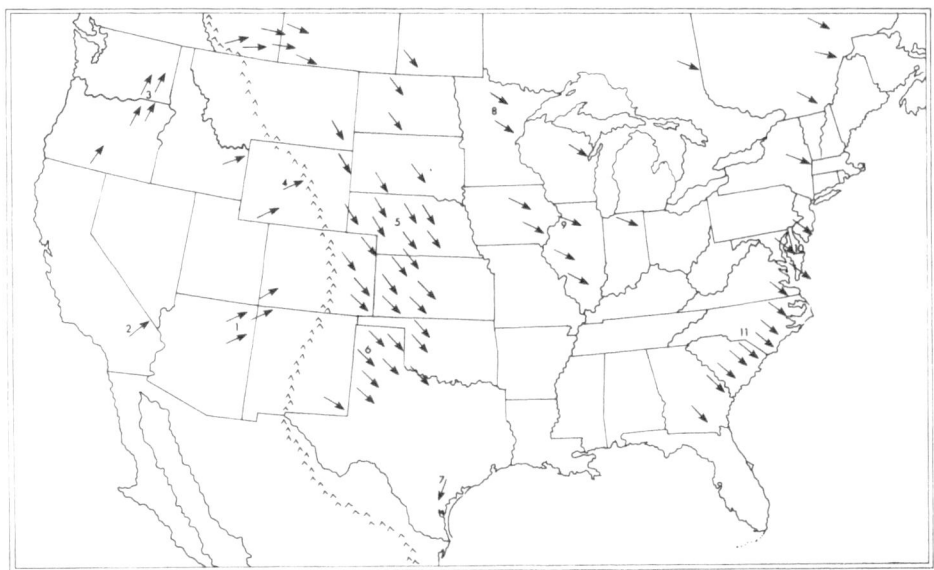

Fig. 5.   Trends of late-glacial features in central North America.
          Sites: 1, Moenkopi Plateau; 2, Kelso area of Mojave
          Desert; 3, Cheney-Palouse loessial ridges; 4, Casper area
          (Wyoming); 5, Nebraska Sand Hills; 6, Llano Estcado;
          7, coastal plain of south Texas; 8, Minnesota glacial
          outwash plain; 9, Illinois and Iowa loessial ridges;
          10, Delmarva Peninsula; 11, Carolina Bays.

an accurate representation of the averaged circulation pattern in
the upper troposphere circulation and of the boundary between
southeasterly tropical airflows carrying moisture from the
Atlantic and Gulf of Mexico and the northwesterly tracks of the
Arctic and Pacific fronts.

     The circulation data in Figures 3 and 4 represent long-term
averages that are not seasonally differentiated. The sand rose
resultants, from which the planetary wave regime and airstream and
precipitation boundaries are inferred, are compiled from meteorological records covering a few decades at most, while the sand
dune trends result from sand movement over periods ranging from
several decades to centuries depending upon the dune size and area.

3.   REGIONAL SUMMARIES OF LATE-GLACIAL AEOLIAN FEATURES

     Circulation patterns persisting for thousands of years are
recorded by relict aeolian features dating from the late stages of
the Wisconsin Glaciation (Figure 5).

3.1 Southwestern United States

The southwest region is alone in supplying little evidence
of any change of surface wind direction. Although the vegetated
sand dunes on the Moenkopi Plateau, Arizona (site 1) date from
100 000 yr BP with some possibly being older (Breed & Breed 1979),
all are aligned S60°W as are adjacent modern dunes as well as the
sand rose resultant for Winslow, Arizona. In southern California,
the Kelso area of the Mojave Desert (site 2) has relict dunes that
trend S60°W and modern active dunes oriented N85°W (Sharp 1966).
This 35° shift is of inconclusive significance for the region as a
whole, owing to the intermontane setting and lack of dating control
for this dune field.

3.2 The Pacific Northwest

The Pacific Northwest of Oregon and Washington provides evidence of a 35° southerly change of surface wind direction.
Initially formed during episodes of the Bull Lake glaciation
(200 000 to 125 000 yr BP), with additional material deposited
along an identical trajectory during the recession of the subsequent
Pinedale stage at *ca.* 13 000 yr BP (Baker 1978), the Cheney-Palouse
loessial ridges (site 3) are aligned S30°W. Superimposed over the
western perimeter of the loess terrain are lobes of modern sand
dunes trending S65°W. In the Casper area of central Wyoming
(site 4), a slight southerly wind shift - displayed by stabilized
sand dunes formed prior to 10 000 yr BP by S50°W winds - has been
identified by Kolm (1974), whereas active dunes in the area trend
S70°W. Other aeolian sites in the Pacific Northwest, including
vegetated dunes in the Snake River Plain of Idaho and the
'Driftless Corridor' of Alberta match the modern circulation pattern.

3.3 The High Plains

The North American continent's most dramatic aeolian landscape, which provides the most conclusive evidence for a major
shift of surface wind direction, is to be found in the High Plains.
During parts of the late Pleistocene, the region was a huge erg
composed of parabolic dunes (see Figure 1(b)) covering river flood-
plains, with megabarchan and transverse ridge dunes overlying most
of Nebraska. Features found in the Dakotas and Montana are aligned
parallel to the direction of modern surface winds, but the trends
of late Pleistocene landforms begin to diverge by 20° to 40° over
Nebraska. Large dunes in the Alliance area of the Nebraska Sand
Hills (site 5) are oriented N35°W, while the local sand rose
resultant is N75°W. A radiocarbon date of 12 600 yr BP has been
associated with megadune stabilization within the Sand Hills (Watts
& Wright 1966). Holocene dates have also been recorded in the Sand
Hills in interdunal lake basins (Bradbury 1980), leading to the
speculation that aeolian activity also occurred during the

Altithermal warm period between 8 000 and 4 000 yr BP (Ahlbrandt
& Fryberger 1980). However, the dynamics of megadune creation
and barchanoid ridge accretion in this area tend to preclude the
formation of these large sand dunes in such a brief interval,
though aeolian activity may have extended from the late Pleistocene
into the early Holocene in some parts of the central High Plains.

To the south in Kansas, Oklahoma, New Mexico and Texas,
present-day surface winds are 70° to 90° out of phase with the
late-glacial circulation pattern. The northwesterly orientation of
relict features gradually shifts from N35°W in Nebraska to N65°W
in west Texas. Small lake basins on the Llano Estacado of west
Texas and New Mexico (site 6) preserve a long history of aeolian
activity (Reeves 1966). Most of these basins are fringed by a
series of three lunette dunes that were created successively by
deflation during pre-, mid- and late-Wisconsin arid intervals
with some dune accumulation during the Altithermal warm period.
The dunes constructed after the Tahokan pluvial period (14 000 yr
BP) are oriented N50°-60°W, approximately 70° out of phase with
modern dune-building winds. On the Pleistocene coastal plain of
south Texas (site 7), lunette dunes were formed during the late-
glacial period by N20°E winds that are 110° out of phase with the
local sand rose resultant and with the trend of currently active
inland sand dunes.

3.4 Midwest

The Midwest continues the pattern of surface wind direction
changes. Dunes (dated at 12 000 yr BP) in Minnesota (site 8),
derived from glacial outwash, are oriented 40° more to the north
than the local sand rose resultants. In Illinois and Iowa,
Woodfordian (22 000 to 12 500 yr BP) dunes and loessial ridges
(site 9) trend N65°W in contrast to the S75°W sand rose resultant
for the region (Flemal et al. 1972). There are many locations
(not shown in Figure 5) where deep (5-20 m) Peorian loess deposits
extend to the southeast of several rivers in the Mississippi
system. Loessial isopachs from the region provide additional
evidence for strong northwesterly winds at the end of the
Wisconsin Glaciation (William & Frye 1970).

3.5 Southeastern Coastal Plain

Late-glacial sand dunes on the Delmarva Peninsula of Maryland
and Delaware (site 10) were formed by N40°W winds that are 30° out
of phase with the N70°W resultant of local sand roses. Radiocarbon
dating of loess deposits in the same area confirms that aeolian
activity took place at about 10 500 yr BP (Foss et al. 1978). The
Carolina Bays of Virginia, North and South Carolina and Georgia
(site 11) trend N30°W to N50°W and are between 50° and 90° out of
phase with modern surface winds over the region. Regardless of the
processes involved in the creation of these elliptical basins (for

which many hypotheses have been offered), the existence of lunette
rims on their southeastern margins indicates deflation by northwesterly winds. The bays were formed during the Wisconsin
Glaciation (Thom 1970), several radiocarbon dates implying bay
development between 40 000 yr BP and the beginning of the Holocene.
That northwesterly winds created many late Pleistocene landforms
along the southeastern Atlantic coast from Maryland to Georgia
demonstrates a crucial change in the prevailing circulation pattern
over the region from the modern tropical airstream moving inland
from the Gulf of Mexico and Atlantic Ocean to a continental airflow crossing the Appalachian Mountains from the northwest.

4.   LATE-GLACIAL CIRCULATION OVER NORTH AMERICA

     Late-glacial aeolian activity occurred during several periods
in different regions. The interval from 17 000 to 12 000 yr BP
encompasses most of the radiocarbon dates, and these cluster
around 14 000 yr BP in a number of regions.
     During the waning stages of the Wisconsin Glaciation, the
south westerly airflow followed a more northward course than today
along the western Rocky Mountains of the Pacific Northwest before
turning sharply eastwards over Alberta and Saskatchewan. Northwesterlies dominated the eastern two-thirds of the continent. The
orientation of sand dunes in relation to the retreating Laurentide
and Cordilleran ice sheets indicates that katabatic winds blowing
off the ice were ineffective in shaping the final form of the dunes.
In the continental area south of $55°N$, there is no evidence for the
formation of dunes by surface winds associated with a glacial
anticyclone centered over the Hudson Bay region, though further
north, in the Lake Athabasca region of Saskatchewan, late-glacial
sand dunes created by easterly winds have been identified
(David 1981). Strong northwesterly winds skirting the southern
perimeter of the Laurentide ice sheet comprised the principal
feature of the circulation regime. Although it is not possible to
derive an absolute estimate of palaeowind velocities based upon
the morphology of sand dunes, the degree of unidirectionality of
the airflow can be judged. Present-day surface winds in the
Alliance, Nebraska area (see Figure 2) blow from opposing northerly
and southerly quadrants. An obtuse biomdal distribution of surface
winds of this kind would probably result in the formation of star
dunes (see Figure 1(g)) if the Nebraska Sand Hills were active at
present (Ahlbrandt & Fryberger 1980). The late-glacial megabarchans
preserved by vegetation in this region reflect a much more uni-
directional northwesterly glacial airflow.
     The dominance of westerly surface winds over the southeastern
USA leads to the conclusion that westerly circulation was greatly
strengthened during the glacial period. This intensification would
require an increased wavelength for the Rossby waves in the upper
troposphere over middle latitudes. The glacial period circulation
pattern can be reconstructed from the trends of late-glacial aeolian

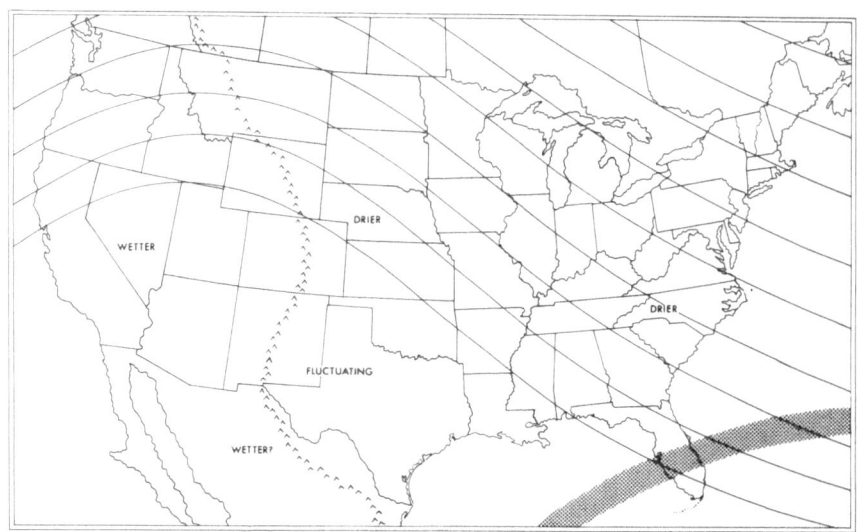

Fig. 6. Late-glacial period circulation pattern and associated
climatic conditions. Possible boundary zone of the
westerly and tropical airstreams shown by the shaded arc.

features (Figure 6). Approximately one-third of a wavelength
(stretching 3250 km) is represented between latitudes 35°N and
50°N over central North America, beginning at 49°N, 113°W in
Alberta and ending at 37°N, 76°W in Virginia. A total wavelength
of about 10 000-11 000 km would require three Rossby waves to
complete the global circulation pattern at mid-latitudes.

With the present circulation, Rossby waves numbering three
or less are found in only 2.4% of observations (5.5% for winter
months) at latitude 50°N (Winston 1960). A four- or five-wave
configuration is found in 69.8% of observations. The trends of
relict aeolian features over central North America signal a transition from the current, interglacial four- or five-wave pattern to
a dominant regime of three planetary waves in the upper troposphere
at mid-latitudes during the last glacial period.

5. PALAEOCLIMATIC IMPLICATIONS

Numerous implications for the nature of late-glacial
climate arise from the persistence of three planetary waves at
mid-latitudes (Figure 6). The vast orographically-induced trough
in the lee of the Rocky Mountains would lead to more arid conditions over the central USA from the High Plains to the Eastern
Seaboard, as is suggested by palaeoecological studies. The lack

of evidence for a mixed mesophytic forest in the southeastern
United State during the glacial maximum (Wright 1981), coupled with
the occurrence of low lake levels and the existence of dry oak-
hickory stands with local prairie vegetation in northern Florida
from 18 500 to 14 000 yr BP (Watts & Stuiver 1980), supports the
view that late-glacial aridity extended from the eastern Rocky
Mountains to the Atlantic coast.

Intensified zonal airflow might have supplied the increased
moisture to the Basin and Range Province required by the presence
of large Late Pleistocene lakes in the closed basins of California,
Oregon, Nevada and Utah. Stronger zonal circulation combined with
the possibility of deflected tropical airflow around the topo-
graphic barrier of the Sierra Madre Oriental from $22°N$ to $27°N$
might have caused greater precipitation in the interior of
northern Mexico (Street-Perrott et $al.$ 1982). Late-glacial
fluctuations in the climate of the Llano Estacado of Texas and
New Mexico (Wendorf 1961, Wendorf & Hester 1975, Reeves 1973) and
Florida (Watts & Stuiver 1980) can be explained by a periodic
weakening of the surface westerly circulation during deglaciation
allowing the reinjection of moisture into these regions by the
tropical airstream.

The presence of large late-glacial sand dunes can also serve
as a direct guide to regional changes in water balance. The
megabarchan, transverse ridges and large parabolic dunes in South
Dakota, Nebraska, Wyoming, Colorado and Kansas require a hyperarid
environment for their formation. In no areas with an annual
precipitation greater than 100 mm are there currently-active dunes
of this size (Breed et $al.$ 1979). In the central High Plains,
annual precipitation ranges from 300 mm to nearly 700 mm, so the
late-glacial contribution of meteoric water to the regional water
balance must have been only 15 to 30% of modern levels. Similarly,
the lunette dunes located on the southeastern margins of many
Carolina Bays arose from deflation under arid conditions; modern
annual precipitation averages more than 1100 mm in the area of
North and South Carolina where the lunette dunes are best preserved.
At the time the lunette dunes were formed, late-glacial annual
precipitation was probably ⩽15% of this value.

6. PLANETARY WAVES AND ICE SHEET DYNAMICS

The existence of a three-wave pattern in the circumpolar
vortex over middle latitudes has implications for the growth and
stabilization of the Laurentide ice sheet. Lamb and Woodroffe
(1970) proposed a model of the early Wisconsin circulatory
conditions required to trigger a rapid glacial advance (Figure 7).
In their reconstruction, cyclonic circulation sweeps across central
North America, dipping into the Gulf of Mexico and carrying moisture
along the Eastern Seaboard to cause greatly increased snowfall over
the eastern Canadian Arctic. Impressive support for the proposed

Fig. 7   Atmospheric circulation for a rapid glacial onset (after
         Lamb & Woodroffe 1970) showing paths of surface cyclones
         (large arrows) and surface winds (small arrows).

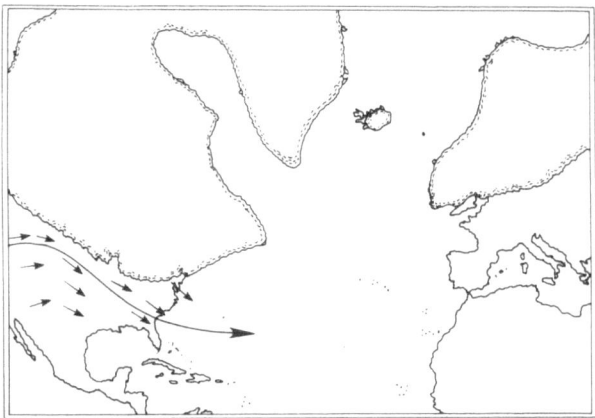

Fig. 8. Atmospheric circulation terminating ice sheet advance
         showing path of surface cyclones (large arrows) and
         surface· winds (small arrows).

circulation pattern comes from the analysis of ocean cores
recovered from the North Atlantic (Ruddiman & McIntyre 1981). The
late-glacial aeolian record demonstrates that this circulation
broke down as the ice sheet reached its maximum extent.
Intensified zonal circulation, averaging three planetary waves
over middle latitudes, would end the growth of the Laurentide ice
sheet by preventing moisture from the Gulf of Mexico and the
tropical Atlantic from reaching it (Figure 8).

A similar autodynamic termination of glacial advance appears to have occurred during earlier glaciations. On at least four occasions, Pleistocene ice sheets have covered central North America to within 150 km of latitude $38°N$. As the glaciers spread across Canada and the northern USA, they gradually induced an intensified zonal circulation pattern characterized by a low planetary wave number, thereby blocking the transport of moisture to the ice sheets from middle latitudes and bringing glacial expansion to a halt.

REFERENCES

Ahlbrandt, T.S. & Fryberger, S.G. 1980 Eolian deposits in the Nebraska Sand Hills. *U.S. Geol. Surv. Prof. Paper* 1120A, pp. 1-24.

Bagnold, R.A. 1941 *The Physics of Blown Sand and Desert Dunes,* 265 pp. London: Methuen.

Baker, V.R. 1978 Quaternary geology of the Channeled Scablands and adjacent areas. In: *The Channeled Scabland* (ed. V.R. Baker & D. Nummedal), pp. 17-35. Washington, D.C.: NASA.

Bradbury, J.P. 1980 Late Quaternary vegetation history of the central Great Plains and its relationship to eolian processes in the Nebraska Sand Hills. *U.S. Geol. Surv. Prof. Paper* 1120A, pp. 29-36.

Breed, C.S. & Breed, W.J. 1979 Dunes and other windforms of central Australia (and a comparison with linear dunes on the Moenkopi Plateau, Arizona). *NASA Spec. Publ.* 412, pp. 319-358.

Breed, C.S., Fryberger, S.G., Andrews, S., McCauley, C., Lennarta, F., Gebel, D & Horstman, K. 1979 Regional studies of sand seas, using LANDSAT (ERTS) imagery. *U.S. Geol. Surv. Prof. Paper* 1052, pp. 305-397.

Bryson, R.A. 1966 Air masses, streamlines, and the boreal forest. *Geogrl. Bull.,* 8, 228-269.

David, P.P. 1981 Stabilized dune ridges in northern Saskatchewan. *Can. J. Earth Sci.,* 18, 286-310.

Flemal, R.C., Odom, I.E. & Vail, R.G. 1972 Stratigraphy and origin of the paha topography of northwestern Illinois. *Quat. Res.,* 2, 232-243.

Foss, J.E., Fanning, D.S., Miller, F.P. & Wagner, D.P. 1978 Loess deposits of the eastern shore of Maryland. *J. Soil Sci. Soc. Am.*, 42, 329-334.

Fryberger, S.G. 1979 Dune forms and wind regime. *U.S. Geol. Surv. Prof. Paper* 1052, pp. 141-169.

Gates, W.L. 1976 Modelling of ice-age climate. *Science, N.Y.*, 191, 1138-1144.

Kolm, K.E. 1974 ERTS MSS imagery applied to mapping and economic evaluation of sand dunes in Wyoming. NAS Contract No. 5-21799: Nat. Tech. Inf. Services, 31 pp. Springfield, Virginia: US Department of Commerce.

Lamb, H.H. & Woodroffe, A. 1970 Atmospheric circulation during the last ice age, *Quat. Res.*, 1, 29-58.

Manabe, S. & Hahn, D.G. 1977 Simulation of the tropical climate of an ice age. *J. geophys. Res.*, 82, 3889-3911.

Reeves, C.C., Jr 1966 Pluvial lake basins of West Texas. *J. Geol.*, 74, 269-291.

Reeves, C.C., Jr 1973 The full-glacial climate of the southern High Plains, West Texas. *J. Geol.*, 81, 693-704.

Ruddiman, W.F. & McIntyre, A. 1981 Oceanic mechanisms for amplification of the 23 000-year ice-volume cycle. *Science, N.Y.*, 212, 617-627.

Sharp, R.P. 1966 Kelso Dunes, Mojave Desert, California. *Bull. Geol. Soc. Am.*, 77, 1045-1074.

Street-Perrott, F.A., Roberts, N. & Metcalfe, S.E. 1982 Late Quaternary climatic fluctuations in the northern hemisphere tropics - some geomorphic implications. In: *Geomorphology and Environmental Change in Tropical Latitude* (ed. I. Douglas). London: Allen and Unwin. [In the press.]

Thom, B.G. 1970 Carolina Bays in Horry and Marion Counties, South Carolina. *Bull. geol. Soc. Am.*, 81, 783-814.

Watts, W.A. & Stuiver, M. 1980 Late Wisconsin climate of northern Florida and the origin of species-rich deciduous trees. *Science, N.Y.*, 210, 325-327.

Watts, W.A. & Wright, H.E., Jr. 1966 Late-Wisconsin pollen and seed analysis from the Nebraska Sand Hills. *Ecology*, 47, 202-210.

Wendorf, F. 1961 *Palaeoecology of the Llano Estacado*, 144 pp. Santa Fe, New Mexico: Museum of New Mexico Press.

Wendorf, F. & Hester, J.J. 1975 Late Pleistocene environments of the southern High Plains. *Fort Burgwin Res. Center Publ.* 9, 290 pp. Santa Fe, New Mexico: Museum of New Mexico Press.

Williams, J., Barry, R.G. & Washington, W.M. 1974 Simulation of the climate of the last glacial maximum using the NCAR global circulation model. *J. appl. Met.*, 13, 305-317.

Willman, H.B. & Frye, J.C. 1970 Pleistocene stratigraphy of Illinois. *Illinois State Geol. Surv. Bull.* 94, 204 pp.

Winston, J.S. 1960 Some new data on the longitudinal dimensions of planetary waves. *J. Met.*, 17, 522-531.

Wright, H.E., Jr 1981 Vegetation east of the Rocky Mountains 18,000 years ago. *Quat. Res.*, 15, 113-125.

FLUCTUATIONS IN CLOSED-BASIN LAKES AS AN INDICATOR OF PAST ATMO-SPHERIC CIRCULATION PATTERNS

F. Alayne Street-Perrott

School of Geography
Mansfield Road
Oxford OX1 3TB, UK

Neil Roberts

Department of Geography
Loughborough University of Technology
Loughborough LE11 3TU, UK

ABSTRACT. At the present day, maxima in the areal extent of lakes are associated with the latitudinal mean positions of the equatorial trough and the mid-latitude westerlies. This association is applied to variations in the relative extent of lakes during the last 18 000 years in order to deduce past changes in atmospheric circulation. The analysis is based on a large number of ^{14}C-dated lake-level sequences from Africa and western Eurasia. We infer that the period following the glacial maximum at $ca.$ 18 000 yr BP was characterized by increasing aridity in the tropics, culminating at around 14 000 - 12 500 yr BP when atmospheric moisture convergence and runoff reached minimum values over almost the entire study area. Between 12 500 and 5000 yr BP, a broad belt of expanded lakes developed in Africa and Arabia. It migrated irregularly northwards to a latitude of 16.5°N and then disintegrated rapidly after 5000 - 4000 yr BP. This early post-glacial intensification of the continental hydrological cycle is attributed primarily to an enhanced monsoonal circulation over Africa and Arabia, probably accompanied by a northward shift of the mean ITCZ. It was interrupted by two episodes of prolonged drought centred on 10 200 and 7400 yr BP. The causes of these abrupt climatic events are discussed.

A. Street-Perrott et al. (eds.), Variations in the Global Water Budget, 331–345.
Copyright © 1983 by D. Reidel Publishing Company.

1.   INTRODUCTION

Lake-level fluctuations are one of the most abundant and
widely distributed sources of palaeohydrological and palaeoclimatic information for low-latitude continental areas over the
late Quaternary. The spatial distribution and relative extent of
lakes at different times have been used to reconstruct atmospheric
circulation regimes for the period during and after the last
global ice-volume maximum (*ca*. 18 000 yr BP), with particular
reference to Africa and western Eurasia (20°W - 60°E and 35°S -
40°N). This time span and geographic area were selected because
they have yielded the best coverage of sites with reliable
chronologies and good stratigraphic resolution; indeed, the
results appear to be representative of the tropics and subtropics
as a whole (Street-Perrott *et al*. 1983).

The analysis is based on the present-day relationship between
the global distribution of lakes and the major features of the
general circulation. The distribution of lakes as a proportion of
continental surface area (Figure 1) shows distinct latitudinal
maxima at the equator and near 45°N and S (Schuiling 1977). These
coincide with zonal peaks in mean annual runoff, P-E, over the
continents, where P is mean annual precipitation and E is evapotranspiration (Baumgartner & Reichel 1975). The three runoff
maxima correspond to the equatorial trough and to the mid-latitude
westerlies. However, the amplitude of the peak in lake area at
45°N is anomalously great relative to the mid-latitude rainfall
and runoff maxima in the northern hemisphere (Jaeger: this
volume). This is partly attributable to the existence at this
latitude of the very large closed catchments of the Caspian and
Aral Seas, but may also reflect the abundance of lake basins
created by northern hemisphere ice sheets. Zonal minima in the
extent of lakes occur between 20° and 35°N and S in the subtropical anticyclone belts and are related to the northern and
southern hemisphere subtropical runoff minima. Closed lakes,
which can only occur where open-water evaporation exceeds precipitation, are concentrated today on either side of the two
subtropical P-E minima.

Analysis of the water balance of the atmosphere (Peixóto &
Oort: this volume) indicates that the observed spatial variations
in the relative extent of lakes (Figure 1) ought also to be
related directly to the atmospheric moisture convergence over the
continents, through the relationship of the latter to P-E (Peixóto
& Oort: this volume).

At the present day, the intensity and mean latitudes of the
equatorial trough and subtropical anticyclone belts vary seasonally
as functions of the tropospheric equator-pole temperature gradient
in each hemisphere and the thermal contrast between the two hemispheres (Korff & Flohn 1969, Flohn 1978). On the glacial/inter-
glacial time scale, comparable adjustments of the two hemispheric
circulations to fluctuations in the extent of ice and snow at

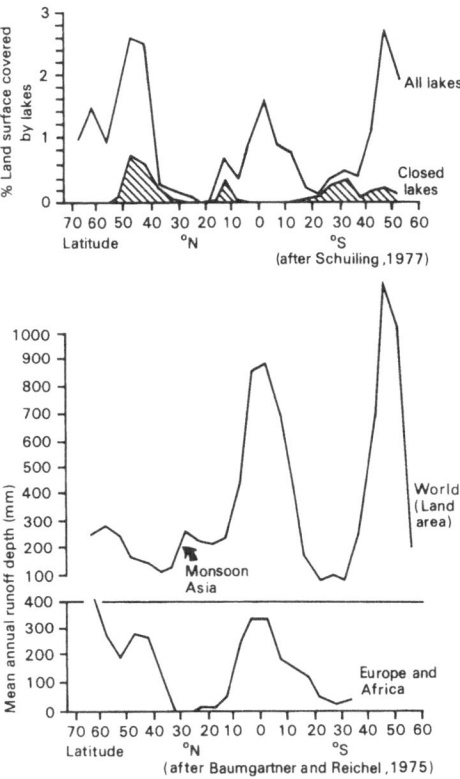

Fig. 1.   Latitudinal variations in: (a) Areal extent of lakes as
a proportion of land surface (after Schuiling 1977);
(b) Mean annual runoff for the global land area and for
Europe and Africa (after Baumgartner & Reichel 1975).

high latitudes seem likely to have influenced both the strength
and mean position of the major circulation features which, in turn,
should have resulted in changes in the amplitude and spacing of
the lake-area maxima and minima shown in Figure 1.

2.   METHODS USED

Following previous work by Street and Grove (1979), published
data have been compiled on water levels, at 1000-yr intervals, in
lakes which have been closed, i.e., in water bodies that were
particularly sensitive to climatic variations during all or part
of their late Quaternary history. Relative water depth, which is
easier to evaluate from stratigraphic and palaeoecological data,
is used as a surrogate for the climatically-determined variable,
lake area. Only ^{14}C-dated lake-level sequences have been included

in the analysis. Following Street and Grove (1979), the relative
water depth in each basin has been divided into three classes,
each of which has a broadly similar frequency of occurrence in
the data set, namely:

low: 0 - 15% of total vertical range of fluctuation,
                  including dry lakes
intermediate: 15 - 70%
high: 70 - 100%, including overflowing lakes.

The analysis was made more precise than in our earlier studies
by evaluating lake-level status for each basin at specific times
(e.g., at 1000 yr BP or 2000 yr BP) rather than by taking the
highest water level recorded within each 1000-yr interval. A
further improvement was the inclusion of data on the direction of
change in the level of a lake, henceforth referred to as lake-
level trend. Lake-level trend, like lake-level status, is
assessed for specific time horizons and is classified as either
up, stable, down or unclassifiable (no data).

Data on lake-level status and trend at 1000-yr intervals have
been stored in a computer data bank, together with about 1500
associated radiocarbon dates, from nearly 200 lake basins - an
increase of about 50% compared with the previous compilation
(Street & Grove 1979).

3.   SPATIAL AND TEMPORAL STRUCTURE OF LAKE LEVELS SINCE 18 000 YR
      BP

The lake-level record since the last global ice-volume
maximum, at *ca.* 18 000 yr BP, can be divided into phases of
relative stability separated by rapid transitions that were
approximately synchronous in time over wide areas, although the
direction of change varied with latitude and, sometimes, with
elevation. Throughout Africa and western Eurasia, five major time
divisions (called A-E) can be recognized. In some parts of intertropical Africa, phase B (10 000 - 5000 yr BP) can be subdivided
into phases B1 and B2 separated by a marked drop in lake levels.
The pattern of lake levels in the respective phases A-E is
described below. (For a more complete series of maps, see Street-
Perrott *et al.* 1983.)

3.1 Phase E (pre-17 000 yr BP)

At 18 000 yr BP (Figure 2a), lakes stood at high levels around
the southern and eastern margins of the Mediterranean and in the
Arabian peninsula. In contrast, water levels in Africa were low
or intermediate and falling, apart from a few equatorial and
montane lakes. The dearth of data points available for the
18 000 yr BP horizon reflects the intensity and widespread nature
of intertropical aridity.

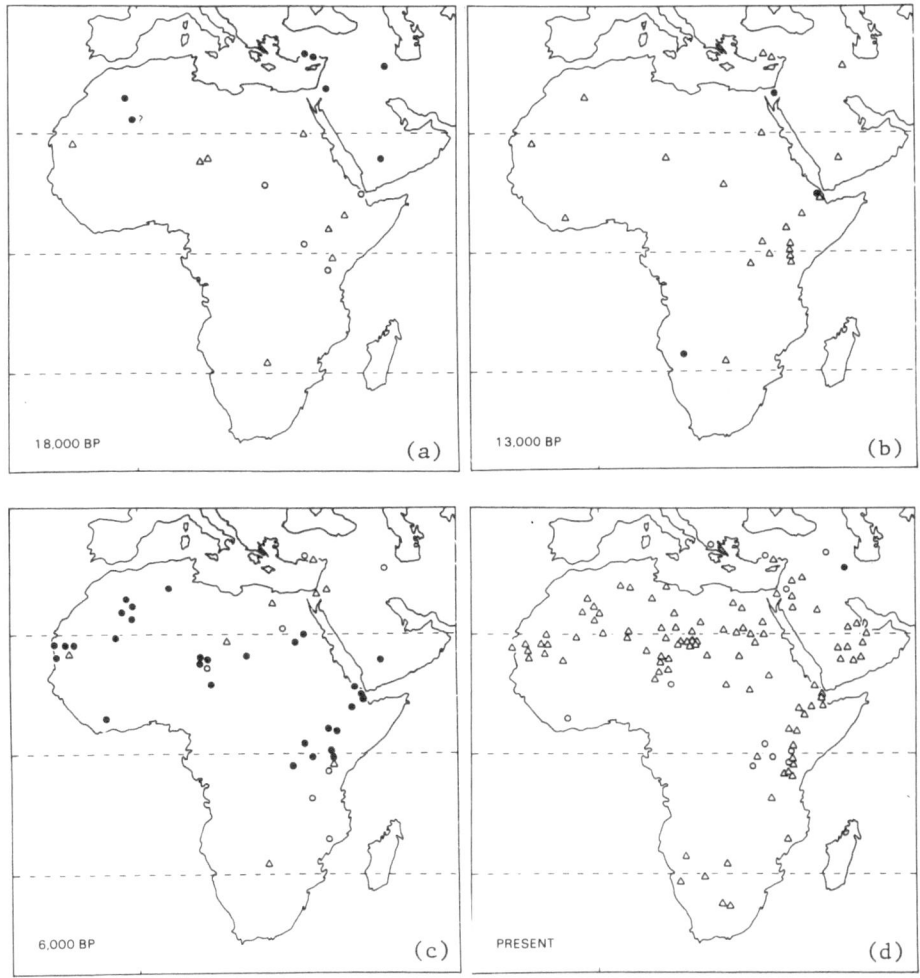

Fig. 2.   Lake-level status at (a) 18 000 yr BP; (b) 13 000 yr BP;
          (c) 6000 yr BP; (d) at the present day: ●, high; ○,
          intermediate; △, low.

3.2   Phase D (17 000 - 12 500 yr BP)

     After *ca.* 17 000 yr BP, the Mediterranean belt of high lake-
levels narrowed and moved southward into North Africa and the
southern Levant. Within the intertropical zone, a prolonged
period of minimal lake levels commenced, with the important
exception of two equatorial lakes (Manyara and Mobutu Sese Seko).
Around 16 000 - 15 000 yr BP, indications of enlarged lakes appear

briefly south of 23°S. The pattern of lake levels at this juncture
shows the most striking degree of hemispheric symmetry to be found
within the whole period of the record. By 14 000 - 13 000 yr BP,
low lake-levels were recorded throughout almost the whole study
area, except for Sinai, Tibesti and southwestern Africa (Figure
2b). The interval 14 000 - 12 500 yr BP represents one of the
most arid periods in the entire late Quaternary, drier even than
at the glacial maximum.

3.3 Phase C (12 500 - 10 000 yr BP)

Phase C was a transitional period in the lake-level record.
Many lakes began to rise between 12 500 and 11 500 yr BP, only to
fall again from 10 800 to 10 200 yr BP. The greatest concentration
of lakes experiencing this terminal Pleistocene water-level maximum
was in the equatorial zone, but lake basins as far apart as 20°S
and 23°N were affected. The African belt was separated from
another area of higher water levels in Turkey (ca. 37°N) by a zone
of lakes with low water levels.

3.4 Phase B1 (10 000 - 7500 yr BP)

There was a major change in the pattern of lake-level status
around 10 000 yr BP, almost all the intertropical lakes having
reached high levels by 9000 yr BP. Maps of lake-level trend
indicate that water levels responded first near the equator and
subsequently rose progressively northwards towards the central
Sahara. At its maximum, the belt of lakes with high water-levels
extended from 4°S to 33°N. However, east of 25°E and north of 26°N,
there was a southward-projecting cusp of lakes with low water-
levels, which clearly defines the northern limit of the early Holocene wet phase in the eastern Mediterranean. The interval 8000 -
7500 yr BP witnessed an abrupt regression of many African lakes.

3.5 Phase B2 (7500 - 5000 yr BP)

The levels of the intertropical lakes which fell at the end
of phase B1 rose again between ca. 7300 and 6800 yr BP and then
generally remained high until around 5000 yr BP. An extensive
belt of lakes with high water-levels developed across Africa,
stretching from 2°S to 32°N, but once more failed to reach the
eastern Mediterranean sector. This pattern, and the zone of
dominantly low or intermediate lake levels south of 2°S, are well
illustrated by the map for 6000 yr BP (Figure 2c).

3.6 Phase A (5000 yr BP to present)

The African belt of high lake-levels disintegrated rapidly
after 5000 yr BP, a modern pattern of lake levels being
established by 2500 yr BP. The present pattern (Figure 2d)

includes no tropical lakes with high water-levels and only a few
of intermediate status, clustered in a narrow band between 2°S
and 13°N which spans the present meteorological equator (ca. 6°N).
All other African lakes are now at low levels (or dry), although
some lakes with high or intermediate levels have persisted in the
eastern Mediterranean region.

4.   ANALYSIS OF THE PATTERN OF LAKE HIGH STANDS AND REGRESSIONS
      (12 500 - 5000 YR BP)

      It is clear that the terminal Pleistocene to mid-Holocene
phases of high lake-levels in Africa and Arabia were the product
of climatic conditions significantly different from those of the
present day (cf. Kutzbach: this volume). Analysis of the 'fine
structure' of the period 12 500 - 5000 yr BP allows the changes
of mode in the atmospheric circulation responsible for the
observed palaeohydrological patterns to be identified with some
precision.
      Each of the three phases C, Bl and B2 included a major lake
high stand which can be differentiated from the other two on the
basis of its latitudinal range and the relative response of
different lake basins. The thirteen African lakes known to have
experienced high levels between 12 500 and 10 000 yr BP (phase C)
have a mean latitude of 4.7°N (Figure 3). This value is not
significantly different from the centre of gravity of the modern
belt of intermediate-status lakes (4.1°N), although there are no
data for phase C from 5 - 15°S. Figure 4 shows that all of the
basins in which phase C was the most important post-glacial high
stand lie south of 1°N.
      Basins in Africa and western Eurasia with high lake-levels
during phases Bl and B2 were not only more numerous, but their
distribution was centred significantly further north than in phase
C (Figure 3). Even after allowance is made for bias due to the
variation in land area with latitude, the mapped patterns still
imply a substantial broadening as well as a northward shift of the
intertropical runoff peak between 12 500 and 5000 yr BP. This
conclusion is supported by the fact that many presently closed
lakes between 2°S and 18°N overflowed between 9000 and 8000 yr BP,
and again around 6000 - 5000 yr BP.
      If the two Holocene phases Bl and B2 are compared, the former
seems to have been the more strongly marked hydrological event.
Most of the lakes reached their highest post-glacial levels during
phase Bl (Figure 4). There is some, albeit inconclusive, evidence
that phase B2 had its greatest impact slightly further north than
phase Bl. Although the mean latitude of the belt of lakes with
high levels in phase B2 (16.5°N) is not significantly different
from the average for phase B2 (14.2°N), all of the sites where B2
was the most important of the three phases are located north of
7°N (Figure 4).

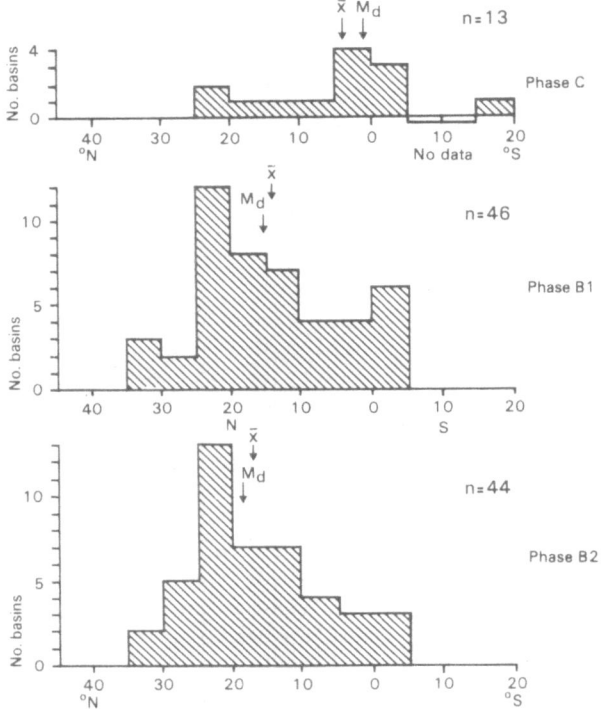

Fig. 3.   Latitudinal distribution of lakes with high levels in
Africa and western Eurasia during phases C, B1 and B2,
showing the mean ($\bar{x}$) and median (M_d) of each distribution.

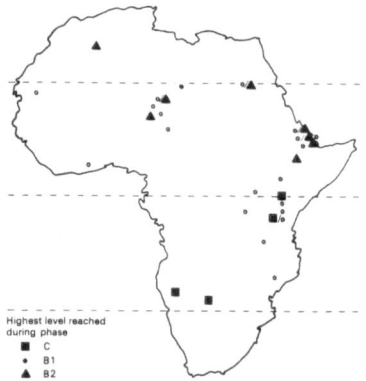

Fig. 4.   Ages of the most important post-glacial high stand in
African basins.

As noted above, the three major high stands were separated by short periods in which the belt of enlarged lakes temporarily collapsed. Close examination of the dating evidence from a number of well-studied basins shows that these dramatic regressions culminated at *ca*. 10 200 yr BP (C/B1) and *ca*. 7400 yr BP (B1/B2) (Figure 5). The distribution of lake basins affected covers the entire northern intertropical zone and probably part of the southern intertropical zone (Figure 6), suggesting that the regressions represent a response of *both* hemispheric circulations to short-lived global fluctuations in climate.

5.   COMPARISON OF PAST ATMOSPHERIC CIRCULATION PATTERNS BASED ON LAKE-LEVEL EVIDENCE

5.1   Glacial (18 000 - 12 500 yr BP)

During the last glacial maximum, the northern mid-latitude runoff peak (Figure 1) appears to have migrated southwards to affect the zone between 20 and 40°N. This shift can be attributed largely to a more southerly position of the westerly depression tracks, but may also partly reflect lower temperatures and, hence, lower evaporation during glacial times than at the present day. From 18 000 yr BP onwards, the scarcity of lakes with high water-levels suggests that the westerly rainfall belt was increasingly weakened and 'pinched'. We estimate that its mean position at the end of phase D was about 30°N. Lake-level evidence from southern Africa suggests that cyclonic rains penetrated as far north as 23°S around 16 000 - 15 000 yr BP.

The intertropical runoff peak appears to have been reduced in width or intensity from 21 000 to 12 500 yr BP, and probably occupied a position at, or just south of, the equator. This displacement can be attributed to a southerly shift or narrowing of the upward-moving branch of the Hadley circulation. High lake-levels persisted relatively late (till 17 000 yr BP) in areas fed by runoff from the highlands, adjacent to the Red Sea, that are affected by cyclonic winter rains today.

5.2   Early Post-glacial (12 500 - 5000 yr BP)

From *ca*. 12 500 yr BP onwards, there was an unsteady north-ward migration and broadening of the intertropical runoff peak, suggesting an enhanced monsoonal circulation over the African and Arabian continents, probably accompanied by a northward shift in the mean ITCZ. The northern mid-latitude runoff peak was also displaced further north during the early and mid-Holocene. The driest part of the study area north of the equator during phases B1 and B2 was the NE margin of the present subtropical arid zone, around 30 to 40°N. A possible explanation for this 'cusp' of low lake-levels is given by Kutzbach (this volume). However, aridity

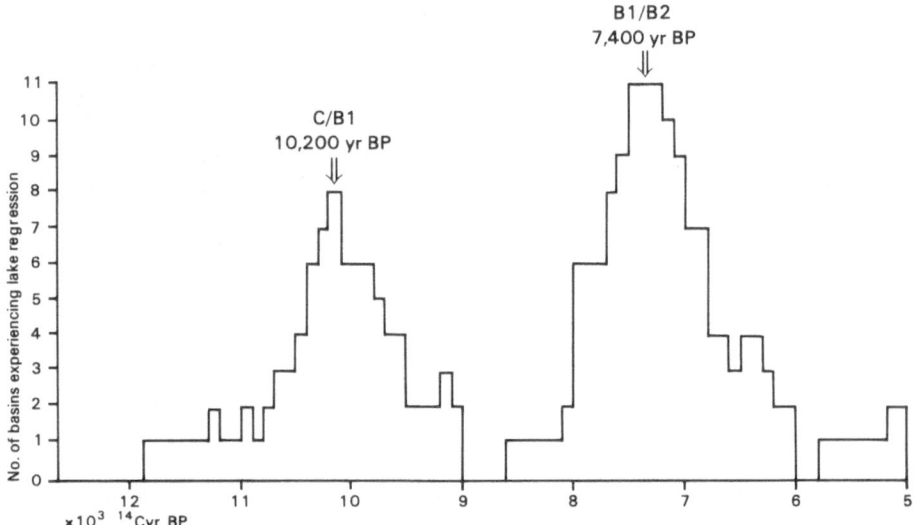

Fig. 5.   Chronological distribution of African lakes experiencing
regressions between 12 000 and 5000 yr BP.

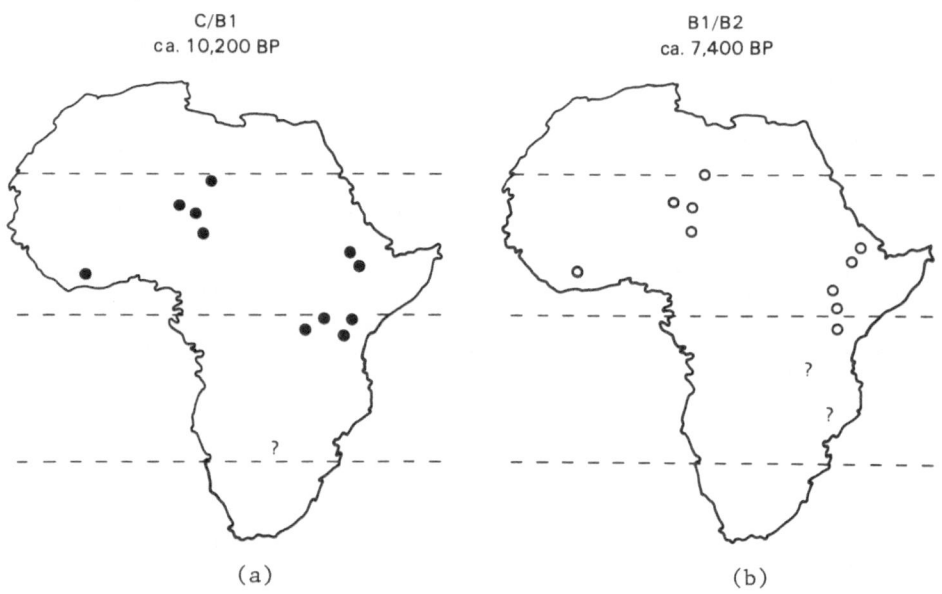

Fig. 6.   Geographical distribution of African lakes experiencing
regressions at: (a) *ca.* 10 200 yr BP (phases C/B1);
(b) *ca.* 7400 yr BP (phases B1/B2).

was nowhere as intense as in today's hyper-arid core to the south.
Evidence from southern African lakes is insufficient in both
quantity and quality to infer Holocene changes in the southern
hemisphere circulation.

6.   CAUSES OF THE LAKE-LEVEL REGRESSIONS: SOME SPECULATIONS

     During the interval 12 500 - 5000 yr BP, the otherwise
orderly northward migration of the African belt of lakes with high
levels was twice interrupted by episodes in which the pattern
abruptly disintegrated. A similar, and even more general,
expansion of arid conditions occurred during the period 14 000 -
12 500 yr BP, when moisture convergence over the entire Afro-
Arabian landmass must have been strongly suppressed. According
to Flohn (1979), abrupt climatic events of this type, on a time
scale of 10^2 - 10^3 yr, are likely to result from (a) solar events,
(b) clustering of explosive volcanic eruptions or (c) surges of
major ice sheets. To this list one could add (d) pulses of meltwater release into the oceans during glacial retreat (Berger et al.
1977, Flohn & Nicholson 1980, Ruddiman & McIntyre 1981).
     Of the four mechanisms cited above, oceanic meltwater 'spikes'
and episodic volcanism appear to provide the most plausible
explanation for the observed lake-level minima. The evidence for
cyclic variations in solar emission with a frequency of 2500 yr
(Denton & Karlen 1973, Pisias et al. 1973), and for Antarctic
surges (Hollin 1969, Wilson 1978) is still controversial and does
not show any convincing relationship to the events in question.
On the other hand, the detailed examination of foraminiferal and
oxygen-isotope records from deep-sea cores, allowing for sediment
mixing by bioturbation, has produced clear evidence of strong
pulses of meltwater release into the northern North Atlantic
during ice-sheet break-up (Ruddiman & McIntyre 1982). The most
important of these occurred around 16 000 - 13 000 yr BP, when
the entire North Atlantic south to at least 50°N was flooded by
meltwater and icebergs in summer and probably covered by sea ice
in winter.
     A weaker pulse of meltwater release at ca. 11 000 - 10 000 yr
BP (= Younger Dryas) has been attributed, by Ruddiman and McIntyre
(1982), to the break-up and outflow of large ice shelves in the
Arctic. A third peak of meltwater release might be expected to
have accompanied the rapid disintegration of the Laurentide ice
sheet after ca. 7900 yr BP, when the sea invaded Hudson's Bay
(Denton & Hughes 1981). However, the magnitude of the change in
ice volume remains controversial and the impact on the North
Atlantic was apparently slight and limited to the Labrador Sea
(W.F. Ruddiman, personal communication).
     It seems, therefore, that both the arid period culminating
around 14 000 - 13 000 yr BP and the temporary collapse of the
belt of high lake-levels between 10 800 and 10 200 yr BP coincided

with episodes of strong cooling of the North Atlantic by icebergs and meltwater, while the abrupt recession centred on 7400 yr BP may possibly have been linked to the evacuation of ice from Hudson's Bay.

These lake-level regressions also appear to have been preceded by important volcanic episodes. In a recent compilation of late Quaternary eruptions, Bryson and Goodman (1980) concluded that major peaks of volcanism occurred around 14 500, 11 500, 8500 - 8300 and 3700 yr BP, and that there was a sudden increase in the general level of volcanic activity at about 4900 - 4800 yr BP. All these episodes, except possibly that at 3700 yr BP, were followed by a major fall in lake levels in tropical Africa. However, the location, magnitude and explosivity of the eruptions concerned (Simkin *et al.* 1981) needs to be examined in more detail in order to explain the lack of agreement with Bray (1974) and Hammer *et al.* (1980) who did not find the same clusters of volcanic events in the record from the southern continents and Greenland.

7.   CONCLUSIONS

Lake-level data for the period 18 000 yr BP to the present day reveal large fluctuations in the water balance of Africa and western Eurasia on a time scale of $10^2 - 10^3$ yr. The last glacial maximum, at *ca.* 18 000 yr BP, was characterized by arid conditions over most of tropical Africa, despite the substantial cooling demonstrated by Livingstone (1980). Our reconstruction agrees quite well with the modelling experiments by Gates (1976) and Manabe and Hahn (1977), but only poorly with the pioneering GCM simulation by Williams *et al.* (1974). The maximum extent of aridity lagged behind the global ice-volume maximum by *ca.* 5000 years. Indeed, the driest conditions of the whole late Quaternary occurred around 14 000 - 12 500 yr BP. This arid episode followed a peak of explosive volcanism at *ca.* 14 500 yr BP and may also have been linked to the spread of a meltwater layer across the mid-latitude oceans. Flohn (this volume) has suggested a feedback mechanism by which such perturbations in the global energy budget might induce marked fluctuations in atmospheric moisture inflow and runoff in tropical and subtropical landmasses.

The lake-level data provide strong support for a northward migration of the mean ITCZ during the period 12 500 - 5000 yr BP, as proposed by Nicholson and Flohn (1980). However, the actual displacement may have been less than the apparent latitudinal shift of 10 - 12° in the centre of gravity of the zone of enlarged lakes, because of two important climatic feedbacks. First, modelling experiments with a low-resolution GCM suggest that the maximum of summer radiation around 9000 yr BP in the northern hemisphere would have significantly enhanced the summer monsoon rains over northern intertropical Africa and Eurasia (Kutzbach:

this volume). Secondly, the northward shift in the mean ITCZ
inferred here would also have led to more frequent interaction
between the low-latitude circulation and the troughs in the upper
westerlies, thus giving rise to increased rainfall over North
Africa (Maley 1977, Nicholson & Flohn 1980).

The late-glacial to early post-glacial phase of high lake-
levels was interrupted by two dramatic regressions which
culminated around 10 200 and 7400 yr BP. These were apparently
preceded by important volcanic episodes and by pulses of meltwater release from the major ice sheets.

The data available at present provide, therefore, considerable
support for the hypothesis that both the ITCZ and the mid-latitude
westerlies have undergone significant changes in mean latitude and
intensity over late Quaternary time. Second-order spatial
patterns evident in the lake-level data may reflect the influence
of orography and changes in the location and intensity of
individual pressure centres.

ACKNOWLEDGEMENTS. We thank R.A. Bryson, S.P. Harrison, J.E.
Kutzbach, S.E. Nicholson, W.F. Ruddiman and T. Webb III for their
helpful advice and comments. Financial support for this research
was provided by U.S. Department of Energy contract No.
DE-ACO2-79EV10097 to T. Webb III at Brown University.

REFERENCES

Baumgartner, A. & Reichel, E. 1975 *The World Water Balance – Mean
    Annual Global, Continental and Maritime Precipitation*, 179 pp.
    Amsterdam: Elsevier.

Berger, W.H., Johnson, R.F. & Killingley, J.S. 1977 "Unmixing" of
    the deep-sea record and the deglacial meltwater spike.
    Nature, Lond., 269, 661–663.

Bray, J.R. 1974 Volcanism and glaciation during the past 40
    millenia. , *Nature, Lond.*, 252, 679–680.

Bryson, R.A. & Goodman, B.M. 1980 Volcanic activity and climatic
    changes. *Science, N.Y.*, 207, 1041–1044.

Denton, G.H. & Hughes, T.J. 1981 *The Last Great Ice Sheets*, 484 pp.
    New York: Wiley-Interscience.

Denton, G.H. & Karlen, W. 1973 Holocene climatic fluctuations;
    their pattern and possible cause. *Quat. Res.*, 3, 155–205.

Flohn, H. 1978 Comparison of Antarctic and Arctic climate and its relevance to climatic evolution. In: *Antarctic Glacial History and World Palaeoenvironments* (ed. E.M. Van Zinderen Bakker Sr), pp. 3-13. Rotterdam: Balkema.

Flohn, H. 1979 On time scales and causes of abrupt palaeoclimatic events. *Quat. Res.*, 12, 135-149.

Flohn, H. & Nicholson, S. 1980 Climatic fluctuations in the arid belt of the 'Old World' since the last glacial maximum; possible causes and future implications. *Palaeoecol. Africa*, 12, 3-21.

Gates, W.L. 1976 The numerical simulation of ice-age climate with a global general circulation model. *J. atmos. Sci.*, 33, 1844-1873.

Hammer, C.U., Clausen, H.B. & Dansgaard, W. 1980 Greenland ice sheet volcanism and its climatic impact. *Nature, Lond.*, 288, 230-235.

Hollin, J.T. 1969 Ice-sheet surges and the geological record. *Can. J. Earth Sci.*, 6, 903-910.

Korff, H.C. & Flohn, H. 1969 Zusammenhang zwischen dem Temperatur-Gefalle Aquator-Pol und den planetarischen Luftdruckgurteln. *Annln. Met., Hamburg*, NF4, 163-164.

Livingstone, D. 1980 Late Quaternary pollen from Lake Bosumtwi, Ghana. *Abstracts, 5th Int. Palynol. Conf. Cambridge, UK.*

Maley, J. 1977 Palaeoclimates of Central Sahara during the early Holocene. *Nature, Lond.*, 269, 573-577.

Manabe, S. & Hahn, D.G. 1977 Simulation of the tropical climate of an ice age. *J. geophys. Res.*, 82, 3889-3911.

Nicholson, S.E. & Flohn, H. 1980 African environmental and climatic changes and the general atmospheric circulation in Late Pleistocene and Holocene. *Clim. Change*, 2, 313-348.

Pisias, N.G., Dauphin, J.P. & Sancetta, C.D. 1973 Spectral analysis of late Pleistocene-Holocene sediments. *Quat. Res.* 3, 3-9.

Ruddiman, W.F. & McIntyre, A. 1981 Oceanic mechanisms for amplification of the 23 000-year ice-volume cycle. *Science, N.Y.*, 212, 617-627.

Ruddiman, W.F. & McIntyre, A. 1982 The North Atlantic during the last deglaciation. *Palaeogeogr. Palaeoclimatol. Palaeoecol.* [In the press.]

Schuiling, R.D. 1977 Source and composition of lake sediments. In: *Interactions between Sediments and Freshwater* (ed. H.L. Golterman), pp. 12-18. Wageningen: PUDOC.

Simkin, T., Siebert, L., McClelland, L., Bridge, D., Newhall, C. & Latter, J.H. 1981 *Volcanoes of the World: A Regional Directory, Gazetteer, and Chronology of Volcanism during the Last 10 000 years*, 232 pp. Stroudsburg, Penn: Hutchinson Ross.

Street, F.A. & Grove, A.T. 1979 Global maps of lake-level fluctuations since 30 000 BP. *Quat. Res.*, 12, 83-118.

Street-Perrott, F.A., Roberts, N. & Metcalfe, S.E. 1983 Geomorphic implications of Late Quaternary hydrological and climatic changes in the northern hemisphere tropics. In: *Geomorphology and Environmental Change in Tropical Latitudes* (ed. I. Douglas). George Allen and Unwin. [In the press.]

Williams, J., Barry, R.G. & Washington, W.M. 1974 Simulation of the atmospheric circulation using the NCAR global circulation model with ice age boundary conditions. *J. appl. Met.*, 13, 305-317.

Wilson, A.T. 1978 Past surges in the West Antarctic ice sheet and their climatological significance. In: *Antarctic Glacial History and World Palaeoenvironments* (ed. E.M. Van Zinderen Bakker Sr), pp. 33-39. Rotterdam: Balkema.

PRESENT-DAY AND EARLY-HOLOCENE EVAPORATION OF LAKE CHAD

G. Tetzlaff and L.J. Adams

Institut für Meteorologie und Klimatologie
Universität Hannover
Federal Republic of Germany

ABSTRACT. A simple model combining the surface energy balance and water balance of a closed lake was used to compute the mean water temperature and evaporation, assuming complete mixing of the water body. Lake Chad, in the Sahel zone of Africa, was taken as an example as it has experienced large climatic variations. The lake level was successfully simulated for the period of the Sahel drought. The calculations show that the fall in the level of Lake Chad resulted mainly from a substantial decrease in runoff rather than from increased evaporation. A data set for the early Holocene (9000 yr BP) indicated a 15% decrease in annual evaporation compared with the present-day value of 2200 mm yr^{-1}. A sensitivity analysis, used to test the compatibility of the data, showed an increase of about 1200% in the total annual inputs from runoff and precipitation during the early Holocene (compared to the present).

1. INTRODUCTION

Closed lakes have long been recognized as good indicators of climatic change (Langbein 1961, Richardson 1969). Most approaches coupling lake level with climatic conditions make use of the water balance of the lake and its surface heat balance (Kutzbach: this volume). The individual terms in the water balance equation are time dependent:

$$R/A + (P-E) - I = \Delta V \tag{1}$$

In Equation (1), the total runoff R from the catchment area is regarded as a point source, whereas the precipitation P onto the lake and the evaporation E depend on the magnitude of the lake's

347

A. Street-Perrott et al. (eds.), Variations in the Global Water Budget, 347–360.

surface area A. In the case of Lake Chad, the loss by seepage I
is assumed to be proportional to the area of the lake. Hence,
changes of the water volume ΔV enable conclusions to be drawn
about climatic processes, provided that the transfer function is
known.

If no changes in water volume occur, Equation (1) can be used
to define the equilibrium area A_E, where

$$A_E = R/(E+I-P) \tag{2}$$

For monthly mean values of R, E, I and P, the equilibrium area A_E
differs from the real surface area A, whereas both A_E and A are of
the same order of magnitude for annual means close to the long-term
average of the period considered.

The surface energy balance of a lake is given by:

$$\Delta S = Q+H+LE+B-\varepsilon\sigma T^{*4}_L \tag{3}$$

where ΔS = net heat storage in the water body,
    Q = absorbed incoming radiation,
    H = flux of sensible heat,
    LE = flux of latent heat,
    L = heat of vaporization of water,
    B = heat flux into the ground,
$\varepsilon\sigma T^{*4}_L$ = long-wave outgoing radiation from the lake surface at
         temperature T^*_L in K, with

    ε = emissivity of the water = 0.97,
    σ = Stefan-Boltzmann constant = 5.67×10^{-8} W m^{-2} K^{-1}.

These two fundamental equations were applied to determine the
sensitivity of a lake to climatic variations. For this purpose,
Lake Chad was selected. Equations (1) and (3) contain more than
two unknown parameters. In most cases, no direct information on
the variables are obtainable. Therefore, these unknowns have to
be replaced by available atmospheric and hydrological data.

2. LAKE CHAD

Lake Chad is situated in the Sahel zone (Figure 1), which is
very sensitive to climatic variations (Nicholson: this volume).
From October to April, dry air masses from the interior of the
Sahara prevail with northeasterly winds while, during the rest of
the year, moist southwesterlies bring summer rains to the area
south of the monsoonal front (Figure 2). In the early 1960s, Lake
Chad reached a volume of nearly 8×10^{10} m^3 and covered an area of
about 20 000 km^2 (Rognon 1980) with a mean depth of about 4 m (Maley
1973a). Because of the shallowness of the lake and the many swamps
and islands, the bathymetric curve is rather complex. The simpli-

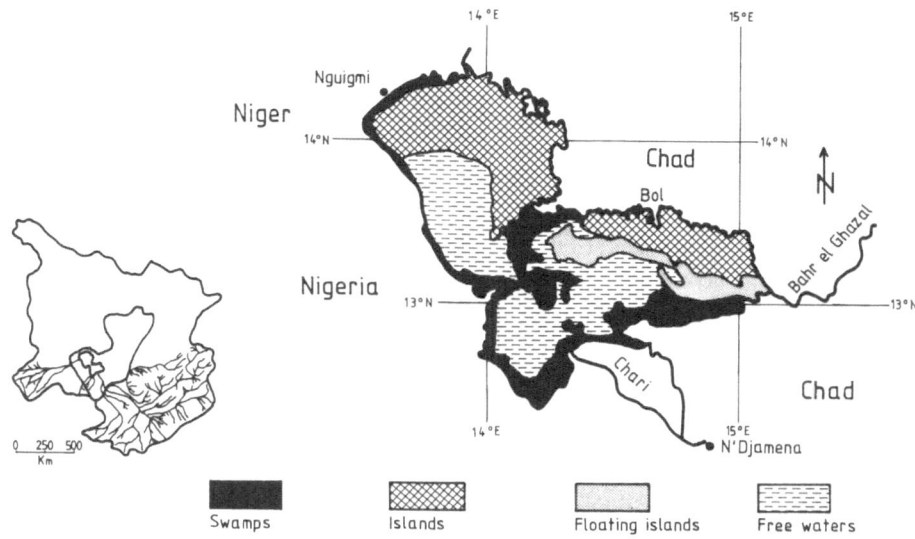

Fig. 1.   The Chad Basin and (inset) Lake Megachad (after Rognon
1980); Lake Chad at the height of 281.50 m a.s.l. (after
Chouret *et al.* 1974).

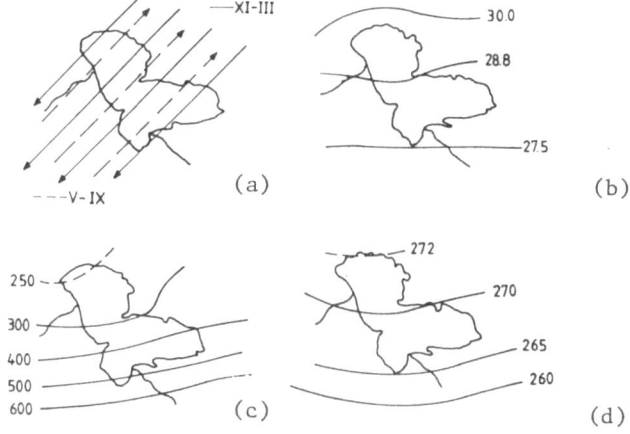

Fig. 2.   Meteorological conditions in the Chad Basin: (a) wind
directions ——, November to March; ---, May to September;
(b) air temperature T_a (°C); (c) precipitation (mm yr^{-1});
(d) global radiation G (W m^{-2}).

fied empirical relationship below was used therefore in the
computations:

$$A = 2.35 \times 10^9 \, h_a - 6.433 \times 10^{11}$$
(4)

where A is the area of the lake in m^2 and h_a is the height of its
surface in m a.s.l. Equation (4) is valid only for a narrow range
of height h_a and is based on data in Bouchardeau and Lefèvre (1957)
and Touchebeuf de Lussigny *et al*. (1969).

3. BRIEF CLIMATIC HISTORY OF THE CHAD BASIN

The climatic history of the Chad Basin has been investigated
by numerous workers (Servant & Servant 1970, Fairbridge 1976,
Rognon 1976, Nicholson & Flohn 1980, Maley 1981, Street & Grove
1979, Adams 1980, Sarnthein *et al*. 1981, Kutzbach: this volume).
There is general agreement on the major events. For our purpose,
it was important to select periods with different climatic patterns,
particularly with respect to precipitation and evaporation.

For example, during the early Holocene from 10 000 yr BP to
8000 yr BP, the enormous Lake Megachad covered an area of about
320 000 km^2 (Schneider 1969). After a short dry period at about
7500 yr BP (Street-Perrott & Roberts: this volume), the water
level reached the same height at about 6000 yr BP. From 5000 yr BP
onwards, the climate became more and more arid, so that the lake
shrank to its present extent.

4. BRIEF DESCRIPTION OF THE PHYSICAL MODEL

Computation of evaporation was conducted in two steps, both
based on the energy budget of the lake surface (Figure 3). The
first facilitates the calculation of the water temperature T_L
(Edinger *et al*. 1968) and the second, the evaporation E (Penman
1948).

In order to calculate T_L, it was necessary to parameterize
the terms in Equation (3). An aerodynamic bulk formula was used
to derive an approximation for the flux of latent heat LE:

$$LE = L\rho_a \ f(\bar{u})(q_L - q_a) \tag{5}$$

where ρ_a = density of air,
      q_L = mixing ratio at the water surface,
      q_a = mixing ratio of air,
    $f(\bar{u})$ = function of the average wind velocity $\bar{u}$ at 2 m height
           = $c_D \bar{u}$, with
      c_D = drag coefficient.

The flux of sensible heat H was similarly parameterized:

$$H = -c_p \rho_a \ f(\bar{u})(T_L - T_a) \tag{6}$$

where c_p = specific heat of air,
      T_L = water temperature in $^\circ C$,
      T_a = air temperature in $^\circ C$.

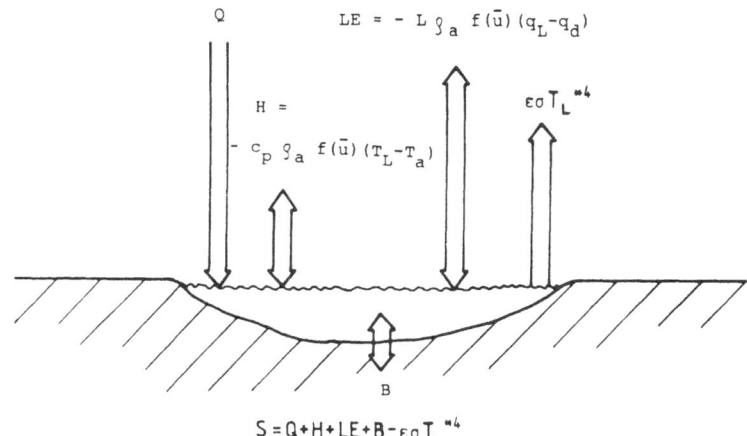

Fig. 3. The energy balance of a shallow lake illustrating the
relationships between absorbed incoming radiation Q, flux
of sensible heat H (Equation (6)), flux of latent heat LE
(Equation (5)), long-wave outgoing radiation $\varepsilon\sigma T_L^{*4}$, and
heat flux to the ground B, where ΔS represents net heat
storage (Equation (3)).

According to Fraedrich (1972) the evaporation from a lake can be
described by Equation (5) using data from the warm, dry environment upstream of the lake. For the evaluation of the water vapour
flux across the internal boundary layer between moist and dry air
above the lake, exchange coefficients were determined by Eggers
and Tetzlaff (1978), using computed and satellite-measured water-
surface temperatures for Lake Chad. Their description makes
formal use of the wind velocity function $f(\bar{u})$, because the product
$c_D \bar{u}$ is relatively constant with height.
      Following Edinger et $al.$ (1968), the relationship between the
water vapour gradient and the temperature gradient above the lake
was also simplified. In our model, the heat flux into the ground
B was neglected. Furthermore, the formula for the long-wave
radiation from the lake was simplified. Finally, an equation was
derived in which the net energy storage in the lake depended only
on the incoming radiation, the water temperature and the temperature and humidity of the advected air. In order to simplify this
equation further, an additional term was introduced, namely the so-
called equilibrium temperature T_e at which no net heat flux occurs
through the water surface so that the water temperature does not
change. It then can be shown that the net energy storage ΔS is
dependent only on the temperature difference between the equilibrium temperature T_e and the mean lake temperature T_L, and on the
transfer coefficient M:

$$\Delta S = M(T_e - T_L) \tag{7}$$

The coefficient M depends on the diffusion coefficient at the top of the moist boundary layer, the density of air, the mean wind velocity and the dew point temperature:

$$M = 4\varepsilon\sigma T_o^{*3} + \rho_a c_D \bar{u}(L\alpha + c_p) \tag{8}$$

where $T^* = 273$ K,

$\alpha^o = (q_L - q_a)/(T_L - T_d)$.

T_d = dew point temperature.

On the other hand, the net storage of energy is proportional to the change of the water temperature with time t, i.e.

$$\Delta S = \rho_L c_L h_L(\delta T_L/\delta t) \tag{9}$$

where ρ_L = density of water,

c_L = specific heat of water,

h_L = mean depth of the lake.

By combining Equations (7) and (9), a simple differential equation for T_L is obtained. The solution used here, derived by Tetzlaff (1979), allows the water temperature of a shallow lake to be obtained using the approach outlined if it is adjusted properly to the problem.

As an error analysis proved (Adams 1980), the evaporation should be computed by modifying Penman's approach as follows after evaluation of the water temperature T_L:

$$LE = \{[(Q-\varepsilon\sigma T_L^{*4})(q_L-q_a)/(T_L-T_a)]-[(0.622/p)f(\bar{u})(q_s-q_a)]\}/$$

$$[(q_L-q_a)/(T_L-T_a)+(c_p/L)] \tag{10}$$

where q_s = the saturation mixing ratio at the air temperature T_a,

p = air pressure.

The gradients and the wind function $f(\bar{u})$ were again taken above and below the boundary layer. The net incoming radiation Q was computed using the empirical Brunt formula with data on global radiation from Touchebeuf de Lussigny et al. (1969).

The first data set used to test the model consisted of measured values of air temperature, dew-point temperature, wind speed, global radiation and the hydrological variables of runoff, precipitation and net seepage to groundwater (Table I). The latter term was assumed to be 100 mm yr^{-1} (Roche 1980). The data originate from Chouret et al. (1974), Touchebeuf de Lussigny et al. (1969), ORSTOM (1974) and Lebedev (1970).

5.   RESULTS

The model was applied to modern data (specified above) and the results are given in Table II. The equilibrium temperature T_e and the calculated water temperature T_L are almost equal

Table I.　Hydrological and meteorological data set for an average
year. R, runoff (m^3 s^{-1}) for 1958-63; P, precipitation
on lake (mm) for 1954-67; G, global radiation (W m^{-2})
for 1958-63; T_a, air temperature (°C) for 1946-60;
T_d, dew point temperature (°C) for 1950-54; $\bar{u}$, wind
velocity at 2 m height (m s^{-1}) for 1956-62.

Month	R	P	G	T_a	T_d	$\bar{u}$
J	840	0	247	20.4	3.5	1.9
F	468	0	282	22.8	4.3	2.0
M	282	0	299	27.0	6.8	2.1
A	202	0	294	29.7	11.5	2.1
M	202	7	282	32.2	16.8	1.9
J	298	17	269	30.7	20.0	2.2
J	558	85	250	28.6	21.6	2.4
A	1330	185	228	26.8	22.5	2.0
S	2516	52	255	28.0	22.5	1.7
O	3468	5	274	28.8	17.4	1.7
N	3463	0	265	24.4	5.8	2.1
D	1995	0	245	22.0	4.5	2.0
Mean	1300	–	266	26.8	13.1	2.0
Total	–	351	–	–	–	–

Table II.　Simulated water balance of Lake Chad for an average
year. T_e, computed equilibrium temperature (°C);
T_L, computed water temperature (°C); T, measured water
temperature (°C); E, evaporation (mm); h_a, water level
of lake (m a.s.l.); A, surface area (km^2); A_E, equilibrium area (km^2).

Month	T_e	T_L	T	E	h_a	A	A_E
					282.50	22040	14430
J	19.8	20.0	20.2	148	282.45	21910	6290
F	22.1	21.4	21.0	172	282.32	21610	3110
M	24.4	23.8	24.0	235	282.11	21120	2120
A	27.0	26.3	27.7	239	281.89	20600	2240
M	30.7	29.8	29.5	240	281.67	20090	3820
J	30.9	30.8	30.4	211	281.51	19700	13920
J	30.3	30.4	29.4	184	281.48	19630	∞
A	30.8	30.7	29.2	138	281.70	20150	59390
S	33.3	32.6	30.0	153	281.91	20650	49810
O	31.1	31.7	29.0	183	282.17	21270	49370
N	22.0	24.6	24.5	174	282.41	21840	31920
D	20.2	20.8	21.8	159	282.50	22040	14430
Mean	26.9	26.9	26.4	–	282.01	20850	20850
Total	–	–	–	2236	–	–	–

throughout the year. The monthly mean values of T_L agree rather
well with the results of water temperature measurements (T) taken
in 1956–1960 (Roche 1980) on the NE shore in Bol (see Figure 1).
The evaporation from Lake Chad was calculated to be 2236 mm y^{-1}.
This value agrees with estimates by Touchebeuf de Lussigny *et al.*
(1969), but not so well with the value of about 1300 mm yr^{-1}
calculated by Kutzbach (1980). The simulated annual lake-level
cycle shows an amplitude of 1.02 m between 1 January (282.50 m
a.s.l.) and 1 August (281.48 m a.s.l.), thus agreeing with data in
Maley (1973*b*). The starting point of h_a = 282.50 m a.s.l. was
derived by an iterative process in order to obtain $\Delta V = 0$ and
identical annual values of A and A_E. During the year, the
equilibrium area A_E exhibits a far larger variation than the
surface A itself. For example, if August conditions were extended
throughout the year, Lake Chad would steadily grow to infinity.

6.   THE SENSITIVITY ANALYSIS

In order to clarify the influence of various meteorological
and hydrological parameters on the area of the lake, a sensitivity
analysis was undertaken. Figure 4(a-d) shows the magnitude of the
open-water evaporation E as a function of the meteorological
parameters, each parameter being varied independently of the
others. Figure 4(a) shows that the influence of the wind velocity
$\bar{u}$ is large. In contrast to the other variables, the reaction of
the model is complex. On the one hand, an increase of $\bar{u}$ causes a
decrease in water temperature which would tend to reduce evaporation; on the other hand, an increase of $\bar{u}$ directly increases E.
The second effect is much larger than the first. Another important
factor is global radiation. As can be seen from Figure 4(b), a 10%
increase leads to a rise in water temperature by about 1 K and an
8% increase in evaporation (180 mm yr^{-1}). The influence of air
temperature T_a is similar to that of global radiation (Figure 4(c)).
Changes in mixing ratio act in the inverse direction and seem
likely to have been less effective than changes in wind velocity.
Figure 4(d) demonstrates that a higher mixing ratio leads to
reduced evaporation and *vice versa*. The main hydrological parameter influencing the lake level is runoff (Figure 4(e)); nearly
90% of the total input come from tributaries like the Chari.
According to Equation (2), a variation in runoff of about ± 10%
leads to a change in lake area of about ± 10%; in Figure 4(e), this
is expressed as a change in the equilibrium area A_E. In contrast
to the runoff, local precipitation P on the lake causes only small
variations in its surface area (Figure 4(f)).

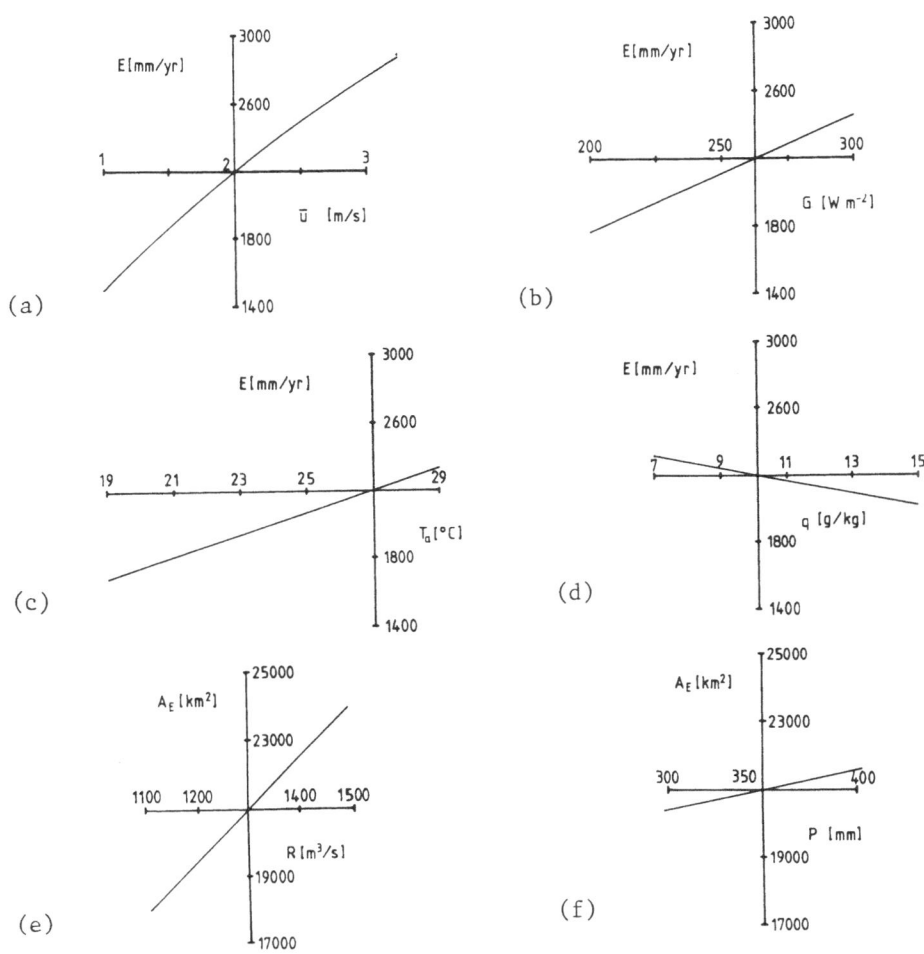

Fig. 4.   Evaporation E (mm yr^{-1}) over Lake Chad as a function of:
(a) wind speed $\bar{\mu}$ (m s^{-1}); (b) global radiation G (W m^{-2});
(c) air temperature T_a upstream of the lake ($^{\circ}$C);
(d) mixing ratio q (g kg^{-1}). The surface area of Lake
Chad A (km^2) as a function of: (e) runoff R (m^3 s^{-1});
(f) precipitation P (mm). The values at the origin of
the coordinates are present-day annual means.

7.   APPLICATION OF THE MODEL

Both the variations in the level of Lake Chad during the
drought period 1968-1973 and the maximum extent of the early
Holocene Lake Megachad were simulated. The data set used for the
Sahel drought was obtained from ORSTOM (1974) and Stranz (1975).
Table III shows the annual values of the meteorological parameters
for the period 1968-1973. It is remarkable that the four years

Table III.   Annual mean values of meteorological parameters for
             Lake Chad for 1968-1973 and computed mean water
             temperature T_L (°C).

             For key, see Tables I and II.

Year	G $(W\ m^{-2})$	T_a (°C)	T_d (°C)	$\bar{u}$ $(m\ s^{-1})$	T_L (°C)	E $(mm\ yr^{-1})$
1968	262	26.1	14.4	1.9	27.6	2052
1969	261	27.2	15.5	2.0	28.2	2158
1970	272	26.9	12.0	2.2	26.1	2452
1971	273	26.9	11.2	2.3	25.3	2542
1972	269	27.4	12.8	2.4	25.9	2534
1973	265	28.2	11.1	2.3	25.2	2661

1970-1973 were characterized by higher global radiation (except in
1973), higher air temperatures, lower humidities and higher wind
speeds over the Chad Basin than in an average year.

     The calculated mean values of water temperature and evapora-
tion are also given in Table III. For the first two years of the
drought, the simulated surface temperature of the lake was above
normal due to higher humidities; conversely, evaporation was below
average. In subsequent years, the influence of increased wind
velocities on the water temperatures predominated over the effects
of increased global radiation and higher air temperatures. An
increase in $\bar{u}$ of about 10 to 20% combined with higher global
radiation and lower humidities (= decreased influence of the ITCZ),
would lead to 10 to 20% more evaporation overall. In Figure 5, the
observed changes in lake level (Chouret *et al*. 1974) are compared
with the computed curve. As the data were obtained in Bol, it is
important to note that wind stress reduces the maximum level
recorded in January and increases the minimum value in August.

     In making these calculations, the hydrological data set was
kept constant, because of the limitations of the bathymetric curve
and the lack of actual hydrological data. From these results, it
can only be said that the increase in evaporation during the Sahel
drought would account for a drop in lake level of nearly 0.8 m
between January 1968 and January 1974. The discrepancy between
the observed and calculated curves therefore implies that both runoff and precipitation were also reduced during this period.

     For the model of Lake Megachad, the conservative assumption is
made that the northward shift in the mean annual position of the
ITCZ around 9000 yr BP was about 5°. Based on a comparison with
data from other African stations near 10°N in the Chad Basin
(Lebedev 1970), the modern meteorological data set was modified
in the following way (Adams 1980). The annual curves of T_a and
T_d were smoothed by a decrease of the monthly values from April
to October by 1.5 K and an increase from December to March by

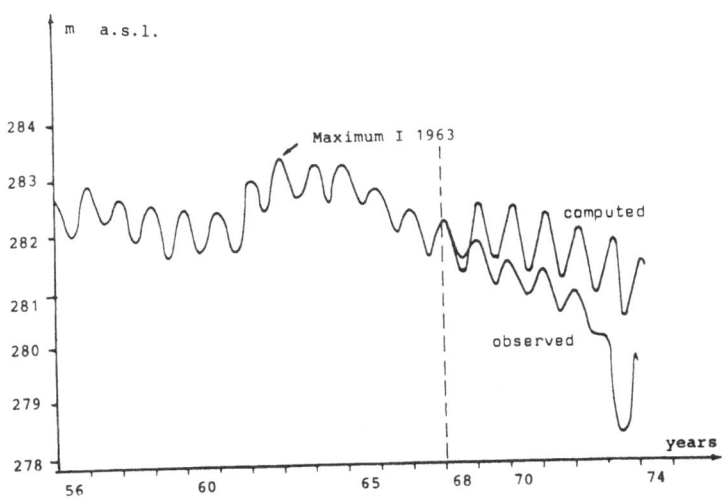

Fig. 5.   Observed changes in the level of Lake Chad between 1956
and 1974 (after Chouret *et al.* 1974), together with the
computed curve for 1968 to 1973.

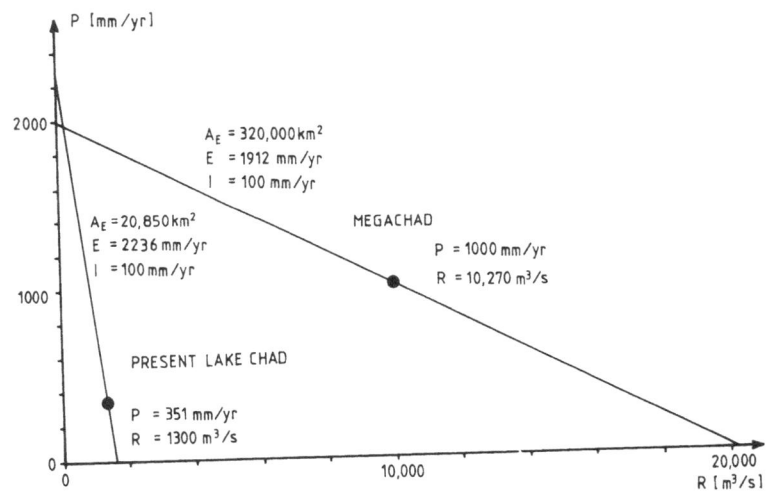

Fig. 6.   Relationship between precipitation P and runoff R for the
modern Lake Chad and for Megachad: E, evaporation;
I, seepage; A_E, equilibrium area.

2.5 K.   The rainy season was extended from the end of March to
early November, resulting in an increase of T_d by 5 K in October
and April and by 2 K in May.   The intensified influence of the

monsoonal air masses would reduce the global radiation by an
average value of about 5%. Furthermore, it is assumed that air
temperatures at about 9000 yr BP were generally lower than today
(Jäkel 1978), perhaps by as much as 2 K.

With an estimated average water temperature of 24.4°C and an
evaporation rate of 1912 mm yr^{-1}, Lake Megachad would have been
about 2.5 K cooler and would have lost about 15% less water by
evaporation than today. Assuming modern values for runoff, precipitation and seepage losses, the increase in the equilibrium
area A_E would be only $\sim$ 4000 km^2. If the palaeolake extended over
320 000 km^2 and seepage losses were the same as today, a relation
between P and R can be deduced from Equation (2). Figure 6 shows
that the values of P and R fall on a straight line. In the case
of an increase in precipitation up to 1000 mm yr^{-1} (10°N), the
runoff from the catchment area would have been about 8 times larger
than the present-day value of 1300 m^3 s^{-1}). These calculations are
nevertheless underestimates of the early Holocene water budget of
the Chad Basin, because Lake Megachad had a southern outlet into
the Benue and Niger Rivers at Bongor. Consequently, the equilibrium area A_E must have been even larger than 320 000 km^2. We
therefore conclude that the total annual input of water into Lake
Megachad during the early Holocene was at least 1200% greater
(i.e., 6.4 × 10^{11} m^3) than today (4.8 × 10^{10} m^3).

ACKNOWLEDGEMENTS. We thank Dr F.A. Street-Perrott for her valuable
comments.

REFERENCES

Adams, L.J. 1980 Der Wasserstand des Tschadsees in Abhängigkeit von
    meteorologischen und hydrologischen Grössen. Thesis: Institut
    für Meteorologie und Klimatologie, Universität Hannover.
    [Unpublished].

Bouchardeau, A. & Lefèvre, R. 1957 *Monographie Hydrologique du Lac
    Tchad*, 112 pp. ORSTOM.

Chouret, A. *et al.* 1974 Les effets de la sécheresse actuelle en
    Afrique sur le niveau du Lac Tchad. *Cah. ORSTOM, ser. Hydrol.*,
    <u>11</u>, No. 1, 35-45.

Edinger, J.E., Duttweiler, D.W. & Geyer, J.C. 1968 The response of
    water temperatures to meteorological conditions. *Water Resour.
    Res.*, <u>4</u>, 1137-1143.

Eggers, K.A. & Tetzlaff, G. 1978 A simple model for describing the
    heat balance of a shallow lake with application to Lake Chad.
    Bound.-Layer Met., <u>15</u>, 205-214.

Fairbridge, R.W. 1976 Effects of Holocene climatic change on some tropical geomorphic processes. *Quat. Res.*, 6, 529-556.

Fraedrich, K. 1972 On the evaporation from a lake in warm and dry environment, *Tellus*, 24, 116-121.

Jäkel, D. 1978 Eine Klimakurve für die Zentralsahara. In: *Sahara: 10 000 Jahre zwischen Weide und Wüste*, pp. 382-396. Köln.

Kutzbach, J.E. 1980 Estimates of past climate at Palaeolake Chad, North Africa, based on a hydrological and energy-balance model. *Quat. Res.*, 14, 210-223.

Langbein, W.B. 1961 Salinity and hydrology of closed lakes. *US geol. Survey Prof. Pap.*, No. 412, 20 pp.

Lebedev, A.N. 1970 *The Climate of Africa, Part I*, 482 pp. Jerusalem: Israel Program for Scientific Translations.

Maley, J. 1973*a* Mechanisme des changements climatique aux basses latitudes. *Palaeogeogr. Palaeoclimatol. Palaeoecol.*, 14, 193-227.

Maley, J. 1973*b* Les variations climatiques dans le bassin du Tchad durant le dernier millénaire; essai d'interpretation climatique de l'Holocène africain. *C. r. hebd. Séanc. Acad. Sci. Paris, D*, 276, 1673-1675.

Maley, J. 1981 Etudes palynologiques dans le bassin du Tchad et paléoclimatologie de l'Afrique nord-tropicale de 30 000 ans à l'époque actuelle. *Trav. et Docs ORSTOM*, No. 129, 586 pp.

Nicholson, S.E. & Flohn, H. 1980 African environmental and climatic changes and the general atmospheric circulation in Late Pleistocene and Holocene. *Climatic Change*, 2, 313-348.

ORSTOM 1974 *Données Climatologiques Mensuelles 1964-1973*. N'Djamena.

Penman, H.L. 1948 Natural evaporation from open water, bare soil and grass. *Proc. R. Soc. Lond., A*, 193, 120-145.

Richardson, J.L. 1969 Former lake-level fluctuations - their recognition and interpretation. *Mitt. Int. Verein. Limnol.*, 17, 78-93.

Roche, M.A. 1980 Tracage naturel salin et isotopique des eaux du système hydrologique du Lac Tchad. *Trav. et Docs ORSTOM*, No. 117, 383 pp.

Rognon, P. 1976 Essai d'interpretation des variations climatiques au Sahara depuis 40.000 ans. *Rev. Géogr. Phys. Geol. Dynam. (2)*, 18, 251-282.

Rognon, P. 1980 Pluvial and arid phases in the Sahara: The role of non-climatic factors. *Palaeoecol. Africa*, 12, 45-62.

Sarnthein, M., Tetzlaff, G., Koopman, B., Wolter, K. & Pflaumann, U. 1981 Glacial and interglacial wind regimes over the eastern subtropical Atlantic and north-west Africa. *Nature, Lond.*, 293, 193-196.

Schneider, J.L. 1969 Evolution du dernier lacustre et peuplements préhistoriques aux pays bas du Tchad. *Bull. Inst. fr. Afr. noire, A*, 31, 259-263.

Servant, M. & Servant, S. 1970 Les formations lacustres et les diatomées du Quaternaire Récent du Fond de la Cuvette Tchadienne. *Rev. Géogr. Phys. Géol. Dynam. (2)*, 12, 63-76.

Stranz, D. 1975 Über den Regen in Afrika und die Trockenheit der letzten Jahre im Sahel (1967-1974). *Deutscher Wetterdienst, Seewetteramt, Einzelveröffentlichungen*, Nr. 88, Hamburg.

Street, F.A. & Grove, A.T. 1979 Global maps of lake-level fluctuations since 30 000 yr BP. *Quat. Res.*, 12, 83-118.

Tetzlaff, G. 1979 The daily cycle of the water temperature of a shallow lake. In: *Developments in Water Science*, Vol. 11: *Hydrodynamics of Lakes* (ed. W.H. Graf & C.H. Mortimer), 325-330 pp. Amsterdam: Elsevier.

Touchebeuf de Lussigny *et al*. 1969 *Monographie Hydrologique du Lac Tchad*, 169 pp. ORSTOM.

MARINE SHORELINES IN ESTUARIES AS PALAEOPRECIPITATION INDICATORS

J.-Y. Gac

Office de la Recherche Scientifique
et Technique d'Outre-Mer (ORSTOM)
B.P. 1386 Hann, Dakar, Sénégal

J. Monteillet

Institut Fondamental d'Afrique Noire (IFAN)
B.P. 206, Dakar, Sénégal

H. Faure

Laboratoire de Géologie du Quaternaire (CNRS)
Case 907, Faculté des Sciences Luminy
13288 Marseille Cédex 9, France

ABSTRACT. The presence of fossil marine faunas in estuarine
environments may be due to sea-water incursions during periods of
low river discharge following diminished precipitation and
increased evaporation over the associated catchment. Thus, on
low-lying tropical coasts, the development of marine fauna in
estuaries or lagoons may indicate periods of drought lasting for
periods from a few years to centuries.
    Ecological studies of present-day marine and brackish-water
fauna and flora in the Senegal River were used to establish the
salinity tolerances of fossil faunas as well as the minimum annual
duration of marine conditions required by each specific assemblage.
In addition, the annual penetration and retreat of the salt-water
wedge has been followed by hydrological and conductivity measurements since the beginning of the century, and is shown to be
related to the river's discharge rates. The mean position of the
salt-water limit for a known annual discharge provides a fairly
good indication of the precipitation pattern. Using this relationship for fossil marine faunas from the lateral lakes in the lower
Senegal basin, palaeoprecipitation totals can be estimated and,

A. Street-Perrott et al. (eds.), Variations in the Global Water Budget, 361–370.

hence, the timings in the mid-Holocene of marine or brackish-water invasions. If these incursions were not due to rises in sea level, they may have resulted from prolonged periods of reduced precipitation in the upper Senegal basin. These dry periods were comparable to the droughts of 1913, 1941 and 1975, and probably lasted for several decades.

1.   INTRODUCTION

     Marine faunas associated with fossil shorelines are usually attributed to relative movements of sea level. During the Holocene, such oscillations occurred in many parts of the world. The development of marine faunas in estuaries may also result from sea-water incursions during phases of low river discharge, i.e., when precipitation over a drainage basin is reduced and evaporation increased. In this way, the presence of a marine fauna in estuaries and lagoons along low-lying tropical coasts may indicate periods of prolonged drought lasting for several years or centuries.
     In the Sahel, the reconstruction of the palaeoprecipitation records has been attempted with the aid of a conceptual model which was calibrated using data for the whole of the Senegal River basin in west Africa (Figure 1). We have endeavoured to specify (either quantitatively or semi-quantitatively) as accurately as possible the relationships between the precipitation over the basin, the discharge of the river, the degree of penetration of the saline wedge from the estuary mouth, and the presence of fresh-water, brackish or marine faunas[1].

2.   MODERN CLIMATE AND HYDROLOGY OF THE SENEGAL BASIN

     The catchment of the Senegal River extends from $10°30'$ to $17°30'N$ and covers an area of about 350 000 km^2 (Figure 1). Formed by the junction of the Bafing and the Bakoye, the Senegal rises in the Fouta Djalon massif and enters the Atlantic via a large delta after flowing 1800 km across the tropical dry zone. The decrease in precipitation northwards (Figure 2) makes it possible to distinguish five major climatic belts: the Guinean, Soudano-Guinean, Soudanian, Sahelian and Saharan zones (see also Nicholson: this volume). When precipitation is abundant and the river discharge high, the various zones are displaced northwards. On the other hand, when aridity becomes pronounced, as at present, the climatic belts recede far to the south (Figure 2). Between these two extremes, the position of the isohyets may be displaced through several degrees of latitude; for example, the 100 mm isohyet by up to 700 km, the 300 mm isohyet by 350 km, and so on.
     Observations of the hydrological regime of the Senegal River date from 1903 to the present day (Faure & Gac 1981) without a

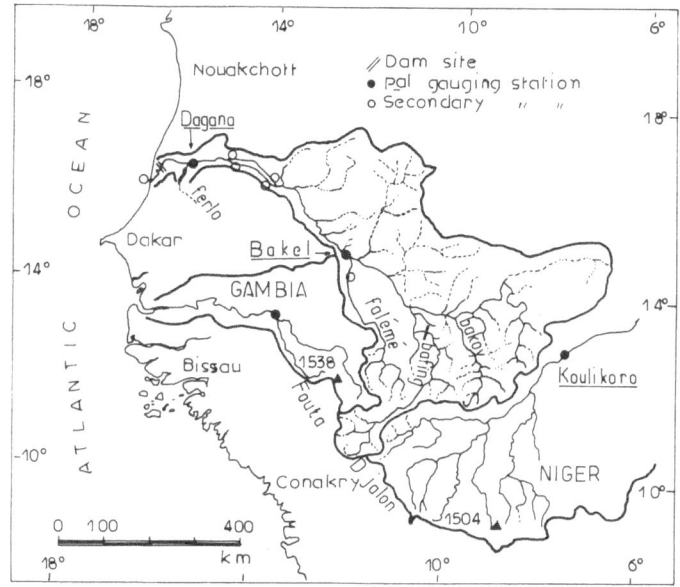

Fig. 1. Location map of the Senegal River basin.

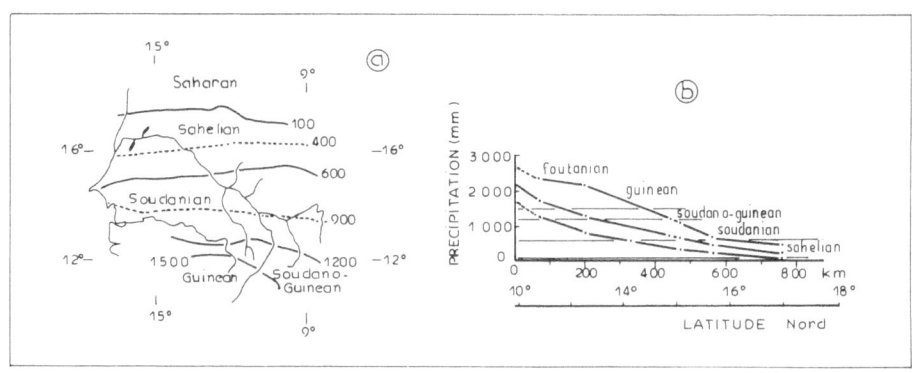

Fig. 2. Present-day climatic data for the Senegal River basin:
         a, average position of the various climatic belts;
         b, latitudinal extent of the various climatic belts as a
         function of precipitation amounts.

break. Between 1903 and 1980, the annual discharge averaged
733 m^3 s^{-1}. The flow records since the turn of the century
(Figure 3) exhibit extreme values of 263 m^3 s^{-1} in 1972 and
1244 m^3 s^{-1} in 1924 and can be classified in six major climatic
episodes: humid (before 1909, 1919-1938, 1949-1968) and arid
(1909-1919, 1938-1949, 1968-1982) (Sircoulon 1976). Only in 1975

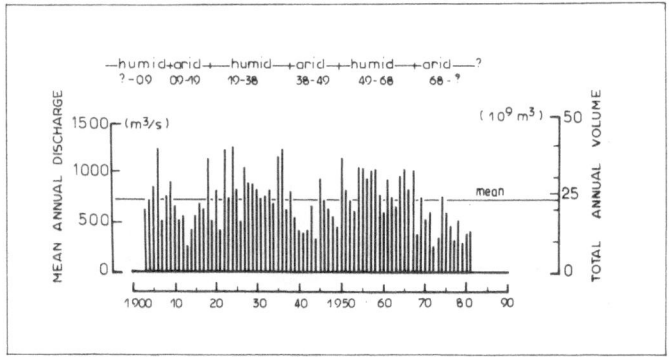

Fig. 3. Variations in mean annual discharge of the Senegal River, 1903-80; the average over the period is 733 m³ s⁻¹.

was the discharge slightly above average over the Sahel zone as a whole after 1968.

The Senegal River and its estuary exhibit the peculiarity that the river bed descends significantly below sea level for more than 350 km. Despite the small tidal range (Faure *et al.* 1980), this unusual situation results in the development of two important phenomena during each period of low flow: a significant estuarine tide and the penetration of salt water into the interior of the continent (Rochette 1964). Four principal reasons can be advanced to explain this marine incursion:
(a) The markedly protracted period of low flow in the river, that lasts for a large part of the year;
(b) The morphology of the bed and its location below sea level;
(c) The energy of the swell off the estuary mouth, which, at 110 J, is among the highest in the world (Coleman & Wright 1975);
(d) The exceptionally steep gradient of the continental slope.

The annual marine incursion into the heart of the Senegal basin can be divided into four phases (Figure 4). Phase 1 corresponds to the flood peak of the river, when fresh water dominates throughout the basin. Phase 2 begins during the recession stage when the discharge falls below 600 m³ s⁻¹; saline water penetrates into the estuary, but its advance into the interior is moderated while the river discharge remains above 50 m³ s⁻¹. When it falls below this value, Phase 3, representing the marine incursion in the strict sense, begins and lasts until the arrival of the next flood; the new flood wave expels the salt water. Phase 4 begins when the saline wedge begins to retreat and ends when the cumulative volume of fresh water discharged reaches 900 × 10⁶ m³.

Every year, the flood peak sets up the initial conditions which govern the duration of each phase. If the precipitation over the catchment is abundant, as in 1953 and 1959 (Figures 5 to

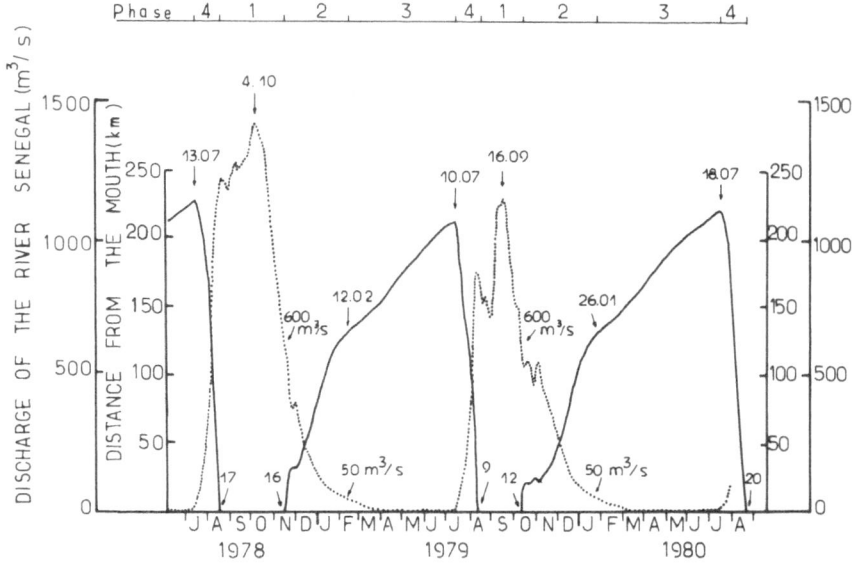

Fig. 4.   The different phases of the annual marine incursion, based
          on data for 1978-80: ——, position of the saline front;
          •••, variations in river discharge (D). Events shown
          during 1978 and 1979: (i) salinity maximum on 13 July
          1978; (ii) retreat of the salt-water wedge, 13 July to
          17 August 1978 (phase 4); (iii) fresh water at the mouth,
          17 August to 16 November 1978 (phase 1); (iv) marine
          advance accompanying decrease in discharge, 16 November
          1978 to 12 February 1979 (phase 2); (v) marine invasion
          (D < 50 m³ s⁻¹), 20 February to 10 July 1979 (phase 3);
          (vi) salinity maximum on 10 July 1979; (vii) retreat of
          the salt-water wedge, 10 July to 9 August 1979 (phase 4).

7) there is a large amount of runoff, the fresh-water stage at the
mouth lasts for five months and the saline wedge remains below the
town of Richard-Toll, 150 km inland. If, however, rainfall is
sparse (as it was between 1974 and 1980), the flood peak is small,
fresh water only persists for two months at the estuary, and sea
water penetrates above Fanaye Oualo, 225 km from the coast (Figure
7: point 9). The latest measurements by ORSTOM of the salinity
of the Senegal River seem to indicate that the marine incursion in
1981 will be the greatest observed this century. By 6 June 1981,
the saline front had already passed the confluence of the Senegal
and the Doué, a point more than 260 km upstream from the mouth.

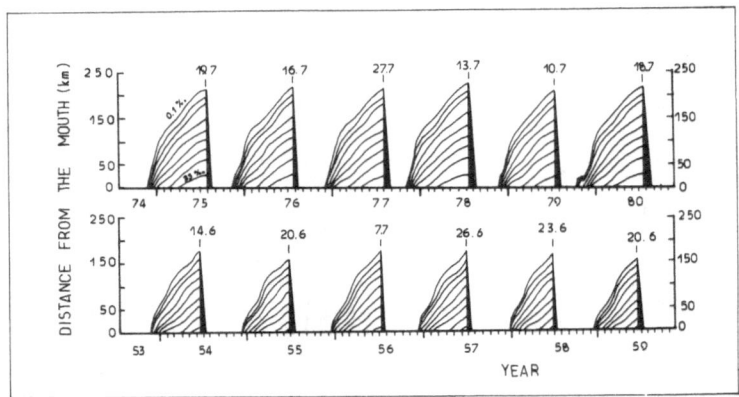

Fig. 5.   Comparison of marine incursions during: a, arid (1974–
80); b, humid (1953–59) periods. The curves correspond
to different salinity levels: 0, 2.5, 5, 10, 15, 20,
25, 30 and 35‰.

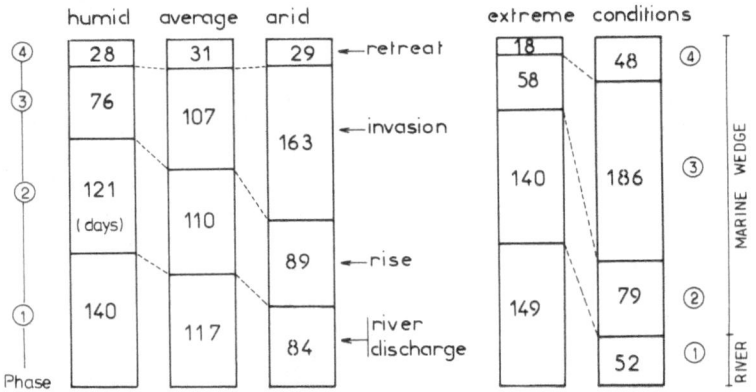

Fig. 6.   Duration (days) of the marine incursion under humid,
average, arid and extreme conditions.

3.   THE PALAEOECOLOGICAL MODEL

In association with this investigation of the precipitation
and discharge patterns of the river regime and different aspects
of the annual marine incursion, detailed ecological studies were
carried out in the saline and brackish-water environments of the
conditions for the survival, development and proliferation of the
modern flora and fauna (Monteillet & Rosso 1977). These studies
made it possible to establish the salinity toleration (in terms of
both concentration and duration) of modern species, and to appre-

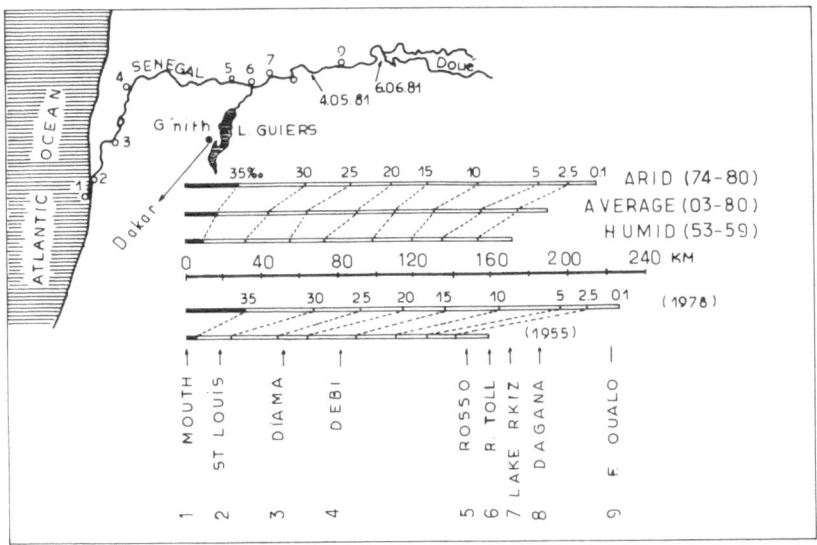

Fig. 7. Penetration of the salt-water wedge under arid (1974-80),
average (1903-80) and humid (1953-59) conditions.

ciate the adaptive capacity of the organisms to the instability of
the lagoonal and estuarine environments. The characteristic
salinity tolerances of each faunal association are linked to the
pattern of river discharge, which is itself a function of precipitation over the upper basin.

Using the inverse argument, this model can be utilized to
estimate the Holocene precipitation record corresponding to the
fossil assemblage collected in the field and ^{14}C age-dated
(Monteillet 1979, Monteillet et al. 1981b), its distance from the
sea along the most likely palaeochannel can be determined, taking
account of the available palaeogeographic information, i.e.,
evolution of the channel and its probable eventual fate (e.g.,
becoming blocked by silt). By entering this distance on the
abscissa of Figure 8, it is possible to calculate the palaeodischarge rate corresponding to the ecological requirements of
the fossil fauna in question. The stratigraphic data and the
radio-carbon dates, together with a knowledge of the rate of
sedimentation and the size and thickness of the fossil shells,
then make it possible to suggest a minimum and maximum duration
for the palaeodischarge. The estimated palaeodischarge gives the
latitudinal position of the isohyets and, hence, the precipitation
pattern (Figure 9).

If the marine or brackish-water incursions identified at 6700,
6200, 5600, 3400, 1850 and 1550 yr BP (Monteillet et al. 1981a)
were not due to rises in sea level, they may have corresponded to

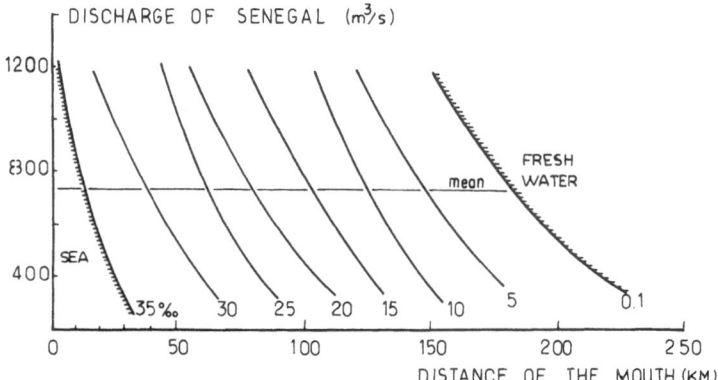

Fig. 8.   Penetration of the salt-water wedge (km) as a function of
          river discharge (m³ s⁻¹). The different salinity levels
          (‰) have been plotted for a duration of 15 days. The
          diagram can be used as a nomograph to estimate the palaeo-
          discharge of the river as a function of the distance from
          the sea of any given fossil fauna for which the ecological
          requirements are known.

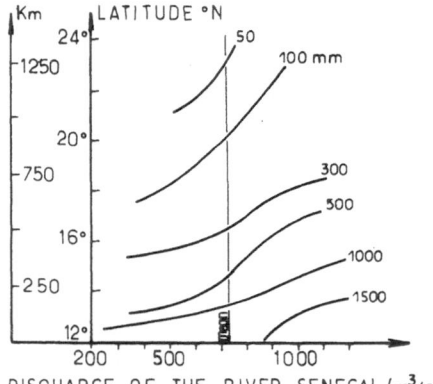

DISCHARGE OF THE RIVER SENEGAL (m³/s)

Fig. 9.   Relation between the latitudinal displacement of the
          annual isohyets (mm) and the average discharge of the
          Senegal River (m³ s⁻¹). The diagram may be used as a
          nomograph to estimate palaeoprecipitation as a function
          of river discharge.

prolonged decreases in precipitation over the upper part of the
Senegal basin. In that event, the palaeoprecipitation levels
would have been of the same order of magnitude as those of the
recent droughts in 1913, 1941 and 1975, though persisting for 50
years or more.

This model constitutes a first methodological approximation. It necessarily rests on a number of assumptions that are acknowledged to be partially false. These include the constancy of such variables as sea level and the mode of life and growth rate of the larvae and molluscs. The model also runs up against the difficulty of translating into a single value the complex ecological requirements of the flora and fauna at different developmental stages. At one and the same time, account must be taken of the duration of the salt-water incursion, of salinity and of other environmental factors. The model needs, therefore, to be improved by means of studies of modern conditions. Such studies are being continued jointly by ORSTOM, CNRS and IFAN within the framework of the CYCLARID project.

NOTE

[1]   Based on observations of present-day climate, the river regime (Faure & Gac 1981) and the penetration of the saline front (Rochette 1964), ecological studies (Monteillet & Rosso 1977) have established the geochemical requirements for the development and growth of the fauna and flora in the Senegal estuary. Conversely, dated fossil assemblages observed in the field (Monteillet 1979, Monteillet *et al.* 1981*b*) may be used to provide an indication of the salinity, hydrology and, probably, the precipitation pattern in the past.

REFERENCES

Coleman, J.M. & Wright, L.D. 1975 Modern river deltas: variability of processes and sand bodies. In: *Deltas* (ed. M.L. Broussard), pp. 99-149. Houston Geological Society.

Faure, H., Fontes, J.C., Hébrard, L., Monteillet, J. & Pirazzoli, P.A. 1980 Geoidal change and shore level tilt along Holocene Estuaries: Sénégal river area, West Africa. *Science, N.Y.*, 210, 421-423.

Faure, H. & Gac, J.-Y. 1981 Will the Sahelian drought end in 1985? *Nature, Lond.*, 291, 475-478.

Monteillet, J. 1979 Le Quaternaire du delta du Sénégal: synthèse des connaissances actuelles avec de nouvelles données sur l'Holocène. *Bull. Inst. fond. Afriq. noire*, A41, 1-20.

Monteillet, J. & Rosso, J.-C. 1977 Répartition de la faune testacée
    actuelle (Mollusques et Crustacés Cirripèdes) dans la basse
    vallée et le delta du Sénégal. *Bull. Inst. fond. Afr. noire,*
    A39, 788-820.

Sircoulon, J. 1976 Les données hydropluviométriques de la
    sécheresse récente en Afrique intertropicale. Comparaison
    avec les sécheresses '1913' et '1940. *Cah. ORSTOM, sér.
    Hydrol.,* 13, 75-174.

Monteillet, J., Faure, J. & Gac, J.-Y. 1981a Variations du niveau
    de la mer ou crues fluviales. IGCP Project 61/INQUA Meeting
    Columbia (South Carolina) USA, April 1981: Abstract.

Monteillet, J., Faure, H., Pirazzoli, P.A. & Ravise, S.A. 1981b
    L'invasion marine du Ferlo (Sénégal) à l'Holocène supérieur
    (1 900 B.P.). *Palaeoecol. Africa,* 13, 205-215.

Rochette, C. 1964 Remontée des eaux marines dans le fleuve Sénégal.
    Monographie Hydrologique, 1, 78 pp. Paris: ORSTOM.

MONSOON RAINS OF THE LATE PLEISTOCENE AND EARLY HOLOCENE: PATTERNS, INTENSITY AND POSSIBLE CAUSES OF CHANGES

J.E. Kutzbach

Center for Climatic Research and Department of Meteorology
University of Wisconsin-Madison
Madison, Wisconsin 53706, USA

ABSTRACT. Field evidence for marked environmental changes in the regions subject to monsoon climate has long been available: ancient lake shorelines, fossil dunes, evaporites and Alpine moraines (Brooks 1926, Zeuner 1946). However, a clear picture of the spatial patterns of these features and an accurate radiocarbon-dated chronology of events has been lacking. Without information on the pattern and timing of events, it has been difficult to characterize and interpret these environmental changes in terms of circulation patterns and climate. Moreover, it has been difficult to test hypotheses that attempt to explain the changing monsoon climates. This paper summarizes recent studies that have provided sufficient information about the nature, location and timing of climatic events to facilitate reconstruction of the changes in monsoonal circulations and estimation of the magnitude of precipitation changes. The possible role of changes in the Earth's orbital elements (obliquity, precession and eccentricity) in causing the observed climatic changes is examined. Some early ideas on the astronomical theory of climatic change, as it relates to tropical climates, are traced.

1.   INTRODUCTION

     During the past decade, a broad outline of the pattern and timing of late Pleistocene and early Holocene climatic events in the monsoon lands has begun to emerge (Fairbridge 1962, Butzer *et al.* 1972, Rognon & Williams 1977, Street & Grove 1979). One important conclusion is that the intertropical belt was drier than today during the last glacial maximum and for several thousand years thereafter. In contrast, the interval beginning around

A. Street-Perrott et al. (eds.), Variations in the Global Water Budget, 371–389.

12 500 yr BP, especially from 10 000 to 5000 yr BP, was wetter than
today (Street & Grove 1979, Street-Perrott & Roberts: this volume).
During most of this period, there were high lake-levels in parts of
North Africa, East Africa, Arabia, and northwest India (Street &
Grove 1979, Street-Perrott & Roberts: this volume - Figure 2).
These higher levels indicate larger values of precipitation-minus-
evaporation and, hence, increased runoff into the lakes from their
catchment areas (see § 2). Referring to the early interglacial
(12 500-5000 yr BP), Street-Perrott and Roberts (this volume) note
"there was an unsteady northward migration and broadening of the
intertropical runoff peak, suggesting an enhanced monsoonal circulation over the African and Arabian continents, probably accompanied
by a northward shift in the mean ITCZ."

Rognon and Williams (1977) reached somewhat similar conclusions having summarized climatic evidence from Australia and Africa
for the period from 10 000 to 5000 yr BP; they concluded that "in
all probability the subtropical high-pressure centres were at their
weakest during this interval of high world temperatures, and the
greater summer warming of the tropics would have favoured the more
active flow of the rain-bearing monsoonal air-masses." Concentra-
ting primarily on the interpretation of pollen analyses from the
central Sahara, Maley (1977) concluded that the increased rains of
the early Holocene were of two kinds: (i) rains associated with
'tropical depressions' triggered by the southward extension of
upper-level troughs from the mid-latitudes and (ii) rains associated
with a northward extension of the monsoon especially after about
6500 yr BP. Flohn and Nicholson (1980) also looked to a combination of extratropical and tropical sources to explain the increased
rains across the Sahara.

In summary, there is now general agreement that broad areas of
Africa, the Middle East and Australia experienced greater rainfall
during the terminal Pleistocene and early Holocene than they do
today. Strengthened monsoonal circulations are generally inferred
as the explanation for increased rainfall; a contributory role has
been suggested for the middle-latitude troughs in the northern
hemisphere which extend far enough south to interact with the
tropical moisture sources. These hypotheses are examined in detail
in the following sections.

2.   QUANTITATIVE ESTIMATES OF PAST RAINFALL

Given the evidence of an enhanced monsoon circulation during
the early Holocene, a further question is the magnitude of the
increase in precipitation or precipitation-minus-evaporation.

Several approaches can be used to derive quantitative estimates
of past rainfall from the water budget of closed lakes. A
pioneering study of palaeolakes in the southwestern USA was make
by Leopold (1951). In Africa, the palaeohydrology of Lake Chad
was studied by Grove and Pullan (1963) and water-budget techniques

were applied to the study of East African palaeolakes (Washbourn 1967, Butzer *et al.* 1972).

The annual water balance of a closed lake at equilibrium, neglecting net subsurface flow, is given by

$$A_B(P_B - E_B) = A_L(E_L - P_L) \tag{1}$$

where A_B is the basin (catchment) area, A_L is the lake area, P is precipitation and E is evaporation (Street 1979). In other words, the runoff from the basin into the lake must balance the net loss of water from the lake $(E_L - P_L)$. If the runoff coefficient K is defined as $K = (P_B - E_B)/P_B$, the equation may be written

$$A_B P_B K = A_L(E_L - P_L) \tag{2}$$

P_B can be calculated from Equation (2) if A_L (the area of the palaeolake) and A_B are known, if E_L and K are estimated, and if P_L is assumed to bear a constant relationship to P_B (either $P_L = P_B$ or $P_L = cP_B$ where c is an empirical constant derived from present-day data).

The estimation of past values of evaporation is required explicitly for E_L and implicitly for basin runoff K, since K depends upon basin evapotranspiration. Evaporation is usually expressed as a function of temperature, in which case large values of precipitation-minus-evaporation may result from either increased precipitation or decreased temperatures (for a discussion of these alternatives, see Brackenridge 1978). In the case of the African lakes, Butzer *et al.* (1972) noted that "whereas the so-called pluvial lakes of higher latitudes were probably due primarily to reduced evaporation ... many or most of the high tropical lake levels were associated with a modest but significant increase in precipitation."

In order to avoid the problems associated with the determination of past evaporation rates from palaeotemperature data, Kutzbach (1980) used an alternative approach involving both the water- and energy-balance equations. This approach permitted the explicit estimation of E_B and E_L from the radiation balance (R) and Bowen ratio (B). With certain assumptions, the Bowen ratio can be expressed as a function of R and P; the average precipitation over a lake and its catchment can then be specified uniquely as a function of the net radiation over both areas (Kutzbach 1980). The numerical values of certain radiation variables (albedo, emissivity) are well known for water, various land surfaces and particular vegetation types, and can therefore be estimated accurately for past conditions. The variability of cloud cover in the past is not well known, but, since it modifies both solar and terrestrial radiation, the errors in estimates of cloudiness only lead to relatively small errors in the estimate of net radiation. The latter is also relatively insensitive to errors in temperature estimates. Additional aspects of past and present water budgets of closed lakes are described by Tetzlaff and Adams (this volume).

The values of precipitation obtained using either simple
water-budget or combined water- and energy-budget techniques are
summarized for three lakes in East Africa, one lake in North Africa
and one lake in northwest India (Table I). These estimates are
minimum values because there is evidence that all five lakes overflowed. For the precipitation estimates derived from hydrological
budgets in which evaporation was estimated from palaeotemperature
data, the assumed temperature change is also indicated in Table I.

Table I. Comparison of palaeoprecipitation estimates for the
          early Holocene from palaeolakes. The increase in pre-
          cipitation, compared with the present, is given by ΔP.
          The inequality (>) indicates that all estimates are
          minimum values because overflow was not taken into
          account. Under Remarks: W refers to the simple water-
          budget approach where a temperature difference from the
          present (ΔT) was used to estimate past evaporation; W/E
          refers to the combined water- and energy-budget approach.

	Ref.	% Present Precipitation	ΔP (mm)	Remarks
Eastern Africa				
Ziway-Shala	a	>147	>450	W; $\Delta T = 0$ K
	a	>128	>268	W; $\Delta T = -2$ K
Nakuru-Elmenteita	b	>152	>505	W; $\Delta T = -2$ K
	b	>165	>625	W; $\Delta T = 2$-3 K
	c	>131	>280	W/E
Naivasha	b	>>125	>>225	W; $\Delta T = 0$ K
	c	>>114	>>130	W/E
North Africa				
Chad	d	>186	>300	W/E
Northwest India				
Sambhar	e	>140	>190	W/E

References: a, Street 1979   d, Kutzbach 1980
             b, Butzer et al. 1972   e, Swain et al. 1982
             c, Hastenrath & Kutzbach (In preparation)

The precipitation estimates for Ziway-Shala and Nakuru-Elmenteita illustrate the sensitivity of the water buget method to the
assumed temperature. For Nakuru-Elmenteita, the differences between the results of the two methods are largest when an increase
in temperature is assumed. For Naivasha, the results of the two
techniques differ by less than 100 mm. Discounting the very large
increase in precipitation that was calculated for Nakuru-Elmenteita
with ΔT = 2 to 3 K, the estimates of the early Holocene precipitation increase in eastern Africa range from 14 to 52% and from 130
to 500 mm.

If the estimates from Chad and northwest India are included, then the increases in precipitation range from 14 to 86% and from 130 to 500 mm. It is desirable to study other basins within these regions as a further check on the results, but there seems to be little doubt as to the order of magnitude of the increase. The considerable range of the precipitation values may be related in part to errors of estimation, but may also be explained in part in terms of the climatological patterns; i.e., a uniform increase, either proportionally or in absolute terms, would not be expected over so broad a region.

Because of the many assumptions involved in determining the water budget, it is advantageous to compare the results with climatic estimates derived from independent lines of evidence. Estimates of past precipitation based upon the former range of plants and animals in the Sudan (Wickens 1975) agree roughly with the results for Lake Megachad (Kutzbach 1980). The result for the Sambhar basin in northwest India compares favourably with esti-mates based on pollen records (Singh *et al*. 1972, Bryson & Swain *et al*. 1982) W.L. Prell (personal communication) also inferred a 1981, Swain *et al*. 1982). W.L. Prell (personal communication) also inferred a strengthened southwest monsoon circulation over the Arabian Sea between 10 000 and 4000 yr BP, based on the evidence for monsoonal upwelling along the Arabian coast derived from planktonic foraminiferal assemblages in deep sea cores.

3.   THE POSSIBLE ROLE OF VARIATIONS IN THE EARTH'S ORBITAL PARA-
     METERS IN CHANGES IN MONSOON CIRCULATION PATTERNS

Changes in the Earth's orbital elements (obliquity, precession and eccentricity) during the late Pleistocene and early Holocene may have influenced monsoon circulations through their effect on the seasonal cycle of solar radiation. Kutzbach (1981) tested this hypothesis by substituting solar radiation values for 9000 yr BP in a low-resolution general circulation model (GCM) in place of present-day values.

At 9000 yr BP, obliquity was 24.23° (*cf*. the present-day value of 23.45°), perihelion was July 30 (January 3) and eccentricity was 0.0193 (0.0167); these factors combine to produce an increase in solar radiation in July and a decrease in January. The seasonal radiation changes were $\sim$ 25–35 W m^{-2} (about 7% of modern values) over a broad band of latitudes (Berger 1978, Kutzbach 1981).

The model into which the modified seasonal radiation cycle was inserted is global in extent; it simulates the regional atmo-spheric circulation, including monsoon circulations and surface climates. The increase in solar radiation for June to August at 9000 yr BP warmed the land surface (relative to the ocean) in the model and produced an intensified summer monsoon circulation over the African and Eurasian land masses; in December to February, the decrease in solar radiation produced an intensified winter monsoon. The model's sea-surface temperature distribution for 9000 yr BP

was assumed to be the same as today. This assumption was based
upon both observational evidence and the argument that the large
heat capacity of the ocean would effectively damp the seasonal
temperature response of the upper oceanic layers to the altered
seasonal radiation cycle.

The experiment with 9000 yr BP solar radiation included an
entire year of model simulation. The method of simulation, the
method of calculating solar radiation, and the differences between
the 9000 yr BP experiment and the modern (control) experiment for
the seasonal averages and the annual average have been described in
detail elsewhere (Kutzbach 1981, Kutzbach & Otto-Bliesner 1982).
Only the most important results are given here.

The surface temperature for June to August 9000 yr BP was
2-4 K higher than in the modern simulation over much of Eurasia
(Figure 1a). The surface temperature was slightly lower than at
present in the tropics and subtropics, primarily because cooler
air was advected from over the ocean. In response to the increased
solar radiation and the resulting higher temperature of the African
and Eurasian land surfaces (relative to the surrounding ocean), the
summer monsoon low intensified considerably compared with the
present (Figure 2). The low-level cyclonic inflow of air was
strengthened, as were the Arabian Sea southwesterlies and the cross-
equatorial flow from the southern to the northern hemisphere over
Africa and the western Indian Ocean. In the upper troposphere, the
tropical easterly jet stream was strengthened. Precipitation and
precipitation-minus-evaporation (Table II) were increased over
North Africa (30°N to the equator) and the Middle East and Asia
(east of 30°E and south of 40°N).

In January, the Eurasian land mass was colder than in the
modern simulation (see Figure 1b) and the winter monsoon circulation was slightly stronger than at present. The simulation
results for the entire seasonal cycle are summarized in Table III.
The average annual temperature did not change as much as the
seasonal averages because the seasonal extremes tended to compensate each other at 9000 yr BP.

4.   COMPARISONS OF MODEL SIMULATION WITH OBSERVATIONS

The model simulation provides an estimate of the climate that
would be produced by the change in solar radiation alone at 9000 yr
BP. This is why it is termed a sensitivity experiment. Sea surface temperatures, land albedo and soil moisture were set at
present-day values. Moreover, the North American ice sheet was not
included in the model as a lower boundary condition. At 9000 yr
BP, the Scandinavian ice sheet had disappeared, but the North
American ice sheet still extended across Canada from west of
Hudson's Bay to the Atlantic coast, although it was considerably
reduced in volume compared with the glacial maximum (Ruddiman &
McIntyre 1981, Denton & Hughes 1981).

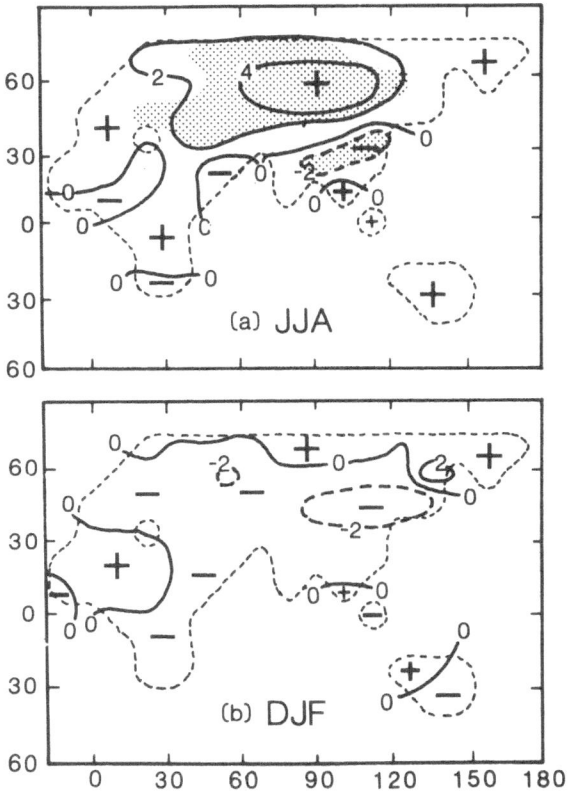

Fig. 1.   Simulated land-surface temperature differences (K) for
          (a) June to August and (b) December to February (9000 yr
          BP minus present-day) for the sector 0-180°E. Grid points
          where temperature differences exceed three standard devia-
          tions are indicated with shading. The temperature
          difference at oceanic grid points is set to zero (see text).

     A GCM experiment incorporating both the North American ice
sheet and the solar radiation regime at 9000 yr BP showed that the
climatic impact of the ice sheet was restricted mainly to the North
American continent (Kutzbach & Otto-Bliesner 1982). With the ice
sheet included, the pattern of the June to August temperature
change for 9000 yr BP over Eurasia was very similar to that shown
in Figure 1(a), although the maximum increase in temperature was
reduced from 5 to 4 K. The Asian summer monsoon circulation at
9000 yr BP was also somewhat weaker in the second experiment,
although it was still strong compared with the modern control case.
Over the North African-Asian sector, the simulated changes in precipitation-minus-evaporation were similar, but smaller (Table IIa).

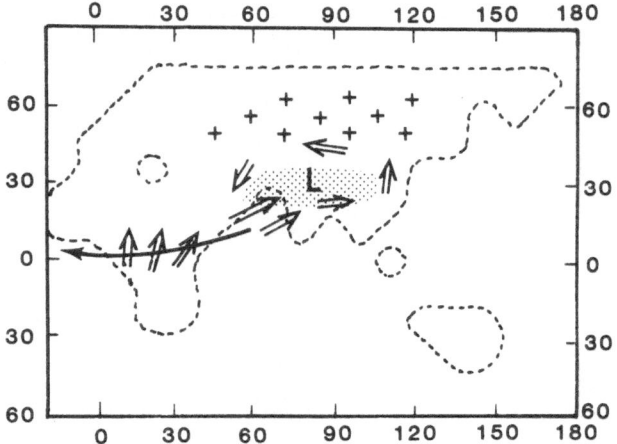

Fig. 2.   Summary of differences between the 9000 yr BP model
          simulation for July and the modern control case: +, higher
          temperature; L, intensified monsoon low; ⇒, stronger low
          level wind; →, stronger upper level wind; shaded area,
          increased precipitation.

Table II. Simulated changes in precipitation and precipitation-
           minus-evaporation for the land areas of North Africa
           (30°N to the equator) and the Middle East and Asia (east
           of 30°E and south of 40°N):   (a) June to August averages;
           (b) annual averages.   Values shown are for 9000 yr BP
           radiation, present-day conditions, and differences.   The
           values in italics in (a) refer to the case of the model
           simulation with a North American ice sheet.

	9000 yr BP (mm)	Present-day (mm)	Differences (mm)	% Change (%)
(a) <u>June to August</u> <u>averages</u>				
Precipitation	546	434	112 *63*	26 *15*
Precipitation-minus-evaporation	28	-55	83 *72*	
(b) <u>Annual averages</u>				
Precipitation	1548	1590	-42	-2.5
Precipitation-minus-evaporation	-154	-241	87	

      Because of the inherent limitations of the low-resolution GCM
(the resolution was approximately that of a 10° latitude and 10°
longitude grid), because the model employed various simplified

Table III. Simulated surface temperature of land in the northern
hemisphere, southern hemisphere and the global average
of land and ocean, for 9000 yr BP compared with present-
day values: (a) June to August averages; (b) annual
averages. The difference between values at 9000 yr BP
and at present are denoted by Δ. The significance
level (S.L.) is determined from the ratio of Δ to the
model standard deviation.

	9000 yr BP (°C)	Present-day (°C)	Δ (K)	S.L. (%)
(a) June to August averages				
Northern hemisphere, land	25.0	23.8	1.2	0.1
Southern hemisphere, land	2.5	1.7	0.8	
Global, land and ocean	17.8	17.5	0.3	1
(b) Annual averages				
Northern hemisphere, land	13.3	13.6	-0.3	
Southern hemisphere, land	7.5	7.2	0.3	
Global, land and ocean	15.9	15.9	0	

parameterizations of physical processes, and because certain
boundary conditions (albedo, soil moisture, ocean temperature) may
have been unrealistic, the 9000 yr BP sensitivity experiments may
include unrealistic features. A detailed comparison between the
simulated and observed palaeoclimate should await therefore the
completion of experiments with models of higher resolution,
improved physics and more accurate palaeoenvironmental boundary
conditions. These experiments are in progress.

Nevertheless, the preliminary results agree with many palaeoclimatic observations (Kutzbach 1981, Kutzbach & Otto-Bliesner
1982). Most significantly, the pattern of increased precipitation
and increased precipitation-minus-evaporation across North Africa
and Asia at 9000 yr BP agrees broadly with the maps of high lake-
level reported by Street and Grove (1979) and Street-Perrott and
Roberts (this volume). The suggestion of these and other authors
that the increased rainfall should be attributed to enhanced
monsoonal circulations are therefore broadly confirmed.

The magnitude of the simulated change in precipitation or
precipitation-minus-evaporation is consistent with the magnitude
of the changes inferred from palaeolake studies (Table I). The
increase in summer precipitation over the North African-Asian
sector exceeded 100 mm (Table IIa), with some grid points
recording increases of over 300 mm. There are increases in precipitation-minus-evaporation of similar magnitude. The corresponding changes were somewhat smaller in the simulation experiment
which included the North American ice sheet as a boundary condition
(Table IIa). For the North African-Asian sector, mean annual
precipitation was slightly less at 9000 yr BP than in the modern

control case, but precipitation-minus-evaporation was increased by
about the same amount as the summer excess (Table II). The zonal
and annual averages of simulated precipitation-minus-evaporation
over land (see Kutzbach & Otto-Bliesner 1982 - Figure 6) show an
increase at 9000 yr BP (wetter conditions) in the northern hemisphere subtropics and a decrease (drier conditions) in the southern
hemisphere subtropics; however, as the model variability is relatively large, these results are not statistically significant.

Eventually, certain details of the palaeolake evidence may
also be explained in terms of these simulated monsoon circulations.
For example, Street-Perrott and Roberts (this volume) refer to a
"southward-projecting cusp" of low lake-levels to the east of 25°E
and north of 26°N, contrasting with the high lake-levels across
Africa, southern Arabia and northwest India. In the model simulation, the region in question lies in the zone of increased
northerly (drier) flow to the west of the intensified monsoon low
(Figure 2).

The results of the simulation experiment suggest increased
rains in parts of China, a region of "no data" on the lake-level
charts of Street and Grove (1979). From pollen records in central
Taiwan, Tsukada (1967) concluded that there was a rapid replacement
of cool-temperate species by subtropical and warm-temperate species
at about 10 000 yr BP. Chang (1968) identified a warm, moist phase
during the early post-glacial in China. His survey included
references to work by Teilhard de Chardin (1936/37) and Andersson
(1947) who reported high water levels of lakes in East Mongolia
and in Honan, Shansi and Shensi. Although these studies lacked
the time control provided by radiocarbon dates, the evidence was
believed by Chang to date from the early Holocene.

Pittock and Salinger (1982) summarized evidence for generally
warmer and wetter conditions in Australia and New Zealand during
the early Holocene. As the low-resolution GCM (Kutzbach 1981,
Kutzbach & Otto-Bliesner 1982) includes relatively few grid points
in these areas, no attempt was made here to compare the model
simulation with observations from this region.

At higher latitudes, there is considerable evidence for a
climate with a warmer growing season from 10 000 to 5000 yr BP in
parts of Eurasia. On the basis of pollen and macrofossil analyses
that reflect primarily conditions during the summer half-year and
that indicate northward shifts of 200-400 km in the principal
vegetation zones, Khotinskii (1973) concluded that a pronounced
warming occurred around 10 000-9000 yr BP in both western and
eastern Siberia, followed by cooling around 5000 yr BP.

Further support for the hypothesis that orbital variations are
related to variations in monsoon climates stems from a comparison
of a time series of the orbitally-induced radiation changes and the
climatic changes. The peak in summertime insolation (compared to
the present) occurred around 11 000 yr BP and the largest radiation
anomalies were confined to the period from 16 000 to 6000 yr BP.
The time and duration of this interval is in fair agreement with

the corresponding interval of high lake-levels, although Street-
Perrott and Roberts (this volume) stress that a period of maximum
aridity occurred in Africa until about 13 000 yr BP and was followed
by a rapid rise in lake levels. The lakes returned to low levels
similar to today around 5000 yr BP. Prell (personal communication)
attributed the monsoonal increase in upwelling in the Arabian Sea
between 10 000 and 4000 yr BP to the increased summer insolation
over southern Asia resulting from changes in the Earth's orbit.

Street and Grove (1979) reported high lake-levels prior to
25 000 yr BP. This occurrence of high lake-levels might possibly
have coincided in time with the previous occurrence of perihelion
in the northern hemisphere summer at *ca.* 30-35 000 yr BP. The
detailed study of lake-level fluctuations also identifies features
in the palaeoclimatic record that cannot be explained simply in
terms of gradual variations in radiation associated with orbital
changes over a period of 20 000 years or longer. Street-Perrott
and Roberts (this volume) report that there were short but dramatic
lake-level regressions around 10 000 and 7500 yr BP. They mention
that the 10 000 yr BP event corresponded to a time of major meltwater release, and also suggest the possible role of meltwater in
causing the extreme aridity of the period 14 000-12 500 yr BP.
Thus, while radiation changes associated with long-period orbital
variations may contribute to an explanation of the changes in
monsoon climates during the early Holocene, they could not have
been the sole cause. Shorter-period orbital influences, different
levels of volcanic or solar activity or internal processes must
also have played a role.

5.   ORBITAL VARIATIONS AND TROPICAL CLIMATES - A BRIEF HISTORICAL
     REVIEW

In 1842, Adhémar published an astronomical theory of climatic
change. Basing his arguments on the 22 000 yr precession cycle, he
theorized that glacial climates would occur every 11 000 years,
alternating in the northern and southern hemispheres. Croll's
work on the astronomical theory during the 1860s and 1870s likewise
concentrated on glacial climates (see Imbrie & Imbrie 1979). With
certain exceptions, Adhémar's and Croll's emphasis on climatic
changes at middle and high latitudes, i.e., on glacial-interglacial
climatic fluctuations, has been common to many of the subsequent
elaborations and advancements of the astronomical theory.

The work of Milankovitch (1920, 1930, 1938, 1941) was primarily concerned with high-latitude climates. He did include
tabulated values of summer half-year and winter half-year insolation for all geographical latitudes (5°, 15°, 25° ... 65°, 75°;
see Milankovitch 1941 - Table XXV or Milankovitch 1938 - Table 92),
but he focused his discussion on the glacial-interglacial flunctua-
tions in high latitudes. He mentioned lake-level changes only
briefly, and only in connection with middle-latitude lakes such as
Bonneville (Milankovitch 1920 - p. 284). Milankovitch (1938),

commenting on the radiation chronology for 5°N, 15°N, 5°S and 15°S
noted that (i) there were 28 cycles (Wellen) in the past 600 000
years – reflecting the predominant influence of the 21 000 yr
precession cycle and (ii) the variations in the summer half-year
insolation were very large, sufficiently so as to explain vertical
variations in tropical mountain snowlines of as much as ±1200 m,
equivalent to land-surface temperature changes of ±8 K. At the
time Milankovitch made these points, estimates of ice-age snowline
and temperature depressions of about this magnitude were available
(Flint 1971). Köppen and Wegener (1924), with whom Milankovitch
corresponded and who used his calculations, also seem to have
concentrated almost exclusively on the middle- to high-latitude
implications of Milankovitch's ideas. They mentioned evidence for
tropical pluvials, but concluded that the data then available were
not sufficient to permit detailed analysis.

An interesting line of research on the relation between the
astronomical theory and the monsoons was pursued by Spitaler (1921).
Milankovitch referred to early work by Spitaler, dating from 1907,
dealing with the astronomical theory and the ice ages, but noted
that the calculations contained technical errors. Working at the
same time as Milankovitch, Spitaler clearly stated that important
land-sea temperature contrasts would be produced in response to
fluctuations in orbital parameters:

"Von sehr zu beachtender Bedeutung fur die Klimaänderungen
während des Umlaufes des Perihels scheint mir die Veränder-
lichkeit des Temperaturgegensatzes zwischen Land und Meer zu
sein. Durch diesen werden barometrische Gradienten geschaffen,
welche die Monsunwinde und monsunartigen Winde hervorrufen.
Diese für das Klima bedeutungsvollen jahreszeitlich veränder-
lichen Luftströmungen vom Meere zum Lande und umgekehrt ändern
im Laufe der Zeit ihre Stärken, weil die Ursache, der Tempera-
turgegensatz Land-Meer, sich ändert."

Spitaler presented tables of the estimated temperature
differences over land and ocean areas between the cases of perihelion at the June solstice and the January solstice. For example,
Spitaler (1921 – Table XXVI, p. 43) indicated that at 40°N for land:
$\Delta T_{summer} = 4.2$ K, $\Delta T_{winter} = -2.6$ K, and for ocean: $\Delta T_{summer} = 0.4$ K, $\Delta T_{winter} = -0.3$ K. These values are very close to those
obtained from the GCM experiment described in the previous section
(Kutzbach 1981). Spitaler's subsequent work (Spitaler 1934a, b)
included a discussion of the latitude of least variation of seasonal
radiation. Both Spitaler and Wundt (1938) considered that secular
changes in this latitude, which was named the 'caloric equator' by
Milankovitch (1938), played an important role in tropical climatic
changes, although this was not discussed by Milankovitch.

Zeuner (1946) also emphasized the possible importance of the
Milankovitch low-latitude insolation chronology for explaining
tropical pluvials. He presented a graph of the latitudinal variations of the 'caloric equator' during the past 600 000 years and
indicated that the timing of pluvials might well be linked to

these variations. Milankovitch (1938 – Table 93, 1941 – Table XXVI)
showed that the 'caloric equator', currently near 3°N, has varied
between about 8°N and 8°S over the past 600 000 years. The
'caloric equator' marks the latitude where the values of summer
half-year and winter half-year radiation are equal. Between the
'caloric equator' and the geographic equator, the 'summer' radiation is less than the 'winter' radiation. In other words, the
zone between the caloric and geographic equators behaves, in
terms of half-year radiation totals, as the opposite hemisphere.
At 10 000 yr BP (see, for example, Milankovitch 1941 – Table XXVI),
the 'caloric equator' was at 3°S and the zone north of 3°S had its
radiation maximum in the northern hemisphere summer.

In 1940, Sir George Simpson, the President of the Royal
Meteorological Society, concluded a series of his own calculations
of possible temperature changes based upon the Milankovitch radiation chronology. Simpson (1940) wrote "we can therefore say that
the changes of temperature which can be brought about by changes
in the Earth's orbit are of the order of 1°C and seldom exceed this
amount in any part of the world. In view of these calculations I
think we are justified in saying that there is no relationship
between the zig-zags on the curve of summer radiation and the
glacial and interglacial epochs of the Pleistocene Period; and
further, that the changes of solar radiation due to changes in the
Earth's orbit are always too small to be of practical importance."
This result by Simpson has often been quoted.

In contrast to this general negative conclusion, Simpson's
own paper contained results that pointed towards a more significant
role for orbital changes. Simpson calculated zonal average temperature departures (from the present) for 11 100 yr BP, the so-called
Climatic Optimum. He found $\Delta T = -1.6$ K for January, $+1.6$ K for
July, and 0 for the annual average. These values were considerably
smaller than those he quoted from Milankovitch (-6.6 K, 6.6 K, 0,
respectively). Simpson wrote that Milankovitch's values for zonal
average temperature changes were about four times too large because,
among other things, Milankovitch considered only the response of a
land surface to the radiation anomalies. Simpson noted that his
own zonal average values were for the appropriate average of land
and ocean at each latitude, and that they might have to be doubled
in "exceptional cases, ... for example, in the middle of Asia."
Simpson's doubled values would be in fair agreement with the
values obtained from the GCM simulation for central Asia (Kutzbach
1981). However, Simpson did not elaborate on the differential
response of land and ocean, and therefore did not relate the monsoons and the astronomical theory.

Zeuner (1959), while noting Simpson's general reservations,
was apparently not dissuaded from his own interest in the astronomical theory; he expanded upon his earlier ideas about climatic
changes (pluvials) and radiation changes in the tropics.

In the early 1960s, Bernard addressed directly the problem of
the astronomical theory and tropical pluvials. In a key paper

(Bernard 1962), he described the temporal fluctuations of the
'caloric equator', previously tabulated by Milankovitch, and concluded that there should be alternations of pluvials and inter-
pluvials in response to the solar radiation changes. He identified
two types of pluvials. *Displuvials* would occur when perihelion was
at the June solstice and when the extreme position of the 'caloric
equator' was far to the south of the geographic equator. Then, the
equator and northern latitudes would undergo large seasonal variations of radiation so that there would be "a contrasting tropical
rainfall regime with strong summer rainfall and dry winters."
Isopluvials would occur near the 'caloric equator' where the seasonal variation of radiation would be small. Bernard (1963)
concluded "One can thus deduce that the last pluvial of the African
Quaternary corresponds to a displuvial installed between the
equator and the northern tropic between 16 500 and 6000 yr BP."

Bernard's treatise, covering some 230 pages, cannot be
examined here in detail. In addition to the exposition of his
astronomical theory, he developed hydrological budget equations for
closed lakes and showed how precipitation depended upon catchment/
lake area ratios. Bernard also considered the problem of moisture
recycling and intensification of the hydrological cycle in response
to an increase in solar radiation.

It is beyond the scope of this review to survey the more
extensive group of papers (Broecker 1966, Broecker *et al.* 1968,
Broecker & Van Donk 1970, Chappell 1974, Ruddiman & McIntyre 1981)
that have suggested that glacial-interglacial climates at high
latitudes may be influenced by low-latitude solar radiation changes,
since these papers did not deal directly with monsoon climates.
Similarly, various model experiments involving the effects of solar
radiation changes on climate (Saltzman & Vernekar 1971, Mason 1976,
Suarez & Held 1976, 1979, Schneider & Thompson 1979, Imbrie &
Imbrie 1980) are not reviewed because they did not treat monsoon
climates explicitly.

6.   CONCLUDING REMARKS

Monsoon climates have been changing from wetter to drier and
back again for a long time. Our understanding of global-scale
monsoonal fluctuations is currently improving because of simultaneous developments in several areas of research: more detailed
maps of the spatial patterns of environmental change, more precise
time control, improved methods for the quantitative estimation of
palaeoclimate from geological indicators and enhanced capabilities
for the numerical simulation of climate have all contributed. A
challenge for the future will be to take full advantage of the
various fields of inquiry that may contribute to the solution of
this long-standing puzzle.

ACKNOWLEDGEMENTS. Research grants to the University of Wisconsin-Madison from the National Science Foundation's Climate Dynamics Program (NSF Grants ATM79-16443 and ATM79-26039) supported this work. The author thanks Peter Guetter for performing the general circulation model computations, and Melanie Woodworth and Elizabeth Seehawer for preparing the manuscript. The computations were made at the National Center for Atmospheric Research (NCAR), Boulder, Colorado, which is sponsored by the National Science Foundation, with a computing grant from the NCAR Computing Facility. This paper has also been presented as the Richard Foster Flint Lecture in Quaternary Geology at the Department of Geology and Geophysics of Yale University, New Haven, Conneticut, on 24 March 1982.

REFERENCES

Andersson, J.G. 1947 Prehistoric sites in Honan. *Bull. Mus. Far East. Antiquities Stockholm,* 19, 1-124.

Berger, A.L. 1978 Long-term variations of caloric solar radiation resulting from the earth's orbital elements. *Quat. Res.,* 9, 139-167.

Bernard, E.A. 1962 Theorie astronomique des pluviaux et inter-pluviaux due Quaternaire African. *Acad. Roy. Sci. Outre-Mer (Brussels), Classe Sci. Tech., Mem. in 8°,* 1, 232 pp.

Bernard, E.A. 1963 The laws of physical palaeoclimatology and the logical significance of palaeoclimatic data. In: *Problems in Palaeoclimatology* (ed. A.E.M. Nairn), pp. 309-321. London: Wiley-Interscience.

Brackenridge, G.R. 1978 Evidence for a cold, dry full-glacial climate in the American Southwest. *Quat. Res.,* 9, 22-40.

Broecker, W.S. 1966 Absolute dating and the astronomical theory of glaciation. *Science, N.Y.,* 151, 299-304.

Broecker, W.S. & von Donk, J. 1970 Insolation changes, ice volumes, and the O^{18} record in deep-sea cores. *Rev. Geophys. Space Phys.,* 8, 169-198.

Broecker, W.S., Thurber, D.L., Goddard, J., Ku, T.L., Matthews, R.K. & Mesolella, K.J. 1968 Milankovitch hypothesis supported by precise dating of coral reefs and deep-sea sediments. *Science, N.Y.,* 159, 297-300.

Brooks, C.P. 1926 *Climate Through the Ages* (Rev. 1949 edn.), 395 pp. New York: Dover.

Bryson, R.A. & Swain, A.M. 1981 Holocene variations of monsoon
    rainfall in Rajasthan. *Quat. Res.*, <u>16</u>, 135-145.

Butzer, K.W., Isaac, G.L., Richardson, J.A. & Washbourn-Kamau, C.
    1972 Radiocarbon dating of East African lake levels. *Science,
    N.Y.*, <u>175</u>, 1069-1076.

Chang, Kwang-chih 1968 *The Archaeology of Ancient China*, 483 pp.
    New Haven: Yale Univ. Press.

Chappell, J. 1974 Relationships between sea levels, O^{18} variations
    and orbital perturbations during the past 250 000 years.
    Nature, Lond., <u>252</u>, 199-202.

Denton, G.H. & Hughes, T.J. (ed.) 1981 *The Last Great Ice Sheets*,
    484 pp. New York: Wiley-Interscience.

Fairbridge, R.W. 1962 New radiocarbon dates of Nile sediments.
    Nature, Lond., <u>196</u>, 108-110.

Flint, R.F. 1971 *Glacial and Quaternary Geology*, 892 pp. New York:
    Wiley-Interscience.

Flohn, H. & Nicholson, S. 1980 Climatic fluctuations in the arid
    belt of the 'Old World' since the last glacial maximum;
    possible causes and future implications. *Palaeoecol. Africa,*
    <u>12</u>, 3-21.

Grove, A.T. & Pullan, R.A. 1963 Some aspects of the Pleistocene
    palaeogeography of the Chad Basin. In: *African Ecology and
    Human Evolution* (ed. F.C. Howell & F. Bourliere) No. 36,
    pp. 230-245. Chicago: Adeline.

Imbrie, J. & Imbrie, J.Z. 1980 Modeling the climatic response to
    224 pp. Short Hills, NJ: Enslow.

Imbrie, J. & Imbrie, J.Z. 1980 Modelling the climatic response to
    orbital variations. *Science, N.Y.*, <u>207</u>, 943-953.

Khotinskii, N.A. 1973 Transkontinental'naia Korreliatsiia etapov
    istorii rastitel'nosti i klimata severnogo; Evrazii v. Golot-
    sene. *Problemy Palinologii Trudy III Mezhdunarodnoi Palino-
    logicheskoi Konferentsii* (ed. M.I. Neishtadt), pp. 116-123.
    Moscow: Nauka (English translation by G.M. Peterson).

Köppen, W. & Wegener, A. 1924 *Die Klimate der Geologischen Vorzeit*,
    255 pp. Berlin: Gebruder Borntraeger.

Kutzbach, J.E. 1980 Estimates of past climate at Paleolake Chad, North Africa based on a hydrological and energy-balance model. *Quat. Res.*, 14, 210-223.

Kutzbach, J.E. 1981 Monsoon climate of the early Holocene: climate experiment with the Earth's orbital parameters for 9000 years ago. *Science, N.Y.*, 214, 59-61.

Kutzbach, J.E. & Otto-Bliesner, B.L. 1982 The sensitivity of the African-Asian monsoonal climate to orbital parameter changes for 9000 yr B.P. in a low resolution general circulation model. *J. atmos. Sci.* [In the press.]

Leopold, L.B. 1951 Pleistocene climate in New Mexico. *Am. J. Sci.*, 249, 152-168.

Maley, J. 1977 Palaeoclimates of central Sahara during the early Holocene. *Nature, Lond.*, 269, 573-577.

Mason, B.J. 1976 Towards the understanding and prediction of climatic variations. *Q. Jl R. met. Soc.*, 102, 473-498.

Milankovitch, M. 1920 *Théorie mathématique des phénomenes thermiques produits par la radiation solaire*, 338 pp. Paris: Gauthier-Villars.

Milankovitch, M. 1930 Mathematische klimalehre und astronomische theorie der klimaschwankungen. In: *Handbuch der Klimatologie*. I(A) (ed. W. Köppen & R. Geiger), pp. 1-176. Berlin: Gebruder Borntraeger.

Milankovitch, M. 1938 Astronomische mittel zur erforschung der erdgeschichtlichen klimate. In: *Handbuch der Geophysik*, 9, (ed. B. Gutenberg), pp. 593-698. Berlin: Gebruder Borntraeger.

Milankovitch, M. 1941 Kanon der Erdbestrahlung und seine Anwendung auf das Eiszeitenproblem. *Royal Serb. Acad. Spec. Publ. (Belgrade)*, No. 133, pp. 1-633. (English translation published in 1969 by Israel Program for Scientific Translations. Available from U.S. Dept. of Commerce, Washington, D.C.)

Pittock, A.B. & Salinger, M.J. 1982 Toward regional scenarios for a CO_2-warmed earth. *Clim. Change*, 4, 23-40.

Rognon, P. & Williams, M.A.J. 1977 Late Quaternary climatic changes in Australia and North Africa: A preliminary interpretation. *Palaeogeogr. Palaeoclimatol. Palaeoecol.*, 21, 285-327.

Ruddiman, W.F. & McIntyre, A. 1981 Oceanic mechanisms for amplification of the 23 000-year ice-volume cycle. *Science, N.Y.*, 212, 617-627.

Saltzman, B. & Vernekar, A.D. 1971 Note on the effect of earth
    orbital radiation variations on climate. *J. geophys. Res.*,
    76, 4195–4197.

Schneider, S.H. & Thompson, S.L. 1979 Ice ages and orbital varia-
    tions: some simple theory and modelling. *Quat. Res.*, 12,
    188–203.

Simpson, G.C. 1940 Possible causes of changes in climate and their
    limitations. *Proc. Linn. Soc. Lond.*, 152, 190–219.

Singh, G., Joshi, R.D. & Singh, A.B. 1972 Stratigraphic and radio-
    carbon evidence for the age and development of three salt lake
    deposits in Rajasthan, India. *Quat. Res.*, 2, 496–505.

Spitaler, R. 1921 *Das Klima der Eiszeitalterns*, 138 pp. Prague:
    R. Spitaler.

Spitaler, R. 1934*a* Der verschiebung der Kalmen in der Vorzeit.
    Met. Z., 51, 206–209.

Spitaler, R. 1934*b* Zur Bestrahlung der Erde durch die Sonne.
    Met. Z., 51, 209–212.

Street, F.A. 1979 Late Quaternary precipitation estimates for the
    Ziway-Shala Basin, Southern Ethiopia. *Palaeoecol. Africa*,
    11, 135–143.

Street, F.A. & Grove, A.T. 1979 Environmental and climatic impli-
    cations of late Quaternary lake-level fluctuations in Africa.
    Nature, Lond., 261, 335–390.

Suarez, M.J. & Held, I.M. 1976 Modelling climatic response to
    orbital parameter variations. *Nature, Lond.*, 263, 46–47.

Suarez, M.J. & Held, I.M. 1979 The sensitivity of an energy balance
    climate model to variations in orbital parameters. *J. geophys.
    Res.*, 84, 4825–4836.

Swain, A.M., Kutzbach, J.E. & Hastenrath, S. 1982 Monsoon climate
    of Rajasthan, India, for the Holocene: estimates of precipi-
    tation based on pollen and lake-levels. *Quat. Res.*
    [In the press.]

Teilhard de Chardin, P. 1936/37 Notes on continental geology.
    Bull. geol. Soc. China, 16, 195–220.

Tsukada, M. 1967 Vegetation in subtropical Formosa during the
    Pleistocene glaciations and the Holocene. *Palaeogeogr.
    Palaeoclimatol. Palaeoecol.*, 3, 49–64.

Washbourn, C.K. 1967 Late Quaternary lakes in the Nakuru-Elmenteita
    Basin, Kenya. Ph.D. Thesis: University of Cambridge, 358 pp.
    [Unpublished.]

Wickens, G.E. 1975 Changes in the climate and vegetation of the
    Sudan since 20 000 B.P. *Boissiera,* 24, 43-65.

Wundt, W. 1938 Die Verschiebung der Klimagurtel seit dem Ausklang
    der Eiszeit. *Petermanns Geogr. Mitt.,* 84, 332-337.

Zeuner, F.E. 1946 *Dating the Past: An Introduction to Geo-
    chronology,* 444 pp. London: Methuen.

Zeuner, F.E. 1959 *The Pleistocene Period: Its Climate, Chronology,
    and Faunal Successions,* 447 pp. London: Hutchinson.

SEA-LEVEL CONTROL OF ICE SHEET DISINTEGRATION

James L. Fastook

Institute of Quaternary Studies
University of Maine
Orono, Maine 04469, USA

ABSTRACT. A simplified model of the Laurentide (North American) Ice Sheet shows that an initial rise in sea level, produced by melting along its landward margin, would trigger an instability and, hence, a major retreat of its sea-level controlled marine margin. Recession of the seaward margin would contribute further to the rise in sea level, leading to retreat rates up to four times faster than on land, which is primarily controlled by ablation.

Of the major late Quaternary ice sheets, only those in the northern hemisphere had extensive terrestrial margins with ablation zones which would have been sensitive to changes in insolation. Marine ice-sheet margins, which predominated in the southern hemisphere and in the high latitudes of the northern hemisphere, are less sensitive to variations in ablation. Indeed, few have ablation zones, mass removal being primarily by iceberg calving. However, marine margins are extremely sensitive to changes in sea level and can only be stabilized if they are buttressed by floating ice shelves pinned to the sea floor at islands and shoals.

Using as an input the retreat history of the terrestrial margin, as deduced from dated moraines, a sea-level curve representing the changing volume of the Laurentide Ice Sheet was produced. This duplicated the essential features of the CLIMAP sea-level curve based on oxygen-isotope ratios. Output from the ice-sheet model includes cross-sections showing the positions of the terrestrial and marine margin and the major ice domes as a function of time, as well as the extent of actively buttressing ice shelves and the velocity of the ice discharge across marine grounding lines.

A. Street-Perrott et al. (eds.), Variations in the Global Water Budget, 391–401.

1.   RESPONSE OF ICE SHEETS TO ORBITAL VARIATIONS

    It has long been recognized that Milankovitch's (1941) theory
of orbital variations provides a means of explaining the periodic
waxing and waning of the world's major ice sheets over the last few
million years. A detailed comparison of the variations in incident
radiation during the summer half-year between 45° to 60°N with
oxygen-isotope curves from deep-sea cores and sea-level records for
isotope stages 5 and 7 (Mesolella *et al.* 1969, Hays *et al.* 1976,
Ruddiman & McIntyre 1981) reveals a close relationship. Outside
this latitude band, a weaker correlation is found, since the insolation signal is strongly dependent on latitude with 180° phase
shifts between its three components in different hemispheres. The
41 000-yr cycle in the tilt of the Earth's axis, which is dominant
at high latitudes, occurs in phase in both hemispheres, whereas
the superimposed 23 000-yr precession cycle, which is dominant at
middle latitudes, has an opposite effect in the northern and
southern hemispheres. Superimposed on, and modulating, both of
these signals is the 96 000-yr variation in the eccentricity of
the Earth's orbit.
    Figure 1 shows summer half-year solar insolation curves for
40° to 90°N for the last 250 000 years expressed as the departure
of insolation from 1950 values (Vernekar 1972). Also shown is the
oxygen-isotope record from deep-sea core V19-29, which is used
here as an index of ocean volume and, hence, of ice volume
(Ninkovich & Shackleton 1975). The correlation between the mid-
latitude radiation curves and the isotope curve is most pronounced
in stages 5 and 7 where the peaks and troughs in the insolation
curve correspond remarkably well with similar features in the
isotope curve. A weaker correlation with the high-latitude insolation signal is evident in stages 3, 5 and, to some extent, 7
where the maxima and minima of the insolation curve do not coincide
with those of the isotope curve.
    The problem of defining the specific mechanism linking the
Milkanovitch curves to the response of the ice sheets stems from
the apparent absence of a direct reaction of the world's cryosphere
to local variations of insolation. For example, dated expansions
of the grounded ice in the Ross Sea in Antarctica correlate well
with the late Quaternary maxima of the northern hemisphere
Laurentide Ice Sheet, although independent alpine glaciers in
Taylor Valley in Antarctica simultaneously experienced their
minimum extent, presumably in response to local insolation or
precipitation changes (Drewry 1980, Stuiver *et al.* 1981, Mayewski
et al. 1981). Thus, when northern hemisphere ice sheets responded
to insolation variations, as reflected by changes in world sea
level, there seems to have been little lag in the response of the
geographically-isolated southern hemisphere ice sheets.
    Looking at the world cryosphere, it can be seen that the most
sensitive latitude band apparently spans the area occupied by the
major terrestrial margins of the northern hemisphere ice sheets

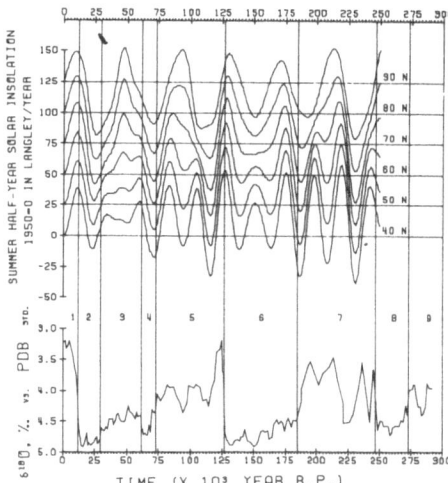

Fig. 1.   Solar insolation (langley yr^{-1}) during the summer half-
          year at latitudes 40-90°N (Vernekar 1972) compared with
          the oxygen-isotope record from core V19-29 (Ninkovich &
          Shackleton 1975).   Isotopic stage boundaries are indicated
          by vertical solid lines.

during their maximum phases. Previous models (Thomas & Bentley
1978, Stuiver *et al.* 1981, Fastook & Hughes 1981) have shown that
the instability of oceanic ice sheets can be triggered by rising
sea level. Marine margins are relatively insensitive to local
temperature changes in contrast to terrestrial margins where
climatically-controlled ablation is solely responsible for ice
removal. This insensitivity stems from the strongly-concentrated
pattern of the ice flow into ice streams, the dynamics of which
control the positions of the margins and the movement of the ice
sheets. While an ablation zone is a necessary condition for a
terrestrial margin to be in a steady state, a marine margin can
reach a steady state without an ablation zone, since the necessary
mass balance is maintained by the calving of icebergs. Thus, a
land-based margin with a positive mass balance must advance until
a sufficient area exists below the level of the 0° isotherm to
provide the necessary ablation zone. Since climatic variations
are likely to manifest themselves as changes in the elevation of
the 0° isotherm, a profile where the surface intersects this
critical datum is more responsive than one with a surface lying
well above the 0° isotherm.
     The mechanism whereby the world's cryosphere responds in
concert to a localized change of insolation can be postulated.
The climatically-sensitive terrestrial margins (found predominantly
around the northern hemisphere ice sheets) melt and retreat in

response to variations of solar insolation. Assuming that long-term changes in groundwater storage are insignificant, this melting would increase world sea level, thereby setting up the retreat of the marine margins on the poleward size of the northern hemisphere ice sheets as well as of the Antarctic Ice Sheet with its largely marine margins. As these margins respond, they too contribute to the rise in sea level, thus initiating a positive feedback mechanism that greatly amplifies the effect of the local insolation variations in the critical latitude-band $45°-60°N$. This treatment ignores the possibility of a local sea-level lowering as the ice masses diminish (Clark *et al*. 1978). The degree of complexity required to include this isostatic effect was beyond the scope of this model, while the rheology used was also considered to be inadequate (Weertman 1978, Hughes 1981*a*). The major ice sheets remaining today (in Greenland and East Antarctica) are less susceptible to the marine type of instability, due to the generally-high bedrock sills around their peripheries, even though large portions of their interiors lie below sea level. West Antarctica, the one remaining marine ice sheet, is bounded by extensive, buttressing ice shelves.

2.   DESCRIPTION OF THE GLACIOLOGICAL MODEL

The dynamic instability, which it is suggested provides the critical feedback mechanism necessary to amplify the effects of the insolation variations occurring at $45°-60°N$, involves the response of the grounding line (the position at which ice begins to float as it flows into the sea) to changes in both sea level and ice thickness at the grounding line. Figure 2 illustrates this response, the position of the grounding line being defined by the flotation characteristics of ice in sea water; the dashed line mirroring the sea bed indicates the level below which ice would float. If the ice thickness diminishes, as is postulated in Figure 2, the 'new surface' will be partly below the flotation line. Consequently, the position of the grounding line will retreat to the point where the new surface and the flotation line intersect. Similarly, a rise in sea level will raise the flotation line and, in its turn, produce a new grounding line position. These arguments can be expressed as follows:

$$v = [\dot{h} + v_{sl}(\rho_w/\rho_i)]/[\alpha - \beta(1 - \rho_w/\rho_i)] \qquad (1)$$

where v is the retreat rate of the grounding line, v_{sl} is the rate of rise of sea level, $\dot{h}$ is the thinning rate at the grounding line, ρ_w and ρ_i are the densities of water and ice respectively, and α and β are the ice surface slope and bedrock slope at the grounding line. Initially, given the appropriate bed and surface slopes, a retreat of the grounding line can be produced simply by raising sea level. However, this dynamic instability will be

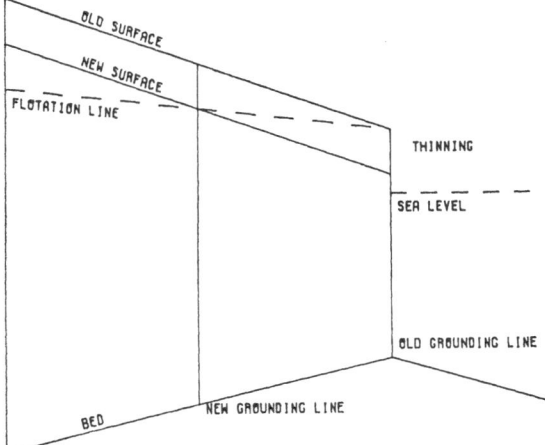

Fig. 2.   Diagram of the grounding line of an ice sheet indicating
its response to a change in ice thickness.

amplified where the bed slopes downwards inland from the grounding
line, thereby increasing the ice thickness at the grounding line
as the retreat progresses. The thinning rate at the grounding
line ($\dot{h}$ in Equation 1) depends on the ice thickness at the
grounding line and can be expressed as:

$$\dot{h} = u(\alpha-\beta)-[(\dot{h}\rho_i g\dot{h})/16A](1-\rho_i/\rho_w)-(2s\tau_s n/WA) + \dot{a} \qquad (2)$$

where u is the velocity of ice across the grounding line, $\dot{h}$ is the
ice thickness at the grounding line, A and n are flow-law parameters (Glen 1955), g is the acceleration of gravity, s, W, and τ_s
are the length, width and shear stress along the sides of the ice
tongue as it flows into the buttressing ice shelf, and $\dot{a}$ is the
accumulation rate at the grounding line. If the conditions at the
grounding line change, the difference between the original thinning
rate and the new thinning rate can be expressed as:

$$\dot{h}'-\dot{h} = \Delta u(\alpha-\beta)-Ch'^{n+1}(1-B'/h')^n-Ch^{n+1}(1-B/h)^n \qquad (3)$$

where $C = n[\rho_i g(1-\rho_i/\rho_w)]/16A$ and B and B' $= 4s\tau_s/[W\rho_i g(1-\rho_i/\rho_w)]$
for the initial and the new configurations, h' is the new ice
thickness, and Δu is the excess (above mass balance) ice velocity
for the ice stream.

     Thus, there are two separate contributions to the global rise
in sea level due, respectively, to melting along the landward ice
margins and to the increased discharge of ice from the streams as
their grounding lines move back along the seaward margins. It was
assumed in the model that the retreat history of the landward
margin (Figure 3: A1) is known adequately from such geological
data as dated moraines (Andersen 1981, Mayewski *et al*. 1981) and

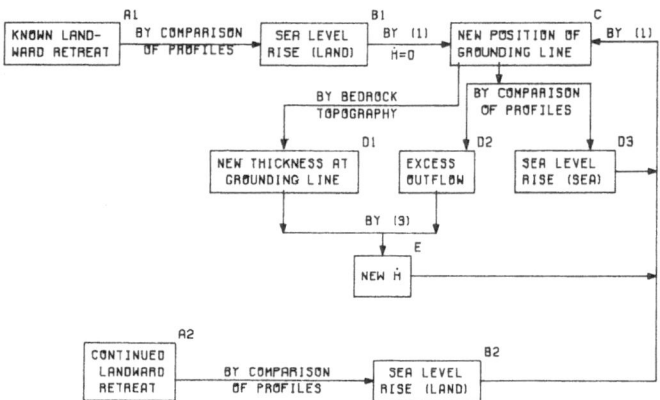

Fig. 3. Diagram outlining the steps in the glaciological model.

it was taken to follow the pattern of the retreat along the Hudson
River Valley in New York. The change in ice volume and the consequent rise in sea level (Figure 3: B1) for a given retreat of the
landward margin were obtained by comparing cross profiles generated
using the CLIMAP finite-difference reconstruction scheme (Hughes
1981b). Briefly, this scheme involves the relationship between
the surface slope α_j, the basal shear stress τ_j, and the ice thickness h_j, at finite difference step j; thus

$$\alpha_j = \tau_j / \rho_i g h_j \tag{4}$$

The basal stresses were obtained from an analytical expression for
an ice sheet in mass-balance equilibrium on a flat bed, and then
applied to the variable-bed topography encountered in a typical
transect across the Laurentide Ice Sheet. Provision was made for
varying degrees of ice-bed coupling, in order to reflect the transition from wet beds dominated by basal sliding to frozen beds
dominated by deformation flow, as well as the highly-lubricated
areas such as ice streams or ice lobes where excessive meltwater
greatly reduces the coupling.
     To minimize computational complexity, two representative flow-
lines were chosen, one terminating on a landward margin and the
other in an ice stream on the seaward margin. The position of
the dome was determined internally by matching the dome heights on
flowlines of differing lengths, but of fixed marginal positions.
The change in ice volume for a given extent of retreat of either
margin was obtained by comparing profiles, assuming a roughly circular shape for the whole ice sheet. A given fraction (1/3) of the
ice-sheet margin was taken to be land-based and the rest marine.
Although variations in this fraction affect the detailed results
from the model, the general result, namely the significant retreat
of the marine margin, remains the same. Once a sea-level change
due to landward retreat was obtained, Equation (1) with the

thinning term set at zero was used to derive the retreat rate of
the grounding line and its new position (Figure 3: C). From this
new position of the grounding line and the known topography of the
bed, the new ice thickness at the grounding line was found (Figure
3: D1). In addition, the same technique of comparing profiles
was used to ascertain the change in ice volume and the resulting
sea-level rise, due to retreat of the seaward margin (Figure 3:
D3). Knowing the amount of excess ice discharged via the ice
streams in a given time, and making a reasonable assumption as to
the width and spacing of average ice streams along the seaward
margins, the excess velocity, Δu (in Equation 3), was derived
(Figure 3: D2). Using the new ice thickness at the grounding
line and the change of velocity, and assuming the thinning rate $\dot{h}$
to be zero initially, Equation (3) was used to obtain a new
thinning rate $\dot{h}'$ at the grounding line (Figure 3: E). At this
point, the continued landward retreat, as defined by the geological
data and the consequent rise in sea level (Figure 3: B2), was
combined with the new thinning rate and the sea-level rise due to
retreat of the seaward margin in Equation (1) to yield the next
position of the grounding line (Figure 3: C). The process was
then repeated.

3.   RESULTS

     The model was applied to an initially-circular ice sheet with
dimensions approximating to the size of the Laurentide Ice Sheet
of 18 000 yr BP (1.32×10^7 km^2 area, 2050 km radius, simulated by
50 km finite-difference steps). In Equation (2), the term $s\tau_s$
describes the compressive back stress from the buttressing ice
shelf and provides the major negative feedback necessary to
prevent a runaway instability. As the rate of retreat of the
grounding line exceeds the rate of calving, the buttressing ice
shelf grows, increasing the compressive back-stress term and
reducing the thinning term in Equation (3). Using this back-stress
term as an adjustable parameter, a close match to a sea-level
curve derived from oxygen isotope data (Matthews 1977, reported
in Stuiver *et al*. 1981) was obtained (Figure 4). The calibration
of this sea-level curve involved the use of a calculated value of
0.11 ‰ per 10 m of sea-level change. The reconstructed sea level,
adjusted to indicate the effect of the collapsing Laurentide Ice
Sheet, is compared in Figure 4 with the sea-level curve calculated
with the model.
     Figure 5 shows the difference between the steady-state
velocity (the value of which is unknown due to uncertainties in
the accumulation rate during the last glaciation) and the velocity
necessary to account for surface lowering during the retreat of
the grounding line. The excess velocity Δu increased slowly to a
peak value of 1.5 km yr^{-1} and then declined rapidly. Figure 5
also shows the compressive back-stress term $s\tau_s$ which represents

J. L. FASTOOK

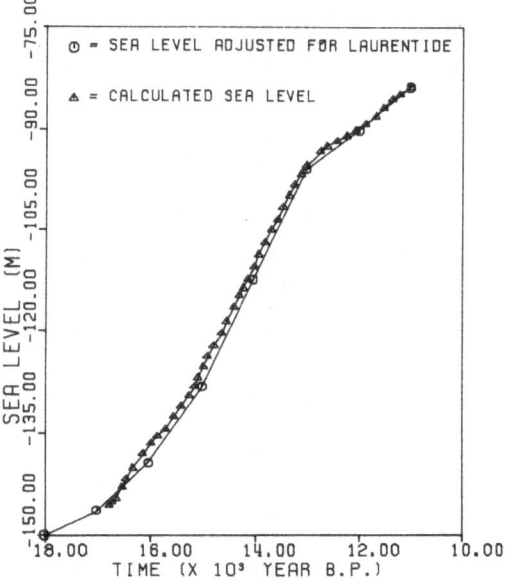

Fig. 4.   Comparison between sea-level curves obtained from oxygen-
          isotope data (○) and calculated with the model (Δ).

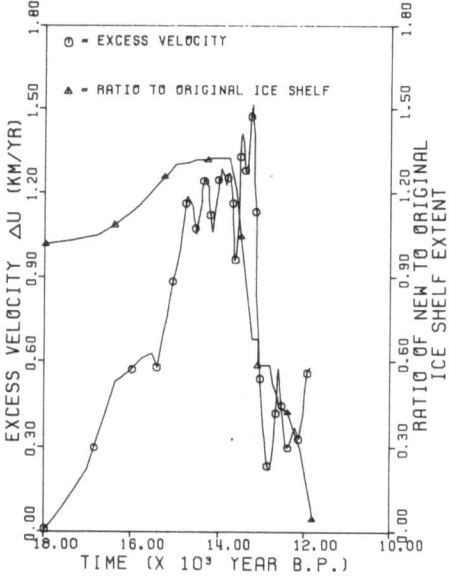

Fig. 5.   Excess ice-stream velocity, namely the difference between
          mass-balance velocity and the velocity necessary to
          account for surface lowering as the grounding line
          retreats (○), and the ratio of the ice-shelf extent (Δ)
          necessary to fit the sea-level curve to the original ice-
          shelf extent of 100 km at a shear stress of 0.55 b.

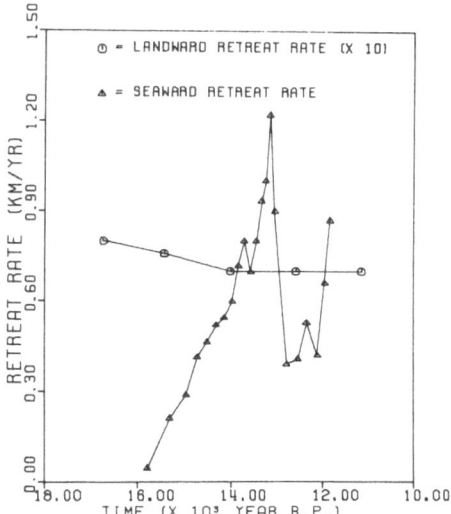

Fig. 6.   Retreat rate ($\times$ 10 km yr^{-1}) for the margins of an ice
          sheet: O, landward-margin velocity (from geological
          data); Δ, seaward-margin velocity (from the model.

the buttressing ice shelf, expressed as a ratio with respect to
its original value. The ratio began at 1.0, climbed slowly to
about 1.3, and then declined rapidly after about 14 000 yr BP.
This does not mean that the ice shelf never exceeded 1.3 times its
original length of 100 km with a shear stress of 0.55 b, rather
that the product of the ice-shelf length s and the shear stress
τ_s attained this value in the model calculation.
          The retreat rates of the landward (as defined by dated moraine
sequences along the Hudson River Valley in New York) and marine
margins (as generated by the model) are shown in Figure 6. The
marine retreat rate peaked at just over 1.2 km yr^{-1} about 13 000
yr BP.
          Cross-sections through the ice sheets at intervals of *ca.*
1300 yr are given in Figure 7, which shows the initially-slow
retreat of the marine margin, and its subsequent rapid acceleration once the grounding line moved into deep water. The movement
of the ice divide, which also would have some effect on the
dynamics of the retreating ice margins, may be noted.
          This model was devised to demonstrate the feasibility of a
connection between the behaviour of the climatically-controlled
landward ice margins and the sea-level controlled marine margins.
Gross over-simplifications severely limit the predictive capacity
of the model, but it serves nonetheless to demonstrate the
possibility that such a mechanism acts as the critical link
between the Milankovitch variations in insolation and observed
fluctuations in sea level.

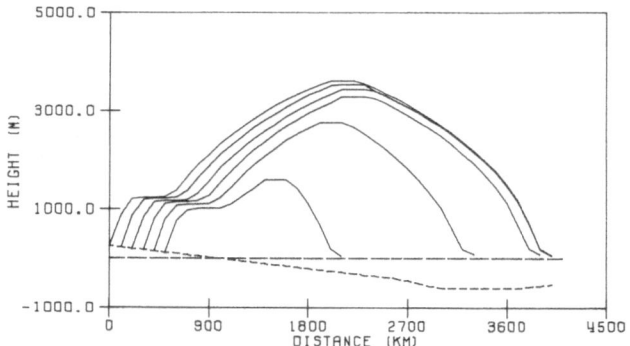

Fig. 7. Cross-sections of an ice sheet with an ice lobe on the
         landward margin (left) and an ice stream on the seaward
         margin (right). Successive profiles are approximately
         1300 yr apart.

ACKNOWLEDGEMENTS. The help of Terry Hughes and George Denton,
whose discussions helped to clarify the concepts in this paper, is
acknowledged. This research was supported by the National Science
Foundation through its grant #DPP8017072.

REFERENCES

Anderson, B.G. 1981 Late Weichselian ice sheets in Eurasia and
     Greenland. In: *The Last Great Ice Sheets* (ed. G.H. Denton &
     T.J. Hughes), pp. 3-54. New York: Wiley-Interscience.

Clark, J.A., Farrell, W.E. & Peltier, W.R. 1978 Global changes in
     post-glacial sea level: A numerical calculation. *Quat. Res.*,
     9, 265-287.

Drewry, D.J. 1980 Pleistocene bimodal response of Antarctic ice.
     Nature, Lond., 287, 214-216.

Fastook, J.L. & Hughes, T.J. 1981 A numerical model for recon-
     struction and disintegration of the late Wisconsin glaciation
     in the Gulf of Maine. In: *Late Wisconsin Glaciation of New
     England* (ed. G. Larson). Debuque, Iowa: Kendall/Hunt Publ.
     Co.

Glen, J.W. 1955 The creep of polycrystalline ice. *Proc. R. Soc.
     Lond.*, A, 228, 519-538.

Hays, J.D., Imbrie, J. & Shackleton, N.J. 1976 Variations in the
     Earth's orbit: Pacemaker of the ice ages. *Science, N.Y.*,
     194, 1121-1132.

Hughes, T.J. 1981*a* Lithosphere deformation by continental ice
    sheets. *Proc. R. Soc. Lond, A,* __378__, 507-527.

Hughes, T.J. 1981*b* Numerical reconstruction of paleo-ice sheets.
    In: *The Last Great Ice Sheets* (ed. G.H. Denton & T.J.
    Hughes), pp. 222-260. New York: Wiley-Interscience.

Matthews, T.K. 1977 The unmixed δO^{18} volume signal for the last
    220 000 years: Model I. [Unpublished CLIMAP Ms.]

Mayewski, P.A., Denton, G.H. & Hughes, T.J. 1981 Late Wisconsin
    ice sheets of North America. In: *The Last Great Ice Sheets*
    (ed. G.H. Denton & T.J. Hughes), pp. 67-170. New York:
    Wiley-Interscience.

Mesolella, K.J., Matthews, R.K., Broecker, W.S. & Thurber, D.L.
    1969 The astronomical theory of climatic change: Barbados
    data. *J. Geol.,* __77__, 250-274.

Milankovitch, M. 1941 Kanon der Erdbestrahlung und seine Anwendung
    aug das Eiszeitenproblem. *Royal Serb. Acad. Spec. Publ.
    (Belgrade),* No. 133, pp. 1-633. (English translation published
    in 1969 by Israel Program for Scientific Translations.
    Available from U.S. Dept. of Commerce, Washington, D.C.)

Ninkovich, D. & Shackleton, N.J. 1975 Distribution, stratigraphic
    position and age of ash layer "L", in the Panama Basin region.
    Earth Planet. Sci. Lett., __27__, 20-34.

Ruddiman, W.F. & McIntyre, A. 1981 Oceanic mechanism for amplifi-
    cation of the 23 000-year ice-volume cycle. *Science, N.Y.,*
    __212__, 617-627.

Stuvier, M., Denton, G.H., Hughes, T.J. & Fastook, J.L. 1981
    History of the marine ice sheet in West Antarctica during the
    last glaciation: A working hypothesis. In: *The Last Great
    Ice Sheets* (ed. G.H. Denton & T.J. Hughes), pp. 319-435. New
    York: Wiley-Interscience.

Thomas, R.H. & Bentley, C.R. 1978 Model for Holocene retreat of
    the West Antarctic Ice Sheet. *Quat. Res.,* __10__, 150-170.

Vernaker, A.D. 1972 Long-period global variations of incoming solar
    radiation. *Met. Mon.* No. 34. Hartford, Conn.: Am. Met. Soc.

Weertman, J. 1974 Stability of the junction of an ice sheet and ice
    shelf. *J. Glaciol.,* __13__, 3-11.

Weertman, J. 1978 Creep laws for the mantle of the Earth. *Phil.
    Trans. R. Soc. Lond., A,* __288__, 9-26.

A CLIMATE FEEDBACK MECHANISM INVOLVING OCEANIC UPWELLING,
ATMOSPHERIC CO_2 AND WATER VAPOUR

H. Flohn

Meteorologisches Institut
Universität Bonn
Federal Republic of Germany

ABSTRACT. Two major problems of climatic history on time scales of
$1-10^3$ years have commanded recent attention, namely the abrupt
transitions (of less than 100 years duration) between warm and cold
phases, and the apparently simultaneous global increase or decrease
of sensible heat flux and evaporation during glacial (cold) and
interglacial (warm) phases, respectively. A global geophysical
feedback mechanism is proposed, based on recent evidence relating
to oceanic upwelling (and downwelling). These data indicate a
simultaneous increase or decrease in the atmospheric content of
carbon dioxide and water vapour, both of which are responsible for
changes in the 'greenhouse effect' in response to changes in the
intensity of upwelling. Quantitative estimates are given of the
variations in oceanic evaporation during the transition between the
late-glacial and the Holocene. This feedback mechanism probably has
caused significant changes of atmospheric CO_2 and H_2O on a time
scale of about 100 years, leading to abrupt climatic fluctuations
and marked variations in the heat budget of the ocean-atmosphere
system.

1. INTRODUCTION

One of the most intriguing events in the Earth's recent climatic history is the occurrence of a practically ubiquitous arid
phase around 18 000 to 12 500 yr BP, during and after the peak of
of the last glaciation (Street-Perrott & Roberts: this volume).
During that interval, the equatorial rain forests of South America,
Africa and, possibly, Indonesia were reduced to about one third of
their present extent (Shackleton 1977). Subsequently, during the

403

A. Street-Perrott et al. (eds.), Variations in the Global Water Budget, 403–418.

early Holocene (10 000-8000 yr BP), the available evidence from all
continents (except North America where large areas remained ice-
covered) suggests a 1°-2°C warming and a climate more humid than at
present. By the peak of the Holocene, around 6000 yr BP, several
areas had become desiccated, and the higher southern latitudes were
cooler. The Eem interglacial (isotope substage 5e, about 125 000
yr BP) was also about 2°C warmer and more humid than today (Frenzel
1968), at least in Eurasia. These facts refer exclusively to the
continents, though the first two events were well distributed over
all climatic belts and therefore can be considered representative
of the globe as a whole. Since global evaporation at present takes
up about 78% of the net radiation available at the Earth's surface,
the question arises as to how far a general increase or decrease in
both sensible heat and evaporation is compatible with the assumption of a constant 'solar constant'.

   Another challenging problem in Quaternary palaeoclimatology
is the occurrence of very abrupt climatic changes on a time scale
of $\leqslant$ 100 years, with an amplitude of $\sim$ half the glacial-interglacial
difference (Müller 1979, Woillard 1979, Flohn 1979, Kukla 1980,
Street-Perrott & Roberts: this volume). Examples are the transitions between isotope stages 5 and 4, and the oldest Dryas-Bølling
and Allerød-Younger Dryas transitions in the late-glacial that
lasted no more than 100-150 years (Eicher & Siegenthaler 1976,
Coope 1977, Eicher *et al*. 1981). The available evidence indicates
variations in annual temperature of about 3°C per century. Such
climatic catastrophes, although quite rare (about one per 10^4 years),
require a physical interpretation or at least a reasonable working
hypothesis that can be tested. Their occurrence under purely
natural conditions sheds an ominous light on the possibility of
future climatic variations stemming from increasing human interference with the climatic system (Kellogg 1977, Flohn 1980).

2.   A LINK BETWEEN THE GLOBAL CO_2 BUDGET, OCEANIC UPWELLING AND
     CLIMATIC CHANGE

   A climate in the past tending to greater aridity or greater
humidity than today implies changes in the global water budget,
i.e., in global evaporation and precipitation. According to
recent general circulation model (GCM) experiments, such changes
are correlated significantly with changes in the CO_2 budget; there
is a positive correlation between changes in global temperature,
atmospheric CO_2 content and the hydrological balance (see Manabe &
Wetherald 1980). Long-term variations of atmospheric CO_2 under un-
questionably natural conditions have been demonstrated by Delmas
et al. (1980) and Berner *et al*. (1980), who independently studied
ice cores from Greenland and Antarctica and found that the atmospheric CO_2 content has varied between about 180 ppm (last-glacial)
and about 350 ppm during the early Holocene. This result intimately
links the problem of a future CO_2-induced climatic change, which

has aroused so much public interest, with the more academic problem of past climatic variations.

The search for an appropriate mechanism leads to a well-known short-term climatic fluctuation with a time scale of only a few months or years: the switch from oceanic upwelling to downwelling in the equatorial Pacific and Atlantic, usually called 'El Niño' after its local occurrence along the Peruvian coast. A remarkable positive correlation between the equatorial Pacific sea surface temperature (SST) and the interannual rate of increase of atmospheric CO_2 content at Mauna Loa has been found (Keeling & Bacastow 1977, Angell 1981). Newell et al. (1978) and Baes (1981) suggested that organisms in the nutrient-rich cold upwelling water consume more atmospheric CO_2 than those in barren warm water, and thus reduce the expected CO_2 increase in the atmosphere. The average CO_2 increase during a composite of five years (between 1958 and 1974) with prevailing upwelling was only 0.57 ppm yr^{-1}; during a different composite of five years with dominantly warm water, CO_2 rose by 1.11 ppm yr^{-1}. This difference is equivalent to $\sim$ 1 GT carbon yr^{-1} (1 GT = 10^{12} kg), i.e., about 40% of the annual input of CO_2 into the atmosphere arising from fossil fuel. Thus, under purely natural conditions (i.e., with a stationary CO_2 budget), atmospheric CO_2 should decrease during upwelling and increase during downwelling episodes. Such observed variations need to be taken into account when the geophysical mechanisms of climatic change at a time scale of 10^3 yr are discussed (see § 6).

3.   UPWELLING AND OCEANIC EVAPORATION

The frequent occurrence of coastal fog in regions of oceanic upwelling (e.g., off the coasts of California, Peru and Angola/ Namibia) is well-known since A. von Humboldt first described the phenomenon along the Peruvian coast in the early 19th century. In these cases, the low temperature of water upwelling from intermediate depths reverses the vertical humidity gradient and replaces evaporation by condensation at the sea surface. In all oceanic locations, the saturation deficit of atmospheric water vapour (q_s-q_a, where q = mixing ratio) is reduced when the SST drops below the air temperature (T_a), i.e., when the flux of sensible heat is reversed. Such thermal stability also tends to reduce the vertical exchange of momentum and thus the speed of surface winds. Hence, both terms of the well-known aerodynamic bulk formula (Tetzlaff & Adams: this volume) contribute to a reduction in evaporation. Upwelling is used here only in the regional sense of cool water from intermediate depths appearing at the surface, giving a SST below the latitudinal average (broader aspects of the ocean circulation are not considered).

Many earlier evaluations of sea surface observations resulted in the smoothing of horizontal gradients of the saturation deficit, mostly because of the low resolution of the data used. These gra-

dients are relatively intense (of the order of several g kg^{-1} per
100 km). Recent heat-budget atlases (Hastenrath & Lamb 1978, 1979)
with a 1° × 1° resolution and similar studies (Trempel 1978, Henning
& Flohn 1980, Weber 1981) indicate a remarkable reduction of
evaporation in upwelling areas. In the vicinity of the equator,
the vertical component of motion at the bottom of the oceanic Ekman
layer w = $(\rho f)^{-1}$ curl$_z$ $\vec{\tau}$ (where ρ = density) changes its sign with
the Coriolis parameter f, f^{-1} approaching ± ∞ in the immediate
vicinity of the equator. However, this only applies to those frequent cases in which the vorticity of the wind stress vector curl$_z\vec{\tau}$
has the same sign on both sides of the equator, where f and, therefore, w change sign. Data from the eastern Pacific and eastern
Atlantic indicate that, during the northern summer when both Hadley
circulations reach their northernmost (and thus most asymmetric)
position and the southern trades cross the equator with a clockwise
curvature, the average seasonal evaporation increases from less
than 1 mm per day at 0-2°S to more than 5 mm per day at 4-6°N,
i.e., in a distance of no more than 600 km.

Weber (1981) has evaluated maritime observations for a 10-year
period (Anon 1961-1970) in two selected areas of the Atlantic: at
1-4°S, 5-8°W, and 3-5.5°N, 1-5°W – the isolated upwelling region
in the Gulf of Guinea south of Ghana where southerly onshore winds
prevail during the whole year. Figure 1 gives the seasonal
variations in the average saturation deficit (expressed as the
vapour pressure deficit e$_s$-e$_a$ (in mb)), relative humidity at the
10 m level, thermal stability and T$_s$ (= SST), together with the
standard deviation of the individual months, for the Gulf of
Guinea field. The significant anomaly during the short upwelling
season (July-September) is clearly shown. In this area, the
relative humidity depends strongly on the thermal stability; it
increases from about 70% where T$_s$-T$_a$ = +1.6°C to about 93% where
T$_s$-T$_a$ = -1.2°C, with a correlation coefficient r = -0.83 for 120
independent pairs of monthly averages. More representative are
the data from the other field south of the equator, where the
evaporation LE increases linearly with T$_s$ (r = 0.78) from 45 W m^{-2}
(1.6 mm per day) at 22°C to more than 110 W m^{-2} (3.9 mm per day)
at 29°C (Figure 2). This is in good agreement with the findings
of Weare (this volume). For a discussion of systematic errors in
the evaluation of the bulk formula using data available from
merchant ships, see Weber (1981); for individual monthly averages,
the errors may amount to 15%.

The effect of wind speed is most striking in the summer monsoon
area of the Indian Ocean. Coastal upwelling controls the climate
off the Somali coast (0-12°N) where there are remarkably strong
winds averaging up to 17 m s^{-1} (Weber 1981), as well as along the
SE coast of the Arabian peninsula. Here, the seasonal increase in
wind speed actually outweighs the drop in saturation deficit and
leads to a primary (or secondary) maximum of evaporation during the
upwelling period. This phenomenon results in very high average
values of evaporation, ⩾ 320 W m^{-2} or 11 mm per day; since such

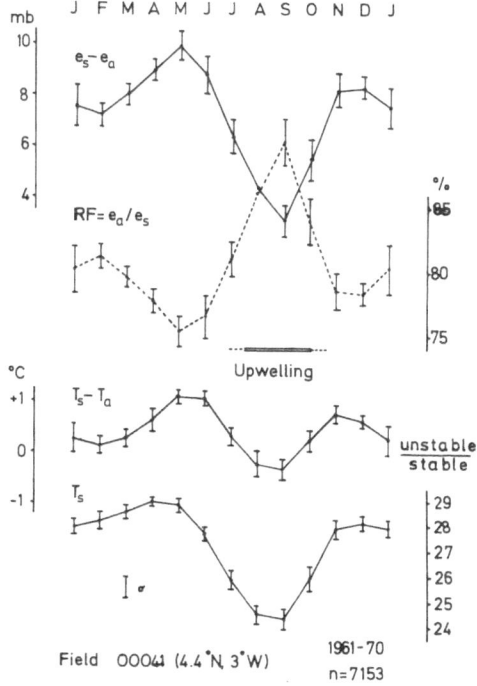

Fig. 1. Seasonal variation of air-sea interaction parameters in
the Gulf of Guinea at 4.4°N, 3°W, 1961-70: T_a, air
temperature; T_s, sea surface temperature; e_s, saturation
vapour pressure at T_s; e_a, vapour pressure; σ, standard
deviation of monthly averages. Upwelling season: July
to September or October.

values exceed the available net radiation, additional energy must
be contributed by advection. The same is true in marginal seas
such as the Red Sea and the Persian Gulf, where the saturation
deficit is increased by advection of hot dry air from the desert.
     Interannual fluctuations in the global water balance have
almost certainly been affected by equatorial and coastal upwelling.
In the Atlantic, interannual fluctuations of SST are mainly
restricted to a narrow belt between the equator and 4°S (Henning &
Flohn 1980). A tendency for a negative (spatial) correlation
between anomalies in the Pacific and Atlantic was found by Doberitz
(1969); coastal upwelling also fluctuates significantly. A reasonable estimate of the interannual fluctuations in evaporation is
unlikely therefore to exceed $6-10 \times 10^3$ km^3 yr^{-1}, i.e., about 2%
of present-day average evaporation from the oceans; the same should
be valid for precipitation. Thus, any attempt to detect such
variations in the available data would be futile, in view of gaps
and inaccuracies in precipitation and evaporation data.

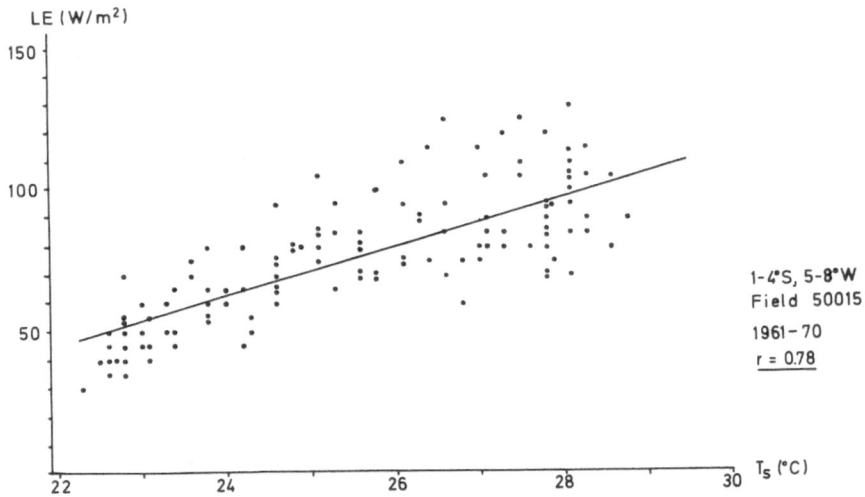

Fig. 2. Correlation between sea surface temperature (T_s) and the
         energy needed for evaporation (LE in $W\ m^{-2}$) in the upwel-
         ling region of the Atlantic south of the equator (1-4°S,
         5-8°W), 1967-70, after Weber (1981): ·, number of indivi-
         dual months in a T_s-LE diagram; ——, regression equation,
         correlation coefficient r = 0.78 (120 independent pairs).

4.   ESTIMATES OF THE GLOBAL WATER BUDGET DURING THE LAST
      GLACIATION AND EARLY HOLOCENE

      Using the best data available, the present-day global water
budget (evapotranspiration ≃ precipitation) has been estimated to
average 970 mm yr^{-1} by Baumgartner and Reichel (1975) and 1030 mm
yr^{-1} by L'vovich (1979). Taking the average latent heat of
condensation as 590 cal g^{-1}, these estimates correspond to energies
of 76 or 80 $W\ m^{-2}$, respectively, compared with the average net
surface radiation of about 100 $W\ m^{-2}$ (96-102 $W\ m^{-2}$). Since about
86% of the global evaporation (equivalent to 2/3 of the net radiation) stems from the oceans, oceanic evaporation must be considered
first in seeking the geophysical causes of past changes in the
global water balance.
      Although recent short-lived anomalies in the global water
balance (on a time scale of a few months or years) are unlikely to
be detectable, there are notable indications of changes on a geological time scale. Preliminary results from the CLIMAP project
(CLIMAP Project Members 1976) indicate a large increase in equatorial upwelling during the last glaciation, with seasonal SST
values as low as 14-16°C; these data are representative of the equa-

torial Pacific and Atlantic. In the Pacific, upwelling occurred
from the Ecuador coast as far west as Nauru (167°E) and the Ontong-
Java Plateau (160°E), i.e., over a 120° longitude span. In contrast, the equatorial Indian Ocean today is practically free of
upwelling – this was also the case during the last glaciation
(Prell *et al.* 1980).

The most recent data from the coastal upwelling region off NW
Africa (Pflaumann 1980) suggest a temperature lowering of about
3-4°C during both summer and winter of the last glaciation. The
highest SST values, which lagged significantly behind the so-called
'Holocene climatic optimum', were 2-3°C higher than today in summer
and about 1°C higher in winter. Significantly, positive correlations exist between upwelling in the equatorial Pacific and
coastal upwelling along the west coasts of both North and South
America. Similar correlations between the equatorial Atlantic and
the coasts of NW and SW Africa are still to be investigated.

Using the reconstructed SST data for the peak of the last
glaciation and for the Holocene climatic optimum (Hypsithermal),
some estimates of the changes (Δ) in oceanic evaporation have been
made (Table I). These values only represent an 'educated guess'.
They depend strongly on the critical assumption that wind speeds
have not changed significantly in equatorial latitudes, although
the palaeoclimatic evidence (Sarnthein & Koopman 1980) suggests
that such changes may in fact have occurred.

Table I. Estimated changes in oceanic evaporation between the last
glacial and the early Holocene (10^3 km^3 yr^{-1}), updated
from Flohn (1980).

	Area (10^6 km^2)	Present- day	ΔGlacial	ΔHolocene
Pacific, Atlantic 0-10°S	25	35	-27*	+8
Pacific, Atlantic 0-10°N	27	34	-20	+2
Coastal upwelling zones	2-3	2	- 2	+1
Atlantic sea ice	14+	11	- 8	+1
Sub-Antarctic sea ice	12-15+	9	- 5	+4
Effect on sea-level change	20+	24	-15	+2§
Net change			-77	+16

* Modern values are probably too high (∿ 30 × 10^3 km^3 yr^{-1});
  in which case, Δglacial value is *ca.* -22 × 10^3 km^3 yr^{-1};
+ Valid for last glaciation;
§ Valid for last interglacial, with a sea-level rise of 5-7 m
  (Hollin 1980).

The estimated changes in oceanic evaporation for the last
glacial and for the Holocene amount to -18% and +4% of the modern
total (425 × 10^3 km^3 yr^{-1}, according to Baumgartner & Reichel 1975),

respectively. Since these changes were concentrated mainly in the
tropics (including the vast shelf areas between Indonesia and
Australia, which were exposed by the sea-level fall during the last
glaciation), their influence on the global water balance cannot be
neglected. As mentioned above, during the peak of the last glaciation, the tropical rain forests were reduced to a few refuge
areas (Shackleton 1977). Simultaneously, boreal and temperate
forests were either buried by continental ice or reduced to a steppe
tundra or, at best, forest tundra. It is justifiable therefore to
reduce the continental evaporation in the $10°N-10°S$ and $40°-70°N$
latitude belts by approximately 50%; this yields a further reduction
in global evaporation of about $11+8 = 19 \times 10^3$ km^3. Thus the total
reduction in global evaporation (and, hence, in precipitation)
during the last glaciation can be estimated to be about 96×10^3 km^3
or nearly 20% of the present-day global value of 496×10^3 km^3 yr^{-1};
$cf.$ L'vovich (1979) who gives 525×10^3 km^3 yr^{-1}. During the Holocene optimum, the global value may have been 3-4% higher than today,
and perhaps as much as 5-6% higher during the last interglacial[1].

How far is such a remarkable change in the global water balance
consistent with the assumption that the solar radiation flux has
remained constant? A decrease in global evaporation by 96×10^3 km^3
would be equivalent to about 15 W m^{-2}. Certainly, the surface
energy balance during the last glaciation must have been significantly different from the present-day balance: the estimated change
in the surface albedo solely due to increased ice cover, from about
0.14 today to about 0.18 (Flohn 1973), would be equivalent to a
global cooling of 4-5°C. If not counter-balanced by a decrease in
cloudiness (as would be expected with decreased evaporation), this
increase in albedo would probably lead to a reduction in available
radiation at the surface. In fact, increased upwelling would
probably occur, leading to a regional increase of low-level stratus
clouds. In this situation, the surplus energy not needed for evaporation in the tropical oceans would be taken up in heating the
increased amount of upwelling cool water. This process may have
been responsible for the remarkable near-constancy (or even slight
increase) of SST in the centre of the subtropical gyres (CLIMAP
Project Members 1976), and may also have played a role in the
poleward heat transport by ocean currents ($cf.$ the hypothesis
(Newell 1974) of a substantial decrease in this transport at higher
latitudes during the glaciations). The role of upwelling regions
in the net heat gain of the oceans can clearly be seen in the Heat
Budget Atlas of the tropical Atlantic and eastern Pacific
(Hastenrath & Lamb 1978).

Sonntag et $al.$ (1980) have shown that the stable isotope content of fossil groundwater in the Sahara indicates a comparatively
high relative humidity of the atmosphere over the ocean ($cf.$ Merlivat & Jouzel 1979). Since these waters have an average ^{14}C-age
of around 25 000 yr BP, they mainly represent waters from an earlier
interstadial of the last glaciation, when the tropical circulation
was stronger than now, but not as strong as during the aridity peak

of the last glacial - as is shown by data on the transport of
Saharan dust (Sarnthein & Koopmann 1980). A relative humidity of
85-90% is clearly compatible with intensified upwelling over large
parts of the equatorial Atlantic including the entire Gulf of Guinea.
The reduction in evaporation due to this increase may have been
compensated by a higher wind speed, as is now the case off the
Somali coast. Even so, the isotopic composition and age of Saharan
fossil groundwater can be considered additional evidence for an
increase of upwelling during a prolonged part of the last glacial.

5.   AIR-SEA INTERACTION AND ITS IMPLICATIONS FOR CLIMATIC
     MODELLING

     Advanced climate models capable of investigating the effects
of a rising CO_2 level (Ramanathan 1981) have shown that atmospheric
water vapour contributes greatly to the CO_2 'greenhouse effect' in
these models. Two related processes are responsible: warming
causes an increase of evaporation and thus of the H_2O content of
the atmosphere thereby leading to a substantially-enhanced greenhouse effect; condensation and precipitation of additional H_2O
liberate more latent heat. Other models (Newell & Dopplick 1979,
Gates *et al.* 1981), which isolate the CO_2 effect by excluding any
change in H_2O content, yielded no significant warming. However,
the results described herein indicate that equatorial upwelling is
strongly correlated with low contents of both H_2O and CO_2, while
the atmospheric content of both gases increases significantly during
episodes of downwelling. Thus air-sea interaction favours a simultaneous rise or fall in the atmospheric content of both infrared
radiation-absorbing gases. This emphasizes the need to introduce
a positive interaction of both gases into climate models. Interactive ocean-atmosphere models should include, as a matter of high
priority, the role of equatorial and coastal upwelling.

6.   A PROPOSED HEMISPHERIC FEEDBACK MECHANISM FOR CLIMATIC CHANGE

     In the preceding sections, a system of interrelated relation-
ships between oceanic upwelling and downwelling, the intensity of
the trades (representing the Hadley circulation) and the CO_2- and
H_2O-content of the atmosphere has been described (see also Kellogg
1981). In this section, an attempt is made to use these relationships to arrive at a physical interpretation of the rare and
enigmatic, abrupt climatic changes on a time scale of $\leqslant 100$ years
(see § 1). To this end, the scale of this investigation has to be
extended from the tropical Hadley cell to a whole hemisphere.
     The intensity of the Hadley cell should be related to its
width, i.e., the latitude difference between the Intertropical
Convergence Zone (ITCZ) and the subtropical anticyclonic belt (STA).
The latter can be approximated by the subtropical maximum of surface

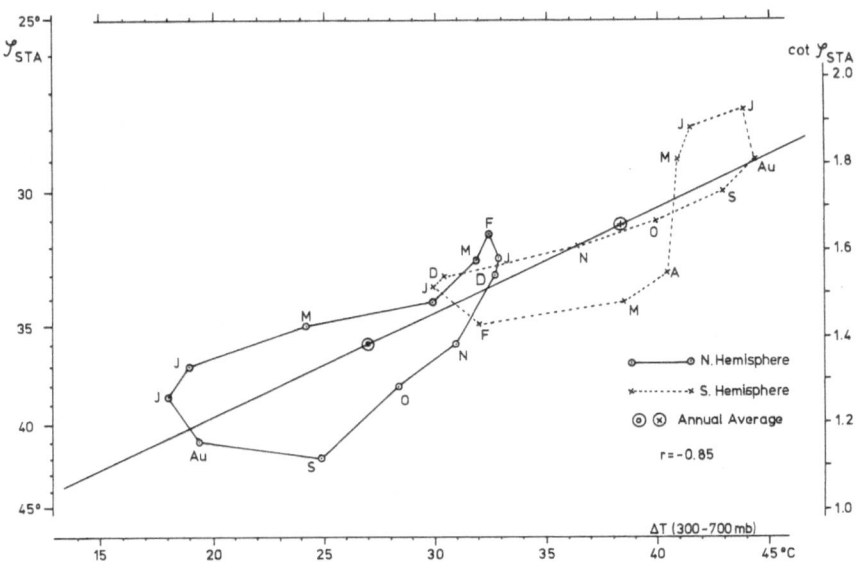

Fig. 3. Correlation between monthly averages of the latitude Ψ of
         the subtropical anticyclonic belts (STA), expressed as
         $\cot\Psi_{STA}$, and the temperature difference in the 700-300 mb
         layer ($ca.$ 3-9 km) between the equator and the pole in
         each hemisphere (after Korff & Flohn 1980).

pressure (slightly distorted by monsoonal summer heat lows over
central Asia) and by the subtropical jet stream near 200 mb.
Korff & Flohn (1969) found a strong, empirical, negative correlation
between the seasonal variations of the latitude Ψ_{STA} (derived from
surface data) and the meridional tropospheric temperature gradient
Γ between the equator and the pole (Figure 3). The latter parameter will be affected by CO_2-induced warming because of the well-
known ice- and snow-albedo temperature feedback (Kellogg 1973):
with increased global surface temperatures, Arctic surface temperatures should increase even more strongly (by a factor of about 3,
as verified empirically), and the equator-pole temperature gradient
should drop significantly. The relation between Γ and Ψ_{STA} therefore predicts a poleward shift of Ψ_{STA}, i.e., a slackening of the
Hadley cell resulting in an increasing CO_2- and H_2O-content of the
atmosphere. If an initial cooling at high latitudes is assumed,
e.g., following a cluster of large volcanic eruptions, a reversed
sequence of linkages should be observed. This model disregards a
number of short time lags, of the order of a few months' duration
(Wyrtki 1975, Reiter 1978, Bacastow et $al.$ 1980, Angell 1981, Weare:
this volume, and others).
     Together, the correlations and linkages described above suggest
a rare, but effective, hemispheric-scale climatic feedback mechanism,

Table II. Natural hemispheric climatic feedbacks.

Initial change in polar regions	cooling	warming
Meridional temperature gradient	stronger	weaker
Latitude subtropical anticyclones (Ψ_{STA})	down	up
Intensity of trades (Hadley cell)	strong	weak
Equatorial SST	cold	warm
Atmospheric content of CO_2 and H_2O	down	up
producing further	cooling	warming

summarized in Table II (see also Flohn 1981, Kellogg 1981).

Such a hemispheric feedback could change drastically the frequency, duration and intensity of equatorial (and coastal) upwelling and downwelling, which at present are more or less in balance, at least on a time scale of several decades. A prolonged CO_2-induced warming might initiate downwelling and thus intensify further warming, whereas a volcanogenic cooling could initiate upwelling and thus be self-intensifying. This effect may be highly significant over several decades or even a century or more. However, since the oceans are largely closed basins and since the turnover time of the deep oceans seems to be $\sim$ 500-1000 years (Broecker *et al.* 1980), the efficiency of these processes is limited in time; it should decrease substantially after a few centuries (or even less).

Further studies are required to check the validity of the suggested mechanisms that involve an extrapolation from the period of instrumental records to much longer time scales. The concept may prove useful for the physical interpretation of abrupt transitions and cold periods on a time scale of a few decades or centuries, whereas the sequence of glacial and interglacial periods with time scales of 10^4 to 10^5 yr, seems to be controlled mainly by the Earth's orbital elements (Hays *et al.* 1976, see also Kutzbach: this volume). It is suggested nevertheless that the essential distinction between an abortive glaciation (or stadial), caused by this type of feedback mechanism within the climatic system, and a full glaciation (with a time scale of 10^4 yr) is due to the selective role of the orbital elements. The same may also be true for short warm periods (or interstadials) with a duration of $\sim 10^2$-10^3 yr, in contrast to interglacials (10^4 yr).

While the accumulation or decay of the northern hemisphere continental ice sheets lasted for 8000-10 000 yr, the abrupt transitions between stadials and interstadials may last for only 50-150 yr, triggered by an internal mechanism of the type outlined above. The continuation of this mechanism during an abrupt transition period of 50-150 yr may depend on the maintenance of a wide-spread snow cover on the northern continents, especially during the summer melting phases. However, its duration may be limited by the effectiveness of the shallow circulation of the upper ocean layers in tropical latitudes, which is unlikely to be much more extensive, either horizontally and vertically, than that described by Wyrtki (1981).

7. CONCLUSIONS

In this paper, a geophysical feedback mechanism of hemispheric
scale is proposed as a rational interpretation of two major
problems of climatic history at a time scale of less than 1000
years, namely:
(a) The abrupt changes between warm and cold phases (on a time
scale of *ca*. 100 years);
(b) A simultaneous world-wide decrease or increase of sensible
heat and evaporation during glacial (cold) and interglacial (warm)
phases, apparently at variance with the tacit assumption of a
constant 'solar constant'.
The proposed mechanism involves a simultaneous decrease (or
increase) of atmospheric carbon dioxide and water vapour, both of
which are responsible for changes in the 'greenhouse effect'.
Such a decrease can be produced by intensified ocean upwelling
along the equator (in the Pacific and Atlantic) and in adjacent
coastal areas, while an increase of like scale can be produced by
intensified downwelling of warm water. These processes are part
of an effective positive-feedback mechanism involving both the
atmospheric and oceanic circulation (Table II). They may lead
also to significant changes in the global water budget.

NOTE

(1) The positive spatial correlation of SST throughout the
tropical Pacific (20°S-20°N) (Weare: this volume) suggests
even higher deviations of the global water budget, perhaps
as much as -22 to -24% for the last glacial and +7% for the
last interglacial.

REFERENCES

Angell, J.K. 1981 Comparison of variations in atmospheric quantities
with sea surface temperature variations in the Equatorial
Eastern Pacific. *Mon. Weath. Rev.*, 109, 230-243.

Anon 1961-70 *Marine Climatological Summaries, 1961-70*. Hamburg:
Seewetteramt.

Bacastow, R.B., Adams, J.A., Keeling, C.D., Moss, D.J., Whorf, T.P.
& Wong, C.S. 1980 Atmospheric carbon dioxide, the southern
oscillation and the weak 1975 El Niño. *Science, N.Y.*, 210,
66-68.

Baes, C.F. Jr 1981 The response of the oceans to increasing atmospheric carbon dioxide. Institute of Energy Analysis, Oak
Ridge Associated Universities. ORAU/JEA-81-6(M).

Baumgartner, A. & Reichel, E. 1975 *The World Water Balance: Mean Annual Global, Continental and Maritime Precipitation, Evaporation and Runoff*, 179 pp. Amsterdam: Elsevier.

Berner, W., Oeschger, H. & Stouffer, B. 1980 Information on the CO_2 cycle from ice core studies. *Radiocarbon,* 22, 227-235.

Broecker, W.S., Peng, T.H. & Engh, R. 1980 Modelling the carbon system. *Radiocarbon,* 22, 565-598.

CLIMAP Project Members 1976 The surface of the ice-age Earth. Modelling the ice-age climate. *Science, N.Y.,* 191, 1131-1144.

Coope, G.R. 1977 Fossil coleopteran assemblages as sensitive indicators of climatic changes during the Devonian (Last) cold stage. *Phil.Trans R.Soc. Lond., B,* 280, 313-340.

Delmas, R.J., Ascencio, J.M. & Legrand, M. 1980 Polar ice evidence that atmospheric CO_2 20 000 yr BP was 50% of present. *Nature, Lond.,* 284, 155-157.

Doberitz, R. 1969 Cross spectrum and filter analysis of monthly rainfall and wind data in the tropical Atlantic region. *Bonner Met. Abhandl.,* 11, 53 pp.

Eicher, U. & Siegenthaler, U. 1976 Palynological and oxygen isotope investigations on Late Glacial sediment cores from Swiss lakes. *Boreas,* 5, 109-117.

Eicher, U., Siegenthaler, U. & Wegmüller, S. 1981 Pollen and oxygen isotope analyses on Late- and Post-Glacial sediments of the Tourbière de Chirens (Dauphiné, France). *Quat. Res.,* 15, 160-170.

Flohn, H. 1973 Globale Energiebilanz und Klimaschwankungen. *Bonner Met. Abhandl.,* 19; *Rhein. Westfäl. Ak. Wiss.* No. 234, pp. 75-117.

Flohn, H. 1979 On time scales and causes of abrupt palaeoclimatic events. *Quat. Res.,* 12, 135-149.

Flohn, H. 1980 Possible climatic consequences of a man-made global warming. *Int. Inst. Appl. Systems Anal. Rept* RR-80-30, xi + 80 pp. Laxenburg (Austria): IIASA.

Flohn, H. 1981 Klimaänderungen als Folge der CO_2-Zunahme? *Phys. Blätt.,* 37, 184-190.

Frenzel, B. 1968 Grundzüge der pleistozänen Vegetationsgeschichte Nordeurasiens. In: *Erdwissenschaftliche Forschungen,* Vol. 1, 326 pp. Wiesbaden: Steiner Verlag.

Gates, W.L., Cook, K.H. & Schlesinger, M.E. 1981 Preliminary analy-
    sis of experiments on the climatic effects of increased CO_2
    with an atmospheric general circulation model and a climato-
    logical ocean. *J. geophys. Res.*, 86, 6385-6393.

Hastenrath, S. & Lamb, P. 1978 *Heat Budget Atlas of the Tropical
    Atlantic and Eastern Pacific Oceans*, 90 pp. University of
    Wisconsin Press.

Hastenrath, S. & Lamb, P. 1979 *Climatic Atlas of the Indian Ocean,
    Part II*, xvii + 93 pp. University of Wisconsin Press.

Hays, J.D., Imbrie, J. & Shackleton, N.J. 1976 Variations in the
    Earth's orbit: Pacemaker of the ice ages. *Science, N.Y.*,
    194, 1121-1132.

Henning, D. & Flohn, H. 1980 Some aspects of evaporation and sen-
    sible heat flux of the Tropical Atlantic. *Contr. atmos. Phys.*,
    53, 430-441.

Hollin, J.D. 1980 Climate and sea level in isotope stage 5: an East
    Antarctic ice surge at ∿ 95 000 BP? *Nature, Lond.*, 283, 629-
    633.

Keeling, C.D. & Bacastow, R.B. 1977 Impact of industrial gases on
    climate. In: *Energy and Climate*, pp. 72-95. Washington, D.C.:
    US Nat. Acad. Sciences.

Kellogg, W.W. 1973 Climatic feedback mechanisms involving the polar
    regions. In: *Climate of the Arctic, 24th Alaska Science
    Conference* (ed. G. Weller & S.A. Bowling), pp. 111-116.

Kellogg, W.W. 1977 Effects of human activities on global climate.
    WMO Tech. Note 156, xviii + 47 pp. Geneva: WMO.

Kellogg, W.W. 1981 Feedback mechanisms in the climate system
    affecting future levels of carbon dioxide. In: *Analysis and
    Interpretation of Atmospheric CO_2 Data* (World Climate Research
    Program WCP-14), pp. 243-251.

Korff, H.Cl. & Flohn, H. 1969 Zusammenhang zwischen dem Temperatur-
    gefälle Äquator-Pol und den planetarischen Luftdruckgürteln.
    Annl. Met. (N.F.), 4, 163-164.

Kukla, G. 1980 End of the last interglacial: a predictive model of
    the future? *Palaeoecol. Africa*, 12, 395-408.

L'vovich, M. 1979 *World Water Resources and Their Future*, viii +
    415 pp. Washington, D.C.: Am. Geophys. Un.

Manabe, S. & Wetherald, R.T. 1980 On the distribution of climatic change resulting from an increase in CO_2-content of the atmosphere. *J. atmos. Sci.*, 37, 99-118.

Merlivat, L. & Jouzel, J. 1979 Global climatic interpretation of the deuterium oxygen-18 relationship for precipitation. *J. geophys. Res.*, 84, 5029-5033.

Müller. H. 1979 Climatic changes during the last three interglacials. In: *Man's Impact on Climate* (ed. W. Bach, J. Pankrath & W. Kellogg): *Devl. atmos. Sci.*, 10, 29-41.

Newell, R.E. 1974 Changes in the poleward energy flux in the atmosphere and ocean as a possible cause for ice ages. *Quat. Res.*, 4, 117-127.

Newell, R.E., Navato, A.R. & Hsiung, J. 1978 Long-term global sea surface temperature fluctuations and their possible influence on atmospheric CO_2 concentrations. *Pure appl. Geophys.*, 116, 351-371.

Newell, R.E. & Dopplick, T.G. 1979 Questions concerning the possible influence of anthropogenic CO_2 on atmospheric temperature. *J. appl. Met.*, 18, 822-825.

Pflaumann, U. 1980 Variations of the surface water temperatures along the eastern North Atlantic continental margin (sediment surface samples, Holocene climatic optimum, and Last Glacial maximum). *Palaeoecol. Africa*, 12, 191-212.

Prell, W.L., Hŭtson, W.H., Williams, D.F., Bé, A.W.H., Geitzenaŭer, K. & Molfino, B. 1980 Surface circulation of the Indian Ocean during the Last Glacial maximum, approximately 18 000 yr B.P. *Quat. Res.*, 14, 309-336.

Ramanathan, Y. 1981 The role of ocean-atmosphere interactions in the CO_2 climate problem. *J. atmos. Sci.*, 38, 918-930.

Reiter, E.R. 1978 The interannual variability of the ocean-atmosphere system. *J. atmos. Sci.*, 35, 349-370.

Sarnthein, M. & Koopman, B. 1980 Late Quaternary deep-sea record on Northwest Africa dust supply and wind circulation. *Palaeoecol. Africa*, 12, 239-253.

Shackleton, N.J. 1977 Carbon-13 in Uvigerina: tropical forest history and the Equatorial Pacific carbonate dissolution cycles. In: *The Fate of Fossil Fuel CO_2 in the Oceans* (ed. N.R. Andersen & A. Malakoff): *Mar. Sci.*, 6, 401-427.

Sonntag, C., Thornweihe, U., Rudolph, J., Löhnert, E.P., Junghans, Chr., Münnich, K.O., Klitzsch, E., El Shazly, E.M. & Swailem, F.M. 1980 Isotopic identification of Saharian groundwaters, groundwater formation in the past. *Palaeoecol. Africa*, 12, 159-171.

Trempel, U. 1978 Eine klimatologische Auswertung der meteorologischen Beobachtungen deutscher Handelsschiffen vor der Westküste Südamerikas im Zeitraum 1869-1970. Diploma Thesis: Univ. Bonn. [Unpublished.]

Weber, K.-H. 1981 Abschätzungen des Energieaustausches an der Meeresoberfläche im Arabischen und Roten Meer. Diploma Thesis: Univ. Bonn. [Unpublished.]

Woillard, G.M. 1979 Abrupt end of the last interglacial s.s. in NE-France. *Nature, Lond.*, 281, 558-562.

Wyrtki, K. 1975 El Niño – the dynamic response of the Equatorial Pacific Ocean to atmospheric forcing. *J. phys. Oceanogr.*, 5, 572-584.

Wyrtki, K. 1981 An estimate of equatorial upwelling in the Pacific. *J. phys. Oceanogr.*, 11, 1206-1214.

ILLUSIONS AND PROBLEMS IN WATER-BUDGET SYNTHESIS

Nils-Axel Morner

Geological Institute
Stockholm University
S-11368 Stockholm, Sweden

ABSTRACT. Evidence is offered to demonstrate that past changes in sea level (eustasy) and oceanic ^{18}O content are not direct measures of corresponding global ice volume changes and causal palaeoclimatic episodes. Geoidal eustasy is known to have caused non-synchronous eustatic changes in sea level over the globe. The hypothetical glacial expansion at around 115 000 yr BP, inferred from oceanic ^{18}O records, is seriously questioned. It is proposed that expansions and contractions of the Arctic Ocean caused by geoidal eustasy significantly influenced the oceanic ^{18}O records from time to time and were responsible for the isotopic substage 5d signal. Similarly, groundwater fluctuations in Africa and other equatorial areas affect the carbonate content of the oceanic records. Obviously, different impulses may lead to similar signals in records of supposed indicators of past climates. Such considerations must be borne in mind in analysing past climatic fluctuations as well as predicting future climatological events.

The major changes in the water budget of the Earth during late Cenozoic time involved the climatically-controlled balance between the volume of the oceans and the volume of continental ice (Figure 1). It is generally believed that these changes can be quantified fairly precisely by recording past sea-level changes and changes in the oxygen-isotope composition of the oceans. Unfortunately, the truth is not nearly so simple and, thus, previous estimates of water-budget changes may be serious in error, not to say illusory.

Sea-level records for the late Pleistocene and Holocene are known now to have been substantially, often even predominantly, influenced by changes in the geoid (i.e., by geoidal eustasy) (Morner 1976, 1981a, 1982a). These changes represent redistribu-

419

A. Street-Perrott et al. (eds.), Variations in the Global Water Budget, 419–423.
Copyright © 1983 by D. Reidel Publishing Company.

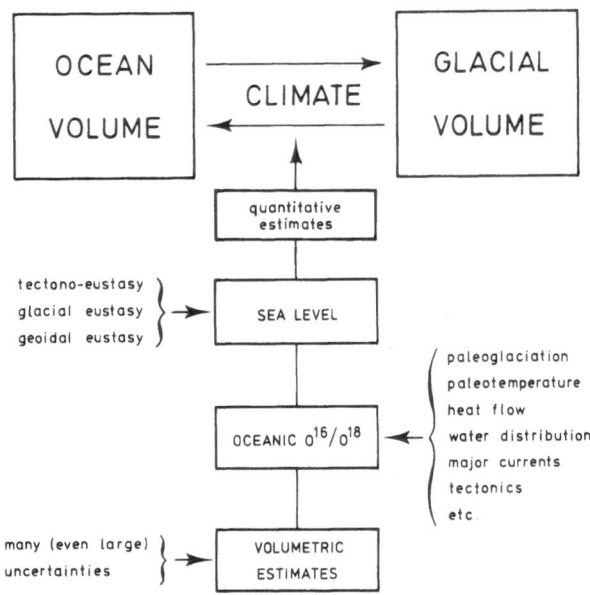

Fig. 1. The water budget of the oceans and cryosphere showing
         the major linkages (each of which includes multiple
         variables), demonstrating that neither eustatic sea
         levels nor oceanic oxygen-isotope records are a direct
         measure of changes in glacial volume.

tions of the oceanic water masses with respect to the gravitational and rotational potentials, i.e., the equipotential surface
of the geoid. Hence, eustatic rises and falls in sea level can
no longer be taken as *a priori* evidence of corresponding changes
in the overall glacial volume. The eustatic oscillations during
the last 6500 years seem to have been caused totally, or at least
in most cases, by palaeogeoid changes and to have been irregular
and variable in magnitude and sign over the globe (Morner 1981*a*,
b). However, within each region, there seems to have been a close
correlation between eustatic, climatic and palaeomagnetic changes
(Morner 1978*a*, 1980*a*). The reason for this seems to be that
a factor controlling the Earth's non-dipole magnetic field is
also linked to gravitational changes affecting both the geoid
(eustasy) (Morner 1976, 1980*b*) and equipotential surfaces in the
atmosphere (hence affecting weather and climate). For example,
the 500 mb level exhibits a correlation with geomagnetic field
changes (King 1974).
     Although the ^{18}O content of the oceans is often claimed to
be a fairly perfect measure of palaeoglaciation (Dansgaard &
Tauber 1969, Shackleton & Opdyke 1973), this may be seriously
questioned (Morner 1981*c*, *d*). Contrary to the above assumption,

the important isotopic event at 3.2 Myr has proved not to represent the "initiation of northern hemisphere glaciation" which occurred much later, namely at about 2.5 Myr BP (Morner 1980*b*). The "expansion of glaciation" at 115 000 yr BP implied by isotope substage 5d (see Fastook: this volume) is not substantiated anywhere in the world by real records (or even traces) of glacierisation or cooling; on the contrary, there is evidence for full interglacial conditions and significant geoidal changes at this time (Morner 1981*c*, *d*). The "glaciation peak" of isotope stage 6 at 135 000 yr BP has the same age as the maximum sea level in New Guinea and the most temperate $\delta^{18}O$ values in the shells from the corresponding high-level marine terraces (Aharon *et al*. 1980, Morner 1981*c*, *d*). The isotopic records in deep-sea cores often differ considerably over the globe (Erez 1979, Duplessy *et al*. 1980). Moreover, the recorded composition of ocean water is a combined effect of many different variables that may have varied in relative importance through time (*cf*. Morner 1981*c*, *d*). As indicated in Figure 1, these factors include: the global ice volume, ocean temperature, heat flow, the distribution of oceanic mass (reflecting the palaeogeoid), the pattern of major currents, tectonic movements. Expansion and contraction of the Arctic Ocean due to geoidal eustasy are likely to have had a significant impact on oceanic ^{18}O records and may well explain deviations between the records of glacial volume and $\delta^{18}O$. Furthermore, such geoidal changes may be caused by the Milankovitch variables (Morner 1981*c*, *d*) and, hence, be in accordance with the frequency analysis of oxygen-isotope curves by Hays *et al*. (1976).

Groundwater level under the continents is not only determined by precipitation, evaporation, the local pattern of subsurface flow and runoff, but also by the geoid level (Morner 1978*b*, 1982*b*) which is known to change with time (involving both the distribution of mass and rotation). Thus humidity and aridity (including desert expansions) may also be caused by factors other than climatic ones (Morner 1978*b*). A lowering of the geoid under a continent would affect the groundwater table and, therefore, could lead to a period of significant aridity (Morner 1978*b*, 1982*b*). Because an episode of rapid deforestation is likely to cause "intense carbonate dissolution" on the ocean floors (Shackleton 1977), the rapid drop in the carbonate content of oceanic sediments off Africa during isotope substage 5d (Hays & Perruzza 1972) should represent a period of continental aridity, which is easily explained in terms of a geoidal lowering rather than a glacial expansion (Morner 1978*b*: Figure 9).

In conclusion, neither eustatic fluctuations in sea level nor the oceanic ^{18}O content are a direct measure of global changes in ice volume (Figure 1). It follows that over optimistic generalizations and evaluations of global systems, that may be much more complex than generally assumed, are to be avoided. This, of course, is especially important if predictions of future climatological events are to be made.

REFERENCES

Aharon, P., Chappell, J. & Compston, W. 1980 Stable isotope and
    sea-level data from New Guinea supports Antarctic ice-surge
    theory of ice ages. *Nature, Lond.*, 283, 649-651.

Dansgaard, W. & Tauber, H. 1969 Glacier oxygen-18 content and
    Pleistocene ocean temperatures. *Science, N.Y.*, 166, 499-502.

Duplessy, J.-C., Moyes, J. & Pujol, C. 1980 Deep water formation
    in the North Atlantic Ocean during the last ice age. *Nature,
    Lond.*, 286, 479-481.

Erez, J. 1979 Modification of the oxygen-isotope record in deep-
    sea cores by Pleistocene dissolution cycles. *Nature, Lond.*,
    281, 535-538.

Hays, J.D. & Perruzza, A. 1972 The signification of calcium
    carbonate oscillations in eastern equatorial Atlantic deep
    sea sediments for the end of the Holocene warm interval.
    Quat. Res., 2, 355-362.

Hays, J.D., Imbrie, J. & Shackleton, N.J. 1976 Variations in the
    Earth's orbit: pacemaker of the ice ages. *Science, N.Y.*,
    194, 1121-1132.

King, J.W. 1974 Weather and the Earth's geomagnetic field. *Nature,
    Lond.*, 247, 131-134.

Morner, N.-A. 1976 Eustasy and geoid changes. *J. Geol.*, 84, 123-
    151.

Morner, N.-A. 1978a Paleoclimatic, paleomagnetic and paleogeoidal
    changes: interaction and complexity. In: *Evolution of
    Planetary Atmospheres and Climatology of the Earth (Colloque
    Int. CNES, Toulouse)*, pp. 221-232.

Morner, N.-A, 1978b Paleogeoid changes and paleoecological changes
    in Africa with respect to real and apparent paleoclimatic
    changes. *Palaeoecol. Africa*, 10/11, 1-12.

Morner, N.-A. 1980a Eustasy and geoid changes as a function of
    core/mantle changes. In: *Earth Rheology, Isostasy and
    Eustasy* (ed. N.-A. Morner), pp. 535-553. New York: Wiley-
    Interscience.

Morner, N.-A. 1980b Earth's movements, paleoceanography, paleo-
    climatology and eustasy: major events in the Cenozoic of the
    North Atlantic. *Geol. Foren. Stockh. Forh.*, 102, 261-268.

Morner, N.-A. 1981*a* Space geodesy, paleogeodesy and paleogeophysics. *Ann. Geophys.*, 37, 69-76.

Morner, N.-A. 1981*b* Eustasy, palaeogeodesy and glacial volume changes. *Int. Ass. Hydr. Sci. Publ.* 131, pp. 277-280.

Morner, N.-A. 1981*c* Eustasy, paleoglaciation and paleoclimatology. *Geol. Rund.*, 70, 691-702.

Morner, N.-A. 1981*d* Ocean/land misfits and the 115 000 BP events. *Palaeogeogr. Palaeoclim. Palaeoecol.*, 34,

Morner, N.-A. 1982*a* Sea levels. In: *Large-Scale Geomorphology* (ed. R.A.M. Gardiner & H. Scoging). Oxford: University Press. [In the press.]

Morner, N.-A. 1982*b* Paleogeoid changes and their possible impact on the formation of natural resources in Africa. *Geoexpl.*, 20. [In the press.]

Shackleton, N.J. 1977 The oxygen isotope stratigraphic record of the Late Pleistocene. *Phil. Trans. R. Soc. Lond.*, B280, 169-182.

Shackleton, N.J. & Opdyke, N.D. 1973 Oxygen isotope and palaeomagnetic stratigraphy of equatorial Pacific core V28-238: oxygen isotope temperature and ice volumes on a 10^5 and 10^6 scale. *Quat. Res.*, 3, 39-55.

Modelling and Prediction

INTRODUCTION TO MODELLING AND PREDICTION

R.P. Pearce

Department of Meteorology
University of Reading
Berkshire, England

M.A. Beran

Institute of Hydrology
Wallingford
Oxfordshire, England

In the classic view of the scientific method, modelling and
prediction are steps in the validation and adjustment procedure
following the primary formulation of a hypothesis. In the past,
even simple hypotheses concerning pathways in the water cycle were
not directly testable. This was so because the sheer complexity
of the system rendered even the simplest of models impossible to
construct and test, by virtue of the daunting computations
involved. Thus, the advent of the electronic computer revolu-
tionized the ability of the meteorologist and hydrologist to study
complex processes, not only in promoting models as a means of
testing hypotheses, but also in suggesting ever more realistic
formulations of the controlling processes within the atmospheric
and terrestrial phases of the hydrological cycle.

In meteorology, it is now possible to integrate numerically
the basic equations of motion, continuity, thermodynamic energy
and moisture conservation, together with the appropriate boundary
conditions, for the whole atmospheric domain. Programs now exist
which, when supplied with global distributions of wind, pressure,
temperature and moisture as initial data, can be run on a modern
computer to yield estimates of the future values of these variables
to cover the whole globe and the full depth of the atmosphere for
long periods of time.

The impact of computers on hydrological modelling has been
similarly profound. The partial differential equations governing
flood waves in a river, fluid motion in a porous medium and heat
transfers within, and across the boundaries of, a snowpack, for
example, can all be solved for practical as well as ideal spatial
units. The focus of attention for the hydrologist is the drainage
basin, the aquifer unit or, sometimes, the agricultural plot,
depending on the ultimate application of the model. This contrast

A. Street-Perrott et al. (eds.), Variations in the Global Water Budget, 426–428.

in spatial scale is one very obvious barrier to the marriage of existing atmospheric and hydrological models.

Clearly, computer speed and storage capacity will always impose limits on the resolution which can be achieved in both space and time. In meteorology, these limitations, together with the assumptions that must be made about small-scale processes, lead to differences between model predictions and the actual behaviour of the atmosphere. Nevertheless, models are capable nowadays of predicting rainfall and evaporation, averaged over areas of, say, 100 km × 100 km, up to a few days ahead with a useful level of accuracy. They can also be used to carry out numerical experiments to investigate the *sensitivity* of different aspects of the hydrological cycle to variations in initial and boundary conditions such as ocean temperatures, soil moisture or surface albedo. Hydrological models are similarly employed in forecasting hydrological outputs, as well as in simulating the effects on the water cycle of land-use changes. Such sensitivity experiments often suggest new avenues of research and new priorities for measurement and monitoring.

Mitchell describes the atmospheric general circulation model (GCM) used by the UK Meteorological Office for experimental investigations and provides a valuable insight into the capabilities of a modern, sophisticated GCM in simulating the hydrological cycle. Rowntree and Bolton then describe the results of experiments carried out with this model on the effects of changing the initial soil moisture distribution; these were found to be quite substantial and, in addition to drawing attention to the need for great care in the inclusion of soil moisture in these models, suggest that the atmosphere is itself occasionally highly sensitive to this factor.

A current problem exercising many hydrologists is how to incorporate land-surface processes and feedbacks into the atmospheric general circulation models in a manner that is not only consistent with known hydrological laws, but that also makes explicit use of measurements of hydrological variables. It is necessary to find a way to parameterize these processes, bearing in mind the computational constraints exerted by the discrete time steps and latitude-longitude grid used in atmospheric models. The following elements need to be included:
(a) state and areal extent of snow and ice cover;
(b) type, density and roughness of vegetation;
(c) state of soil moisture, which is controlled by the infiltration of precipitation and by losses through evapotranspiration. The importance of these particular variables is rooted in their controlling influence on land-surface albedo and on the return of moisture to the atmosphere from the ground.

Apart from the non-linearity of the linkages between albedo and hydrological variables as well as the complexity of the relationship between transpiration and soil moisture, other problems concern the treatment of sub-grid-scale variability and

the fact that computations of soil moisture are effected most
naturally at a drainage-basin scale in order to include the runoff
term explicitly. The recent Greenbelt Conference identified the
following kinds of data as central requirements for a fuller
merging of the surface and atmospheric components of the water
cycle: monthly means of precipitation, evapotranspiration, runoff
and soil-moisture content, vegetation type and canopy density,
surface roughness, hydraulic and thermal properties of the soil,
and monthly mean surface albedo and snow cover.

The two hydrological modelling papers are addressed to the
interface between the atmospheric and land phases of the hydrological cycle. Lockwood and Sellers' model can be regarded as
analogous to an atmospheric model in that it treats the processes
within vegetation canopies of different types in a layered fashion.
The model provides answers to intriguing questions about energy
and water exchanges between the various layers over the course of
the seasons. Such a model is classic in its capacity to suggest
further lines of enquiry and to reveal the need for more detailed
monitoring of hydrological variables corresponding to 'node'
points in the model.

In contrast, Němec's paper deals with the philosophical and
semantic problems involved in bridging the gap between the
hydrological and meteorological modelling traditions. The most
important point is that, to a hydrologist, runoff is a tangible
and fundamental quantity, with the result that he tends to resist
its use as a closing term in a soil-moisture or aerological
balance. A second theme of the paper is the way in which hydrological models can be used to quantify the effects of changes in
climatic boundary conditions. The author is able to demonstrate
the strong multiplier effect of a change in annual rainfall on
river runoff in the semi-arid zone.

THE HYDROLOGICAL CYCLE AS SIMULATED BY AN ATMOSPHERIC GENERAL CIRCULATION MODEL

J.F.B. Mitchell

Meteorological Office
Bracknell
Berkshire, UK

ABSTRACT. The Meteorological Office five layer atmospheric general circulation model has been integrated through three complete annual cycles. The basic model has five layers in the vertical, with a quasi-uniform 330 km horizontal grid, giving 4626 points over the globe. Values of temperature, northward and eastward wind components and specific humidity were predicted at each grid point at each level; surface pressure was predicted at each point.

The model reproduces the main features of the hydrological cycle, including the major regional and seasonal variations in precipitation and evaporation. Over land, the hydrological cycle was more intense than observed, and the model's deserts were less extensive than their real counterparts. The seasonal variation of snow cover was close to that observed, except over central North America where the model was too warm. Although the external forcing in the model was repeated exactly over each annual cycle, there were changes in the monthly mean circulation patterns from year to year, particularly along the model's depression belts in winter. Precipitation varied most where it was heaviest. The fractional change in precipitation was largest over the drier regions in the subtropics.

1. THE MODEL

This paper is based on the final three years of an integration of an atmospheric general circulation model. The integration was carried out using the version of the Meteorological Office atmospheric general circulation model, described by Slingo (1982),

429

A. Street-Perrott et al. (eds.), Variations in the Global Water Budget, 429–446.

which is a development of that of Corby *et al.* (1977). The atmosphere is divided into five layers of equal mass, and the horizontal grid gives a quasi-uniform resolution over the globe with a grid length of approximately 330 km. The model has realistic continents and smoothed topography.

The formulation is based in the usual way on the equations of motion, thermodynamics and continuity of mass and moisture, along with the perfect gas law and the hydrostatic equation. The equations are stepped forward in time from an initial atmospheric state. Temperature, specific humidity and the northward and eastward wind components are predicted at each grid point at each level; surface pressure is predicted at each point.

The radiative fluxes are dependent on the vertical profiles of temperature, cloud and humidity, and the CO_2 concentration (Slingo 1979, 1982). A simple parameterization of radiative heating due to ozone is included in the model's top layer. Cloud amounts and heights are interpolated from four zonally-averaged climatological data sets, one for each season. Sea surface temperatures and sea ice extents are also derived from climatological data, and updated every five days. The surface albedo over land is a function of latitude and, where appropriate, snow depth.

An explicit boundary layer height is carried, and used to control convection and surface exchanges. Supersaturation is removed by condensing excess moisture as rain, and a parameterization of penetrative convection is included. The surface fluxes of heat and moisture are dependent on the surface stability (stable or unstable) and type (sea, land or ice). The heat flux through ice is included.

The treatment of soil moisture is based on that of Manabe (1969). Runoff occurs if the soil moisture content would otherwise exceed 20 cm. The soil moisture content is increased by rainfall, condensation and snowmelt and decreased by evaporation. The evaporation rate E is given by:

$$E = \rho C_D [|V_B| + A(\Delta\theta^\nu/\bar\theta^\nu)^{\frac{1}{2}}] (q_* - q_T) \tag{1}$$

where $\Delta\theta^\nu = (\theta_*^\nu - \theta_B^\nu) = \theta_*(1 + 0.61 q_*) - \theta_B(1 + 0.61 q_T)$

$\bar\theta^\nu = \frac{1}{2}(\theta_*^\nu - \theta_B^\nu)$

and $q_* = 0.8 a q_{SAT}(T_*, P_*) + (1-a) q_B$

ρ = density of air at the surface,
C_D = geostrophic drag coefficient,
q_B, V_B, θ_B = specific humidity, wind and potential temperature at the top of the boundary layer,
T_*, P_*, θ_* = temperature, pressure and potential temperature (adjusted under certain circumstances to allow for condensation in the boundary layer) at the surface,
$q_{SAT}(T_*, P_*)$ = saturation mixing ratio at the surface temperature and pressure.

The second term inside the square brackets allows for convectively unstable situations.

A is 50 m s^{-1} if $\Delta\theta^{v}$ is positive, otherwise it is zero. The ratio of evaporation to potential evaporation is controlled by the parameter a which is a linear increasing function of soil moisture content.

$$a = \text{soil moisture content (cm)}/10, \quad 0 \leqslant q \leqslant 1 \qquad (2)$$

Thus, evaporation is reduced when soil moisture is small. Over snow-free land, evaporation is also limited if the surface temperature falls below freezing point.

If the temperature at the lowest level of the model is less than 267.5 K, then precipitation is assumed to fall as snow rather than rain, snow depth rather than soil moisture content being updated. Both snow and snow depth are represented as an equivalent depth of water. Snow depth is reduced by evaporation and melting. Snow melt occurs if the surface temperature rises above 273 K, heat being extracted from the surface to melt snow until the surface temperature equals 273 K or the snow depth is zero.

2.   THE INTEGRATION

The integration was started from a data set consisting of real data (for 27 May 1977) in the northern hemisphere and data from a previous model integration for July in the southern hemisphere. The initial snow depth was set to zero everywhere except over Greenland and Antarctica where it was given a large value; the soil moisture content was set to 5 cm everywhere, except over deserts where it was set to zero.

The model was integrated for 1192 days. The topmost layer cooled by about 12 K during the first 150 days but, thereafter, the seasonal cycle of global mean atmospheric temperatures was in equilibrium. The soil moisture content increased markedly over the first one to two hundred days, particularly at high latitudes. The model reached quasi-equilibrium by the first northern winter. The results presented here are averaged over the final three years, starting at the beginning of the first September.

3.   ZONAL MEAN CLIMATOLOGY

Some aspects of the hydrological cycle can be illustrated using zonal means. In certain cases, this approach may be misleading as large variations may occur round a latitude circle, and this may obscure cause and effect. For example, in July at 25°N, there was a region of wet land with heavy precipitation (southeast Asia), a desert (the Sahara) and an ocean with meagre precipitation (the eastern Atlantic). In a similar way, time means (particularly those over the complete annual cycle) may be misleading.

3.1   Annual Means

The annual mean of zonally-averaged evaporation reached a
maximum of over 4 mm per day at low latitudes and decreased to
near zero values at the poles (Figure 1c). The variation is due
to the non-linear increase of the saturation vapour pressure with
temperature. At low latitudes, the surface temperature was a
maximum (Figure 1a) as was the atmosphere's capacity to hold water.
Over land, the evaporation was largely determined by the net radiative heating at the surface which was also a maximum in low
latitudes. Over high southern latitudes, there were low surface
temperatures and evaporation rates.

The annual mean precipitation was also largest at low latitudes and small near the poles (Figure 1c). However, there was a
minimum in the subtropics in each hemisphere where precipitation
is less than evaporation, and there were maxima on the equator and
at middle latitudes where precipitation exceeds evaporation. The
minima corresponded approximately to areas where high surface
pressure and downward atmospheric motion ($dP/dt > 0$) were prevalent, and the maxima coincided more or less with the regions where
average surface pressure was low and vertical motion upward
($dP/dt < 0$) (Figure 1b). In regions of ascent, air was cooled
adiabatically. Where upward motion was sufficiently intense the
air became supersaturated, and the excess water vapour condensed
as rain. On the other hand, in regions of descent, air was warmed
adiabatically and the relative humidity was low. Thus, the distribution of zonally-averaged precipitation depends largely on the
prevailing atmospheric circulation. In the subtropics, the surface supplied moisture to the atmosphere, the reverse being true
elsewhere. The water balance was maintained by atmospheric motions
which transported water vapour to low and high latitudes from the
subtropics, as indicated by Figure 1(e) which shows the model's
northward moisture flux in July and January.

The annual mean precipitation rate of 2.83 mm per day is
within the range of climatological estimates given, for example,
by Möller (1951) of 2.20 mm per day, Budyko (1970) of 2.80 mm per
day and L'vovich and Ovtchinnikov (1963) of 3.05 mm per day. The
annual zonal mean precipitation was slightly greater than the
estimates from observations by L'vovich and Ovtchinnikov (1963) in
northern mid-latitudes (Figure 1d), but is less than observed in
much of the southern hemisphere, although here most of the surface
is covered by ocean where estimates from observations are less
reliable. The zonal means of precipitation averaged over land
only (Figure 2) were much higher than the corresponding estimates
of L'vovich and Ovtchinnikov. Manabe and Holloway (1975) also
found that, in their general circulation model, precipitation over
land in the tropics was higher than observed. The observations of
Budyko (1963) suggest that the model's rates of evaporation were
too high in the tropics and northern subtropics (Figure 2).

The model accumulated snow at high latitudes (not shown). As

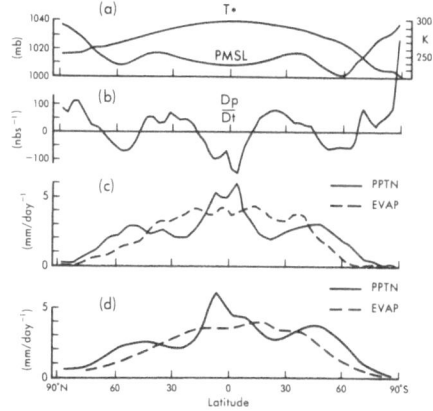

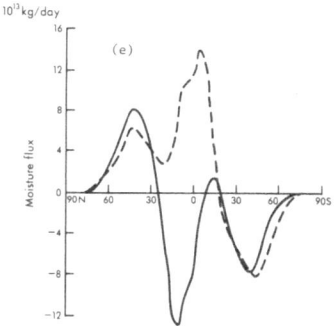

Fig. 1. Annual means of zonally-averaged quantities:
(a) Temperature (K) and surface pressure reduced to mean
sea level (mb); (b) Vertical velocity (nb s^{-1}); (c) Total
precipitation (——) and evaporation (---) from the model
(mm per day); (d) Observed total precipitation: ——, from
L'vovich and Ovtchinnikov (1964); evaporation ---, from
Budyko (1963) (mm per day); (e) Northward moisture flux
(10^{13} kg per day): ---, July; ——, January.

elsewhere the snow disappeared each year, snowfall must be balanced
by sublimation and snowmelt averaged over the year. Over Greenland
and Antarctica, small amounts of snow were removed by numerical
rounding errors. There was a peak in zonally-meaned snowfall near
35°N due to the Himalayas.

3.2 Seasonal Variation over Land

The flux of moisture to the surface from the atmosphere (precipitation-minus-evaporation, or P-E) over land was generally
positive (Figure 3a), so that runoff occurred to complete the water

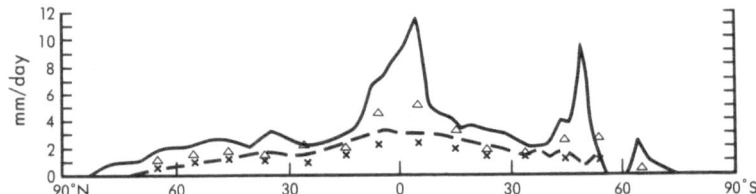

Fig. 2.   Annual means (mm per day) of zonally-averaged total pre-
          cipitation and evaporation (for land points only):
          ——, precipitation from the model; Δ, observed precipita-
          tion (from L'vovich and Ovtchinnikov 1964); ---, evapora-
          tion from the model; ×, observed evaporation (from Budyko
          1963).

balance at the surface. Evaporation exceeded precipitation over
the oceans, and there was a net flux of atmospheric moisture from
ocean to land to complete the hydrological cycle. Precipitation
was a maximum during the summer months (Figure 3b). The continents heated up faster than the oceans during this period as they
have a much smaller heat capacity. The pressure over the continents fell relative to that over the oceans, leading to increased
low-level convergence, ascending motion and, in general, more
precipitation. In winter, this land-sea temperature contrast
reversed and much of the winter land masses were covered by anticyclones with weak downward motion giving little precipitation:
this was most obvious in the outer tropics. For example, at 15°
latitude, the summer precipitation rate was three times that found
in winter. At higher latitudes, the seasonal variation was less
marked.

     The seasonal variation of evaporation was large outside the
tropics (Figure 3a). This followed the net input of energy at the
surface due to solar- and long-wave radiation (Figure 3d), except
where evaporation was limited by low soil-moisture values, for
example, near 20°N in spring (Figures 3c and 6a). At middle and
high latitudes, the seasonal variation of evaporation was greater
than that of precipitation, so that P-E was at a maximum in winter
when evaporation was smallest. However, at low latitudes, the
variation in precipitation was greater, so P-E was at a maximum in
summer corresponding to the maximum in precipitation.

     The seasonal maxima in P-E, which represents the supply of
moisture to the surface, were followed by seasonal maxima in soil
moisture content (Figure 4a). Thus, the maximum at middle latitudes occurred in early spring, and the greatest values at low
latitudes in the autumn. The drying zone around 40° of latitude
in summer coinciding with the minimum in P-E was due to the large
rates of evaporation. Similarly, there was a drying zone near 20°
of latitude corresponding to the winter minimum in P-E. At 50°N,
the change in the land-sea contrast outlined earlier led to drying

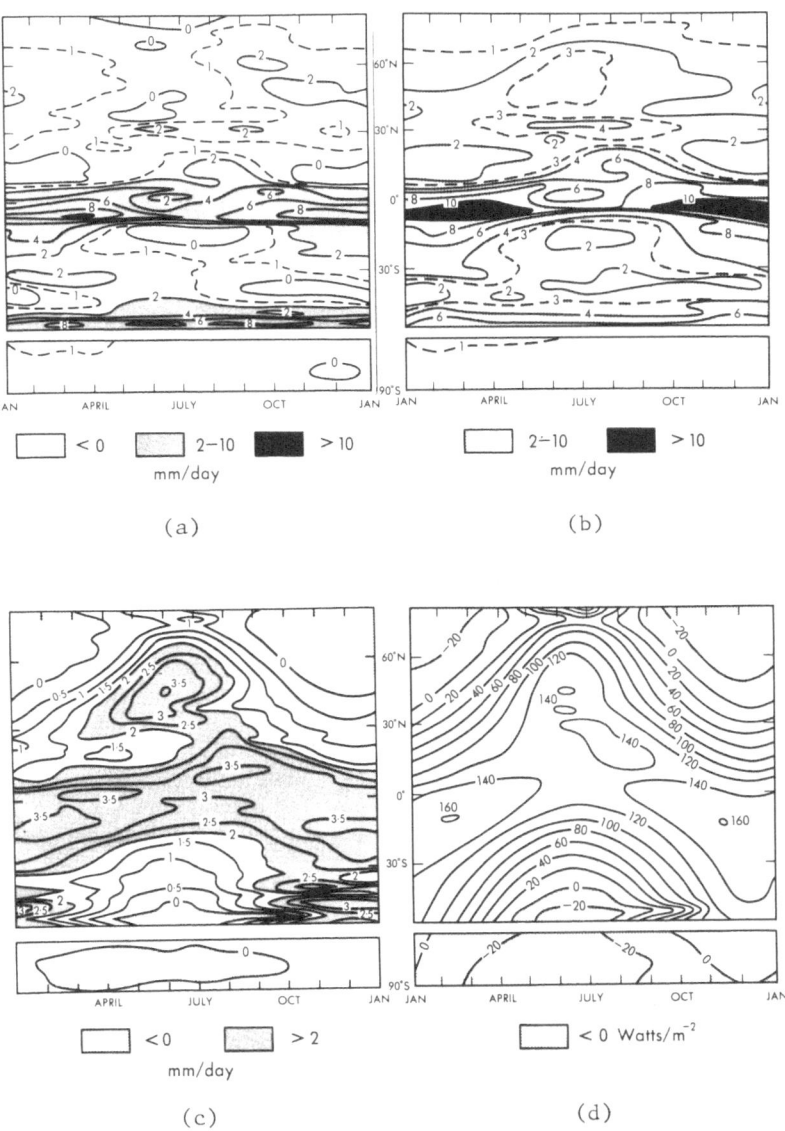

Fig. 3. Time-latitude diagrams, averaged over land points only.
(a) Precipitation minus evaporation: ——, 2 mm per day
contours; ---, 1 mm per day contours; (b) Total precipitation: ——, 2 mm per day contours; ---, 1 and 3 mm per
day contours; (c) Evaporation: ——, 0.5 mm per day contours; (d) Net radiative heating at the surface: ——,
20 W m^{-2} contours.

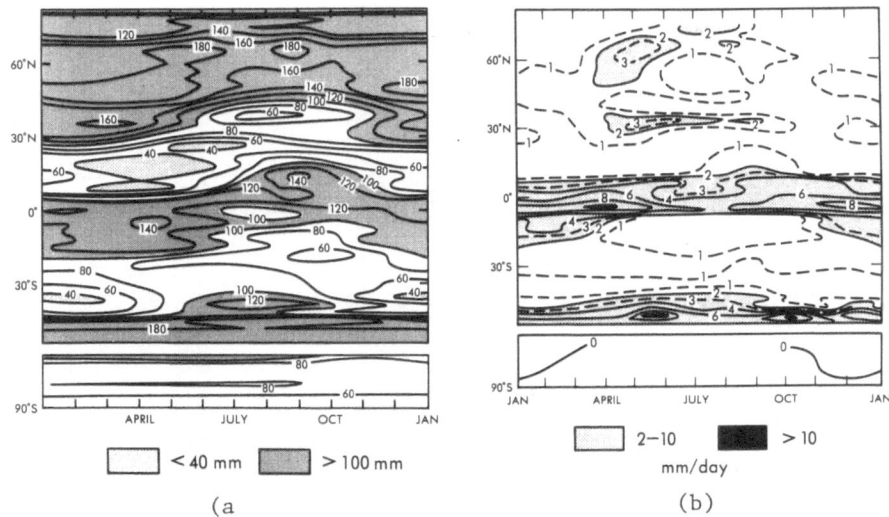

Fig. 4.   Time latitude diagrams, averaged over land points only.
         (a) Soil moisture content: ——, 20 mm contours;
         (b) Runoff: ——, 2 mm per day contours; ---, 1 and 3 mm
         per day contours.

of the surface and a reduction in precipitation over southeast
Asia in winter. (Over the Sahara, low soil-moisture values may
limit both precipitation and evaporation, e.g., see Carson &
Sangster 1980). The low values of soil moisture near 20°N in
spring and 30°N in summer limited evaporation in those regions
(Figure 3c). The large values of soil moisture in high latitudes
in late spring were due to snowmelt. Considerable snowmelt also
occurred over the Himalayas in summer.
       The water balance at the surface is completed by runoff
(Figure 4b). This was greatest near the equator and at middle and
high latitudes where P-E was greatest. Once again, snowmelt was
responsible for the large runoff values at high latitudes in the
northern hemisphere at the end of spring. Runoff was mainly confined to particular regions.

4.   GEOGRAPHICAL DISTRIBUTIONS

4.1   Surface Pressure

       As has been shown, the distribution of precipitation depended
to a large extent on the prevailing atmospheric circulation. The
model produced all the main features of the general circulation as

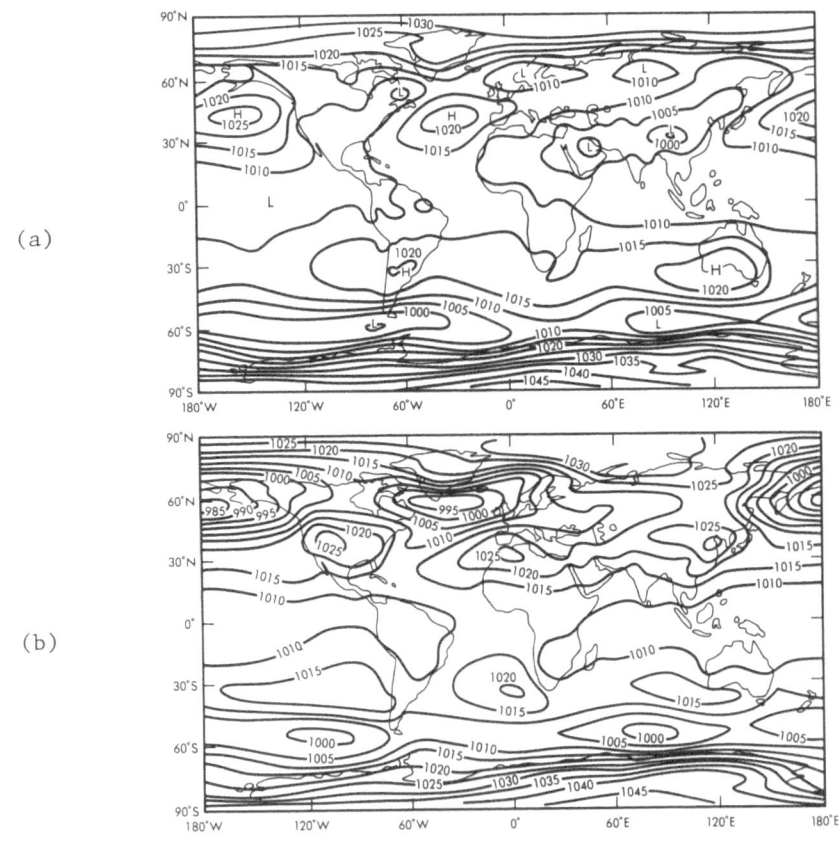

Fig. 5.   Simulated pressure reduced to mean sea level (5 mb con-
          tours):   (a) June, July and August; (b) December, January
          and February.

may be seen in Figure 5, which shows the mean sea level pressure
fields for the northern hemisphere summer and winter, averaged over
the last three years of the integration. The subtropical anticyclones formed predominantly over the continents in winter and
over the oceans in summer. The continental heat lows were present
in the summer hemisphere, notably over North Africa and southern
Asia. The mid-latitude depression belts were more pronounced and
shifted further poleward in winter. Comparison of the simulated
and climatological patterns (Crutcher & Meserve 1970, Taljaard *et
al.* 1969) revealed some systematic differences. The oceanic areas
of high pressure were generally displaced polewards in the model.
The oceanic depressions around Antarctica were shallower and further north than observed. The Aleutian and Icelandic lows were

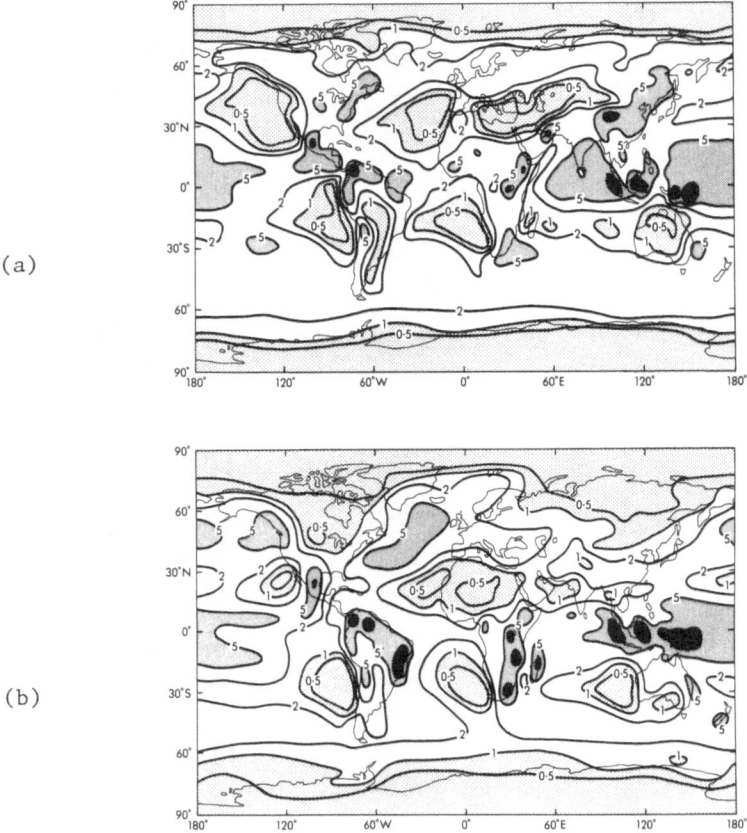

(a)

(b)

Fig. 6.   Simulated total precipitation (0.5, 1, 2, 5 and 10 mm per
          day contours):   (a) June, July and August; (b) December,
          January and February.

deeper in winter and shallower in summer than in the climatological
record.   The simulated Siberian anticyclone was too weak and displaced southwards.   The surface pressure over the poles throughout
the year was much higher than observed.

4.2 Precipitation

     The distribution of precipitation (Figure 6) may be related
to the circulation patterns indicated by the corresponding surface
pressure fields (Figure 5).   The flow, to a first approximation,
was parallel to the isobars, anticlockwise round low pressure areas
in the northern hemisphere and clockwise in the southern hemisphere.   Frictional drag at the surface deflects the flow slightly

across the isobars towards low pressure, so that air near the surface converges in the regions of low pressure and is forced to rise, cool and condense excess moisture as precipitation.

In the northern hemisphere in summer (Figure 6), there was a maximum in precipitation near 60°N due to the passage of mid-latitude depressions. There were further maxima associated with the equatorial low pressure trough from the West Pacific across the Indian Ocean to East Africa and also over Central America and northern South America. Air originating in the subtropical anti-cyclones moved equatorward in the northeast and southeast trade-winds, picking up moisture from the warm tropical oceans. The tradewinds converged along the equatorial trough forcing this moist air to rise, producing areas of heavy rainfall. Other areas of heavy rainfall extended northwards from the tropics along the eastern coasts of continents notably over southeast Asia, the western edges of continents and the adjoining subtropical oceans remaining dry. The enhancement of rainfall along the eastern coasts of the subtropical continents was due to warm moist air travelling poleward around the western edge of oceanic anticyclones and converging into the continental heat lows. This process is discussed in detail by Manabe (1969). Precipitation rates over India and Thailand were smaller than observed, probably because the model's Asian monsoon was too weak. Along the eastern side of the subtropical anticyclones, there was a flow of cool dry air from the north and a descending motion producing little precipitation. Australia and southern South America, where the surface pressure patterns are also anticyclonic, were dry. There was a further belt of precipitation associated with the Antarctic circumpolar trough. The model produced heavy rainfall over southern Africa, the Red Sea and Persian Gulf, contrary to the climatological data (Peixoto: this volume), and rainfall over the southern Sahara extended too far northward. The dry zone from Mongolia to the Mediterranean should have extended right across North Africa and southern Europe.

In the northern hemisphere in winter (Figure 6b), the maxima in precipitation occurred along the oceanic depression tracks, and their extension across western boundaries of the continents. There was a maximum in precipitation where warm moist air was forced to rise over the Canadian Rockies and extension of the North Atlantic maximum across southern Europe. The continental interiors were relatively dry, as were the regions under the subtropical anti-cyclones. The equatorial maxima shifted south with the associated pressure trough, and were found just south of the equator over the western Pacific and Indonesia, East Africa and the north and east of South America. Once again, wet regions extended polewards from the tropics along the eastern coasts of the continents in the summer subtropics. The western edges of the southern continents and the areas under the subtropical anticyclones remained dry. The circumpolar precipitation belt was again evident, but, like the surface pressure trough, was placed to the north of its observed position. The rainfall maxima over Mexico, Venezuela and

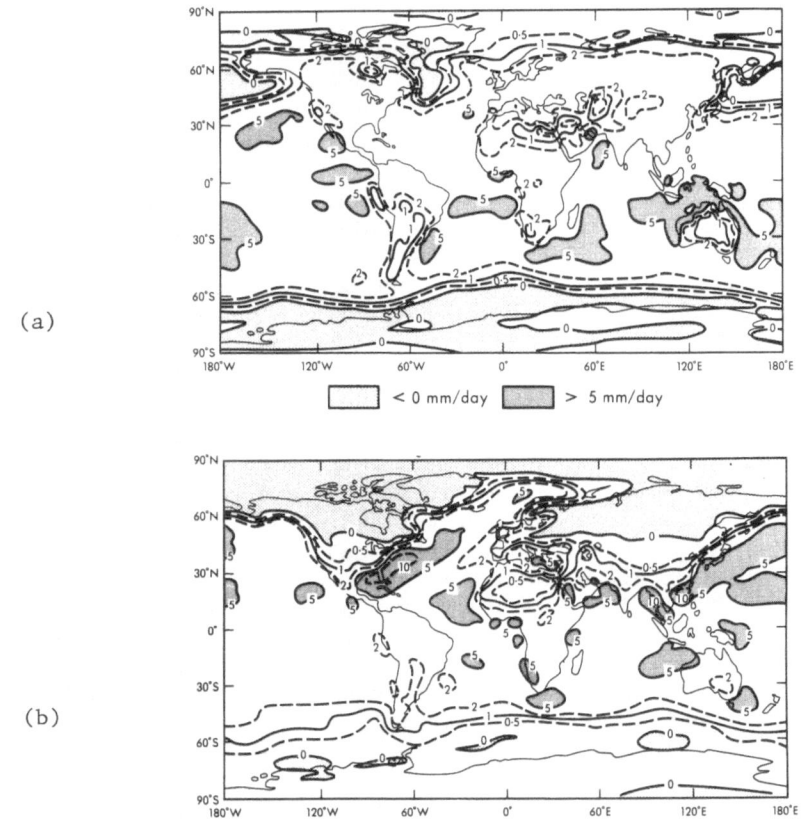

Fig. 7. Simulated evaporation (——, 1, 5 and 10 mm per day contours, ---, 0.5 and 2 mm per day contours): (a) June,
July and August; (b) December, January and February.

East Africa north of the equator were not found in the observed
data, and precipitation over Asia between 30° and 45°N, and much
of southern Africa, was heavier than observed.

4.3 Evaporation and Soil Moisture

Evaporation was generally greater over the oceans than over
land (Figure 7). Rates greater than 5 mm per day occurred over the
tropical and subtropical oceans, particularly in winter. The cool
dry winter tradewinds picked up large amounts of moisture from the
relatively warm ocean (see Equation 1). In winter, small, often
negative, evaporation rates were found over Antarctica, Greenland
and the Arctic, and much of North America and Eurasia, as the sur-

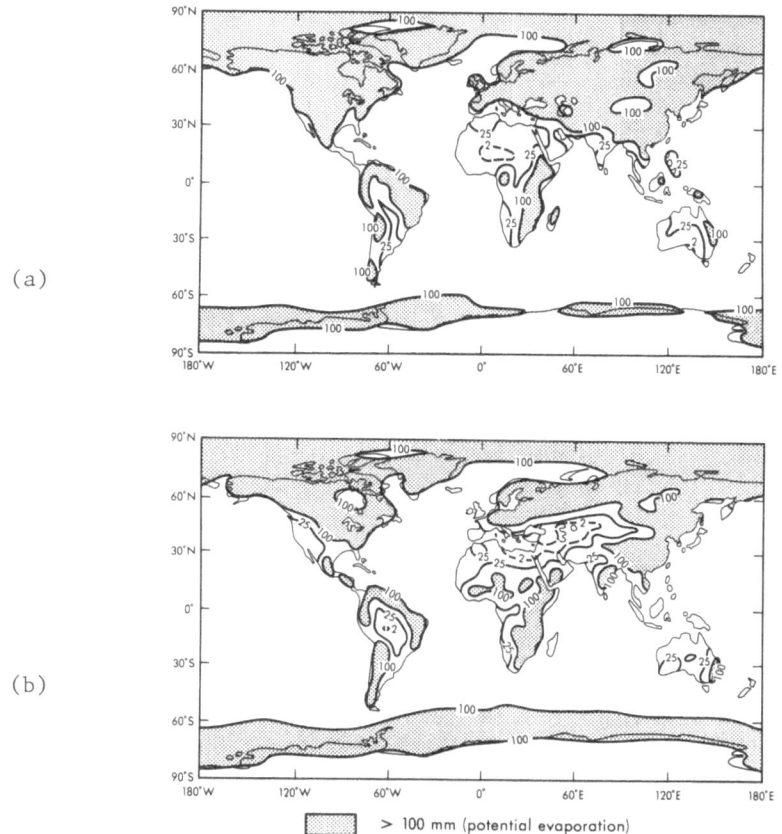

Fig. 8. Simulated soil moisture (2, 25 and 100 mm contours):
         (a) February mean; (b) August mean.

face cooled radiatively and temperatures were low. The areas of
high evaporation to the east of North America and Asia in winter,
evident in Figure 7, were due to cold dry continental air passing
over the warm western boundary currents (the Gulf Stream and the
Kuroshio) and their northward extensions. In summer, the land/sea
temperature contrast was reversed leading to condensation over the
Labrador Sea and the Sea of Okhotsk.

     The soil moisture variable used in the model cannot be com-
pared directly with observations. The treatment of soil moisture
is intended to represent the reduction in evaporation and evapotranspiration which occurs as the surface and subsurface dry out.
Processes were being represented in 330 km grid boxes in which the
terrain, soil and vegetation types would vary considerably. It is
important to make a qualitative assessment of the model's distri-

bution of soil moisture, as this can have an important limiting
effect on evaporation and hence precipitation (see Rowntree and
Bolton: this volume).

The model preserved in part, particularly in winter, the dry
areas over North Africa, the Middle East, southwest Africa,
Argentina and Western Australia, where evaporation was generally
small due to low values of soil moisture (Figure 8).   Elsewhere,
soil moisture content was generally in excess of 10 cm (apart from
the western USA and central southern Asia in summer) and so did not
limit evaporation.   Evaporation was not limited by soil moisture
over snow-covered land or sea ice.   The model's main shortcoming
was its failure to keep the southern Sahara dry in summer.   The
estimates of Budyko (1963) suggest that the zone of large evaporation rates and, hence, probably of soil moisture values over
central Africa extended too far to the north throughout the year,
and that southern Africa was too wet during the southern winter.
These shortcomings are not apparent in a higher resolution model
(Carson & Cunnington 1980) in which the treatment of soil moisture
was similar, but in which the thresholds for potential evaporation
and runoff were 5 and 15 cm rather than 10 and 20 cm used here.
However, as many of the other physical processes were parameterized differently in the higher resolution model, it is not
possible to isolate the factor which led to the improvement.

The main supply of moisture from the surface to the atmosphere was over the subtropical oceans, as shown by the areas of
negative P-E in Figure 9.   In higher latitudes, the atmosphere
supplies moisture to the ocean, except off the eastern coasts of
continents in winter and off western coasts in summer.   Over much
of the land, P-E was positive throughout the year, the surface
moisture balance being completed by runoff.   Negative values,
indicating a drying of the surface, occurred over the western USA,
around the Mediterranean and eastward to Mongolia during summer
(Figure 9a), producing the seasonal change in soil moisture found
in these regions (Figure 8).   The surface also became drier in
central South America at this time of year.   The northern hemisphere drying zone was found nearer the equator in winter, extending from the Sahara across Arabia to India and the extreme southeast of Asia; there was a further drying zone near $35°S$ across
southern Argentina and Australia.   The model's values of P-E may
be compared with values of atmospheric moisture convergence
derived from observations (Peixoto: this volume) as the storage
of moisture in the atmosphere over a season is small.   Regions of
positive P-E should correspond to regions of moisture convergence.

5.   MODEL VARIABILITY

During the final three years of the integration, the specifications for solar heating at the top of the atmosphere and for sea
surface temperatures were repeated exactly.   The simulation did

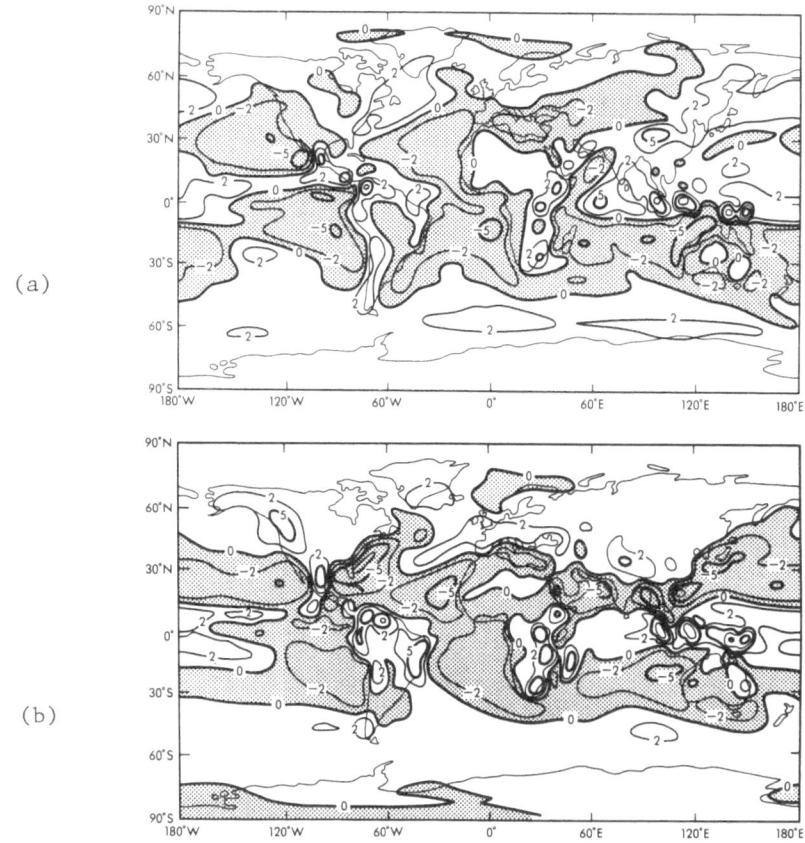

Fig. 9.   Simulated precipitation-minus-evaporation (0, ± 2, ± 5
           and ± 10 mm per day contours, stippled where negative):
           (a) June, July and August; (b) December, January and
           February.

exhibit year-to-year variations which were generated within the
model. The average standard deviation (SD) of monthly mean surface pressure (not shown) for December, January and February
reached a maximum of 9 mb over the east of the North Pacific and
North Atlantic, and a further maximum of 12 mb over Northwest Asia.
These maxima lay along the model's depression tracks. Minimum
values occurred near the east of continents and at low latitudes.
Much of the variability occurred because the model's middle-latitude depression tracks did not reproduce themselves exactly from
year to year. The observed SD of mean January surface pressure
from the 100-yr (1875-1974) mean has a similar pattern, though the
Atlantic maximum is further north and the maximum over Northwest
Asia is less pronounced than in the model.

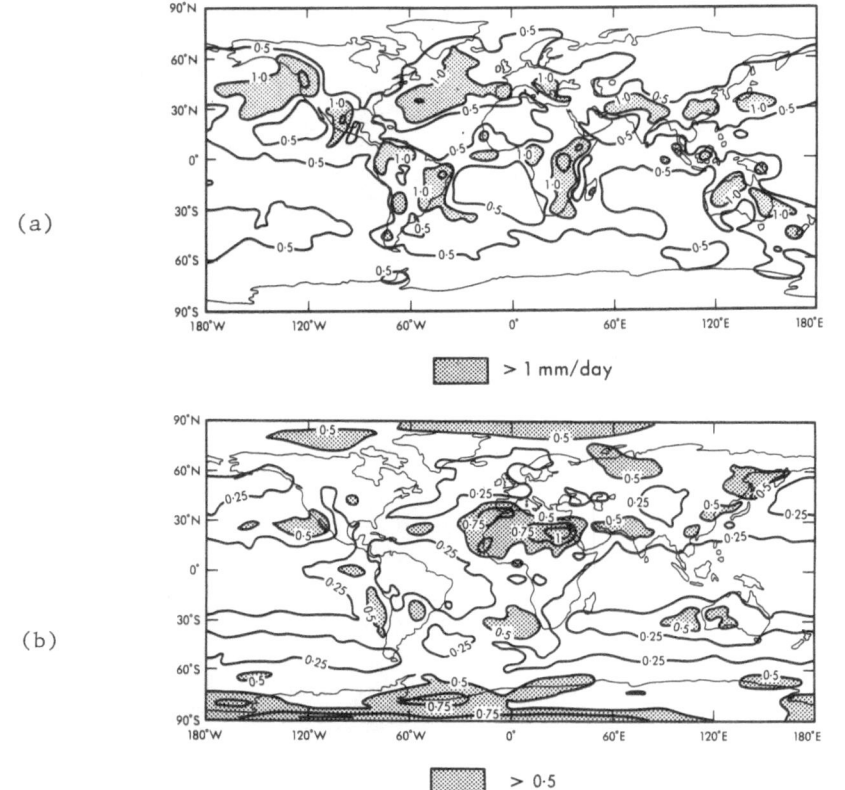

Fig. 10. Monthly mean total precipitation from 3-yr monthly mean
          values: (a) Standard deviations of monthly means averaged
          over December to February (0.5 mm per day contours);
          (b) Coefficient of variation (standard deviation/mean;
          0.25 contours, averaged over the depth of the atmosphere)
          for December to February.

     The year-to-year SD of monthly mean precipitation, averaged
over December, January and February (Figure 10a) was largest where
the mean amounts of precipitation were largest (see Figure 6b).
The main exception was over the tropical oceans, particularly in
the central and western Pacific. Fluctuations in tropical rainfall are closely related to changes in sea surface temperatures
(Doberitz 1968). The sea surface temperatures in the model did
not vary from year to year, so this lack of interannual variability over the oceans in low latitudes is not surprising. In
contrast, there were peaks in variability over the North Atlantic
at higher latitudes resulting from the variability in the model's
simulated depressions noted above. The coefficient of variation

(= SD/mean) tended to be greatest where precipitation amounts
were small (Figures 10b and 6b), that is in the subtropics and at
high latitudes. Although the amplitude of the year-to-year variation was small in these regions, the fractional change and, hence,
the potential impact on the environment was largest.

6.   SUMMARY AND CONCLUSIONS

     The model reproduces the main features of the hydrological
cycle, including the major regional and seasonal variations in
precipitation and evaporation. Over land, the hydrological cycle
was more intense than observed, and the model's deserts were less
extensive than their real counterparts. The seasonal variation of
snow cover was close to that observed, except over central North
America where the model was too warm. Although the external forcing in the model was repeated exactly over each annual cycle,
there were changes in the monthly mean circulation patterns from
year to year, particularly along the model's depression belts in
winter. Precipitation varied most where it was heaviest. The
fractional change in precipitation was largest over the drier
regions in the subtropics.

ACKNOWLEDGEMENTS. I wish to thank Mr J. Bolton for help with the
preparation of material for this paper, Dr P.R. Rowntree for making
useful comments on the manuscript, and Mr N. Birkett and
Mr C. Staunton-Lambert who drew the diagrams.

REFERENCES

Budyko, M.I. (ed.) 1963 *Atlas of the Heat Balance of the Earth*,
        69 pp. (in Russian). Moscow: Glavnaia Geofiz. Obs.

Budyko, M.I. 1970 The water balance of the ocean. In: *Proc. Symp.
        World Water Balance, Reading 1970*, pp. 24-33. Gentbrugge:
        Int. Ass. Sci. Hydrol. (Publ. 92.)

Carson, D.J. & Cunnington, W.M. 1980 General circulation experi-
        ments with an 11-layer model using FGGE data (GARP Numerical
        Experimentation Programme). *Research Activities in Atmo-
        spheric and Ocean Modelling Rept* No. 1, pp. 1.17-1.20.
        Geneva: ICSU/WMO.

Carson, D.J. & Sangster, A.B. 1980 The sensitivity of general cir-
        culation simulations to land-surface properties (GARP
        Numerical Experimentation Programme). *Research Activities in
        Atmospheric and Oceanic Modelling Rept* No. 1, pp. 6.3-6.8.
        Geneva: ICSU/WMO.

Corby, G.A., Gilchrist, A. & Rowntree, P.R. 1977 United Kingdom
    Meteorological Office five-level general circulation model.
    Meth. comput. Phys, <u>17</u>, 67-110.

Crutcher, H.L. & Meserve, J.M. 1970 Selected level heights, tem-
    peratures and dew points for the Northern Hemisphere. NAVAIR
    50-IC-52. Washington: US Naval Weather Service.

Doberitz, R. 1968 Cross-spectrum analyses of rainfall and sea sur-
    face temperatures at the equatorial Pacific Ocean. *Bonner
    Met. Abhandl.*, <u>8</u>.

L'vovich, M.I. & Ovtchinnikov, S.P. 1963 *Physical-Geographical
    Atlas of the World* (in Russian). Moscow: Academy of Sciences
    and Department of Geodesy and Cartography, State Geodetic
    Commission.

Manabe, S. 1969 Climate and the ocean circulation. I: the atmo-
    spheric circulation and the hydrology of the Earth's surface.
    Month. Weath. Rev., <u>97</u>, 361-385.

Manabe, S. & Holloway, J.L., Jr 1975 The seasonal variation of the
    hydrologic cycle as simulated by a global model of the atmo-
    sphere. *J. geophys. Res.*, <u>80</u>, 1617-1649.

Möller, F. 1951 Vierteljahrskarten des Niederschlags für die ganze
    Erde. *Petermanns Geogr. Mitt.*, <u>95</u>, 1-7.

Slingo, Julia M. 1979 A new interactive radiation scheme for the
    5-level model. *Met. O. 20 Tech. Note* No. II/135. Bracknell:
    Meteorological Office, UK.

Slingo, Julia M. 1982 A study of the Earth's radiation budget
    using a general circulation model. *Q. Jl R. met. Soc.*, <u>108</u>,
    379-406.

Taljaard, J.J., Van Loon, H., Crutcher, H.L. & Jenne, R.L. 1969
    Climate of the upper air: 1. Southern Hemisphere, 1.
    Temperatures, dew points and heights at selected pressure
    levels. NAVAIR 50-IC-55. Washington: US Naval Weather
    Service.

EFFECTS OF SOIL MOISTURE ANOMALIES OVER EUROPE IN SUMMER

P.R. Rowntree and J.A. Bolton

Meteorological Office
Bracknell
Berkshire, UK

ABSTRACT. Two global models of differing resolutions of the atmosphere in July have been run for 50 days with anomalies in the initial soil moisture over Europe. The results are similar for both models and show that such anomalies can have a considerable effect on the modelled rainfall, humidity and temperature in the course of the following 50 days not only over the region with the anomaly, but also over a considerable area of the adjacent land. The effects of the anomaly spread to much of Scandinavia and Spain and a large part of North Africa.

  The persistence of the anomaly was dependent on the prevailing flow conditions. With a westerly flow advecting moist air from the Atlantic, the model's equilibrium state was wet over northern Europe with the dry anomaly persisting for only about 20 days before much of it became too weak to affect the evaporation significantly. With weaker circulation patterns, the model equilibrium state was dry over the anomaly area and, although the wet anomaly was slowly being eroded, this process was far from complete after 50 days.

1. INTRODUCTION

  The moisture budget of the atmosphere over a region depends on three components:
(a) The net flux of water vapour and liquid water across the vertical boundaries of the region;
(b) The net upward flux of water vapour from the underlying surface (evaporation less condensation);
(c) The precipitation of rain and snow from the atmosphere to the surface.

A. Street-Perrott et al. (eds.), Variations in the Global Water Budget, 447–462.
© *British Crown Copyright 1983*

The question considered here is the importance of the second
component (mainly evaporation) in determining the third component
(precipitation).

Mintz (1981) noted that studies of the world water balance
(Korzun 1974, Baumgartner & Reichel 1975) show that about 60% of
the rainfall over the land mass of Eurasia evaporates; from an
analysis of the water budget of eastern and central USA, he argues
that much of the continental rainfall is due to evaporation from
the surface of the continent. This is because continental evaporation increases the water vapour available for precipitation and
also provides the energy needed to initiate precipitation.

Experiments with numerical models in recent years generally
have confirmed the importance of evaporation over the land for the
precipitation process. Walker and Rowntree (1977) carried out two
experiments using a tropical model with idealized geography in
which the land surface over an 18°-wide latitude belt was initially
prescribed as either dry or wet. The model experiments showed
that the absence of evaporation in the dry case almost eliminated
the rainfall over the anomaly area whereas, in the wet case, the
initially-dry atmosphere over the anomaly area soon became moist
and, after about 10 days, rainfall balanced evaporation and so
maintained the soil in a wet state. A subsequent series of
experiments by Cunnington (personal communication) used a global
version of this model with realistic topography. The results confirmed the sensitivity of rainfall over the Saharan region to
anomalies in the initial availability of moisture, both in the soil
and in the atmosphere. Initial anomalies persisted for at least
50 days, partly due to a positive feedback in the moisture convergence.

Mintz (1981) reported recent experiments on the effects of
global-scale soil moisture anomalies. He described three rather
similar experiments, undertaken by Suarez and Arakawa, Shukla and
Mintz, and Carson and Sangster, in which global models were run
for northern summer conditions with all the land either dry or
wet. Over much of the land, precipitation was found to be
considerably reduced by the absence of evaporation. Even the
southern (winter) continents equatorward of 40°S were affected.
Shukla and Mintz found increased rainfall over parts of southern
Asia, but this was not evident in the other experiments. Only in
Carson and Sangster's experiment was the soil moisture allowed to
vary with time. Their results are instructive although perpetual
July conditions (radiation, sea surface temperature, etc.), were
applied. By days 21-50 from the start of the integration for
initially-dry land, there was substantial rainfall over most of
the rainy regions of the tropics, though generally less than in
the case of initially-wet land. However, in middle latitudes,
there was still much less rainfall except locally near rainy
eastern coasts. Even after 250 days, there were still substantial
land areas in middle latitudes and the subtropics with more rain
in the initially-wet case.

The experiments discussed above were designed to investigate the effects of either rather large-scale anomalies in the summer tropics and subtropics or of global-scale anomalies. In this paper, the results are reported of experiments made to assess the response to an anomaly of relatively small horizontal scale in middle latitudes in order to approach the question of how large an anomaly needs to be to have a significant effect. This is an important question because most observed anomalies on the inter-annual time scale tend to be on a rather small spatial scale.

2.   THE MODEL

The models used for the experiments are developments of that described by Corby *et al.* (1977). The main changes were the inclusion of interactive soil moisture and snow, as described by Slingo (1982); also the moist rather than the dry gas constant was used in the hydrostatic equation (and elsewhere). For the main series of experiments, the climatological radiation scheme described by Corby *et al.* (1977), in which long-wave radiative cooling was a function only of temperature and latitude, has been retained. For a subsequent experiment with the annual-cycle version of the model, the radiation scheme described by Slingo (1982), in which atmospheric absorption of solar and long-wave radiation depends on moisture as well as temperature, was used.

The models have five layers of equal mass and use a σ-coordinate system (σ = pressure/surface pressure). The horizontal grid gives a quasi-uniform resolution over the sphere. In the medium resolution version used for the main series of experiments, the grid length was approximately 330 km (3° latitude north to south); some results are given for a series of experiments with a low resolution model having a grid length of about 500 km, other features of the model being unchanged. Further details of the model and, specifically, of the annual-cycle version are given by Mitchell (this volume).

There are a number of relevant features of the models which could be made more elaborate. The radiative transfer does not depend on modelled cloud or, in the main series of experiments, on modelled humidity. The albedo is kept independent of soil moisture, although Idso *et al.* (1975) and Norton *et al.* (1979) reported increases in albedo by 0.1 to 0.15 due to a change from wet to dry soil or wet to dry vegetation.

3.   THE EXPERIMENTS

For the main series of experiments, the low and medium resolution models were each integrated for 50 days from an initial global data set compiled from real data for most of the northern hemisphere (for 27 May 1977) and a modelled data set for July for

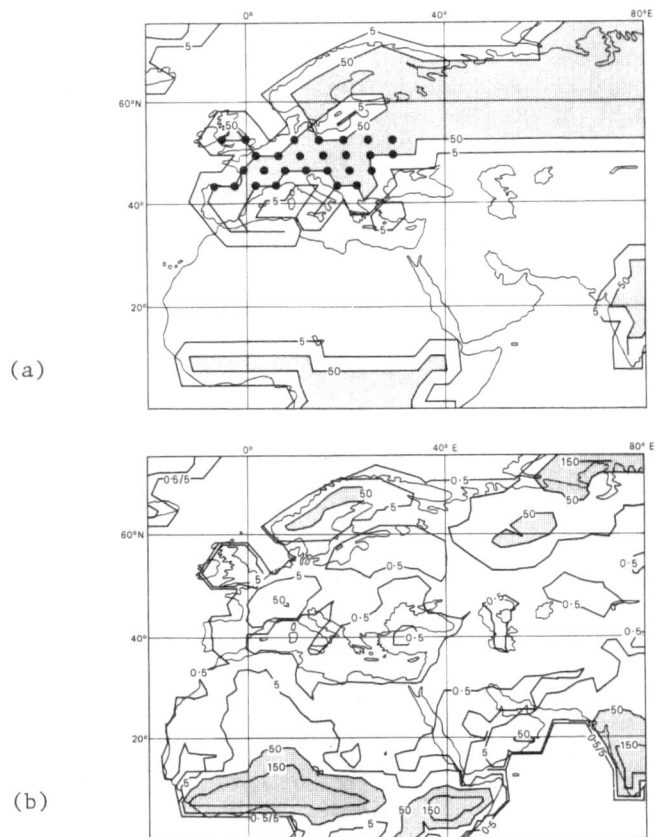

Fig. 1   Soil moisture (mm) for control experiment (MC):   (a) Day 0;
(b) Day 41-50 mean; •, grid points in (a) at which anomalous values of the initial soil moisture were applied.
Areas with 50 mm or more are stippled.

the southern hemisphere. The constants used in the radiation
scheme were appropriate for July, while sea surface temperature
and ice extent were as prescribed from climatology for mid-June.
For the control integration, initial soil moisture was as prescribed in Figure 1 for the medium resolution model, 5 cm being
assigned in all areas except in some arid regions with no soil
moisture. Specifications were generally similar for the low
resolution model, apart from minor changes due to differing locations of grid points.
    In the anomaly experiments, the models were again integrated
for 50 days starting from initial conditions which were identical
to those of the control experiments except for the soil moisture

content for the European area indicated for the medium resolution
models in Figure 1; this area extends from the Atlantic nearly to
the Black Sea. The area was similar in the low resolution model,
the latitudes affected being $42\frac{3}{4}°$, $47\frac{1}{4}°$ and $51\frac{3}{4}°$N compared to
$43\frac{1}{2}°$, $46\frac{1}{2}°$, $49\frac{1}{2}°$ and $52\frac{1}{2}°$N in the medium resolution model. Two
experiments were run with the initial soil moisture over the
anomaly area set to zero (dry) and 15 cm (wet). The nomenclature
of all the experiments is defined in Table I.

Table I. Nomenclature of experiments.

Model	Experiment series	Grid length (km)	Initial soil moisture over the anomaly area		
			Dry (0 cm)	Control (5 cm)	Wet (15 cm)
Low resolution	L	500	LD	LC	LW
Medium resolution	M	333	MD	MC	MW
Annual cycle	A	333	AD		AW*

* Initial soil moistures over Europe generally exceeded 15 cm in
   experiment AW.

The initial data for the anomaly experiments using the annual-
cycle version of the model were taken from day 376 (7 June of the
second year) of the run described by Slingo (1982). The soil
moisture was set to zero at the same points over Europe as for the
medium resolution dry-start experiment (MD). It should be noted
that soil moisture in middle latitudes at day 376 of the annual
cycle run was generally greater than the value of 5 cm specified
for the main series of experiments. In particular, over the
anomaly area (see below), over Scandinavia in the north and over
Russia in the east, soil moisture generally exceeded 15 cm. Since
evaporation is little affected by soil moisture until soil moisture
content is well below 10 cm, this prevents the anomaly over Europe
affecting evaporation in these adjacent regions in experiment AD.
The area adopted for the anomaly experiment is similar to that
affected by the European drought of 1976. For February to June
1976, rainfall was less than 80% of normal over most of the anomaly
area, and below 50% over most of the northwestern quadrant.

4.   THE MEDIUM-RESOLUTION CONTROL EXPERIMENT

The models tended to produce a drying-out over most middle
latitude areas in summer, as is shown by a comparison (Figure 1)
of the mean soil moisture content initially and at days 41-50 for

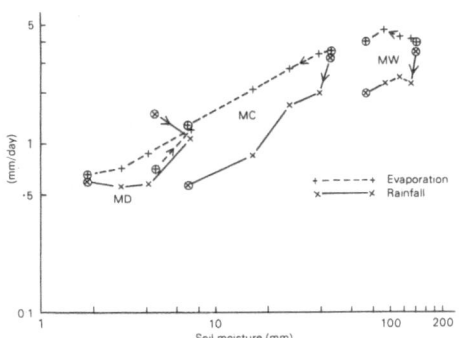

Fig. 2 Variations of evaporation and rainfall compared with soil
        moisture averaged for each of the five 10-day periods of
        the medium resolution experiments (MW, MC and MD). Arrows
        indicate progression from days 1-10 to days 41-50.

Europe, western Asia and North Africa. The only parts of the area
to show an increase were tropical Africa and India and small
regions of Scandinavia and Russia near 60°N.

    Because the middle-latitude 'summer' continents tended to dry
out in the control experiment, rainfall distribution did not reach
an equilibrium state. This is demonstrated (in Figure 2) by the
10-day means of evaporation and rainfall for the anomaly area (in
Figure 1a). In the control experiment, soil moisture, evaporation
and rainfall all decreased monotonically throughout the 50-day
period. The marked drop in precipitation between the first and
second 10-day periods is probably due to the model's self-adjust-
ment towards its own equilibrium from the observed initial state.

That the rainfall for days 11-20 was near the model's equilibrium
with a local evaporation of 3-4 mm per day is confirmed by the wet-
start experiment (MW) wherein rainfall stayed near 2-2.5 mm per day
from day 21 to day 50. As this amount was less than the potential
evapotranspiration, drying out was observed with both the MC and
MW experiment runs. Evidently, over the anomaly area, there was a
moisture flux divergence which decreased as the ground dried out.
A larger decrease in flux divergence might have occurred if the
adjacent land, especially upwind of the anomaly area, did not dry
out at the same time (see Figure 1).

    As soil moisture (SM) decreased from 10 to 0 cm, evaporation
did not decrease linearly as might have been expected from the
model's formulation of evaporation (Mitchell: this volume). This
is because, as a (= SM/10 cm) decreases, the surface temperature
and saturation mixing ratio increase in order to maintain the
balance between net radiation and the upward turbulent fluxes -
ground heat storage being small. From the data used in Figure 2,
the relation between evaporation E and a can be estimated to be
approximately $E \propto a^{\frac{1}{2}}$.

5.   DEVELOPMENT OF DIFFERENCES DUE TO THE ANOMALY

In this section, the development is outlined of the predictions
for the first week of the medium resolution experiments with dry
and wet soil moisture anomalies over the area indicated in Figure
1(a). As described in § 1, the direct effect of a dry anomaly is
to modify the partitioning of the surface turbulent heat flux
between the sensible and latent parts so as to increase the
temperature and decrease the specific humidity of the atmosphere,
particularly of the boundary layer. After 48 h, the temperature
in the lowest layer of the model centred at $\sigma = 0.9$, was 4 K higher
in the dry model run over most of the anomaly area, with local
increases of up to 9 K; specific humidities were lower by 1 to
3 g kg^{-1}.
The differences in rainfall up to this time (48 h after start-
up) were small though, on the second day, rainfall rates exceeding
1 mm per day spread north from a rain area over southern Europe to
about 48°N near 5°E with the wet start, but only to 45°N with the
dry start. More marked differences in rainfall were evident on the
third day. By then, the rainfall area over southern France on
day 2 had disappeared with the dry start, whereas much of France
had over 2 mm per day in the wet-start case. Heavy rain over SE
Europe diminished in area with the dry start and areas with over
1 mm per day outside the anomaly area, over southern Spain and
over and to the west of the British Isles, were absent or much
smaller. At this stage, there was little change in the rainfall
north of the anomaly owing to the mainly northerly flow in these
regions.
With the increase in lower tropospheric temperatures in the
case of the dry start, there were decreases in surface pressure of
2 to 3 mb over most of the anomaly area by day 3, with little
change in the adjacent regions. This association of decreased
pressure and decreased soil moisture and rainfall should not be
surprising, having been evident in the soil moisture experiments
of Walker and Rowntree (1977) for an idealized North Africa; it
can also be related to observed conditions over the Sahara. The
pressure drop led to a more marked trough over France and southern
Britain, but had little evident effect on rainfall distribution:
the rain associated with the trough, though developing in the dry
case, generally yielded < 2 mm per day on days 4 and 5 compared
with peak values of > 5 mm per day in the wet case.
By day 5, the depression which had formed over southern
Britain 24 h earlier in both experiments developed a trough eastwards over northern Germany, with advection of air from central
and eastern Europe into southern Scandinavia between the trough and
an anticyclone over central Scandinavia. Thus the effects of the
soil moisture anomaly were advected northward into Scandinavia as
is evident from Figure 3 which shows that differences in the
specific humidity of the lowest layer exceeded 1 g kg^{-1} well north
of the $52\frac{1}{2}$°N limit of the anomaly. During the following 24 h (with

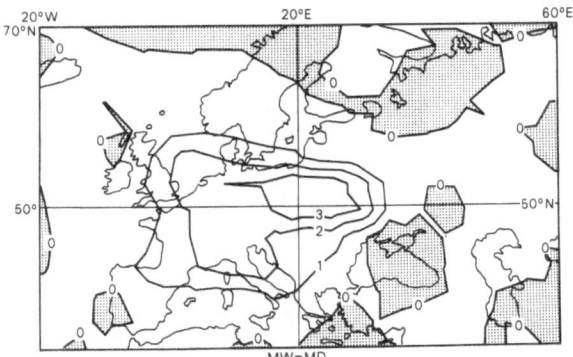

Fig. 3   Difference in specific humidity (g kg^{-1}) on day 6 of medium
         resolution experiments at σ = 0.9 between wet start and dry
         start (MW-MD).

the wet start), substantial rainfall (Figure 4) occurred over
southern Scandinavia ahead of the surface trough as it moved slowly
northeast towards the Baltic. Over 2 mm of rain fell in a belt
linking regions of heavier rain over north Germany and Norway.
Yet in the dry case, even though the pressure patterns were very
similar, there was only a somewhat weakened rain-affected area over
Norway, with less than 0.5 mm over Denmark. Thus, as early as the
sixth day, the effects of the anomaly were evident over land outside the anomaly area. These land areas then progressively dried
out so that the area with reduced evaporation expanded.

6.   TIME-AVERAGED EFFECTS FROM DAY 21 TO DAY 50

     Although the model did not reach an equilibrium state during
the 50 days of the experiments, time-averaging helped to remove
some transient features of the simulations. Averages were calculated, therefore, for days 21-50. The results of the low resolution experiments were generally very similar to those of the medium
resolution experiments and increase confidence in the conclusions.

6.1 Rainfall

     In the medium resolution experiments, there is a striking
contrast between the mean rainfall values for the dry (MD) and wet
(MW) starts, with reductions by a factor of three or more over most
of the anomaly area and much of Sweden (Figure 5). These results
provide a clear answer to the questions (posed in § 1) concerning
the impact and the persistence of an anomaly of this horizontal
scale in middle latitudes, and the possibility of its extension
to affect other areas. In the MW experiment, it is particularly

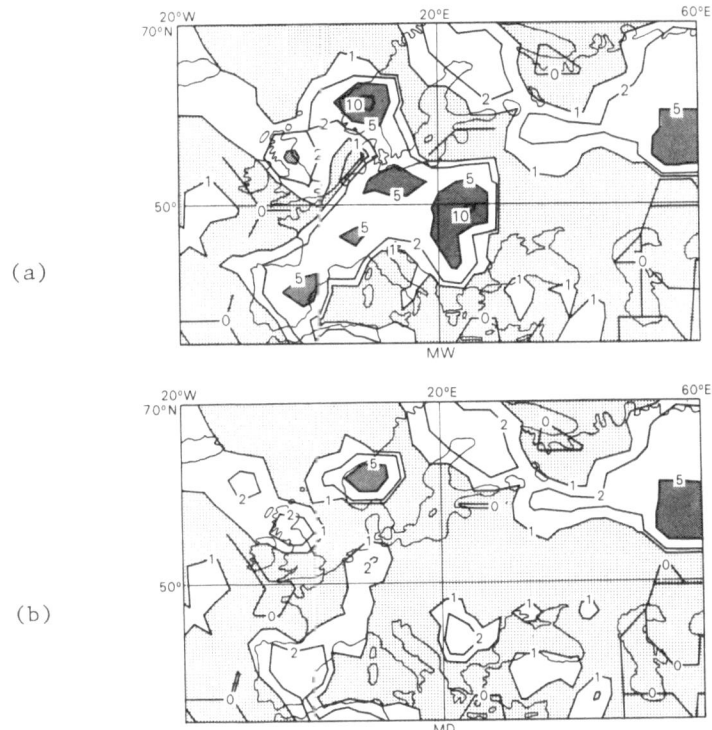

(a)

(b)

Fig. 4   Rainfall (mm) on day 6 of medium resolution experiments:
         (a) wet start (MW); (b) dry start (MD).

interesting that the precipitation in the western central part of
the anomaly area was still sufficient to balance the potential
evaporation of about 4 mm per day. This is confirmed by soil
moisture values (not shown) which exceeded the initial value of
15 cm at four of the rainiest points according to the day 21-50
mean; with the dry start, these points had at most 1.55 mm per day
rainfall over this period. In the eastern part of the anomaly
area, rainfall was well below potential evaporation for both the
MW and MD experiments, and the lifetime of the wet anomaly was
clearly limited to the time taken for soil moisture to decrease
to values at which it restricted evaporation; it is thus a function
of (i) the excess of the initial value (15 cm in this case) above
the critical value (10 cm) and (ii) the excess of evaporation over
precipitation.
        The rainfall difference pattern (Figure 6) delineates more
clearly the regions outside the anomaly area where rainfall had
been affected. As well as the large areas of Scandinavia and the
Iberian peninsula where differences were already developing at

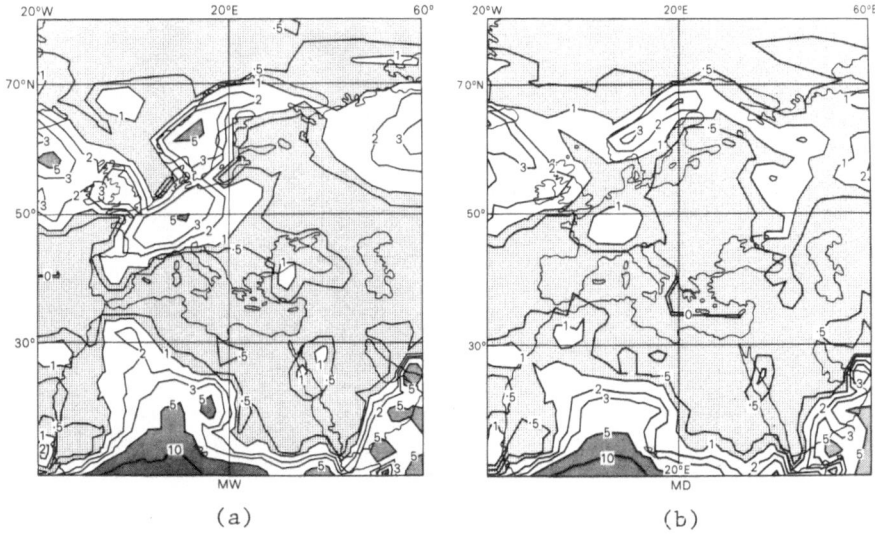

Fig. 5   Rainfall (mm per day) averaged for days 21-50 of medium
         resolution experiments: (a) wet start (MW); (b) dry start
         (MD).

day 6 (Figure 4), with the dry start there were decreases over
Turkey and much of North Africa extending in a rather broken
fashion to the Persian Gulf. The differences over North Africa
were partly in a desert region which the model commonly failed to
keep dry. Although there was still too much rainfall in the MD
experiment, these results indicate that a moisture source to the
north over Europe can contribute to the model's rainfall in this
area.

     The rainfall differences for the low resolution model in
Figure 7 were generally similar to those of the medium resolution
model (Figure 6) with an equally striking contrast between the dry
and wet cases over the anomaly area. In both models, rainfall
over Europe was greatly affected by surface moisture availability.

There were some important differences outside the anomaly area,
notably over northern Scandinavia where rainfall was increased in
the case of the dry start. This increase was associated with lower
surface pressure (not shown) suggesting increased cyclonicity which
was probably a chance occurrence. The maximum differences in rainfall were 6 mm per day over central Europe, southeast of the
maximum difference in the medium resolution experiment (MW-MD).
In the dry-start experiment, the northern regions of the Sahara
had slightly more rainfall than with the wet-start experiment,
but most of North Africa is drier, as in the medium resolution
case.

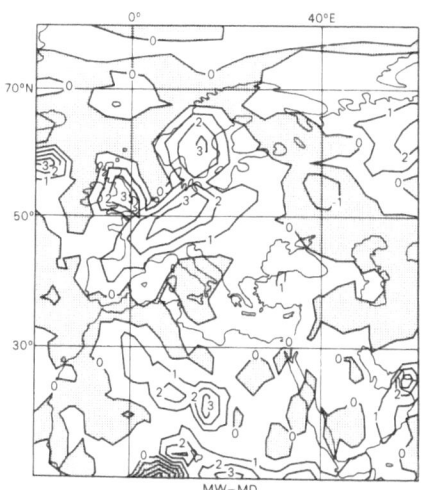

Fig. 6   Difference in rainfall (mm per day) for days 21-50 of
         medium resolution experiments between wet start and dry
         start (MW-MD). Negative values (i.e., rainfall less in
         MW than in MD) are stippled.

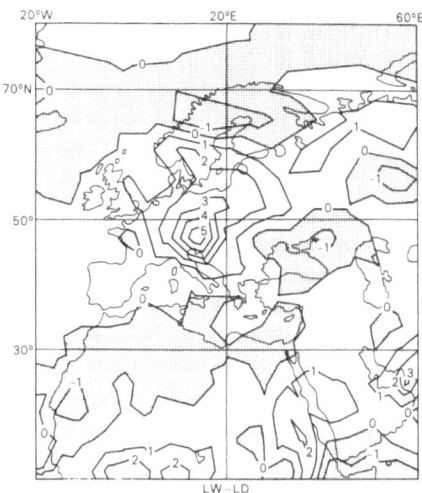

Fig. 7   Difference in rainfall (mm per day) for days 21-50 of low
         resolution experiments between wet start and dry start
         (LW-LD). Negative values (i.e., rainfall less in LW than
         in LD) are stippled.

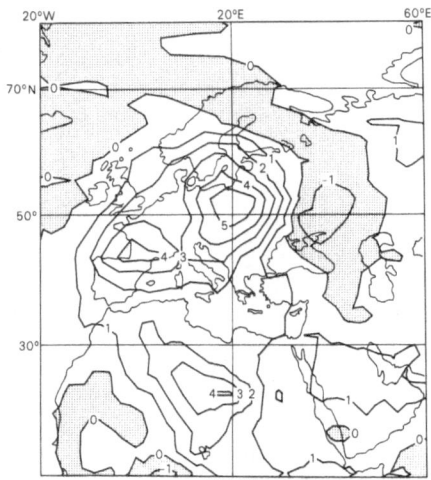

Fig. 8   Difference in specific humidity (g kg^{-1}) for days 21-50 of
medium resolution experiments at σ = 0.9 between wet start
and dry start (MW-MD). Negative values (i.e., humidity
less in MW than in MD) are stippled.

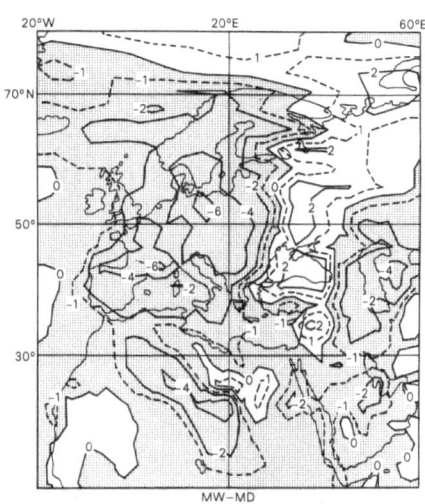

Fig. 9   Difference in temperature (K) for days 21-50 of medium
resolution experiments between wet start and dry start
(MW-MD). Negative values (i.e., temperature less in MW
than in MD) are stippled.

6.2 Atmospheric Moisture

Just as the moisture content of the lowest layer of the model showed a large response within two days of start-up (see § 4), so did a response of similar magnitude appear in the mean of days 21-50 (Figure 8) with differences of up to 5 g kg^{-1} over eastern Europe. This is clearly a direct effect of reduction in the local surface moisture source in the dry-start experiment. In the south, the changes extended well beyond the anomaly area to Africa and Arabia. It is most probable that this was due to advection of drier air, though feedback of local soil moisture can help to maintain the anomalies. It is likely that this happened where the differences were at a maximum in the central Sahara. The large horizontal scale of the differences in atmospheric moisture content is strong evidence for regarding the differences in rainfall over North Africa and Arabia as being due to the European soil moisture anomaly.

6.3 Temperature

The differences in temperature for days 21-50 at $\sigma = 0.9$ (Figure 9) are quite similar in pattern to the differences in specific humidity at that level just as they were after two days. Again the differences over the anomaly area are clearly due to the local soil moisture while the general warming over North Africa and Arabia must be due initially to advection together with some local soil-moisture feedback effects. Relative humidity was, of course, considerably reduced in these areas of increased temperature and decreased specific humidity, so that both were likely to contribute to decreases in rainfall.

6.4 Statistical Aspects

Statistical assessments of the results have been made. In summary, these showed that, using Student's t test on five 10-day mean differences, the differences in most variables discussed above are significant at the 1% level over the anomaly area. The differences in precipitation are significant at the 10% level over southern Scandinavia and Spain.

7.   THE ANNUAL-CYCLE MODEL EXPERIMENT

Sea-level pressure patterns for experiments which included seasonal variation showed a much more westerly flow over northern Europe in summer than that showed by Corby *et al.* (1977) which was typical of medium resolution experiments MD, MC and MW. As would be expected in the real atmosphere, the flow varied somewhat from one 10-day period to another, but a ridge near 45°-50°N

(a)

(b)

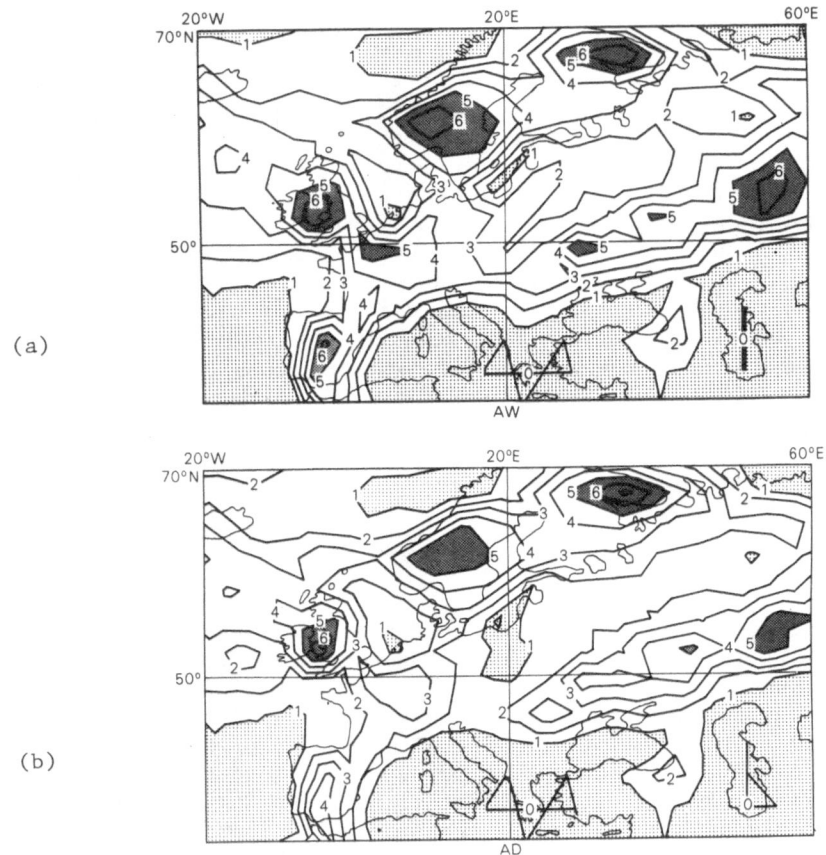

Fig. 10 Rainfall (mm per day) averaged for days 1-10 of annual
        cycle experiments: (a) wet start (AW); (b) dry start (AD).

from the Azores high and troughs at 60°-65°N are characteristics
of most 10-day periods of the annual-cycle experiments AD and AW
over Europe. Troughs and ridges moved eastward from the Atlantic
bringing alternating wet and dry spells, while the occasional
secondary depressions near 50°N brought considerable rainfall, as
shown by the days 1-10 mean in the wet start annual-cycle
experiment (AW) (Figure 10a).
     In the initial absence of soil moisture over western and
central Europe (experiment AD), evaporation must be zero initially
and, from the arguments presented above, rainfall ought to be
reduced over the anomaly area. This expectation was borne out as
rainfall with the dry start (Figure 10b) was typically between 30
and 70% of that with the wet start (Figure 10a). There were also
reductions over southern Scandinavia and central Spain as for the

corresponding medium resolution experiments (see Figure 5). The
moisture content of the starting condition of the annual-cycle
experiments made little difference to the mean pressure pattern
with falls of about 2 mb over the anomaly area. Although, with
the dry start, rainfall exceeded evaporation (which averaged about
1 mm per day over the period) so that soil moisture and evaporation
were increasing, evaporation was still below 2 mm per day over much
of the anomaly area in the day 11-20 mean while generally rainfall
continued to be less than with the wet-start experiment (AW) in
which evaporation exceeded 3 mm per day.

Around day 20 in the dry-start experiment, a depression moved
into central Europe and the day 21-30 mean showed a depression
over central Europe. As might be expected, there was heavy rainfall over much of the anomaly area in association with this cyclonic
flow. (Rainfall was actually less heavy for AW for which the mean
pressure pattern showed a depression further east near the Black
Sea). For the rest of the 50-day integration with the dry-start
experiment, the rain increased evaporation to around 3 mm per day
over the northern half of the anomaly area. This is too similar
to the value for the wet-start experiment (3 to 4 mm per day) for
any significant difference in the rainfall to be detected.

Averaged for days 21-50, the rainfall was heavier with the
dry-start experiment over most northern and eastern parts of the
anomaly area. Evaporation and rainfall remained less over the
southwestern part of the anomaly area and over central Spain. The
differences in this region are statistically significant up to the
1% level.

8.   CONCLUSIONS

These experiments have demonstrated that an anomaly in the
soil moisture over Europe can have a considerable effect on the
modelled rainfall, humidity and temperature over the 50 days
following the start not only over the anomaly, but also over a
considerable area of the adjacent land. The effects of the anomaly
spread to much of Scandinavia and Spain and a large part of North
Africa.

The persistence of the anomaly was dependent on the prevailing
flow conditions. With a westerly flow advecting moist air from
the Atlantic, the model's equilibrium state was wet over northern
Europe, the dry anomaly persisting for only about 20 days before
much of it became too weak to affect evaporation significantly.
With weaker circulation patterns, the model equilibrium state was
dry over the anomaly area; although the wet anomaly was slowly
being eroded, this process was far from complete after 50 days.

REFERENCES

Baumgartner, A. & Reichel, E. 1975 *The World Water Balance: Mean Annual Global, Continental and Maritime Precipitation, Evaporation and Runoff*, 179 pp. Amsterdam: Elsevier.

Corby, G.A., Gilchrist, A. & Rowntree, P.R. 1977 United Kingdom Meteorological Office five-level general circulation model. *Meth. comput. Phys.*, 17, 67-110.

Idso, S.B., Jackson, R.D., Reginato, R.J., Kimball, B.A. & Nakayama, F.S. 1975 The dependence of bare soil albedo on soil water content. *J. appl. Met.*, 14, 109-113.

Korzun, V.I. (ed.) 1974 *World Water Balance and Water Resources of the Earth*. Report of the USSR Committee for the IHD, 663 pp. (English translation 1977-78: *Studies and Reports in Hydrology* No. 25. Paris: UNESCO.)

Mintz, Y. 1981 The influence of soil moisture on rainfall and circulation: a review of simulation experiments. In: *Proc. Study Conf. on Land Surface Processes in Atmospheric General Circulation Models*. Geneva: WMO/ICSU.

Norton, C.C., Mosher, F.R. & Hinton, B. 1979 An investigation of surface albedo variations during the recent Sahel drought. *J. appl. Met.*, 18, 1252-1262.

Slingo, Julia M. 1982 A study of the Earth's radiation budget using a general circulation model. *Q. Jl R. met. Soc.*, 108. [In the press.]

Walker, J. & Rowntree, P.R. 1977 The effect of soil moisture on circulation and rainfall in a tropical model. *Q. Jl R. met. Soc.*, 103, 29-46.

SOME SIMULATION MODEL RESULTS OF THE EFFECT OF VEGETATION CHANGE
ON THE NEAR-SURFACE HYDROCLIMATE

John G. Lockwood and Piers J. Sellers

School of Geography
University of Leeds
United Kingdom

ABSTRACT. It is well established that the principal effects of
forest on the hydrological cycle are in the reception and disposal
of precipitation. Forests provide the greatest surface area for
the interception and re-evaporation of water and are effective traps
for the absorption of solar radiation. Simply altering ground
cover affects surface albedo and runoff, changes the ratio of sensible to latent heat transport, and greatly modifies the surface
winds. A complex multilayer crop model coupled with a simple hydrological description is used to simulate the hydrological cycle in
coniferous forest (pine), deciduous forest (oak), arable land
(wheat) and grassland. Standard meteorological observations made
at an airfield in the UK were used for the model's inputs, after
being adjusted to allow for different surfaces. Allowance was made
for changes in such parameters as varied as the seasons progressed.
     The model was used with hourly data from a typical site in low-
land Britain for 1977. Over this annual period the model simulates
least runoff from pine and greatest from wheat. These changes are
due to differences in the interception and evaporative properties
of the canopies. The interception loss from coniferous forest is
particularly high. The model suggests that the clearance of forest
leads to a generally wetter soil moisture regime and higher runoff
under constant climatic conditions. Experimental evidence from a
number of humid and semi-humid river basins is shown to support
these conclusions.

1.   INTRODUCTION

     A large body of literature emphasizes the importance of vege-
tation cover on the hydrological cycle. Figure 1 summarizes the

463

A. Street-Perrott et al. (eds.), Variations in the Global Water Budget, 463–477.

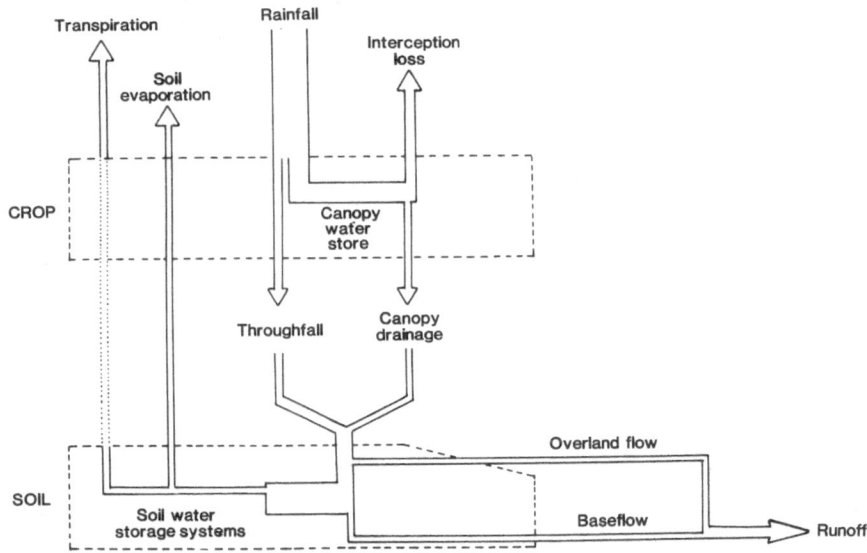

Fig. 1. Generalized water balance of a small, vegetated catchment.

water balance of a small basin covered by a specific vegetation
type and shows the processes involved. The effect of changing a
crop type within a basin is to alter the interception loss, transpiration and, to a lesser extent, soil evaporation rates. Changes
in any of these processes will have feedback effects on the soil
moisture content and the runoff. A numerical model is outlined
which simulates the exchanges shown in Figure 1. The model is
applied to a simple catchment under four different crop types:
grass, pine, oak and wheat. The hydrological performances of the
catchment under these four regimes are contrasted in terms of
interception, transpiration, soil moisture status and runoff. To
evaluate this simulation model, some results are cited from field
experiments dealing with similar crop contrasts.

2. DESCRIPTION OF MODEL

The model used to simulate the processes identified in Figure
1 consisted of three linked sub-models, each being designed to
replicate the processes in a physically realistic manner. As
complete detailed descriptions of the model have been given previously (Sellers 1981, Sellers & Lockwood 1981), only a brief
description is included here. A flow diagram for the basic model
is outlined in Figure 2.

The core of the linked model (termed MANTA) is the Multilayer

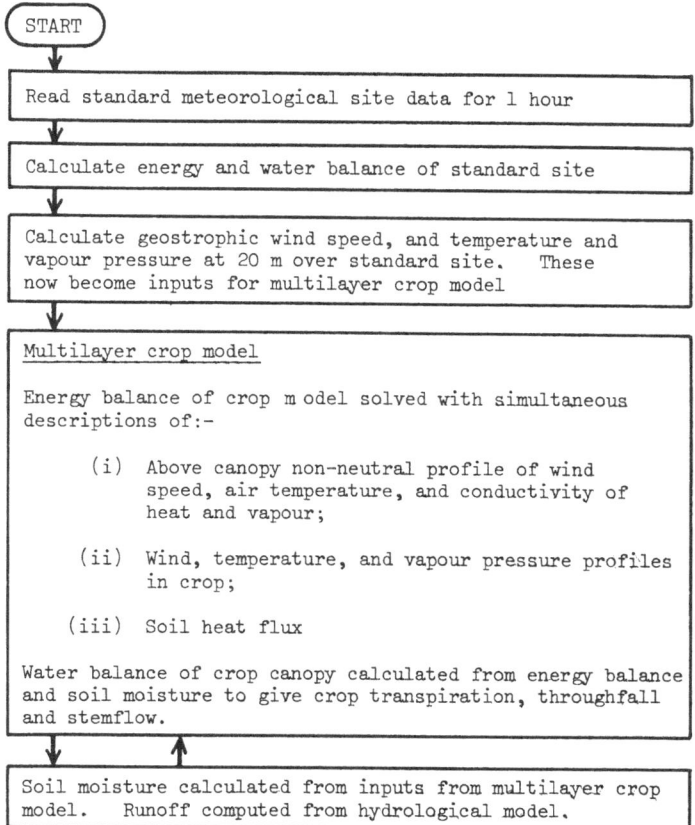

Fig. 2.   Schematic diagram of the multilayer crop, described in
          detail in Sellers and Lockwood (1981).

Crop Model which has the following virtues:
(a) It is capable of simulating changes in the locations of heat
     and water vapour sinks and sources;
(b) It can be used to describe leaf drainage and the redistribution
     of intercepted water between different parts of the canopy;
(c) With the inclusion of a wetted area term, the leaf's energy
     balance can be realistically modelled.
The multilayer approach, however, is demanding in terms of data and
parameterization; in particular:
(i)   The data available for this study were hourly measurements
       of air temperature, wet bulb temperature, wind speed at 10 m,
       total cloud cover, low cloud cover and precipitation, as
       observed at standard meteorological sites in the UK. As the
       sites were mainly grass covered, the data are not directly
       applicable to other vegetation types;

(ii)   Most multilayer models require values from below the canopy
        of humidity and air temperature which are needed as fixed
        points for the solution. They were not available in the
        above data set;

(iii)   Long-term simulation necessitates the inclusion of some kind
        of runoff model since realistic feedback effects of soil
        moisture stress on transpiration are required.

     In the crop model, allowance was made for the changing leaf
area index, albedo, roughness length, moisture tension and stomatal
resistance in each crop as the seasons progressed. For grassland,
a unilayer model was used; for the other vegetation types, crop
models of four or more layers were used. The interception process
is considered using a multilayer formulation.

     Figure 2 also shows the adjustment of the meteorological input
data from standard meteorological sites, e.g., grass-covered air-
fields, to those that would be applicable to a particular crop type.
This procedure was undertaken in the 'Standard Meteorological Site
Submodel'.

     The hydrological model used to estimate runoff represented the
complete catchment as a vertical series of cascading soil moisture
stores, the contents of which determine the runoff. Rainfall, runoff and actual evaporation data from the Grendon Underwood river
basin (of 18.6 km^2 area) in Buckinghamshire, were used to optimize
the parameters for the hydrological model. During wetting and
drying of the vegetation canopies, the model used 12-min. time
steps; when the canopy was dry, the time step was extended to one
hour.

3. SOME RESULTS OF THE NUMERICAL MODEL

     As yet, MANTA has only been run with a 12-month data set, this
being for 1977 from Benson in Oxfordshire; the monthly mean climatological data are shown in Table I. The hydrological runoff model
applies to the nearby Grendon Underwood basin, and this should be
considered to be covered completely by each hypothetical vegetation
type in turn. The simulated development of transpiration, interception loss (including canopy loss plus soil evaporation) and runoff for four crop types are shown in Figure 3. The lowest runoff
values throughout the year are associated with pine and the highest
for much of the year with wheat.

     Simulated annual values of the hydrological components are
shown in Table II, where slight variations in total runoff and
evaporation are due to small variations in soil moisture at the
beginning and end of the year. Changes in vegetation type bring
about profound changes in the operation of the surface hydrological
system illustrated in Figure 1. In particular, there are fundamental changes in the partitioning of water between runoff and
evaporation, and also between evaporation from intercepted water
and from plants. A change from forest to a non-forest vegetation

Table I. Monthly averages of climatological observations for
          Benson, for 1977.

	J	F	M	A	M	J
Daily air temperature (°C)	3.2	6.1	7.4	7.6	10.9	12.3
Monthly total precipitation (mm)	71.1	78.9	54.9	28.6	39.5	77.7
Daily mean hours of bright sunshine	1.30	3.21	3.36	5.37	7.32	4.60

	J	A	S	O	N	D
Daily air temperature (°C)	16.3	15.7	13.7	12.3	6.5	6.5
Monthly total precipitation (mm)	20.8	177.6	20.1	44.3	39.9	70.7
Daily mean hours of bright sunshine	7.07	4.51	3.75	3.57	2.99	1.53

Table II. Annual totals of simulated runoff, interception loss
           and transpiration for four vegetation types (in mm
           rainfall equivalent).

	Runoff	Interception loss	Transpiration
Wheat	417	158	134
Pine	152	253	309
Oak	279	172	263
Grass	318	62	335

type leads to an increase in runoff. This is particularly marked
with a change from coniferous forest (pine) to any other type of
vegetation, or from any other type of vegetation to arable land
(wheat). The change in runoff is less significant with a change
from deciduous forest (oak) to grassland.

3.1 Interception and Transpiration Losses

     The vertical distribution of the interception loss contribution
from a pine canopy predicted by the MANTA pine model is shown in
Figure 4. The most significant prediction is that the topmost leaf
layer is by far the most dominant source of interception loss, providing 42% of the canopy's total annual loss despite being wet for
the shortest time. Simulations using synthetic data (Sellers 1981,
Sellers & Lockwood 1981) indicate that the dominance of the top

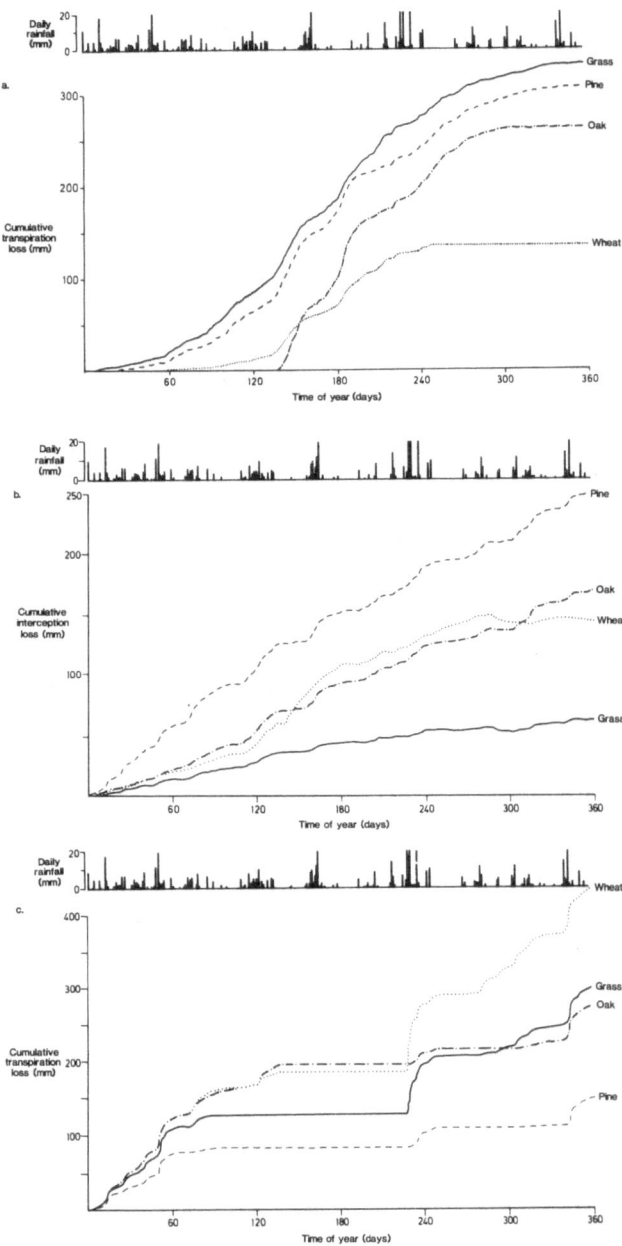

Fig. 3.   Simulated values over one year, for four crop types of
(a) accumulated transpiration loss; (b) accumulated interception loss; (c) accumulated runoff. Meteorological data
set from Benson, Oxfordshire, for 1977.

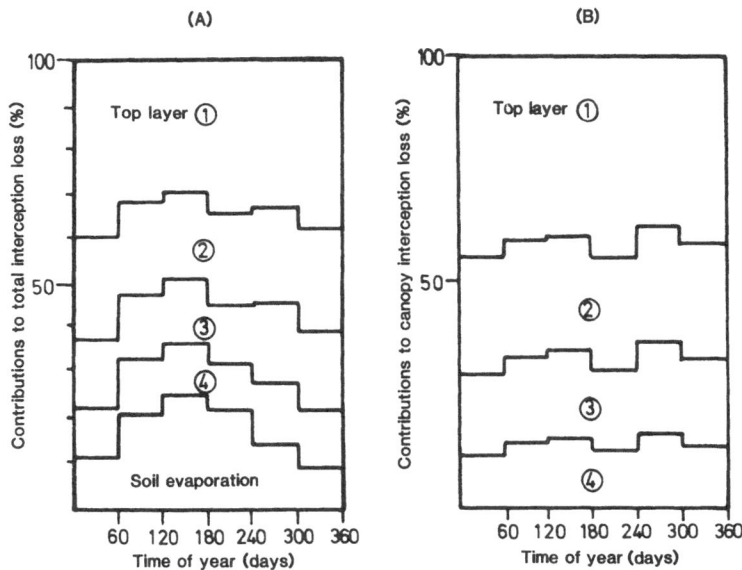

Fig. 4. Predicted variation of interception loss for the MANTA
pine model for 1977: (a) canopy interception and soil
evaporation; (b) canopy interception only. Circled
numbers denote leaf layer position. Meteorological data
set from Benson, Oxfordshire, for 1977.

leaf layer is likely to vary according to weather conditions.
Figure 4 demonstrates that soil evaporation is a relatively large
loss during the summer months, but is suppressed during the winter
by an actively-evaporating canopy which is able to consume a larger
proportion of the available energy. Similarly, if canopy interception loss only is considered, the upper layers are slightly more
dominant during the winter months. This is because, under conditions of low evaporative potential, the topmost leaf layer uses up
a relatively large proportion of the available energy, and in the
process suppresses the evaporation from lower layers.

Figure 5 illustrates the five-day means of predicted daily
transpiration from an oak stand, given assumed values for the leaf
area index, and the simulated transpiration losses over one year
are shown in Figure 3(a). The marked seasonal variations in foliage
cover limits transpiration to a few months of the year though,
according to the simulation, the fully-foliated oak forest tran-
spires at a considerably greater rate than the other three crops
considered; thus, on an annual basis, the transpiration loss from
the oak stand is comparable with grass- and pine-dominated areas.

Figure 6 reveals that, as with transpiration, the annual course
of interception loss from an oak-covered area is closely related to
the state of the crop's foliage. The leafless crop gives rise to

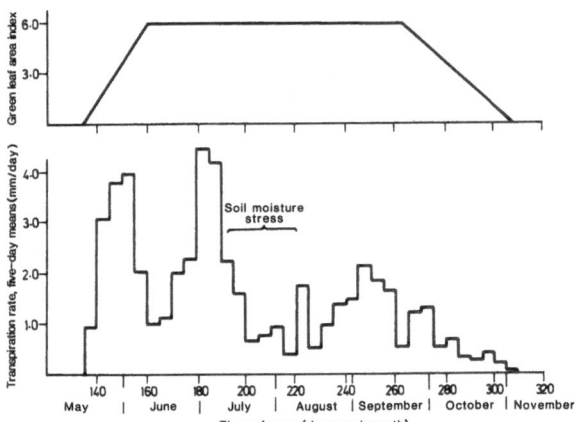

Fig. 5.   Five-day mean values of the predicted daily transpiration
          loss for the MANTA oak model for 1977. Meteorological
          data from Benson, Oxfordshire, for 1977.

a much reduced rate of interception loss due to minimal water
storage in the canopy, high throughfall coefficient, small maximum
exposed wetted area, all tending to a low bulk-vapour source. Over
the year as a whole, the canopy contribution to interception loss
amounted to 53% of the total. However, as this value incorporates
predictions for both **a leafless** and full-foliated canopy, it reveals
little about the physical interactions between a structured crop
and the atmosphere. Between the day 1-60 period (when the crop was
completely leafless) and the day 181-240 period (when the crop was
fully foliated), the canopy contribution to interception loss
increased from 59% to 80% of the total interception losses. As the
leaf-area index increased and rainfall income decreased during the
summer, the canopy contributed a larger proportion of the total
interception loss. When fully foliated, the upper part of the
canopy contributes relatively more to the total interception loss,
the distribution of interception loss from the leafless trees
being related directly to the leaf-area index of each leaf layer.

        The predicted cumulative transpiration loss from the MANTA
wheat model is shown in Figure 3(a). The predictions for the lowest
annual total of transpiration loss of the four crop types may be
underestimates (Sellers 1981). Figure 7 shows the contribution of
leaf layers and soil to interception loss over the year. Soil
evaporation accounts for over 77% of the total interception loss;
even when the canopy is fully developed, the soil evaporation loss
still makes up 65% of the total. The interception loss from wheat
foliage was just over 5% (37 mm) of the annual precipitation. This
low value (compared to the equivalent value of 29% for pine) is the
result of the relatively short period of full plant cover in the

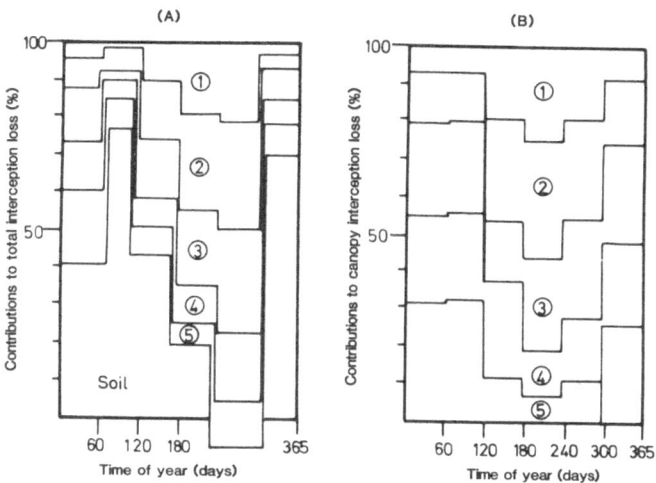

Fig. 6. Predicted variation of interception loss for the MANTA
oak model for 1977: (a) canopy interception and soil
evaporation; (b) canopy interception only. Circled
numbers denote leaf layer position. Meteorological data
set from Benson, Oxfordshire, for 1977.

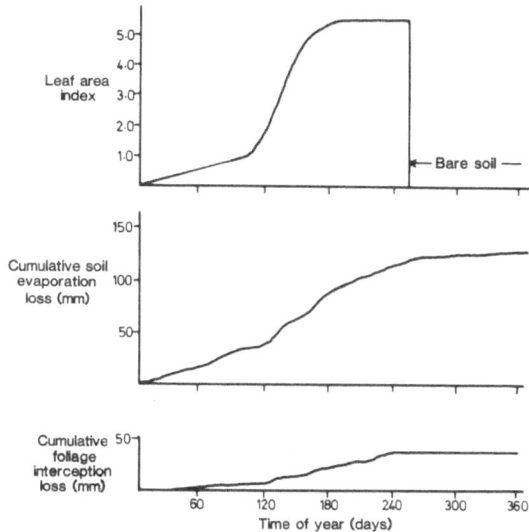

Fig. 7. Interception losses from soil and foliage as predicted by
the MANTA wheat model for 1977. Meteorological data set
from Benson, Oxfordshire, for 1977.

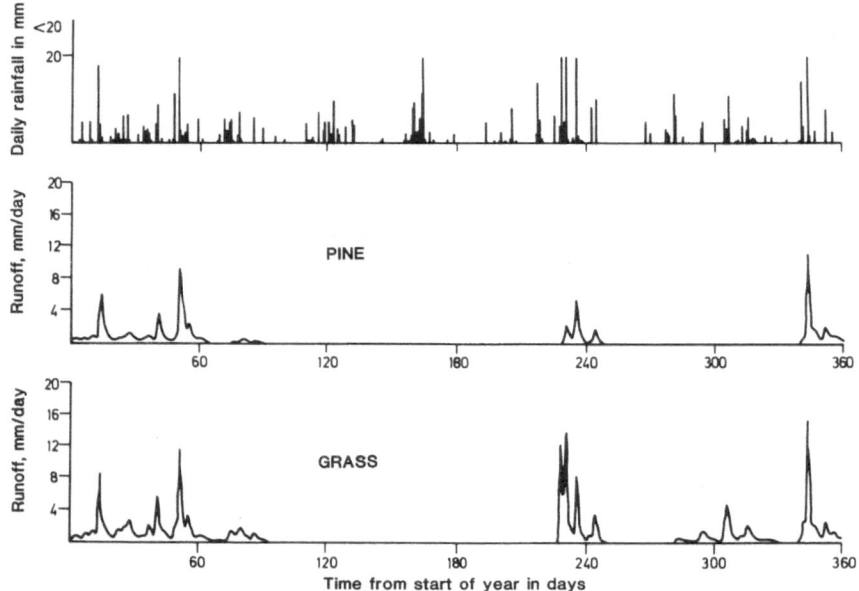

Fig. 8. Predicted daily runoff totals for pine- and grass-covered
         areas. Meteorological data set from Benson, Oxfordshire,
         for 1977.

summer and the complete absence of foliage during other wetter
months. Loss from the soil surface, which is predicted to be 17%
(123 mm) of annual precipitation, is related to the availability of
soil moisture and the seasonal variation of the evaporative potential.

3.2 Soil Moisture and Runoff

     Although the transpiration rate from grassland is relatively
large, the interception loss is small compared with forest. Thus
considerably more rainfall passes through the vegetation canopy of
a grassed area into the soil moisture store than for a forest area.
Consequently, the predicted soil moisture levels are generally
higher under grass than under forest, particularly coniferous
forest. The MANTA oak model predicts the most extreme runoff
regime of the four models as there is no transpiration and little
interception loss during winter and, by contrast, the highest
predicted transpiration loss of the four crop models during the
summer. As a result of the unbalanced annual distribution of the
evapotranspiration loss, the runoff regime of the oakland model is
characterized by high peak flows and high levels of soil moisture

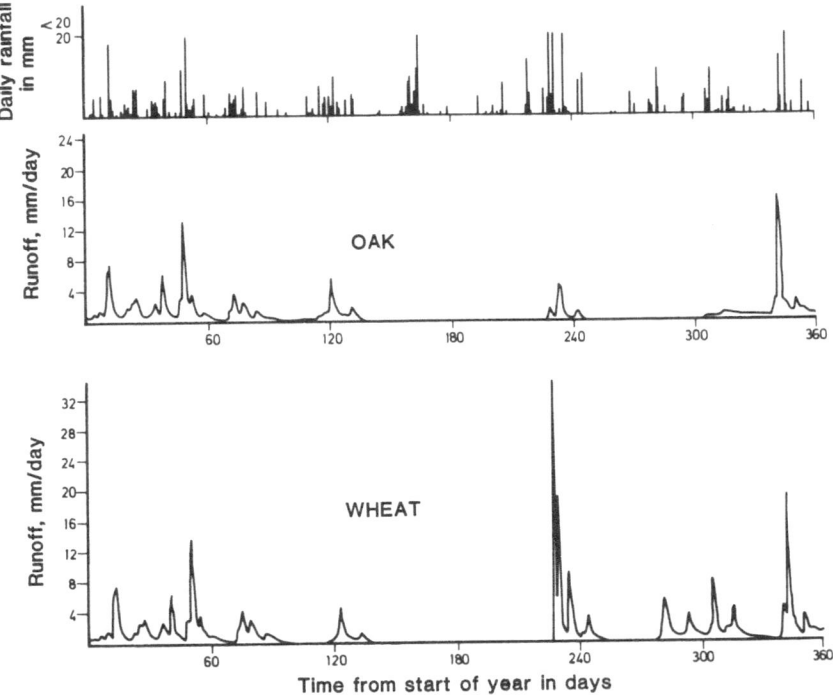

Fig. 9. Predicted daily runoff totals for oak- and wheat-covered
         areas. Meteorological data set from Benson, Oxfordshire,
         for 1977.

in winter with an extended period of low flows during the summer.
For much of the year, the lowest soil moisture totals are associated with pine: this is reflected in the low runoff totals
predicted by MANTA.

    Predicted daily runoff totals for the four vegetation types
are shown in Figures 8 and 9. Statistical analysis of the runoff
values indicates that an increase in the mean discharge over a
period leads to a proportionately greater increase in the frequency
of high flows. In addition, the runoff regime generated by each
crop model is idiosyncratic in terms of the characteristics of its
frequency distribution. Grass maintains the least leptokurtic and
least skewed runoff regime throughout the year, thereby implying
that its runoff frequency distribution is the most even of the
four crops. Oak and wheat show highly skewed and leptokurtic runoff rates during the winter when neither crop is able to regulate
soil moisture levels via evapotranspiration to the extent predicted
for grass and pine. Pine, with the highest predicted annual evapo-

Table III. Interception losses from experimental catchments with coniferous forest cover in the UK

Site	Species	Duration of Study	Rainfall, P (mm)	Interception loss, I (mm)	I/P x 100 (%)	Source
Hodder forest (Yorkshire Pennines)	Sitka spruce	4 June 1955 – 8 June 1956	984	371	37.7	Law 1956
Crawthorne, Berkshire	Scots pine	Oct. 1959 – March 1960 April – Sept. 1958 1958 – 1960	430 366 ≈ 800 p.a.	161 128 –	37.4 34.8 31.7	Rutter 1963
Thetford Chase (Norfolk)	Scots pine	365 days in 1977 (continuous)	595	213	35.8	Gash & Stewart 1977
Upper Severn, Plynlimon (Wales)	Mixed mature conifers	6 Feb. 1974 – 1 Oct. 1976	5444	1580	29.0	Calder & Newson 1979
Hafren forest (Plynlimon)	Sitka spruce	351 (continuous) days	1757	479	27.2	Gash et al. 1980
Roseisle forest Moray Firth (Scotland)	Scots pine	281 (continuous) days	493	209	42.4	Gash et al. 1980
Kielder forest (Cheviot Hills)	Sitka spruce	302 (continuous) days	802	254	31.7	Gash et al. 1980
River Ray (Buckinghamshire)	Pine model	1977	724	252	34.8	MANTA prediction (Sellers 1981)

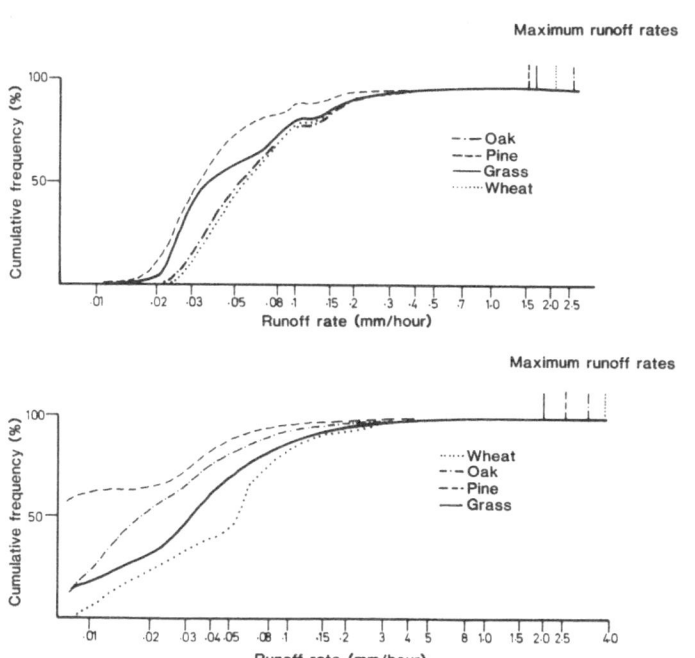

Fig. 10. Predicted runoff rate frequency distribution for the
four crop types: (a) over the period day 1 to 60;
(b) over the period day 301 to 365.

transpiration loss, is predicted to have the highest frequency of
low flows (Figure 10, but not necessarily the lowest peak flows.

4. EXPERIMENTAL STUDIES

Table III compares the interception losses measured in experimental UK catchments having coniferous forest cover with the interception losses predicted by MANTA. The MANTA prediction of
interception loss from a pine-covered area agrees with the results
obtained from coniferous forest sites in the dryer parts of the UK.
The large interception loss from trees has been illustrated in the
field by Clarke and Newson (1978) who compared the water yields
from the upper Wye and Severn river basins which are, respectively,
under hill pasture and largely coniferous forest. For the upper
Wye basin, the mean annual loss (precipitation minus streamflow)

for the years 1970 to 1975 inclusive was 18% of the mean annual precipitation of 2415 mm; for the upper Severn basin, the mean annual loss over the same period was 30% of its mean annual precipitation, of 2388 mm. Moreover, the latter made no allowance for the fact that about one-third of the Severn basin is unforested; when Clarke and Newson made a statistical adjustment to allow for this lack of homogeneity, the mean annual loss from the forested area of the upper Severn basin rose to 38% of mean annual precipitation. Clarke and Newson claim that the reality of these results has been confirmed by plot studies within the Severn forest on 'natural' lysimeters underlain by impermeable boulder clay.

A number of experimental studies have been made of the response of streamflow to a change in vegetation type. Hibbert (1967) reports results from two small river basins at Coweeta in North Carolina (USA) in which oak-hickory forest made up the principal vegetation cover. Flow and yield parameters for two small forested river basins near Coweeta were compared over three years at the beginning of the study period. In 1940, one of the basins was denuded of trees. In the first year after this clear-cutting, annual streamflow from the treated basin increased by 373 mm over the value expected from the regression equation with the control basin. As the deciduous forest regrew during the following 23 years, streamflow from the treated basin decreased logarithmically until, at the end of this period, the flow was only 75 mm greater than the value predicted for a full-forest cover. Clear-cutting again at that juncture resulted in another marked increase in streamflow from the treated basin almost exactly equivalent to the response after the first clearance.

Swift *et al.* (1975) developed a phenomenological model of water exchange between soil, plant and atmosphere which was used to simulate evapotranspiration and annual drainage over two years from a mature oak-hickory forest in the southern Appalachians. In a year of unusually high precipitation, the simulated annual drainage was within 1.5% of measured streamflow. Simulations were also performed using the same two years of meteorological data, but with vegetation parameters changed to represent a young, white pine plantation and a regrowing hardwood forest, one year after clear-cutting. The results were similar to the changes of -200 mm and +300 mm observed in comparable basin experiments near Coweeta. Simulated evapotranspiration during the summer was nearly identical for hardwood and pine forests, but water losses were greater for the coniferous forest in winter and early spring. The simulation suggested that the greater evapotranspiration by pine was due to increased interception in all seasons and increased transpiration in the dormant season. These results are in agreement with those from the MANTA model.

ACKNOWLEDGEMENTS. Piers Sellers acknowledges the receipt of a
NERC studentship during the course of this work. Thanks are due
to the Meteorological Office, Bracknell, and the Institute of
Hydrology, Wallingford, for providing data.

REFERENCES

Calder, I.R. & Newson, M.D. 1979 Land use and upland water resources
    in Britain - a strategic look. *Wat. Resour. Bull., U.S.*, 15,
    1628-1639.

Clarke, R.T. & Newson, M.D. 1978 Some detailed water balance studies
    of research catchments. *Proc. R. Soc. Lond., A*, 363, 21-42.

Gash, J.H.C. & Stewart, J.B. 1977 The evaporation from Thetford
    forest during 1973. *J. Hydrol.*, 35, 385-395.

Gash, J.H.C., Wright, I.R. & Lloyd, C.R. 1980 Comparative estimates
    of interception loss from three coniferous forests in Great
    Britain. *J. Hydrol.*, 48, 89-105.

Hibbert, A.R. 1967 Forest treatment effects on water yield. In:
    Forest Hydrology (ed. W.E. Sopper & H.W. Lull), pp. 536-538.
    Oxford: Pergamon Press.

Law, F. 1956 The effect of afforestation upon the yield of water
    catchment areas. *J. Br. Waterwks Ass.*, 38, 484-494..

Rutter, A.J. 1963 Studies in the water relations of *Pinus sylvestris*
    in plantation conditions 1. Measurement of rainfall and inter-
    ception. *J. Ecol.*, 51, 191-203.

Sellers, P.J. 1981 Vegetation type and catchment water balance:
    a simulation study. University of Leeds: Ph.D. Thesis.
    [Unpublished.]

Sellers, P.J. & Lockwood, J.G. 1981 A computer simulation of the
    effects of differing crop types on the water balance of small
    catchments over long time periods. *Qt. Jl R. met. Soc.*, 107,
    395-414.

Swift, L.W., Swank, W.T., Mankin, J.B., Luxmoore, R.J. & Goldstein,
    R.A. 1975 Simulation of Evapotranspiration and drainage from
    mature and clearcut deciduous forests and young pine
    plantation. *Water Resour. Res.*, 11, 667-673.

THE CONCEPT OF RUNOFF IN THE GLOBAL WATER BUDGET

J. Němec

Hydrology and Water Resources Department
World Meteorological Organization
Geneva, Switzerland

ABSTRACT. Four classes of users of water budget concepts are
identified each with a different perception of the place of runoff within the budget: the global circulation modeller,
climatologist, geographical hydrologist, and basin modeller. The
uses are contrasted in terms of the emphasis placed on runoff
compared with other elements of the budget, the time and space
scales, and whether interest focuses on time averages or on the
details of time variations. Despite fundamental differences, it
is emphasized that, in the present state of water budget science,
each user needs to validate the budget by comparing the runoff
terms with runoff data from stream gauging stations. In many
parts of the world, the existing network is inadequate. One particular new requirement for flow measurement arises in the context
of the new generation of Global Atmospheric Circulation Models.
Problems arise in matching the time lags inherent in the atmospheric and land phases of the water cycle, and in the differing
space scales of a grid square and irregularly-shaped river basins.

1. THE FOUR APPROACHES TO RUNOFF

The concept of runoff in the global water budget is of
interest to the specialists of at least four geophysical fields,
namely: (i) modelling the global atmospheric circulation, (ii)
climatology (or climatonomy), (iii) geographic hydrology and (iv)
modelling the hydrological cycle in a basin. The order of these
fields is not random, for it moves from emphasis on the 'global'
view via emphasis on the 'water budget' to emphasis on 'runoff'.
The modeller of the global atmospheric circulation is
interested less in runoff and more in other elements of the water

479

A. Street-Perrott et al. (eds.), Variations in the Global Water Budget, 479–488.

budget, of which precipitation is the output of the models and
evapotranspiratio n (or a parameterization of the soil moisture) is
an important input into the models, at least over the continents.
Runoff interests the global atmospheric circulation modeller
rather marginally, sometimes being used to verify the performance
of the model. The integrating capacity of the basin with respect
to runoff is an advantage, offset by the effect on the runoff of
surface and sub-surface storages which introduce time delays and
so complicate matters for models with short-time (e.g., hourly or
daily) steps.

Runoff is of concern to the climatologist although his main
interest lies in the water budget as a whole, so all terms of
this budget have almost equal importance. On the global scale,
the actual evapotranspiration is an unobserved (or very inade-
quately observed) term which must be derived as the difference
between precipitation and runoff, the latter terms being much more
precisely measurable than the former term. While the energy
budget of the land surface serves the climatologist as a basis
for the derivation of both potential evapotranspiration and the
so-called 'climatic runoff', the measured runoff is again a useful
verification criterion for his 'models' of the global water budget.
This usefulness derives from the employment of time steps in his
models where only time averages are of importance ranging from,
at the shorter end, the month, through the more frequently calculated seasonal budget to the most frequently used time step, the
year. In this way (particularly for the annual averages), the
storage term of the budget equation almost entirely loses its
importance, and full advantage can be taken of the integrating
capacity of the basin.

The emphasis placed on the budget principle by the geographical hydrologist is as great as by the climatologist, but
the former focuses on the runoff term, since his purpose in
establishing a water balance is to examine the distribution of
water resources in the world, in the continents and in major
basins. The storage term of the balance equation does not
represent a major problem to the geographical hydrologist because,
like the climatologist, he operates in essence with long-term
values averaged over monthly, seasonal or annual periods.

The water budget on a global scale has rather marginal
interest to the basin modeller whose predominant interest is the
runoff term. Perhaps the modeller hopes that his basin models can
be utilized in the other three fields, just as models are employed
in water and other engineering fields. Storage in the basin is of
vital importance to the basin modeller whose time step is similar
to that of the global atmospheric modeller, e.g., an hourly or
daily scale. He only uses time averaged data for verification
and error assessment.

An almost complete circle can be described encompassing these
four fields, with the modellers at the two extremes agreeing with
respect to the time step, the two internal fields being very

similar in almost entirely neglecting the storage aspects of the
runoff element of the water budget, and using global and average
values with large time steps.

2.   CALCULATION OF RUNOFF

It appears from § 1 that it is not possible to consider the
concept of runoff in the global water budget without specifying
(a) for which of the above fields such a concept is conceived,
(b) how it is to be used. It seems necessary, therefore, to
contrast and quantify all aspects of the runoff concept. The
major distinction is between the climatologist for whom the
global water budget terms are rather 'static' (i.e., long-term
average values) and the modellers of the global atmospheric circulation or of the hydrological cycle in a basin who view the terms
as dynamic and time varying. Of course, for computational
purposes, the basin modellers are obliged to work with short-term
averages rather than with the instantaneous and continuously
varying values. Conceptually, however, such averages can be
regarded as instantaneous.

The runoff concept and its use for global circulation models
(GCMs) were discussed in detail at a GARP Study Conference, in
particular by Eagleson (1981) and Dooge (1981). Runoff is one of
the land surface processes to be parameterized for use in GCM
verification and one of the elements which has to be formulated
adequately for inclusion in an homogeneous hydrothermal system at
an appropriate grid scale (62 000 km^2, at least) for GCMs.
Modifications of these formulations are necessary, at the cost of
a sub-grid parameterization, to account for the non-homogeneities
in inputs to and outputs from the hydrothermal system simulated by
the GCM. It was the view of the Study Conference that this is as
yet an unsolved problem.

In this connection, runoff can be represented in a one
dimensional formulation:

$$Y = R_s + p = R_s + R_g - \vec{n}(\delta Z_w/\delta t) = Y(\theta,P,t;\text{soil}) \tag{1}$$

where Y = rate of water yield, R_s = surface runoff rate, p = rate
of downward percolation of soil moisture to the water table,
R_g = rate of net groundwater runoff, $\vec{n}$ = unit normal vector,
Z_w = depth to the water table, P = rate of rainfall, and θ = volu-
metric soil moisture content.

Equation (1), which can be used at any time step, permits the
short-term soil moisture dynamics to be incorporated into the
water balance of the soil-atmosphere system. The incorporation of
soil moisture variation is the essential step which eventually
enables a long-term average water balance to be derived. Eagleson
(1981) proposed a 'statistical dynamic' solution that also
includes the vegetation moisture flux (extraction of soil moisture

by plant roots). Present parameterization of runoff in the
existing GCMs has been reviewed by Carson (1981). Eagleson's
approach and Carson's review reveal how the representation of
runoff in the water budget used in GCMs bears very little
resemblance to the basin modeller's concept of runoff which can
be regarded as the actual river flow. So direct comparisons are
difficult: the terms bear the same name, but carry a different
significance.

The climatological (or climatonomical) concept of runoff in
the global water budget or balance can be best illustrated by the
approaches of Budyko (1977) and Lettau (1973), respectively, as
has been indicated by Němec and Rodier (1979). The Budyko
approach uses the 'Radiation Index of Aridity (RIA)', defined as
R/L_r where R = radiation balance of the land surface, and L_r =
heat necessary to evaporate the amount of precipitation r (all
taken as annual averages per unit of land surface). Budyko
considers RIA as a characteristic of a combined energy and water
budget over land and uses it to calculate the 'climatic runoff'
by Equation (2):

$$f = r - E(R, L_r, r) \tag{2}$$

where f = mean annual climatic runoff, and E = annual geometric
mean actual evapotranspiration as given by Equation (3):

$$E = \{(Rr/L_r) - \tanh(L_r/R)[1 - \cosh(R/L_r) + \sinh(R/L_r)]\}^{\frac{1}{2}} \tag{3}$$

Budyko indicates that, when using actual streamflow and precipitation observations on 29 European rivers for verification, the
standard error of the annual runoff coefficient f/r =
$1 - [E(R, L_r, r)]/r$, calculated by his method, is 0.04, the
coefficient itself varying from 0.14 to 0.64. Budyko's concept
of climatic runoff again gives only a mean value which is, nevertheless, much nearer to the hydrological concept of runoff, as
illustrated by the graphical representation of f and r in Figure 1.
This is similar to the geographical hydrologist's approach.
Lettau (1973) analysed the thermal budget, which is the main
driving force of the water budget, by dividing it into three
simultaneous, but distinct processes according to the relevant
space boundaries (sea-land; intra-continental or regional; local).
This separation of the processes, by climatonomical concept,
results in the following expression for runoff:

$$D = P - E - (\delta m/\delta t) \tag{4}$$

where D = runoff, P = precipitation, E = evapotranspiration, and
m = soil moisture capacity. Lettau defines a 'flushing rate' of
runoff (r_D,) using a threshold value of precipitation (P_D) and
transforms Equation (4) to:

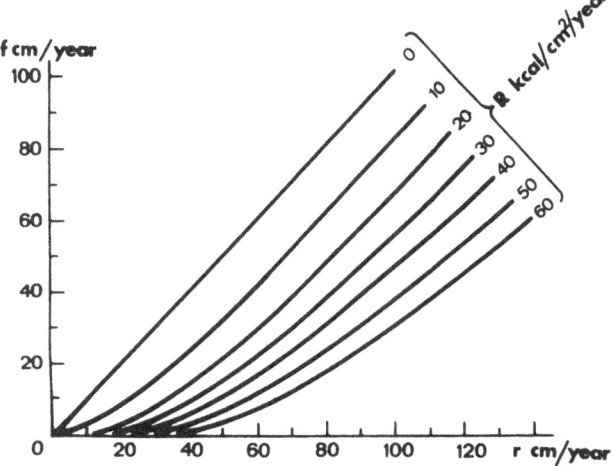

Fig. 1. Annual runoff (r) versus annual precipitation (f) as a
         function of R (kcal cm^{-2} yr^{-1}).

$$D = D' + r_D m \qquad\qquad\qquad\qquad\qquad\qquad (5)$$

where $D' = d_P (P - P_D)$, with $D' = 0$ if $P < P_D$, m = soil moisture
capacity, and d_P = a dimensionless characteristic defining the
input-dependent parts of runoff. The solution of this and of a
similar equation for evapotranspiration assumes non-linear
flushing rates for both runoff and evapotranspiration, but yields
only average values. However, it adds the recycling of regionally-
or locally-evaporated water to the basic concept of the atmo-
spheric- and land-water balance. Thus, the advected water vapour
from the sea will be precipitated on the continents at least twice.

     Geographical hydrology also uses the concept of water
recycling over continents in the computation of the water balance
of the continents, but starts from measured values of runoff and
precipitation. This method for calculating world water balance
was used by Baumgartner and Reichel (1975) and Korzun (1974). The
approach uses an empirical expression relating large-scale average
monthly, seasonal or annual evapotranspiration to air moisture
saturation deficit, wind velocity and temperature (or radiation).
It is thus possible to calculate a coefficient for water recycling
over continents and to establish a 'static' world water balance
giving averages of runoff, corrected for those parts of the
continents where direct measurements of runoff are not available.

     Finally, the basin modeller also uses a water balance concept
as evaluated over a river basin or an elementary, quasi-homogeneous
sub-basin, using as independent inputs the precipitation over the
basin together with some estimate of potential evapotranspiration
obtained indirectly (by energy balance or empirical formula) or
directly (e.g., measurement with evaporimetric pans). The soil

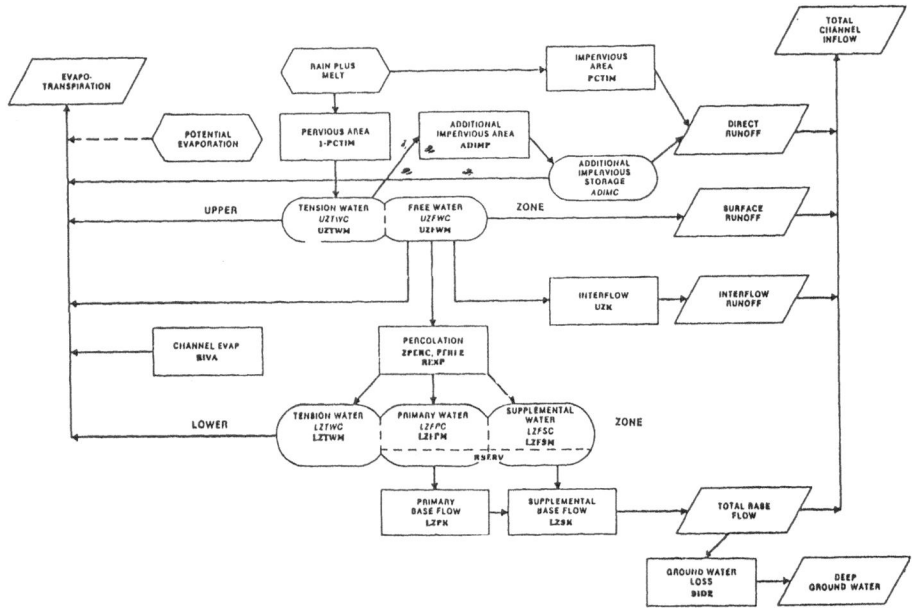

Fig. 2.   Schematic diagram of the NWSRFS (Sacramento) Rainfall-
          Runoff Model.

moisture and groundwater processes within the basin are represented
by schematized storages. The time step of such models ranges from
minutes to months, but a daily time step is the most commonly
found. The Sacramento Rainfall-Runoff Model (Burnash *et al.* 1973)
can be cited as an example of this approach (see Figure 2). It is
one of at least six similar models used by hydrologists to synthesize basin runoff. Their use ranges over real-time forecasting,
design flood computation, dam and reservoir design, and research
into the influence of man on runoff on the scale of the Amazon
basin.

     The Sacramento model is a lumped (spatially-averaged) input,
lumped-parameter type of model. To allow for spatial variability
of inputs and parameter values, a heterogenous basin can be
divided into areas assumed to be uniform in their behaviour,
e.g., on the basis of permeability. Two soil moisture zones,
upper and lower, are identified in the model (Figure 2), each zone
storing moisture in two forms: as 'tension water' and 'free
water'. The amount of water in each of these storage zones
represents a state of the model. The flow of water from upper to
lower zone is controlled, as elsewhere within the model, by parameter values expressing the relationship between the state within
the box and the rates of filling and removal of water. The
acronyms of Figure 2 indicate the complexity of the model and the
number of parameters. A combination of manual and automatic

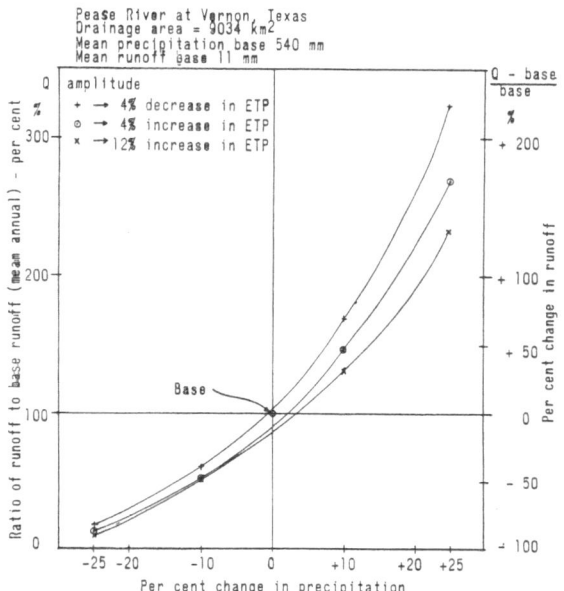

Fig. 3.   Changes (%) in mean annual runoff ($\bar{Q}$) as a function of
          changes in precipitation (P) and potential evaporation
          (PE). Data from Peace River at Vernon, Texas; drainage
          area = 9034 km²; mean precipitation base = 54 cm; mean
          runoff base = 1.1 cm; +, 4% decrease in PE; ⊙, 4% increase
          in PE; ×, 12% increase in PE.

optimization techniques is used to calibrate the model. As with
any conceptual model, considerable hydrological skill is required
to produce a set of parameters that 'best fit' the physical
characteristics of the basin. In general, the data base should
span a sufficient period to include both extremely dry and
extremely wet conditions, and also should reflect current land use
in the basin (Peck *et al.* 1981).

     The model is usually applied to basins ranging in size from
500 km² to 2500 km², but has been applied successfully to a small
research basin of 7.3 km² area and to a basin as large as the
Sanaga River in the Republic of Cameroon (131 500 km²) (WMO 1975).

3.   VERIFICATION OF THE RUNOFF TERM

     Whichever of the four water budget concepts are used to
compute runoff, it appears that in no case can the result be
accepted without verification based on the observed flow in the

rivers that often integrates the surface and sub-surface runoff.
These data are only directly comparable with those in the basin
modelling approach as this is the only one accounting for the lag
in runoff caused by surface and sub-surface storage in the basin.
For the same reason, mean runoff as derived by a 'static balance'
approach is not directly comparable with the GCM or basin modelling approaches which feature the short-time step. This problem
has been addressed recently within the World Climate Programme
(launched by WMO and ICSU among others).

A study by the author (Němec 1981) has addressed the problem
of the influence of a moderate climate change on runoff and,
consequently, on water resources management. One of the conclusions of that study is particularly pertinent to the discussion
here. Indeed, as indicated above, if moderate changes of temperature and precipitation are applied to approaches concerned only with
averages, such as the case in Figure 1 (Budyko's annual average
climatic runoff), the changes of the resulting runoff will be
entirely different from those actually derived from the basin
model in Figure 2. In the first case, a change of yearly precipitation by ±25% would, according to the average model, produce at
most a change of ±25% in runoff (the exact value depending on the
change in the radiation balance and thus of temperature); the same
change in precipitation, simulated by the Sacramento model, could
yield a change of up to 250% in runoff in a very dry catchment as
shown in Figure 3 (Němec 1981). As pointed out earlier, the
source of all verification data for the above approaches and
models can only be the actual stream gauging network in the world,
whether used directly or after adaptation to account for the
storage lag. It is expected that a total of 1000 to 1500 stations
will be needed to provide data for verification of the runoff component of GCMs. To overcome the problems of the mismatch between
basin and grid scale as well as the runoff lag problem, two different methods have been suggested (S.I. Solomon: private communication). The first uses small basins and extrapolates the observations to the entire grid square. The second distributes runoff
measured in large river basins (for example, in the Amazon basin
which covers an area of 4.5×10^6 km^2) in space according to determining factors, in particular, temperature, precipitation and topography. With respect to lag compensation, the first method assumes
that the flow variation of small river basins (encompassing areas
measured in hundreds of km^2) is directly proportional to the daily
direct surface runoff, providing there is no extraordinary surface
or sub-surface storage in the basin (e.g., in reservoirs, lakes or
the karst). The second method is based on the analysis of the
hydrograph and, particularly, of the recession curve which indicates the outflow from storage. The first method is probably more
accurate, but requires the larger number of stations.

5.   CONCLUSIONS

It can be stated that, while the stream gauging and computed runoff are considered as two aspects of the same process, the equivalence is far from being evident when closely analysed. The runoff concept implies different things to different categories of user. On the other hand, no progress in using runoff data for verification or calibration of water budget computations, particularly on the global scale, can be made without increasing the number of stream gauging stations, especially in the developing countries of Africa, Asia and Latin America. Thus, the development of stream gauging networks in these countries will not only assist their economic development, but will also provide data of value to the world's scientific community.

REFERENCES

Baumgartner, A. & Reichel, A. 1975 *World Water Balance: Mean Annual Global, Continental and Maritime Precipitation, Evaporation and Runoff*, 179 pp. Amsterdam: Elsevier.

Budyko, M.I. 1977 *Globalna Ecologya (Global Ecology)*. Moscow: Mysl Publishers.

Burnash, R.J.C., Ferral, R.L. & McGuire, R.A. 1973 A generalized streamflow simulation system: Conceptual modelling for digital computers. Sacramento: US Department of Commerce (NWS) and State of California (Department of Water Resources).

Carson, D.J. 1981 Current parameterizations of land surface processes in AGCM. In: *Proc. Study Conf. on Land Surface Processes in Atmospheric General Circulation Models*. Geneva: WMO/ICSU.

Dooge, J.C.I. 1981 Parameterization of hydrological processes. In: *Proc. Study Conf. on Land Surface Processes in Atmospheric General Circulation Models*. Geneva: WMO/ICSU.

Eagleson, P.S. 1981 Dynamic hydro-thermal balances at macroscale. In: *Proc. Study Conf. on Land Surface Processes in Atmospheric General Circulation Models*. Geneva: WMO/ICSU.

Korzun, V.I. (ed.) 1974 *World Water Balance and Water Resources of the Earth*. Report of the USSR Committee for the IHD, 663 pp. Leningrad: Gidromet. Izdatel. (English translation 1977-78: *Studies and Reports in Hydrology* No. 25. Paris: UNESCO.

Lettau, H.H. 1973 Evapotranspiration climatonomy. *Mon. Weath. Rev. Rev.*, 101, 636-649.

Němec, J. 1981 WCP implications for water-resource management.
    IPM/WCAP Working Document 4. Geneva: WMO.

Němec, J. & Rodier, J. 1979 Streamflow characteristics in areas of
    low precipitation. In: *Proc. IAHS Symp. on the Hydrology of
    Areas of Low Precipitation*, pp. 125-140. (Int. Ass. Hydrol.
    Sci. Publ. 128.)

Peck, E.L., McQuivey, R.S., Keefer, T.N., Johnson, E.R. & Erekson,
    J.L. 1981 Review of hydrologic models for evaluating use of
    remote sensing capabilities, NASA Rept. Greenbelt, Md: NASA.